Proceedings of the 21st European Symposium on Poultry Nutrition

ESPN
2017

8-11 May, 2017
Salou/Vila-seca, Spain

education organization research

Proceedings of the 21st European Symposium on Poultry Nutrition

edited by:

M. Francesch, D. Torrallardona and J. Brufau

May 8-11, 2017

Salou/Vila-seca, Spain

Organized by:

The Working Group 2 Nutrition of the European Federation of Branches of the World's Poultry Science Association (WPSA)

and

The Spanish Branch of the WPSA

EAN: 9789086863068
e-EAN: 9789086868513
ISBN: 978-90-8686-306-8
e-ISBN: 978-90-8686-851-3
DOI: 10.3920/978-90-8686-851-3

The individual contributions in this publication and any liabilities arising from them remain the responsibility of the authors.

First published, 2017

The designations employed and the presentation of material in this publication do not imply the expression of any opinion whatsoever on the part of the European Federation of Animal Science concerning the legal status of any country, territory, city or area or of its authorities, or concerning the delimitation of its frontiers or boundaries.

© Wageningen Academic Publishers
The Netherlands, 2017

Scientific committee:

Dr. Sanna Steenfeldt (Chair)
Dr. Ignacio Badiola
Dr. Ana Cristina Barroeta
Dr. Joaquim Brufau
Dr. Franco Calini
Mr. Federic Castelló
Dr. Ir. Evelyne Delezie
Dr. Károly Dublecz
Dr. Fernando Escribano
Dr. Enric Esteve-Garcia

Dr. Colin Fisher
Dr. Maria Francesch
Dr. Carlos Garcés
Dr. Pierre-André Geraert
Dr. Rene Kwakkel
Dr. Pim Lanhout
Dr. Gonzalo G. Mateos
Dr. Yael Noy
Dr. Lenka Papesova

Dr. Anna Pérez-Vendrell
Prof. Dr. Lidija Perić
Prof. Dr. Markus Rodehutscord
Dr. Gerard Santomà
Dr. Achille Schiavone
Dr. Nizamettin Senkoylu
Dr. Birger Svihus
Dr. Borja Vilà
Dr. Lotta Wallenstedt

The organizers thank the following scientists for reviewing the abstracts submitted to this symposium:

Dr. Ana Cristina Barroeta
Dr. Jiri Broz
Dr. Fernando Escribano
Dr. Enric Esteve-Garcia
Dr. Maria Francesch

Dr. Carlos Garcés
Dr. Michael Grashorn
Dr. René Kwakkel
Dr. Gonzalo G. Mateos
Dr. Anna Pérez-Vendrell

Prof. Dr. Markus Rodehutscord
Dr. Gerard Santomà
Dr. Sanna Steenfeldt
Dr. Borja Vilà

Local organizing committee:

Dr. Joaquim Brufau (Chair)
Dr. Enric Esteve-Garcia
Dr. Maria Francesch
Dr. David Torrallardona

Dr. Anna Perez-Vendrell
Dr. Borja Vilà
Dr. Rosil Lizardo
Ms. Rosa Cubel

Mr. Albert Gurri
Mr. Miquel Portals
Mr. Jordi Sanahuja
Ms. Anna Pinto

Sponsors

Diamond:

BASF
BIOMIN
CHRISTIAN HANSEN
DSM
NOVUS
PERSTORP

Platinum:

ADISSEO
ANDRÉS PINTALUBA
DUPONT
HUVEPHARMA
ZOETIS

Gold:

AB VISTA
CARGILL
DELACON
ELANCO
EVONIK
EW NUTRITION
ITPSA
PATENT CO
PHILEO LESSAFRE ANIMAL CARE
RETTENMAIER & SÖHNE

Silver:

ADIVETER
CCPA
HAMLET PROTEIN
IMPEXTRACO
KAESLER NUTRITION
KEMIN
NUSCIENCE
TROUW NUTRITION
NUTRIAD
PHYTOBIOTICS
ZINPRO

Media partners:

AVINEWS
POULTRY WORLD
SELECCIONES AVÍCOLAS
WATTGLOBAL

Table of contents

Plenary session 05. Optimized use of feed ingredients

Thursday 11 May – 9:00

Plenary session 06. Hot Topics: sustainability on poultry feeding

Thursday 11 May – 14:30

Abstracts 149

Session 01. Feed raw materials

Theatre abstracts

Poster abstracts

Session 02. Protein and amino acids nutrition

Theatre abstracts

Poster abstracts

Session 03. Feed additives

Theatre abstracts

Poster abstracts

Session 04. Feed enzymes

Theatre abstracts

Poster abstracts

Session 05. Nutrition and gut health

Theatre abstracts

Poster abstracts

Session 06. Mineral nutrition

Theatre abstracts

Poster abstracts

Session 07. Other topics

Theatre abstracts

Poster abstracts

Author index

Papers of invited speakers

Physiology of feed intake regulation in chickens: a chemosensory perspective

E. Roura[*] and S. Cho
Queensland Alliance for Agriculture and Food Innovation, The University of Queensland, 4072, Australia; e.roura@uq.edu.au

Summary

Nutritional needs are sensed through sensory cues resulting in nutrient specific appetites. However, the potential impact of individual dietary nutrients on chicken appetite is largely unknown except for the most limiting nutrients (i.e. energy and limiting essential amino acids). In the wild, birds are able to balance their nutritional needs by making appropriate dietary choices. In contrast, chickens raised under standard commercial practices have no choice. Marginal nutrient imbalances may occur under this conditions due to individual variations in feeding behavior and digestive and metabolic efficiencies. Thus, a lack of understanding of individual differences in nutrient needs may have a high impact on poultry production. Advances in nutritional chemosensing have recently uncovered the full repertoire of nutrient sensing mechanisms including those related to taste and the existence of these molecular mechanisms outside the oral cavity, particularly in the gut, orchestrating the hunger-satiety cycle. The study of the avian chemosensory mechanisms offers novel tools to understand the dynamic states of nutrient appetite and feed intake relevant to growth uniformity in broilers and feather pecking in laying hens among other practical issues.

Introduction

Animals in the wild balance their nutritional needs through diet selection. However, production animals (particularly non-grazing animals) are raised in a no-choice feeding environment. Commercial feeds are designed to maximize the efficiency of production by applying the principle that one nutritionally balanced feed formulation will cover the needs of the flock. While this principle is undisputed on a flock bases, on the other hand it overlooks individual needs which, in turn, may have a significant negative impact on the overall flock performance. Thus, no-choice feeding practices may result in marginal over and under-feeding metabolic consequences for specific nutrients which, in turn, may determine feed intake in farm animals (Roura and Navarro, 2017).

The taste and nutrient sensing system has an important role in identifying dietary carbohydrates, protein, fats, and poisonous compounds and seems to play a key role in orchestrating the hunger-satiety cycle (Meyerhof and Korsching, 2009). The emergence of genomic tools have contributed to the understanding on how these chemosensory mechanisms may impact feed intake in birds. Of particular interest to this brief review is how the nutritional status is monitored and the signals conveyed to the brain affecting appetite for specific nutrients.

Nutrient sensing in chickens

It has been argued that birds have a lower taste acuity compared to mammals, because they have a low number of taste bud, lack of mastication, relatively low saliva secretion and fewer taste receptor genes (Shi and Zhang, 2006; Berkhoudt, 1985). The relative small number of taste buds in birds may suggest that they are unable to discriminate between different flavours, however, when expressed relative to the volume of the oral cavity it becomes evident that the taste sensing capacity in birds is not inferior to mammals (Roura et al., 2013). In addition, the small number of taste receptor genes is not synonymous to lower taste sensitivity since some of the avian receptors were found to be more broadly tuned than the heterologous mammalian receptors (Behrens et al., 2014).

Plenary session 01. Feed intake, thermoregulation and heat stress

Most studies referring to nutrient sensing in chickens have been addressed by specific nutrient deficiency challenges. Chickens have been shown to respond to deficiencies of calcium (Wood-Gush and Kare, 1966; Hughes and Wood-Gush, 1971b; Stewart *et al.*, 2011; Wilkinson *et al.*, 2014), zinc (Hughes and Dewar, 1971), sodium (Hughes and Whitehead, 1974; Hughes and Whitehead, 1979; Hughes and Wood-Gush, 1971a), and amino acids (Picard *et al.*, 1993; Holcombe *et al.*, 1976; Newman and Sands, 1983; Steinruck *et al.*, 1990) as well as some micro-minerals and vitamins (Table 1).

Several groups of nutrient sensing genes have been recently identified in chickens following the discoveries in mammalian species (Roura *et al.*, 2013). These are receptors belonging to the super-family of G protein-coupled receptors (GPCR) which have been characterized in chickens based on mammalian gene homologies. In brief, their function as nutritional sensors is described below.

Carbohydrate sensing (related to sweet taste in humans)

The ability of avian species, including chickens, to sense carbohydrates is controversial (Roura *et al.*, 2013; Baldwin *et al.*, 2014). For example, chickens lack the main mammalian sweet taste receptor T1R2. However, chickens, such as it has been shown in mammals, might be able of sensing glucose through a T1R-independent pathway involving the transmembrane transporter SGLT1 (Sukumaran *et al.*, 2016). Results from our group indicated that laying hens may develop high preferences for sucrose (Cho and Roura, 2017).

Table 1. Nutrient specific appetites in chickens

Nutrient Appetite		References
Energy	Metab. Energy	Kaminska, 1979; Shariatmadari and Forbes, 1992; Cerrate *et al.*, 2007
	Fat	Klasing 1998; Sahin and Forbes, 1998; Cerrate *et al.*, 2007
	Long Chain F.A.	Furuse *et al.*, 1996
Protein	Crude protein	Chah and Moran, 1985; Shariatmadari and Forbes, 1993
	L-Lysine	Newmand and Sands, 1983; Picard *et al.*, 1993
	L-Methionine	Steinruck *et al.*, 1990; Picard *et al.*, 1993
	L-Tryptophan	Picard *et al.*, 1993
	L-Glutamic	Moran and Stilborn, 1996
Energy/protein	Ratio (ME/CP)	Kaufman *et al.*, 1978; Shariatmadari and Forbes, 1990; Syafwan *et al.*, 2012
Minerals	Calcium	Wood-gush and Kare, 1966; Hughes andWood-Gush,1971; Stewart *et al.*, 2011; Wilkinson *et al.*, 2014
	Phosphorus	Holcombe *et al.*, 1976; Barkley *et al.*, 2004
	Selenium	Forbes 1995
	Sodium	Duncan, 1962; Hughes and Whitehead, 1974; Karen and Mason, 1986; Balog and Millar, 1989
	Zinc	Hughes and Dewar, 1971; Kirchgessner *et al.*, 1990
Vitamins	Thiamine	Hughes and Wood-Gush, 1971
	Vitamin A	Jukes, 1938
	Vitamin B2	Steinruck *et al.*, 1991
	Vitamin C	Jukes, 1938; Kutlu and Forbes, 1993
Water	Grain hydration	Forbes, 2003; Scott, 2005

Plenary session 01. Feed intake, thermoregulation and heat stress

Amino acid sensing (related to umami/savoury taste in humans)

The umami taste (also known as savoury which relates to glutamate in humans) is basically associated to L-amino acids (L-AA) and peptides in mammalian species (Roura *et al.*, 2008). The two genes coding for the T1R1 and T1R3 heterodimer receptor work together as L-AA sensors (Nelson *et al.*, 2002). In chickens, the T1R1/T1R3 receptor had a high affinity for L-alanine and L-serine in an '*in vitro*' study (Baldwin *et al.*, 2014). Other amino acid and peptone receptors identified in rodents are also present in the chicken genome and are expressed in the gastrointestinal tract (GIT) including GPR92, mGluR1 and 4, GPRC6A and CaSR (Roura *et al.*, unpublished results).

Fatty acid sensing

It has become apparent in recent years that birds and mammals may have the ability to perceive fatty acids (Sclafani, 2015). Fat sensing involves several GPCR receptors expressed on sensory cells in and outside the oral cavity: GPR40 and 120 which sense medium and long chain fatty acids; GPR41 and 43 sensing short chain fatty acids; and the fatty acid translocase CD36 (Fukuwatari *et al.*, 1997).

Bitter compounds sensing

Chickens have only three functional bitter taste receptors (while humans have 25) which may indicate that they have a much lower capacity to sense bitter compounds (Dey *et al.*, 2017). However Behrens *et al.* (2014) have recently shown that the bitter taste receptors in chicken are broadly tuned and may account for the majority of the bitter compounds known in humans. In addition, further evidence of the deterrence of chickens to consume bitter compounds have been published in the past including denatonium benzoate and quinine (Balog and Millar, 1989; Vermaut *et al.*, 1997). Bitter compounds have also been effective in reducing feather pecking in laying hens (Harlander-Matauschek and Rodenburg, 2011).

Nutrient sensing in the gut and the post-prandial regulation of appetite

In recent years it has become apparent that the main groups of nutrient sensing GPCR were not only present in the oral cavity but, also throughout the GIT (Margolskee *et al.*, 2007; Moran *et al.*, 2010; Kitamura *et al.*, 2014; Batchelor *et al.*, 2011; Hofer and Drenckhahn, 1999). Consequently, not only oral but also extra-oral nutrient sensory mechanisms are main players orchestrating the pre and post-prandial regulation of feed intake to guarantee nutritional balance. Chickens swallow foods with little oral manipulation, thus, it is tempting to speculate that gut nutrient sensing may have even higher relevance in birds than it does in mammalian species. Thus, learning about how the GIT senses dietary nutrients will contribute substantially to better understanding diet selection in chickens (Harlander-Matauschek *et al.*, 2008). Taste and nutrient sensors are expressed in intestinal enteroendocrine cells mediating the release of gut hormones CCK and GLP-1 among others (Tanaka *et al.*, 2008; Liou *et al.*, 2011; Hirasawa *et al.*, 2005). To date, very little is known in chickens, but unpublished results from our lab would confirm the presence of a diffused chemosensory system of cells in the GIT similar to what has been described in mammals.

Nutrient specific appetites and diet selection in chickens

'Anything that a chicken needs will be sensed through sensory cues.'
(Forbes and Shariatmadari, 1994)

The concept of nutrient specific appetite relates to the capacity of animals to sense and select feeds or feed ingredients differing in one specific nutrient to balance nutritional deficiencies, meet nutrient requirements and/or ease gastrointestinal discomfort (Villalba and Provenza, 2007; Forbes, 2009; Forbes and Shariatmadari, 1994; Roura and Navarro, 2017). Chicken taste was found to be an important determinant of feed choice and consumption which, in turn, were highly correlated with growth performance and feed efficiencies in broiler chickens (Kare and Medway, 1959; Gentle, 1972; Furuse *et al.*, 1996; El Boushy and van der Poel, 2013; De Verdal *et al.*, 2011). In addition, the avian taste system was found to guide foraging decisions aimed at covering nutritional needs (Forbes and Shariatmadari, 1994; Emmans, 1991; Duncan, 1992; Henuk and Dingle, 2002). For example, chickens have been shown to develop specific amino acid appetites for lysine, methionine, and tryptophan, some of the most limiting essential amino acids in poultry feeding (Newman and Sands, 1983; Steinruck *et al.*, 1990; Picard *et al.*, 1993). Chickens offered a balanced/complete mix of free dietary amino acids showed lower feed intake than a supplement based on soy protein isolate (Siegert *et al.*, 2016). Whilst taste aversion to some free amino acids may explain the later, it is also likely that amino acid sensing in outside the oral cavity may have played an important role.

Broilers and laying hens were able to select among various feed ingredients and taste active compounds such as aspartame, saccharin, citric acid and salt (Duncan, 1992; Emmans, 1991; Gentle, 1972). In addition, chickens have shown preferences for linoleic and oleic acids as well as different plant oils (Sawamura *et al.*, 2015; Furuse *et al.*, 1996). Taste aversions to bitter compounds or some seeds (e.g. Jojoba oilseed) have also been reported (Balog and Millar, 1989; Vermaut *et al.*, 1997; El Boushy and van der Poel, 2013).

Individual differences in taste (nutrient) preference

According to the principle of nutrient specific appetites, individual variations in digestive and metabolic efficiencies on nutrient utilization have an influence on diet selection and intake. Pousga and co-workers reported individual variations in diet selection in chickens which were potentially reflecting different protein and energy requirements and/or efficiencies (Pousga *et al.*, 2005). In a multiple choice experiment Joshua and Muller reported that the variations on calcium preference and intake they found were related to individual differences on calcium requirement (Joshua and Mueller, 1979).

In laying hens, a dietary deficiency of amino acids influences feather pecking behaviour only in a selected group of birds in the flock also refereed as feather peckers (Van Krimpen *et al.*, 2005). Similarly, methionine and lysine specific appetites have been observed in feather eating hens but not in non-feather eating animals (Van Krimpen *et al.*, 2005; Savory, 1995; Cho *et al.*, 2017). In addition, the feathers consumed by feather eating birds were partially digested resulting in supplemental dietary amino acid source (Prescilla *et al.*, 2017).

One of the main research topics in our group is currently based on the hypothesis that broiler chickens showing high growth rates show different appetite for dietary limiting amino acids when compared to slow growing animals. Preliminary results showed that slow growing chickens had a specific appetite for non-limiting amino acids when supplemented to a commercial broiler diet (Niknafs *et al.*, 2017). These findings warrant further investigation.

Part of these individual differences seem to be explained by differences in the level of expression or the sensitivity of nutrient sensors in mammals and birds (Roura *et al.*, 2013). For example, Byerly and co-workers studied a unique model of obesity in chickens by genetically selecting a fat and lean line. Their findings supported the role of the T1R1 in energy metabolism showing a higher level of expression of this gene in the hypothalamus, liver and adipose tissues of the fat group compared to the lean subpopulation (Byerly *et al.*, 2010). However, a potential role of T1R1 gene polymorphisms between the two lines of broilers (fat vs lean) was not assessed. That brings to surface a profound gap in our knowledge in chicken nutrition regarding the genetic polymorphisms in nutrient sensors. It is tempting to speculate that differences in growth and other foraging behaviours (such as feather pecking) may be partially explained by genetic diversity of our genes of interest. That is topic that deserves future research attention.

Conclusions

The nutrient sensory system consists of a network of sensory cells present in the oral cavity and all along the gastro-intestinal tract monitoring the nutritional status of the animal and playing an important role in energy homeostasis and the hunger-satiety cycle. Whilst normal rearing conditions in the poultry industry are based in flock needs, the sensory system responds to individual needs creating a conflict which, in turn, may deter producers from achieving efficient flock productivity. Outside the most limiting nutrients (such as energy and essential limiting amino acids) there is a profound lack of understanding of how dietary nutrients affect feed intake and growth. Many of these nutrients are not even regularly monitored in normal poultry feed formulation hence potentially being supplied in excess to keep an optimal flow of the metabolic pathways. Thanks to genetic and molecular advances it is currently an appropriate time to re-visit nutrient needs from a new perspective (particularly regarding excess nutrients). Examples such as the implication of nutrient sensing and genetic polymorphisms on the incidence of feather pecking in laying hens and lack of growth uniformity in broiler chickens maybe showing a path forward worth to pursue.

References

Baldwin, M.W., Toda, Y., Nakagita, T., O'Connell, M.J., Klasing, K.C., Misaka, T., Edwards, S.V. and Liberles, S.D. 2014. *Science* 345: 929-933.

Balog, J. and Millar, R. 1989. *Poultry Science* 68: 1519-1526.

Barkley, G., Miller, H. and Forbes, J. 2004. *British Journal of Nutrition* 92: 233-240.

Batchelor, D. J., Al-Rammahi, M., Moran, A. W., Brand, J. G., Li, X., Haskins, M., German, A. J. and Shirazi-Beechey, S. P. 2011. *Physiol Regul Integr Comp Physiol* 300: R67-R75.

Behrens, M., Korsching, S. I. and Meyerhof, W. 2014. *Molecular Biology and Evolution* 31: 3216-3227.

Berkhoudt, H. 1985. *Form and function in birds* 3: 463-496.

Byerly, M. S., Simon, J., Cogburn, L. A., Le Bihan-Duval, E., Duclos, M. J., Aggrey, S. E. and Porter, T. E. 2010. *Physiol Genomics* 42:157-67.

Cerrate, S., Wang, Z., Coto, C., Yan, F. and Waldroup, P. 2007. *International Journal of Poultry Science* 6: 713-724.

Chah, C. and Moran, E. 1985. *Poultry Science* 64: 1696-1712.

Cho, S., Kim, J. and Roura, E. 2017. Feather-eating hens show specific essential amino acid appetites in a double-choice model. *Australian Poultry Science Symposium.* Sydney, Australia.

Cho, S. and Roura, E. 2017. Nutrient appetites in feather-eating laying hens. *Australasian Veterinary Poultry Association.* Sydney, Australia.

De Verdal, H., Narcy, A., Bastianelli, D., Chapuis, H., Même, N., Urvoix, S., Le Bihan-Duval, E. and Mignon-Grasteau, S. 2011. *BMC genetics* 12: 59.

Dey, B., Kawabata, F., Kawabata, Y., Yoshida, Y., Nishimura, S. and Tabata, S. 2017. *Biochemical and Biophysical Research Communications* 482: 693-699.

Duncan, C. 1962. *Physiological Zoology* 35: 120-132.

Duncan, I. J. H. 1992. *Poultry Science* 71: 658-663.

El Boushy, A. and Van Der Poel, A. F. 2013. *Handbook of poultry feed from waste: processing and use,* Springer Science and Business Media.

Emmans, G. C. 1991. *Proceedings of the Nutrition Society* 50: 59-64.

Forbes, J. 2009. *Voluntary feed intake in pigs. Wageningen, Wageningen Academic Publishers*: 61-86.

Forbes, J. M. 2003. *Avian and Poultry Biology Reviews* 14: 175-193.

Forbes, J. M. and Covasa, M. 1995. *World's Poultry Science Journal* 51: 149-165.

Forbes, J. M. and Shariatmadari, F. 1994. *World's Poultry Science Journal* 50: 7-24.

Fukuwatari, T., Kawada, T., Tsuruta, M., Hiraoka, T., Iwanaga, T., Sugimoto, E. and Fushiki, T. 1997. *FEBS Letters* 414: 461-464.

Furuse, M., Mabayo, R. T. and Okumura, J.-I. 1996. *Japanese Poultry Science* 33: 256-260.

Gentle, M. 1972. *British Poultry Science* 13: 141-155.

Harlander-Matauschek, A. and Rodenburg, T. B. 2011. *Applied Animal Behaviour Science* 132: 146-151.

Harlander-Matauschek, A., Wassermann, F., Zentek, J. and Bessei, W. 2008. *Poultry Science* 87: 1720-1724.

Henuk, Y. L. and Dingle, J. G. 2002. *World's Poultry Science Journal* 58: 199-208.

Hirasawa, A., Tsumaya, K., Awaji, T., Katsuma, S., Adachi, T., Yamada, M., Sugimoto, Y., Miyazaki, S. and Tsujimoto, G. 2005. *Nature Medicine* 11: 90-94.

Hofer, D. and Drenckhahn, D. 1999. *Histochemistry and Cell Biology* 112: 79-86.

Holcombe, D. J., Roland, D. A. and Harms, R. H. 1976. *Poultry Science* 55: 1731-1737.

Hughes, B. O. and Dewar, W. A. 1971. *British Poultry Science* 12: 255-258.

Hughes, B. O. and Whitehead, C. C. 1974. *Br Poult Sci* 15: 435-9.

Hughes, B. O. and Whitehead, C. C. 1979. *Applied Animal Ethology* 5: 255-266.

Hughes, B. O. and Wood-Gush, D. G. M. 1971a. *Physiology and Behavior* 6: 331-339.

Hughes, B. O. and Wood-Gush, D. G. M. 1971b. *Animal Behaviour* 19: 490-499.

Joshua, I. G. and Mueller, W. J. 1979. *British Poultry Science* 20: 481-490.

Jukes, T. H. 1938. *Poultry Science* 17: 227-234.

Kaminska, B. 1979. *Instytut Zootechniki Prace Bad* 8: 47-57.

Kare, M. R. and Medway, W. 1959. *Poultry Science* 38: 1119-1127.

Kare, M.R. and Mason, J.R. 1986. The chemical senses in birds, in: Sturkie, P.D. (Ed.), Avian Physiology 4th edition. Springer-Verlag Inc., New York, USA, pp. 59-67.

Kaufman, L. W., Collier, G. and Squibb, R. L. 1978. *Physiology and Behavior* 20: 339-344.

Kirchgessner, A. L. and Gershon, M. D. 1990. *J Neurosci* 10: 1626-42.

Kitamura, A., Tsurugizawa, T., Uematsu, A. and Uneyama, H. 2014. *Current Pharmaceutical Design* 20: 2713-2724.

Klasing, K. 1998. *Poultry Science* 77: 1119-1125.

Kutlu, H. R. and Forbes, J. M. 1993. *Livestock Production Science* 36: 335-350.

Liou, A. P., Lu, X., Sei, Y., Zhao, X., Pechhold, S., Carrero, R. J., Raybould, H. E. and Wank, S. 2011. *Gastroenterology* 140: 903-912.e4.

Margolskee, R. F., Dyer, J., Kokrashvili, Z., Salmon, K. S. H., Ilegems, E., Daly, K., Maillet, E. L., Ninomiya, Y., Mosinger, B. and Shirazi-Beechey, S. P. 2007. *Proceedings of the National Academy of Sciences* 104: 15075-15080.

Meyerhof, W. and Korsching, S. 2009. *Preface*, in: Meyerhof, W. and Korshing, S. (Eds.). Chemosensory systems in mammals, fishes, and insects. Springer-Verlag Inc., New York, USA, pp. v-xi.

Moran, A. W., Al-Rammahi, M. A., Arora, D. K., Batchelor, D. J., Coulter, E. A., Daly, K., Ionescu, C., Bravo, D. and Shirazi-Beechey, S. P. 2010. *British Journal of Nutrition,* 104: 637-646.

Moran, E. and Stilborn, H. 1996. *Poultry science* 75: 120-129.

Nelson, G., Chandrashekar, J., Hoon, M. A., Feng, L., Zhao, G., Ryba, N. J. and Zuker, C. S. 2002. *Nature* 416: 199-202.

Newman, R. K. and Sands, D. C. 1983. *Physiology and Behavior* 31: 13-19.

Niknafs, S., Kim, J. and Roura, E. 2017. Fast and slow-growing broiler chickens show different appetite for limiting and non-essential amino acids. *Australian Poultry Science Symposium.* Sydney, Australia.

Picard, M. L., Uzu, G., Dunnington, E. A. and Siegel, P. B. 1993. *British Poultry Science* 34: 737-746.

Pollak, L. and Scott, M. 2005. *Maydica* 50: 247.

Pousga, S., Boly, H. and Ogle, B. 2005. *Livestock Research for Rural Development* 17.

Prescilla, K. M., Cronin, G. M., Liu, S., Hartcher, K. M. and Singh, M. 2017. Measuring feather appetite and feather digestibility in ISA Brown hens. *Australian Poultry Science Symposium.* Sydney, Australia.

Roura, E., Baldwin, M. W. and Klasing, K. C. 2013. *Animal Feed Science and Technology* 180: 1-9.

Roura, E., Humphrey, B., Tedo, G. and Ipharraguerre, I. 2008. *Canadian Journal of Animal Science* 88: 535-558.

Roura, E. and Navarro, M. 2017. *Animal Production Science (in press)*.

Sahin, A. and Forbes, J. 1999. *British Poultry Science* 40: 52-54.

Savory, C. J. 1995. *World's Poultry Science Journal* 51: 215-219.

Sawamura, R., Kawabata, Y., Kawabata, F., Nishimura, S. and Tabata, S. 2015. *Biochemical and Biophysical Research Communications* 458: 387-391.

Sclafani, A. 2015. Flavor Preferences in Animals: Role of Mouth and Gut Nutrient Sensors. *Frontiers in Integrative Neuroscience.*

Sclafani, A. and Ackroff, K. 2012. *American Journal of Physiology-Regulatory, Integrative and Comparative Physiology* 302: R1119-R1133.

Plenary session 01. Feed intake, thermoregulation and heat stress

Shariatmadari, F. and Forbes, J. 1993a. *British Poultry Science* 34: 959-970.
Shariatmadari, F. and Forbes, J. M. 1993b. *British Poultry Science* 34: 959-970.
Shi, P. and Zhang, J. 2006. *Molecular Biology and Evolution* 23: 292-300.
Siegert, W., Wild, K. J., Schollenberger, M., Helmbrecht, A. and Rodehutscord, M. 2016. *British Poultry Science* 57: 424-434.
Skelhorn, J. and Rowe, C. 2010. *Proc Biol Sci* 277: 1729-34.
Sneddon, H., Hadden, R. and Hepper, P. G. 1998. *Physiology and Behavior* 64: 133-139.
Steinruck, U., Roth, F. X. and Kirchgessner, M. 1990. *Archiv fur Geflugelkunde* 54: 173-183.
Stewart, J. R., Ecay, T. W., Heulin, B., Fregoso, S. P. and Linville, B. J. 2011. *The Journal of Experimental Biology* 214: 2999-3004.
Sukumaran, S. K., Yee, K. K., Iwata, S., Kotha, R., Quezada-Calvillo, R., Nichols, B. L., Mohan, S., Pinto, B. M., Shigemura, N., Ninomiya, Y. and Margolskee, R. F. 2016. *Proceedings of the National Academy of Sciences* 113: 6035-6040.
Syafwan, S., Wermink, G. J. D., Kwakkel, R. P. and Verstegen, M. W. A. 2012. *Poultry Science* 91: 537-549.
Tanaka, T., Katsuma, S., Adachi, T., Koshimizu, T.-A., Hirasawa, A. and Tsujimoto, G. 2008. *Naunyn-Schmiedeberg's Archives of Pharmacology* 377: 523-527.
Van Krimpen, M. M., Kwakkel, R. P., Reuvekamp, B. F. J., Van Der Peet-Schwering, C. M. C., Den Hartog, L. A. and Verstegen, M. W. A. 2005. *World's Poultry Science Journal* 61: 663-686.
Vermaut, S., De Coninck, K., Flo, G., Cokelaere, M., Onagbesan, M. and Decuypere, E. 1997. *Journal of Agricultural and Food Chemistry* 45: 3158-3163.
Villalba, J. J. and Provenza, F. D. 2007. *Animal* 1: 1360-1370.
Wilkinson, S., Bradbury, E., Bedford, M. and Cowieson, A. 2014. *Poultry Science* 93: 1695-1703.
Wood-Gush, D. G. M. and Kare, M. R. 1966. *British Poultry Science* 7: 285-290.

Factors affecting broiler volunteer feed intake in practice

H.L. Classen[*]*, E. Herwig, N. Karunaratne, D.D.L.S. Bryan and R.K. Savary*
Department of Animal and Poultry Science, University of Saskatchewan, Saskatoon, SK, S7N 5A8, Canada; hank.classen@usask.ca

Summary

Feed intake plays an important role in determining the performance potential of broiler chickens and is affected by a large range of nutritional and non-nutritional factors. A deficiency or excess of dietary nutrients will negatively impact feed intake, but for most nutrients, levels within commercially acceptable ranges have a limited effect on feed intake. Dietary components that are known to cause anorexia, such as myotoxins, need to be avoided and maximum feed intake is achieved by feeding high quality pelleted diets. Water is a key nutrient and because of the strong relationship between water and feed intake, availability must be ensured. Environmental factors play an important role in regulating feed intake and are highly interrelated. Management practices that minimize stress and provide optimal environmental conditions increase feed intake and broiler productivity. Disease reduces feed intake and therefore strategies that minimize disease potential are warranted. In conclusion, providing broilers with a well-balanced, palatable diet and good husbandry provides an opportunity to increase feed intake and productivity.

Introduction

Feed intake is a key index in broiler chicken production and is impacted by a complex assortment of factors. A close association exists between feed intake and growth rate, but establishing a cause and effect relationship is often not obvious. Therefore, research and production results must be interpreted with some caution. Never-the-less, promoting feed intake is an appropriate strategy to maximize broiler performance and understanding the factors that impact feed intake is fundamental to achieving this goal. This review will examine nutritional and environmental factors that influence or are thought to influence feed consumption.

Nutrition

Nutrition can have a major impact on feed intake, with deficiency and excess of most nutrients both decreasing feed intake. However, these extreme effects are only present after feed mixing errors and are not of practical significance. Therefore, this section will focus on nutritional factors that may impact feed intake under more commercially relevant nutrient levels and processing conditions.

Energy and protein

Of all nutrients, diet energy has been most discussed in relationship to feed intake as it relates to energy homeostasis, which is fundamental to animal well-being. Research in this area is extensive and goes back to before the development of the broiler industry. It has been reviewed on many occasions, including by the senior author (Classen, 2016). Although historically chickens were assumed eat to meet their energy requirements, early research (Lacy *et al.*, 1986) identified that *feed* intake control mechanisms of broilers selected for growth were not the same as more slowly growing birds. Similarly, broiler genotypes responded differently to diet energy with some showing no feed intake response (Pym, 2005). Additional research has convincingly found that diet energy does not accurately predict feed intake in broiler chickens (Classen, 2016). Therefore, manipulating diet energy is not likely to affect feed intake in a positive manner.

When birds are fed well balanced diets at or above their amino acid (AA) requirements for growth and maintenance, increasing AA density does not affect feed intake (Corzo *et al.*, 2005; Hernandez *et al.*, 2012). However, both amino acid deficiency and imbalance promote feed intake responses. If diets are balanced for key AA, but the levels are lower than the bird's growth and maintenance requirement, feed intake increases in an attempt to maintain AA intake to meet the bird's requirement (Dozier *et al.*, 2007; Taschetto *et al.*, 2012). A severe AA imbalance plus low levels of a single essential AA tend to produce two main effects on feed intake. If diets have a severe deficiency in one or two indispensable AA's, broilers will reduce feed intake regardless of dietary AA density (Awad *et al.*, 2014). If the deficiency results in an AA antagonism (e.g. lysine, arginine), birds will consume less diet to reduce the effect of the antagonism (Ghahri *et al.*, 2010).

For both diet energy and balanced digestible amino acid content, good practice in the formulation of a balanced diet is key to maximizing feed intake and bird production. If this occurs, little is to be gained from a feed intake perspective by altering their levels.

Fibre

Fibre in poultry feeds has gained more attention in recent years with emphasis on insoluble and soluble fibre. Recent studies have observed that the presence of a moderate amount of insoluble fibre in the diet increases digesta retention time in the upper gastro-intestinal tract (GIT), which improves gizzard function, facilitates enzyme-substrate interaction and improves digestibility (Mateos *et al.*, 2012). At the same time, insoluble fibre may increase feed passage in the lower digestive tract and thereby increase feed intake (Hetland and Svihus, 2001; Jiménez-Moreno *et al.*, 2010). Of note, the insoluble fibre effect on feed passage rate is dependent on large particle size (Hetland and Svihus, 2001; Sacranie *et al.*, 2012).

In contrast, some types of soluble fibre (e.g. high molecular weight arabinoxylans and β-glucans) increase intestinal digesta viscosity and cause distension of the GIT. In turn, soluble fibre induced viscosity can slow feed passage rate and reduce broiler feed intake (Salih *et al.*, 1991; González-Alvarado *et al.*, 2010). The high water absorbing capacity of soluble fibre increases the bulk of digesta in the GIT and may also reduce feed intake as a result of increased satiety (Mateos *et al.*, 2012). Fortunately, dietary exogenous carbohydrase activity is commonly used to decrease the viscosity of soluble fibre, which negates their impact on feed intake.

Feed palatability

Chickens are highly capable of recognizing differences in feed that impact its consumption. Therefore, changes in the nature of feed or the presence of less palatable ingredients can reduce feed intake. A variety of factors in ingredients can impact diet acceptability, but likely of most importance are mycotoxins. Mycotoxins are a global problem and are well known to affect feed intake (Andretta *et al.*, 2011).

Feed form and processing

Feed form affects broiler performance with pelleted diets resulting in faster and more efficient growth (McKinney and Teeter, 2004; Abdollahi *et al.*, 2013). The increase in productivity is thought to relate to feed intake and feeding behaviour. Mash diets have reduced volumetric density and this affects the bird's ability to consume sufficient feed for maximum production. In addition, fine particle size in mash diets can reduce diet palatability. Broilers fed pellets also benefit from reduced energy use due to less time eating and more time resting (McKinney and Teeter, 2004). Because of the latter energy savings and possibly other effects, pelleting diets provides a higher effective caloric value than mash feed (McKinney and Teeter, 2004). As might be expected, the degree of energy savings is also dependent on pellet quality.

Water

Adequate water intake is required to maintain feed intake in broiler chickens. When water is unavailable or restricted, birds will consume less feed, resulting in reduced growth rate and poorer feed conversion (Marks, 1981). Elements that may influence the amount of water consumed include nutritional and environmental factors. For example, increasing dietary protein (Alleman and Leclercq, 1997), soluble non-starch polysaccharides (Langhout *et al.*, 2000) and minerals (Borges *et al.*, 2004) all increase water intake. The most important environmental factor affecting water intake is ambient air temperature. Heat stressed broilers consume more water and less feed than those maintained at a lower ambient temperature (May and Lott, 1992; Alleman and Leclercq, 1997). Water temperature also affects feed intake with cooler water usually increasing water intake and consequently feed intake (Harris Jr. *et al.*, 1975). Both waterer type (e.g. bell vs nipple drinkers) and management can affect water intake with the impact most important under high temperature conditions (May *et al.*, 1997; Bruno *et al.*, 2011).

Bird management and environment

Temperature

Providing broilers with an appropriate environment plays an important role in maintenance of core body temperature and flock productivity. The bird's perception of its thermal environment is based on the combination of absolute temperature and relative humidity. Providing the appropriate thermal environment during the brooding period is essential for broiler performance to ensure adaptation to the barn environment and maximum feed intake. Both low and high temperatures during this time have been shown to reduce feed intake (Moraes *et al.*, 2002). Heat stress after the brooding period adversely affects feed consumption and bird performance, and is particularly relevant as birds approach marketing. Feed consumption and utilization are associated with heat production, so the reduction in feed intake is an attempt to reduce heat production and maintain homeothermy (Quinteiro-Filho *et al.*, 2012).

Stocking density

Stocking density effects on broiler feed intake, performance and welfare are complex (Bessei, 2005; Estevez, 2007). As stocking density increases, feed intake decreases, and both direct and indirect mechanisms of action have been proposed. Direct effects relate to the stress associated with housing density and are mediated via the hypothalamic-pituitary adrenal axis HPA axis. Indirect effects relate to other aspects of the environment that are influenced by stocking density. Ensuring that the equipment needs (feeders, waterers, ventilation) of broiler production are met is an obvious consideration when increasing stocking density. However, even if sufficient equipment is in place, the increased mass of birds (floor covering) can still affect these parameters. The efficiency of ventilation at bird level is affected (litter drying, temperature control) and more excreta from more birds affects litter quality and microbial activity, which in turn increases the production of ammonia. In the case of wet litter, the feed intake effect is possibly due to an increase in the incidence of foot pad dermatitis and poor skeletal health, which in turn reduces bird mobility and feeder access. Decreased exercise due to high density housing may further exacerbate this effect. In apparent disagreement with this argument is the finding that feeding behaviour is not affected by housing density. The production of gases such as ammonia may also impact feed intake, with the degree of related to ammonia level and time of exposure. Because of the increased number of birds and floor coverage, the temperature immediately surrounding birds is higher than ambient temperature under high density housing (Reiter and Bessei, 2000).

Light

Chickens are diurnal and feed intake takes place during the day with a very characteristic pattern of increased feeding at the beginning and end of the day. The former, is the consequence of the depletion of energy storage during the night and the latter, known as anticipatory feeding, is observed under diurnal lighting regimens in which the dark period is greater than four hours (Savory, 1976). The response of feed intake to daylength is quadratic and in a comparison of 14, 17, 20 and 23 hours of light, the lowest and highest intake were found for the 14 and 20 hour daylengths, respectively (Schwean-Lardner *et al.*, 2012). Variation in light intensity from 1 to 100 lux has shown no effect on feed intake (e.g. Kristensen *et al.*, 2006; Deep *et al.*, 2010). Light spectrum or wavelength affects feed intake although results are inconsistent; broilers reared under green, and blue to a lesser extent, light appear to have a higher feed intake (Rozenboim *et al.*, 1999; Karakaya *et al.*, 2009).

Bird health

In a disease state it is common for chickens to display symptoms such as reduced feed intake, growth rate, and feed efficiency (Klasing *et al.*, 1987; Korver *et al.*, 1998; Takahashi *et al.*, 2002). Up to 70% of the loss in performance can be accounted for by reduced feed intake while the other 30% is attributed to immunological or metabolic changes unrelated to feed intake (Klasing *et al.*, 1987). Although the effect of disease on feed intake is complex, pro-inflammatory cytokines such as IL-1 reduce feed intake.

Conclusions

Consumption of nutrients is essential to achieve the genetic potential of broilers. Providing a well-balanced pelleted diet, an unlimited water supply, a suitable environment and good health management are the basis for achieving maximum feed intake and productivity.

References

Abdollahi, M.R., Ravindran, V., Svihus, B. 2013. *Animal Feed Science and Technology* 179:1-23.

Alleman, F., Leclercq, B. 1997. *British Poultry Science* 38: 607-610.

Andretta, I., Kipper, M., Lehnen, C.R., Hauschild, L., Vale, M.M., Lovatto, P.A. 2011. *Poultry Science* 90: 1934-1940.

Awad, E.A., Fadlullah, M., Zulkifli, I., Farjam, A.S., Chwen, L.T. 2014. *Italian Journal of Animal Science* 13: 3166-3172.

Bessei, W. 2006. *World's Poultry Science Journal* 62: 455-466.

Borges, S.A., Fischer da Silva, A.V., Majorka, A., Hooge, D.M., Cummings, K.R. 2004. *Poultry Science* 83: 1551-1558.

Bruno, L.D.G, Maiorka, A., Macari, M., Furlan, R.L., Givisiez, P.E.N. 2011. *Brazilian Journal of Poultry Science* 13: 147-152.

Classen, H.L. 2016. *Animal Feed Science and Technology* http://dx.doi.org/10.1016/j.anifeedsci.2016.03.004

Corzo, A., Fritts, C.A., Kidd, M.T., Kerr, B.J. 2005. *Animal Feed Science and Technology* 118:319-327.

Deep, A., Schwean-Lardner, K., Crowe, T.G., Fancher, B.I., Classen, H.L. 2010. *Poultry Science* 89: 2326-2333.

Dozier W.A., Corzo, A., Kidd, M.T., Branton, S.L. 2007. *Journal of Applied Poultry Research* 16: 192-205.

Estevez, I. 2007. *Poultry Science* 86: 1265-1272.

Ghahri, H., Gaykani, R., Toloie, T. 2010. *African Journal of Agricultural Research* 5: 1228-1234.

González-Alvarado, J.M., Jiménez-Moreno, E., González-Sánchez, D., Lázaro, R., Mateos, G.G. 2010. *Animal Feed Science and Technology* 162: 37-46.

Harris, Jr., G.C., Nelson, G.S., Seay, R.L., Dodgen, W.H. 1975. *Poultry Science* 54: 775-779.

Hernández, F., López, M., Martínez, S., Megías, M.D., Catalá, P., Madrid. J. 2012. *Poultry Science* 91: 683-692.

Hetland, H., Svihus, B. 2001. *British Poultry Science* 42: 354-361.

Jiménez-Moreno, E., González-Alvarado, J.M, González-Sánchez, D., Lázaro, R., Mateos, G.G. 2010. *Poultry Science* 89: 2197-2212.

Karakaya, M., Parlat, S.S., Yilmaz, M.T., Yildirim, I., Ozalp, B. 2009. *British Poultry Science* 50:76-82.

Klasing, K.C., Laurin, D.E., Peng, R.K., Fry, D.M. 1987. *The Journal of Nutrition* 117: 1629-2637.

Korver, D.R., Roura, E., Klasing, K.C. 1998. *Poultry Science* 77: 1217-1227.

Kristensen, H.H., Perry, G.C., Prescott, N.B., Ladewig, J., Ersbøll, A.K., Wathes, C.M. 2006. *British Poultry Science* 47: 257-263.

Lacy, M.P., Van Krey, H.P., Skewes, P.A., Denbow, D.M. 1986. *Physiology and Behavior* 36: 533-538.

Langhout, D.J., Schutte, J.B., de Jong, J., Sloetjes, H., Verstegen, M.W.A., Tamminga, S. 2000. *British Journal of Nutrition* 83: 533-540.

Marks, H.L. 1981. *Poultry Science* 60: 698-707.

Mateos, G.G., Jiménez-Moreno, E., Serrano, M.P., Lázaro, R.P. 2012. *Journal of Applied Poultry Research* 21: 156-174.

May, J.D., Lott, B.D. 1992. *Poultry Science* 71: 331-336.

May, J.D., Lott, B.D., Simmons, J.D. 1997. *Poultry Science* 76: 944-947.

McKinney, L.J., Teeter, R.G. 2004. *Poultry Science* 83: 1165-1174.

Moraes, V.M.B., Malheiros, R.D., Furlan, R.L., Bruno, L.D.G., Malheiros, E.B., Macari, M. 2002. *Revista Brasileira de Ciência Avícola* 4. doi: 10.1590/S1516-635X2002000100003.

Pym, R.A.E. 2005. *Proceedings of the Australian Poultry Science Symposium* 17: 153-162.

Quinteiro-Filho, W.M., Rodrigues, M.V., Ribeiro, A., Ferraz-de-Paula, V., Pinheiro, M.L., Sá, L.R. M., Ferreira, A.J.P., Palermo-Neto, J. 2012. *Journal of Animal Science* 90: 1986-1994.

Reiter, K., Bessei, W. 2000. *Archiv für Geflügelkunde* 64: 204-206.

Rozenboim, I., Biran, I., Uni, Z., Robinzon, B., Halevy, O. 1999. *Poultry Science* 78: 135-138.

Sacranie, A., Svihus, B., Denstadli, V., Moen, B., Iji, P.A., Choct, M. 2012. *Poultry Science* 91: 693-700.

Salih, M.E., Classen, H.L., Campbell, G.L. 1991. *Animal Feed Science and Technology* 33: 139-149.

Savory, C.J. 1976. *British Poultry Science* 17: 557-560.

Schwean-Lardner, K., Fancher, B.I., Classen, H.L. 2012. *British Poultry Science* 53: 7-18.

Takahashi, K., Kawamata, K., Akiba, Y., Iwata, T., Kasai, M. 2002. *British Poultry Science* 43: 47-53.

Taschetto, D., Vieira, S.L., Angel, R., Favero, A., Cruz, R.A. 2012. *Livestock Science* 146: 183-188.

Physiology of thermoregulation in broilers, and strategies for sustaining performance under hot conditions

S. Druyan[1*] and A. Haron[1,2]

[1]Institute of Animal Science, Agricultural Research Organization, Volcani Center, HaMaccabim Road, Rishon Le Tsiyon P.O. Box 15159, 7528809, Israel; [2]The Robert H. Smith Institute of Plant Sciences and Genetics in Agriculture, The Hebrew University of Jerusalem, Rehovot 7610001, Israel; shelly.druyan@mail.huji.ac.il

Summary

The continuous genetic selection for performance traits resulted in a considerable enhancement of daily feed consumption, leading to alterations in growth mechanisms and development. These developments were not accompanied by the necessary increases in the size of the cardiovascular and respiratory systems, nor sufficient enhancements of their functional efficiency. This has resulted in a relatively low capability for maintaining adequate dynamic steady-state mechanisms in the body that should balance energy expenditure under extreme environmental conditions. Thus, modern broilers have an elevated metabolic rate and consequently elevated internal heat production that leads to insufficient maintenance of dynamic steady-state of thermoregulation processes, resulting in enhancement of body temperature fluctuations. Epigenetic adaptation seems to be a suitable means for achieving the goal of improved broilers resistance to changes in environmental conditions pre- and post-hatch. Environmental manipulations during the critical phases of embryonic development may induce alterations in the thermoregulatory control system. With regard to incubation environmental conditions, it seems that low O_2 partial pressure (hypoxia) can play a role, affecting embryonic development, metabolism, oxygen demands and the available energy for post-hatch growth and development. Hypoxic daily exposures of 12 h or continuously for 48 h to hypoxia of 17% from E16 to E18 (the plateau period), caused a decrease in the metabolic rate of the embryo. This represents a metabolic adaptation characterized by a lower resting metabolic rate. Such alterations affect post-hatch performance and energy allocation between maintenance and growth, especially under heat stress.

Introduction

Body temperature control

In endothermic birds, body temperature (T_b) is the most physiologically-protected parameter of the body; therefore, the thermoregulatory system operates at a very high gain, in order to hold T_b within a narrow range, despite moderate to extreme changes in environmental temperatures. The ability to maintain a stable T_b is dependent upon regulatory mechanisms that maintain a balance between heat production and heat loss. Both processes are permanently activated and regulated by neuronal and hormonal signals (Morrison *et al.*, 2008; Richards and Proszkowiec-Weglarz, 2007). The thermoregulatory response is mediated mainly by the level of metabolism induced or permitted by the thyroid hormones axis (thyroxine, T_4; triiodothyronine, T_3) and the adrenal hormone corticosterone (Dimicco and Zaretsky, 2007). Thermogenesis (heat production) is regulated by the thyroid hormones (Mcnabb, 1995; Silva, 2006), and can be divided into obligatory and facultative thermogenesis (Silva, 2006). Obligatory thermogenesis refers to the energy required to maintain T_b, under ambient temperature (T_a) that is in the thermoneutral zone, the range in which the body is in thermal equilibrium with the environment and produces energy at the resting metabolic rate (RMR) (Gordon, 1993). Facultative thermogenesis refers to prompted production of the energy required when T_a deviates below or, to some extent, above the thermoneutral zone. Facultative thermogenesis comprises the short-term shivering thermogenesis (ST) and a long-term mechanism, the non-shivering thermogenesis (NST). In birds, NST occurs in the skeletal muscles (Dridi *et al.*, 2008). The T_a influences the rate of metabolic activity and, in turn, the amount of oxygen required by the bird for its growth (Buys *et al.*, 1999).

Plenary session 01. Feed intake, thermoregulation and heat stress

The conflict between high productivity and thermotolerance

Since the 1950s, commercial genetic selection programs had led to a dramatic improvement in broilers' production traits. This improvement was mainly due to rapid growth, high feed utilization and higher meat production characteristics (Havenstein *et al.*, 1994a; Havenstein *et al.*, 1994b; Havenstein *et al.*, 2003). Genetic selection for performance traits resulted in considerable enhancement in daily feed consumption, leading to alterations of growth mechanisms and development (Druyan, 2010). Such developments logically necessitate parallel increases in the size of the cardiovascular and respiratory systems, as well as enhancements in their functional efficiency. However, insufficient development of these major systems has led to a relatively low capability for maintaining adequate dynamic steady-state mechanisms in the body, that should balance energy expenditure and body water balance under extreme environmental conditions (Yahav, 2009). Thus, modern broilers have an elevated metabolic rate and consequently elevated internal heat production that leads to insufficient maintenance of dynamic steady-state of thermoregulation processes, resulting in the enhancement of body temperature fluctuations.

Exposure of broilers to hot conditions from 21d onward, either short-term rapid elevation in T_a (from 24 to 32 °C for 5 h on a weekly base) or chronic hot conditions characteristic of the summer season in Israel (cycling temperature between 33 and 24 °C every 12 h) has resulted in major economic losses. The body weight (BW) was found to be lower in broilers exposed to short term heat and significantly lower in broilers exposed to chronic heat, compared to control broilers (Figure 1; 2,474±38, 2,178±36 vs 2,535±37 g for short heat, chronic heat and control broilers, respectively). Under chronic heat conditions, broilers adapt to the heat by reducing feed intake (FI) resulting not only in a lower BW at standard marketing age, but also in poorer feed conversion ratio (FCR). While FCR calculated from 14 to 35 days for the chronically-exposed broilers was 1.62±0.02, FCR calculated for the control broilers was significantly better (1.54±0.02; Figure 2).

To sustain thermotolerance and avoid the deleterious consequences of thermal stresses, three direct responses are elicited by chickens: (1) rapid thermal shock response; (2) acclimation; (3) epigenetic adaptation during the perinatal period.

The epigenetic approach

Epigenetic adaptation is based on the assumption that, during the 'critical' developmental phases, environmental factors substantially affect the determination of the 'set-point' of physiological control systems (Nichelmann and Tzschentke, 2002; Tzschentke and Basta, 2002). Epigenetic adaptation seems to be a suitable means for achieving the goal of improved broilers resistance

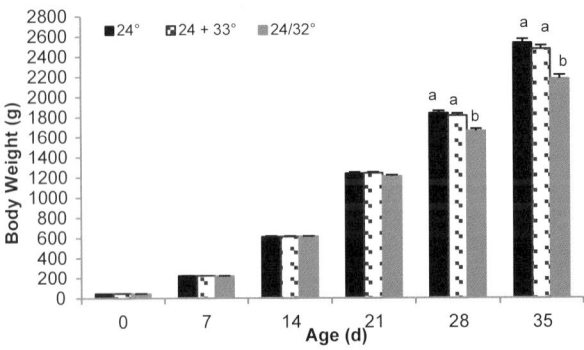

Figure 1. Body Weight (g) of male broiler chickens from hatch until 35d of age raised under either standard conditions, short-term rapid elevation in ambient temperature, or Chronic hot conditions. Different letters indicate differences (P≤0.05) among treatments.

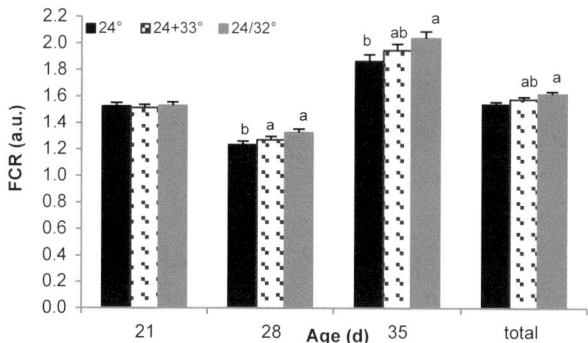

Figure 2. FCR of male broiler chickens from 14d until 35d of age raised under either standard conditions, short-term rapid elevation in ambient temperature, or Chronic hot conditions. Different letters indicate differences (P≤0.05) among treatments.

to changes in environmental conditions pre- and post-hatch. During early development, most functional systems evolve from an open-loop system without feedback to a closed control system with feedback ('transformation rule') (Dorner, 1974). Environmental manipulations during the critical phases of embryonic development may induce alterations in the thermoregulatory control system.

With regard to incubation environmental conditions, it seems that low O_2 partial pressure (hypoxia) can play a role in affecting embryonic development, metabolism, oxygen demand and the available energy for post-hatch growth and development (Haron et al., 2017; Druyan et al., 2012).

Hypoxic exposure during embryonic development

The actual effects of hypoxia on embryonic development depend on the critical period, hypoxia level and duration of exposure. Embryos exposed to hypoxia of 15% oxygen for a period of six days on embryonic days: E1-E6, E6-E12, and E12-E18 were developmentally retarded and smaller compared to Controls (Chan and Burggren, 2005). Hypoxic exposure at an earlier stage, had more severe effects, causing a significant damage to the eyes and beak (Dzialowski et al., 2001), changing the nature of yolk sac consumption (Chan and Burggren, 2005). However, we have reported in 2012 that exposure to moderate hypoxia of 17% for 12 h daily from E5 to E12 had a positive impact on the chorioalantoic membrane (CAM) which exhibited a better ability to deliver oxygen to tissues. Exposure to hypoxia during the plateau period was designed to optimize the function of the embryo circulatory system and affect its metabolic demands and energy allocation between growth and maintenance (Haron et al., 2017). Hypoxic exposure during this time window, forces the embryo to adapt to the new conditions in several ways, including: alteration in cardiac output and redistribution of oxygenated blood from the periphery to vital organs such as the brain, heart and adrenal gland (Mulder et al., 1998). Other measures include increasing blood oxygen-carrying capacity (Dusseau and Hutchins, 1988), modification of hemoglobin (Liu et al., 2009), increasing vascularization (Dusseau and Hutchins, 1989), or combination of those.

Following daily exposures of 12 h or continuously for 48 h (12H, and 48H, respectively) to hypoxia of 17% from E16 to E18, induced an upward trend in haematocrit levels compared to the Control treatment (Table 1). Upon hatch, this trend was pronounced especially in 12H embryos. Another adaptation caused by hypoxic conditions is an elevation of heart rate (HR) and an increase of stroke volume (Mortola et al., 2012). We observed differences in HR between

Plenary session 01. Feed intake, thermoregulation and heat stress

Table 1. Haematocrit level (%), heart rate (beat/min) and oxygen consumption (ml/g·h) of control, 12H and 48H chicks from eggs incubated under differing oxygen regimes from E16 to E18.

	Haematocrit			Heart rate			O_2 consumption		
	Con	12H	48H	Con	12H	48H	Con	12H	48H
E16_0	38.4±0.4			278±2	275±2	280±2	0.94±0.02		
E16_6				282±3[b]	294±3[a]	302±3[a]			
E16_12	38.5±0.4	39.6±0.4		280±2[b]	294±2[a]	299±2[a]	0.91±0.02	0.89±0.02	
E17_0	39.6±0.6	39.4±0.6	39.3±0.5	276±2[b]	271±2[b]	289±2[a]	0.95±0.02	0.9±0.02	0.9±0.02
E17_6				278±3[b]	287±3[a]	290±3[a]			
E17_12	39.9±0.6	41.0±0.5	40.5±0.5	278±3[b]	282±4[a]	293±3[a]	0.84±0.02	0.80±0.02	0.81±0.02
E18_0	39.8±0.5	40.7±0.4	40.8±0.4	273±3	265±3	272±3	0.84±0.02[a]	0.86±0.02[a]	0.74±0.02[b]
E18_6				272±3	257±3	264±3			
E18_12	40.1±0.5[b]	41.4±0.5[a]	41.1±0.5[a]	272±3	272±3	271±3	0.79±0.02	0.82±0.02	0.83±0.02
E19_0	39.1±0.5	40.4±0.6	39.9±0.5	278±4	268±4	275±4	0.81±0.02	0.83±0.02	0.79±0.02
EP	37.4±0.5	39.2±0.6	38.6±0.4						
Hatch	37.4±0.4[b]	39.2±0.5[a]	38.7±0.6ab						

[ab] Different letters indicate differences ($P \leq 0.05$) among treatments.

hypoxic and Control embryos and found them to be correlated to the time and duration of exposure. Hypoxic embryos exhibited significant increases in HR as early as 6 hours after the onset of exposure to low oxygen conditions. When embryos were returned to standard conditions, concomitant with the progress of embryogenesis, the embryonic HR declined. This may be due to a better adaptation of the circulatory system to a recurring hypoxic stress or a decrease in embryonic demands for oxygen.

In general, oxygen consumption of embryos increases with embryonic development and growth during incubation until E17 and then remains approximately constant in the plateau phase until E19 (Druyan, 2010). Following hypoxic exposure, both 12H and 48H embryos exhibited a decrease in oxygen consumption compared to Controls. We interpreted this to be a result of the hypoxic stress exposure causing lower oxygen demand per 1 g of tissue.

Additional support for this metabolic adaptation following hypoxic exposure during the plateau period was obtained from body temperature (Table 2). Body temperature of 12H and 48H chicks was significantly lower at hatch, indicating a lower heat production, which is consistent with lower metabolic rate during incubation. Possible explanations include a metabolic adaptation by decline in RMR. A lower RMR leads to lower heat production and enables embryos to invest less energy in maintenance and allocate the remaining metabolic energy to growth in order to reach

Table 2. Body weight (g), Yolk weight (g) and body temperature (°C) of control, 12H and 48H chicks incubated under differing oxygen regimes from E16 to E18.

	Con (n=211)	12H (n=223)	48H (n=213)
% hatch	94.31%	94.90%	95.52%
Chick weight (g)	46.15	46.45	46.3
Yolk weight	6.77[b]	7.70[a]	7.95[a]
Temperature (°C)	39.78[a]	39.67[b]	39.66[b]

[ab] Different letters indicate differences ($P \leq 0.05$) among treatments.

a similar hatching weight as that of the Control embryos. Such alterations may affect post-hatch performance and energy allocation between maintenance and growth, especially under suboptimal growing conditions when there is an increased energy demand.

Hypoxic Exposure during embryonic development affects post hatch thermotolerance

12H, 48H and control broilers were raised under different T_a treatments (after brooding from 14d) of either standard constant 24 °C, long-term high T_a (32 °C from 18d to 42d) or cycling chronic hot conditions: (cycling temperature between 33 °C and 24 °C 12/12h from 18d to 42d). Hypoxic exposure did not exert a long-lasting effect on BW of either 12H or 48H male chicks raised under standard growing conditions. From hatching to 42 d of age, BW of the 12H and 48H chickens was similar to control chickens. However, when the birds were raised under suboptimal environmental conditions, the effect of hypoxic exposure during the plateau period on broiler performance was pronounced (Figure 3).

Under long-term high T_a, the 12H and 48H broilers' mean body weight at 42d was slightly higher than the control (2,256, 2,317 vs 2,246 g, respectively), while under cycling chronic hot conditions, both hypoxic treatments (12H and 48H) had a significant effect on BW. At 42d, broilers from these treatments had a significantly higher BW than the controls (3,049, 3,009 vs 2,833 g respectively). One could speculate that this was related to a lower metabolic rate. Hypoxic and non-hypoxic treated broilers exhibited similar feed intake under all environmental conditions (data not shown). The FCR of the chickens raised under hypoxic conditions was found to be numerically to significantly lower (Figure 4). Those birds gained more BW than controls and had a higher weekly mean growth rate with a similar portion of feed intake.

Energy consumption in general, and in the domestic fowl in particular, has been shown to be divided between maintenance and production. In endotherms, the ability to maintain body temperature (T_b) depends on daily energy requirement for maintenance. Therefore, lowering the demands for maintenance, while keeping total energy consumption approximately constant, allows the allocation of more energy for growth. An alternative beneficial response may consist of a combination of reduced maintenance energy demands coupled with an overall reduction in energy consumption.

In the present studies, lower metabolic rates, evidenced by lower T_b at hatch, were found in the 12H and 48H broilers, indicating lower energy demands for maintenance. One could speculate that as a result of the lower metabolic rate, the BW of hypoxic broilers would be higher than that of the controls. However, significant differences in BW were found among treatments only when raised under cycling chronic hot conditions with a significantly improved FCR.

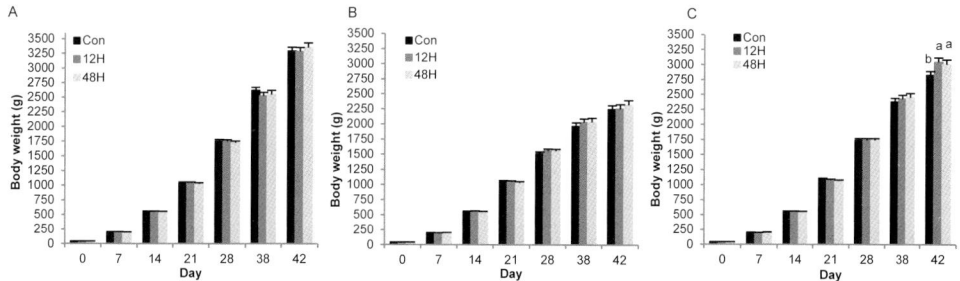

Figure 3. Body Weight (g) of Control, 12H and 48H male broiler chickens from hatch until 42d of age raised under either standard condition (A), long term high ambient temperature (B) or cycling chronic hot conditions (C). Different letters indicate differences (P≤0.05) between treatments.

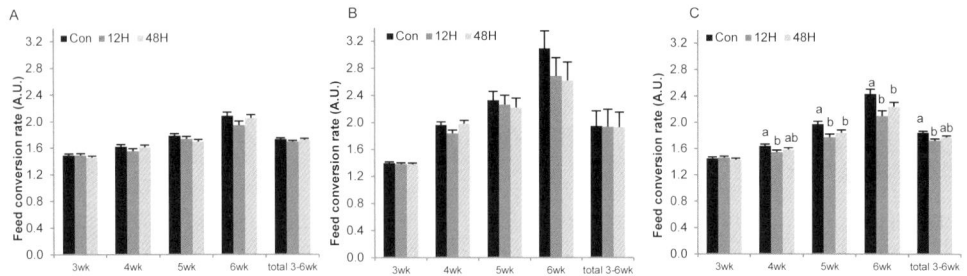

Figure 4. Feed conversion ratio (FCR) of Control, 12H and 48H male broiler chickens from 14 until 42d of age raised under either standard condition (A), long term high ambient temperature (B) or cycling chronic hot conditions (C). Different letters indicate differences (P≤0.05) between treatments.

Conclusions

Hypoxic exposure during the plateau period caused metabolic adaptation characterized by lower RMR. Lower RMR leads to lower heat production and enables the bird to invest less energy in maintenance and to allocate the remaining metabolic energy to growth. This adaptation had a long-lasting effect by improved thermotolerance of broiler chickens exposed to either long term high T_a or cycling chronic hot conditions.

References

Buys, N., Scheele, C. W., Kwakernaak, C. and Decuypere, E. 1999. *British Poultry Science* 40: 140-144.
Chan, T. and Burggren, W. (2005). *Respiratory Physiology and Neurobiology* 145: 251-263.
Dimicco, J. A. and Zaretsky, D. V. (2007). *Am J Physiol Regul Integr Comp Physiol* 292: R47-63.
Dridi, S., Temim, S., Derouet, M., Tesseraud, S. and Taouis, M. (2008). *J Exp Zool A Ecol Genet Physiol* 309: 381-388.
Druyan, S. 2010. *Poultry Science* 89: 1457-1467.
Druyan, S., Levi, E., Shinder, D. and Stern, T. 2012. *Poultry Science* 91: 987-997.
Dusseau, J. W. and Hutchins, P. M. 1988. *Respiration Physiology* 71: 33-44.
Dusseau, J. W. and Hutchins, P. M. 1989. *Microvascular Research* 37: 138-147.
Dzialowski, E. M., Decker, S., Black, J., Burggren, W. W. and Tonhardt, H. 2001. *American Zoologist* 41: 1434-1434.
Gordon, C. J. 1993. New-York: Cambridge University Press.
Haron, A., Dahan, Y., Shinder, D. and Druyan, S. 2017. *Comp Biochem Physiol A Mol Integr Physiol* 203: 32-39.
Havenstein, G. B., Ferket, P. R. and Qureshi, M. A. 2003. *Poultry Science* 82: 1500-1508.
Havenstein, G. B., Ferket, P. R., Scheideler, S. E. and Larson, B. T. 1994a. *Poultry Science* 73: 1785-1794.
Havenstein, G. B., Ferket, P. R., Scheideler, S. E. and Rives, D. V. 1994b. *Poultry Science* 73: 1795-1804.
Liu, C., Zhang, L. F., Song, M. L., Bao, H. G., Zhao, C. J. and Li, N. 2009. *Poultry Science* 88: 2689-2694.
Mcnabb, F. M. A. 1995. *Journal of Nutrition* 125: S1773-S1776.
Morrison, S. F., Nakamura, K. and Madden, C. J. 2008. *Experimental Physiology* 93: 773-797.
Mortola, J. P., Marinescu, D. C., Pierre, A. and Artman, L. 2012. *Respiratory Physiology and Neurobiology* 181: 109-117.
Mulder, A. L. M., van Golde, J. C., Prinzen, F. W. and Blanco, C. E. 1998. *Journal of Physiology-London* 508: 281-287.
Nichelmann, M. and Tzschentke, B. 2002. *Comparative Biochemistry and Physiology – Molecular & Integrative Physiology* 131: 751-763.
Richards, M. P. and Proszkowiec-Weglarz, M. 2007. *Poultry Science* 86: 1478-1490.
Silva, J. E. 2006. T *Physiological Reviews* 86: 435-464.
Tzschentke, B. and Basta, D. 2002. *Comparative Biochemistry and Physiology – Molecular & Integrative Physiology* 131: 825-832.
Yahav, S. 2009. *World's Poultry Science Journal* 65: 719-732.

Update on feed structure and processing: efficiency, digestion and gut development

M.R. Abdollahi* and V. Ravindran
Monogastric Research Centre, Institute of Veterinary, Animal and Biomedical Sciences, Massey University, Palmerston North 4442, New Zealand; m.abdollahi@massey.ac.nz

Summary

The magnitude of pelleting-induced benefits to broiler performance varies depending on physical quality of pellets. However, factors such as nutrient density, the degree to which the grains are ground and pelleting temperature may also influence the pelleting efficiency. Of these, the application of relatively high temperatures during conventional pelleting processes which does not favour high nutrient availability, remains a major concern. Feed texture and physical alterations associated with pelleting process can substantially impact on the development and functionality of digestive tract and nutrient digestion, with the nature of digestibility responses to feed processing being ingredient- and nutrient-dependent. In general, feeding pelleted diets has been reported to either adversely influence energy utilisation and nutrient digestibility in wheat- and sorghum-based diets or have no effect on the digestibility of major nutrients (starch and protein) in maize-based diets. Such effects may stem from the lack of structure in highly processed broiler diets and nutrient overload that coincide with ad libitum feeding of pelleted feed, resulting in a suboptimal foregut development. The implication that feed processing enhances the provision of rapidly digestible starch in feed and results in an accelerated, but not higher, starch digestibility and might have an impact on amino acid utilisation merits further investigation.

Introduction

The progress in the technology of feed manufacture during the past 50 years represents a major development in improving bird performance. Currently, majority of the feed used in the production of broilers is fed in pelleted or crumbled form. With ever-rising cost of poultry feed ingredients, feed is too precious to be wasted. Feed wastage is not only limited to the feed that is not ingested by the bird and spilled during feeding (physical feed wastage), but also includes the feed (more accurately, feed nutrients) which is not digested by the bird and excreted (nutritional feed wastage). Therefore, to maximise pelleting benefits it is critical to identify manufacturing techniques to create highly digestible high quality pellets. There are several mechanisms that underpin the considerable advantages of pellet feeding over mash diets; however, these advantages may be simply a function of increased feed consumption.

Factors influencing pelleting efficiency

The benefits of pellet feeding on broiler performance have been extensively reported (Abdollahi et al., 2013a). Since the introduction of pelleting process, the importance of pellet physical quality in poultry nutrition is being increasingly recognised because of the benefits associated with a better growth response and feed efficiency. McKinney and Teeter (2004) reported that pelleting of maize-based diets contributed 0.78 MJ nitrogen-corrected apparent metabolizable energy (AMEn)/kg diet at 100% pellets (no fines), with this value decreasing with increasing proportions of fines to pellets, and only contributing 0.32 MJ AMEn/kg for 20% pellets. However, other factors such as nutrient density, the degree to which the grains are ground and pelleting temperature are also of great importance in determining the efficiency of pelleting.

Nutrient density

Despite the potential for interactive effects between nutrient density and feed form in poultry, very few studies (Lemme et al., 2006; Brickett et al., 2007; Saldana et al., 2015a) have investigated this possible interaction. In a recent study (unpublished data), we investigated the interaction between five dietary nutrient densities (differing in 100 kcal AME/kg and 0.48 g lysine/kg) and two feed forms (mash vs pellet). Whilst birds fed pelleted diets outperformed those fed mash diets at each density, the pellet-associated benefits were more pronounced at lower nutrient density, confirming an interaction between nutrient density and feed form. Therefore, economic returns must be taken into account to decide which level of nutrient density should be used in broiler diets to achieve the highest efficiency of pelleting. Based on available data, it is reasonable to assume that nutrient requirements of modern broilers may depend on feed form and there is a need to determine nutrient requirements of broilers using pelleted diets.

Degree of grinding

It has been reported that pelleting efficacy in terms of increased feed intake (FI), weight gain and feed efficiency was much higher with fine particles than with coarse particles (Nir *et al.*, 1995). This finding may mostly be attributed to the lower FI and subsequent growth depressions obtained in the finely ground diet compared to the coarse diet fed in mash form. Amerah *et al.* (2007) reported noticeably greater FI increase in medium-ground than in coarse-ground wheat diets as a result of pelleting (64 vs 43%), with corresponding weight gain responses of 84 and 53%, respectively. Chewning *et al.* (2012) reported that pelleting a finely ground maize diet increased FI by 13.6% compared to only 3.3% when the coarse maize diet was pelleted. According to Nir *et al.* (1995), pelleting a finely ground diet with a lack of structural texture, re-texturises the diet, facilitating feed consumption and subsequently enhancing the growth response of broilers.

Pelleting temperature

One of the major issues in the manufacture of pellets is the application of high conditioning temperatures. The need to achieve high pellet quality and to reduce potential levels of feed-borne pathogens (salmonella and campylobacter) for feed safety has led to the application of relatively high conditioning temperatures (between 80 and 90 °C) during conventional pelleting processes, a practice which may not favour high nutrient availability. However, the true impact of conditioning temperature on the nutrient availability of pelleted diets has not been clearly delineated due to the confounding effects of conditioning temperature and feed form or has been neglected due to the focus on physical pellet quality and feed safety. Raastad and Skrede (2003) showed that whilst increasing conditioning temperature from 69 to 78 and 86 °C enhanced pellet quality, it reduced AME of diets from 13.3 to 13.1 and 12.9 MJ/kg, respectively, and depressed weight gain and feed efficiency. By differentiating the effects of conditioning temperature from feed form, Abdollahi *et al.* (2011) showed that application of high conditioning temperatures per se adversely influenced starch digestibility and energy utilisation in wheat-based diets. These researchers reported decreases in starch digestibility from 0.98 in diets conditioned at 60 °C to 0.94 and 0.91 in diets conditioned at 75 and 90 °C, respectively. Increasing conditioning temperatures above 60 °C also reduced the AME of diets from 14.2 MJ/kg in diets conditioned at 60 °C to 13.9 MJ/kg in those conditioned at 75 and 90 °C. In a recent study with maize-soybean meal diet, Loar II *et al.* (2014) found that as conditioning temperature increased from 74 to 85 and 96 °C, digestibility of some amino acids (AA) decreased by 3 to 5%, and feed per gain was impaired by 3 points (1.96 vs 1.99) and 8 points (1.96 vs 2.04), respectively. Heat treatment is currently thought to be the most practical method to achieve satisfying levels of feed safety, but a continued search for other methods which are not detrimental to feed nutrients and bird performance must be undertaken.

Impact of feed processing on nutrient digestion

It is widely believed in the feed industry that one of the factors underpinning the improved performance in pellet-fed birds is an increase in the extent of nutrient digestion. Recent studies in our laboratory, however, have shown that the nature of digestibility responses to pelleting is ingredient- and nutrient-dependent. Feeding pelleted diets, in general, either adversely influenced energy utilisation and nutrient digestibility in wheat- and sorghum-based diets, or had no effect on the digestibility of major nutrients (starch and protein) in maize-based diets (Abdollahi *et al.*, 2011, 2013b; Selle *et al.*, 2012). Besides the over-consumption associated with pellet feeding (Svihus, 2001), the anti-nutritive characteristics of soluble non-starch polysaccharides, which are found in high concentrations in wheat, may also account for the lower nutrient digestibility in birds fed pelleted wheat diets. However, pelleting maize-based diets has been shown to enhance the digestibility of fat, calcium and phosphorus compared to mash diets (Abdollahi *et al.*, 2013b). Naderinejad *et al.* (2016) in a study with maize-based diets reported that pelleting had no effect on the digestibility of dry matter, decreased protein digestibility, but improved that of fat by 7.9%. It appears that, in maize-based diets, some nutrients such as fat, calcium and phosphorus benefit from pelleting, probably caused by the disruption of the aleurone layer of cell walls and greater accessibility of previously encapsulated cellular contents to digestive enzymes. Although pelleting process tends to diminish the grain particle size impact on nutrient digestibility and broiler performance, study by Naderinejad *et al.* (2016) showed that coarse grinding of maize, through enhanced gizzard development and reduction in gizzard pH, was beneficial to nutrient utilisation and growth performance in broilers fed pelleted diets. When the diets were in mash form, different particle sizes offered no advantage; however, medium and coarse grindings appeared to increase starch digestibility and AME in pelleted diets.

One implication is that feed processing may influence growth performance of broiler chickens by manipulating the rate of starch digestion. In the digestive tract of broilers, starch is more rapidly digested compared to AA (Weurding *et al.*, 2003). It has been suggested that the provision of slowly digestible starch (SDS) in the feed not only benefits the starch utilisation but may attenuate the AA loss by providing higher glucose supply to the lower part of small intestine and sparing the AA from being catabolized for energy provision (Weurding *et al.*, 2003; Truong *et al.*, 2016). Selle *et al.* (2013), comparing mash with re-ground steam-pelleted sorghum-based diets, reported that steam-pelleting resulted in 48% increase in jejunal starch digestibility (0.395 vs 0.586), suggesting that feed processing accelerates the rate of starch digestion. Giuberti *et al.* (2012), using *in vitro* assay, found that steam-flaking increased the values of rapidly digestible starch (RDS; g/kg DM) in maize (147 vs 603), barley (151 vs 484) and wheat (181 vs 438) compared to unprocessed counterparts. Steam-flaking whilst decreased resistant starch (RS; g/kg DM) in maize (191 vs 113) and barley (143 vs 32), it generated 30% more RS in wheat grain (53 vs 70 g/kg). Enhanced energy utilisation and consequently better feed efficiency as a result of whole grain feeding to broilers, in addition to generating more developed and functional digestive tract and gizzard in particular, have been proposed to partially stem from the possible generation of SDS following the application of this feeding strategy (Liu *et al.*, 2015). Therefore, as suggested by Weurding *et al.* (2003), the rate of starch digestion as a relevant feed characteristic needs to be taken into account in modern broiler nutrition.

Extrusion cooking (a high-temperature/short-time thermal treatment) of current *faba bean* cultivars, even though they contain low concentrations of tannins, has been shown to improve the nutritional value of these beans in broilers through reducing the phytic phosphorus (PP), trypsin inhibitor activity (TIA) and RS content (Hejdysz *et al.*, 2016a). These researchers reported that extrusion (22% moisture, extrusion time of 10 s, temperature of 135 ± 10 °C, and pressure of 30 kg cm^2) of 5 different *faba beans* decreased the concentration of neutral detergent fibre by 38% (132 vs 213 g/kg), RS by 94% (10 vs 182 g/kg), TIA by 50% (0.3 vs 0.6 g/kg) and PP by 51% (1.9 vs 3.9 g/kg) compared to raw *faba beans*. Extrusion cooking also had a positive effect on starch digestibility (0.970 vs 0.773), fat retention (0.989 vs 0.872) and AMEn (14.95 vs 10.79 MJ/kg). Extruded *faba beans* also had higher dry matter, protein and fat, and all AA (except cysteine and proline) digestibilities than raw beans. In a follow up study (Hejdysz *et al.*, 2016b), the RS, and PP contents and TIA in extruded pea seeds were reduced by 89, 29 and 14%, respectively, compared with raw pea seeds. Extruded pea seeds were also characterised by higher protein (15.6%) and average AA (15.3%) digestibility, and AMEn (2.25 MJ/kg). Comparing an inadequately processed soybean meal characterised by high urease (0.23 pH points) and TIA (16,800 units/g) activities with an extruded SBM (ESBM; temperature of 150 ± 2 °C for 15 s), Jahanian and Rasouli (2016) reported a marked reduction in TIA (16,800 vs 2,400 units/g) and urease activity (0.23 vs 0.0 pH units) in ESBM. Extrusion processing also enhanced ileal digestibility of protein and average AA by 15.6 and 11.3%, respectively. These researchers reported that replacing the inadequately processed soybean meal with ESBM in a 42-d broiler study improved daily weight gain by 9.6% and feed efficiency by 3.8%. Therefore, although extrusion, due to high capital investment costs, has not been used commonly in commercial poultry feed production; it may be a useful processing technique for some feed ingredients with positive advantages in terms of AA digestibility owing to destruction of anti-nutritional factors.

Feed texture and processing: impact on gut development

The gastrointestinal tract (GIT) plays an important role either directly or indirectly on birds' health through various physiological functions. A well-developed gizzard enhances the grinding action, generates stronger reverse peristalsis contractions within the GIT, increases proteolysis by pepsin and stimulates secretion of hydrochloric acid which reduces the pH. Harmful bacteria entering the intestinal tract via the feed have a greater chance of being suppressed in a highly acidic environment. It is being increasingly recognised that the broilers may have a requirement for a certain level of structural components such as coarse particles, insoluble fibre sources and whole grains in their feed to meet their innate feeding behavior development (Ferket and Gernat, 2006). The major motivation for inclusion of structural components in poultry diets is to stimulate gizzard development and functionality which will favourably influence gut health and the bird's ability to better utilise nutrients. The beneficial effects of such practices may also extend to their favourable influence on intestinal morphology and functionality (Amerah *et al.*, 2007) and microbiota profile (Engberg *et al.*, 2004).

Since fine grinding is generally favoured to obtain a high pellet quality and because it is not possible to avoid further particle size reduction during the pelleting process, fine particle are almost inevitable during pelleting; this results in a suboptimal gizzard development with potential negative influence on nutrient digestibility. An elevated gizzard pH and a short digesta retention time, due to an under-developed gizzard, are physiological limits to optimal digestion in poultry. This complex matrix of conditions (pH and retention time) becomes even more limiting when birds are fed pelleted diets (Abdollahi *et al.*, 2013b). According to Liu *et al.* (2015), a negative correlation (r=-0.451) exists between relative gizzard weight and gizzard pH. There is evidence of relatively higher gizzard pH in birds fed pelleted diets compared to those fed mash diets (Huang *et al.*, 2006; Frikha *et al.*, 2009; Saldana *et al.*, 2015a,b), attributed mainly to the pH of feed (5.5 to 6.5), higher FI and possibly lower hydrochloric acid secretion (Svihus, 2011). Under the current system of continuous feeding, the function of the crop as a storage organ appears to be lost. Gizzard is not fully developed when feeding pelleted diets and a less developed gizzard serves as a transit organ rather than a grinding organ, with the implication of reduced retention time. The average retention time in the digestive tract, excluding the caeca, is probably around 3 to 4 h (Svihus, 2011). Of this, digesta possibly spends only 60 to 90 min in the anterior digestive tract, which gives only limited opportunity for enzyme action. Our recent study (Naderinejad *et al.*, 2016), whilst confirmed the higher gizzard pH in pellet-fed birds than those fed mash diets, showed that gizzard pH was responsive to particle size only in pelleted diets. Although pelleting reduced the proportion of coarse particles, it seems that a minimum of 4 to 6% coarse particles of >2,000 μm was sufficient to stimulate hydrochloric acid secretion and reduce the gizzard pH to the same level as mash-fed birds.

The majority of investigations into the effect of feed structure and form on proventriculus and gizzard development have examined the effects on the mass of these organs. Despite gizzard being the organ orchestrating the digestion process in poultry, little attention has been paid to the musculature structure of the gizzard. In an attempt to improve our knowledge of the adjustment of proventriculus and gizzard to changes in diet structure, we investigated the impact of wheat particle size (fine, medium and coarse), fibre source (lignocellulose fibre, oat hulls [OH] and wood shavings [WS]) and whole wheat (WW) inclusion on the musculature of gizzard. Whilst gizzard responses to changes in diet texture were consistent with previous findings; these responses were more emphasised in diets diluted with WS (77%) and OH (70%). Compared to the fine diet, offering coarse and medium ground wheat, OH, WS and WW increased total proventricular glandular height by 11.6, 10.7, 9.6, 8.6 and 6.9%, respectively. Feeding OH and WS diets increased the diameter of gizzard's caudoventral thin muscle, caudodorsal thick muscle and cranioventral thick muscle by an average of 50, 84 and 87%, respectively. Medium and coarse grinding of wheat and WW inclusion resulted in 26, 35 and 29% greater diameter in caudodorsal thick muscle, and 27, 29 and 25% higher diameter in cranioventral thick muscle, respectively. It seems that when the OH and WS are introduced into the broiler diets, these materials even with small proportion of coarse particles that are more resistant to grinding, retain in the gizzard for a longer period, increase the gizzard grinding functionality and reduce digesta passage rate from gizzard to duodenum (Jimenez-Moreno *et al.*, 2009). The gizzard mass and musculature responses to structural components are most likely influenced by the presence of coarse particles and the nature of particles.

The above-mentioned study also showed that different dietary textures did not have any effect on the histological structure of small intestine, a finding which is accordance with those of Amerah *et al.* (2007), Zang *et al.* (2009) and Naderinejad *et al.* (2016) who reported a lack of gut morphology response to feed particle size. Dahlke *et al.* (2003) found that pelleted diets stimulated an increased number of duodenal villi as compared with mash diets. Naderinejad *et al.* (2016) showed an increase in the villus height in the duodenum and jejunum of birds fed pelleted diets compared with mash diets. The increased villus height may increase total luminal villus absorptive area and subsequently result in greater digestive enzyme action and enhanced transport of nutrients at the villus surface (Cera *et al.*, 1988). Increased villus height in pelleted diets can be considered as a general response to the need for greater digestive and absorptive capacity of the proximal small intestine to the greater load of nutrients. Although an increased villus height has been suggested as an indicator of improved digestibility (Chaing *et al.*, 2010), improved nutrient digestibility due to the inclusion of structural components is not necessarily achieved by increased villus height or number. Husveth *et al.* (2015) found no effect of WW incorporation into pelleted broiler diets on villus size and crypt depth but an increase in jejunal activities of α-amylase, lipase and trypsin by 20, 8.7 and 5.3%, respectively. It was suggested that beneficial effects of WW feeding are mediated more by higher digestive enzyme activities than the changes in the tissue structure of the gut.

Conclusions

Whilst the importance of feeding processed feed to broilers is no longer questionable, its efficiency in determining the actual performance responses depends on nutrient availability which is, in turn, influenced by grain type, processing variables such as temperature, feed texture and birds' digestive tract development. Though the nature of digestibility response is dependent on the ingredient and specific nutrient, recent evidence suggests that pelleting has no positive impact on the digestibility of starch and protein in cereal-based poultry diets. The current practice of high degree of processing of feed, which induces particle size reduction and increased FI, and ad libitum feeding does not support the normal development and functionality of the foregut. Incorporation of structural components to poultry diets can impart some benefits to the birds' digestive system, especially in contemporary diets which lack in feed structure.

References

Abdollahi, M.R., Ravindran, V., Wester, T.J., Ravindran, G. and Thomas, D.V. 2011. *Animal Feed Science and Technology* 168: 88-99.
Abdollahi, M.R., Ravindran, V. and Svihus, B. 2013a. *Animal Feed Science and Technology* 179: 1-23.
Abdollahi, M.R., Ravindran, V. and Svihus, B. 2013b. *Animal Feed Science and Technology* 186: 193-203.
Amerah, A.M., Ravindran, V., Lentle, R.G. and Thomas, D.G. 2007. *Poultry Science* 86: 2615-2623.
Brickett, K.E., Dahiya, J.P., Classen, H.L. and Gomis, S. 2007. *Poultry Science* 86: 2172-2181.
Cera, K.R., Mahan, D.C. and Cross, R.F. 1988. *Journal of Animal Science* 66: 574-584.

Chewning, C.G., Stark, C.R. and Brake, J. 2012. *Journal of Applied Poultry Research* 21:830-837.

Chiang, G., Lu, W.Q., Piao, X.S., Hu, J.K., Gong, L.M. and Thacker, P.A. 2010. *Asian-Australian Journal of Animal Sciences* 23: 263-271.

Dahlke, F., Ribeiro, A.M.L., Kessler, A.M., Lima, A.R. and Maiorka, A. 2003. *Brazilian Journal of Poultry Science* 5: 61-67.

Engberg, R.M., Hedemann, M.S., Steenfeldt, S. and Jensen, B.B. 2004. Poultry Science 83: 925-938.

Ferket, P.R. and Gernat, A.G. 2006. *International Journal of Poultry Science* 5: 905-911.

Frikha, M., Safaa, H.M., Serrano, M.P., Arbe, X. and Mateos, G.G. 2009. *Poultry Science* 88: 994-1002.

Giuberti, G., Gallo, A., Cerioli, C. and Masoero, F. 2012. *Animal Feed Science and Technology* 174: 163-173.

Hejdysz, M., Kaczmarek, S.A. and Rutkowski, A. 2016a. *Animal Feed Science and Technology* 212: 100-111.

Hejdysz, M., Kaczmarek, S.A. and Rutkowski, A. 2016b. *Archives of Animal Nutrition* DOI: 10.1080/1745039X.2016.1206736.

Huang, D.S., Li, D.F., Xing, J.J., Ma, Y.X., Li, Z.J. and Lv, S.Q. 2006. *Poultry Science* 85: 831-836.

Husvéth, F., Pál, L., Galamb, E., Ács, K.C., Bustyaházai, L., Wágner, L., Dublecz, F. and Dublecz, K. 2015. *Animal Feed Science and Technology* 210: 144-151.

Jahanian, R. and Rasouli, E. 2016. *Poultry Science* 95: 2871-2878.

Jiménez-Moreno, E., González-Alvarado, J.M., Lázaro, R. and Mateos, G.G. 2009. *Poultry Science* 88: 1925-1933.

Lemme, A., Wijtten, P.J.A., van Wichen, J., Petri, A. and Langhout, D.J. 2006. *Poultry Science* 85: 721-730.

Liu, S.Y., Truong, H.H. and Selle, P.H. 2015. *Animal Production Science* 55: 559-572.

Loar II, R.E., Wamsley, K.G.S., Evans, A., Moritz, J.S. and Corzo, A. 2014. *Journal of Applied Poultry Research* 23: 1-12.

Mckinney, L.J. and Teeter, R.G. 2004. *Poultry Science* 83: 1165-1174.

Naderinejad, S., Zaefarian, F., Abdollahi, M.R., Hassanabadi, A., Kermanshahi, H. and Ravindran, V. 2016. *Animal Feed Science and Technology* 215: 92-104.

Nir, I., Hillel, R., Ptichi, I. and Shefet, G. 1995. *Poultry Science* 74: 771-783.

Raastad, N. and Skrede, A. 2003. WPSA proceedings, 14th European Symposium on Poultry Nutrition 14: 115-116. Lillehammer, Norway.

Saldana, B., Guzmán, P., Cámara, L., García, J. and Mateos, G.G. 2015a. *Poultry Science* 94: 1879-1893.

Saldana, B., Guzmán, P., Safaa, H.M., Harzalli, R. and Mateos, G.G. 2015b. *Poultry Science* 94: 2650-2661.

Selle P.H., Liu S.Y., Cai J. and Cowieson A.J. 2012. *Animal Production Science* 52: 842-852.

Selle P.H., Liu S.Y., Cai J. and Cowieson A.J. 2013. *Animal Production Science* 53: 378-387.

Svihus B., Hetland H., Choct M. and Sundby F. 2002. *British Poultry Science* 43: 662-668.

Svihus B. 2001. *Animal Feed Science and Technology* 92: 45-49.

Svihus B. 2011. *World's Poultry Science Journal* 67: 207-223.

Truong H.H., Liu S.Y. and Selle P.H. 2016. *Animal Production Science* 56: 797-814.

Weurding R.E., Enting H. and Verstegen M.W. 2003. *Poultry Science* 82: 279-284.

Zang J.J., Piao X.S., Huang D.S., Wang J.J., Ma X. and Ma Y.X. 2009. *Asian-Australian Journal of Animal Sciences* 22: 107-112.

Nutritional modulation of microbial signals in the distal intestinal tract and how they can affect broiler health

F. Van Immerseel*, K. Vermeulen, L. Onrust, V. Eeckhaut and R. Ducatelle
Department of Pathology, Bacteriology and Avian Diseases, Faculty of Veterinary Medicine, Ghent University, Belgium; filip.vanimmerseel@Ugent.be

Summary

Traditionally, antimicrobials have not only been used to control intestinal pathogens, but also to increase animal performance in broilers. Different hypotheses have been put forward on the mode of action of the antimicrobial growth promoters, of which modulation of the microbiota composition is one. The quantity and composition of the microbiota of broilers varies along the length of the crop – stomach – intestinal tract, and the highest abundance and diversity can be found in the ceca, with bacterial numbers that outrange the number of eukaryotic cells in the body. As these microbes have the capacity to degrade and ferment (macro-) molecules and thus produce a range of metabolites, the metabolite profile is highly dependent on the nutritional status of the bird. While antimicrobial compounds affect the metabolic profile by inhibiting bacterial populations, one can also stimulate the colonization of specific bacterial groups and thus enrich for specific functions, in order for the metabolic profile of the gut microbial ecosystem to be steered towards certain end products. While the gut lumen contains the majority of bacteria, the mucosa-associated bacteria are likely the most important ones, as they are in close interaction with the epithelial cells, and thus are in a favourable location to send signals to the host. Butyric acid is the most well-known example of a bacterial signal that is affecting gut and systemic health. This simple fermentation acid can be produced by specific species from the *Lachnospiraceae* and *Ruminococcaceae* families, of which some are specifically associated with the mucosa. Degradation products of arabinoxylans (e.g. XOS) and low-particle size wheat bran fractions are examples of feed additives that specifically stimulate colonization of these microbes. This has been shown to result in effects on pathogen colonization and prevention of small intestinal damage.

Introduction

The use of antimicrobial growth promoters (AGPs) is under debate in many regions worldwide, and while some countries (e.g. in EU) have banned these from poultry diets, other countries still use these compounds as in-feed supplements. The use of therapeutic antibiotics is still high in the broiler industry worldwide and although they are essential to maintain animal welfare and to treat birds from diseases, these compounds are also too often used in a prophylactic manner. There have been numerous attempt to develop alternatives to antibiotics, mainly to AGPs. Ideally, to be able to do so, one should understand how the 'old' AGPs exerted their activity, but this is not fully clear. The most well-known proposed mechanism of action is that AGPs have antibacterial action and this favours performance in different ways (reviewed by Huyghebaert *et al.*, 2011): (1) by reducing the incidence and severity of subclinical infections; (2) by reducing the microbial use of nutrients; (3) by improving absorption of nutrients; and (4) by reducing the amount of growth-depressing metabolites produced by Gram-positive bacteria. Niewold (2007) suggested direct anti-inflammatory effects of AGPs on macrophages (Niewold, 2007). Also, indirect anti-inflammatory effects caused by changes in the quantity and composition in the microbiota could affect inflammatory pathway. Shifts in the microbiota composition have already been proven after AGPs were added to broiler feed (Pedroso *et al.*, 2006; Wise and Siragusa, 2007). Indeed, sub-MIC concentrations do not mean that growth-inhibition of certain bacterial species in the gut cannot occur. Additionally effects on microbial metabolic pathways are a potential consequence of the activity of AGPs. Below, emphasis is put on the microbial composition and how this can affect the metabolic profile in the gut, and how we can steer the gut microbiota composition to one that is beneficial for gut health. The example of butyrate as a key metabolite in gut health maintenance is documented.

Plenary session 02. Feeding strategies and gastrointestinal health

The gut microbiota

The major part of the gut microbiota is believed to consist of beneficial microbes that serve the host by protecting against enteropathogenic infections by blocking receptors used for binding of pathogens, production of antimicrobial metabolites and competition for nutrients. These old concepts of the protective mechanisms of the gut microbiota can be true, but in addition, there is new data that indicates that other activities of the microbiota might also be very powerful in preserving gut health. These are activities of end products generated by bacterial metabolism of complex feed components. The digestion of (for the host) indigestible substrates is also a beneficial effect of the microbiota and is thought to support gut health. The term 'gut health' however is not so easy to define, and while in humans gut health means the absence of any gastrointestinal symptoms (pain, diarrhoea, …) related to disease (caused by pathogens, inflammation, …), in production animals performance is often used as an indicator of gut health. While animal performance problems (FCR, BWG) can be the consequence of poor gut health, slower growing animals do not necessarily have gut health problems per se. While the term dysbiosis is used in humans to describe microbiota shifts that are linked with specific gut health disorders (eg. IBD, IBS), in animals the term dysbiosis is often used without any knowledge about the microbiota composition, but merely as an indicator of poor animal performance that can be reversed by antibiotics.

Knowledge on the microbiota composition and the functional activity of the gut microbiota has been strongly influenced by the development of rapid (16S rDNA) DNA sequencing techniques. Metagenome analysis even described the full genomic content of a gut microbiota (the microbiome), yielding valuable information on the metabolic properties of the microbiota. Due to the diversity of functions the gut microbiota can carry out, it is even considered to be an additional organ. It is estimated that the gut microbiota of an adult animal consists out of more than 1000 different bacterial species, belonging to only a handful of phyla. Bacterial species from the phyla *Bacteroidetes* and *Firmicutes* are highly dominant, while less abundant phyla are for example *Proteobacteria* (containing the *Enterobacteraceae* family), *Verrucomicrobia* and *Actinobacteria* (Arumugam *et al.*, 2011; Wei, Morrison, and Yu, 2013; Zoetendal, Rajilic-Stojanovic, and de Vos, 2008). All phyla contain a high diversity in families, genera and species, which can cause a high diversity at this lower taxonomic level when comparing individuals. Despite the high inter-individual diversity in the gut microbiota composition, a functional core microbiome seems to exist. This means that, despite differences in the microbiota composition between individuals, these microbiota can exert similar metabolic actions in the gut (Huttenhower *et al.*, 2012; Lozupone *et al.*, 2012; Qin *et al.*, 2010). In this way, different bacterial species can colonize identical niches in the gut and exert similar functions (e.g. breakdown of specific polysaccharides). In the gut of chickens, the proximal intestinal tract harbors a rather low number of bacteria (about 10^{4-6} per g content) while the distal intestinal tract, specifically the ceca, contains a huge number of bacteria (more than 10^{10} per g). The terminal ileum also contains high bacterial numbers as it is physically very closely located to the ceca. The bacterial diversity is very high in the distal compartment of the gut. There is also a clear difference between the bacterial composition in the gut lumen and the mucus. In the lumen, the numbers and the diversity are very high, while the numbers and diversity of the mucosa-associated microbiota are rather low. The mucosa-associated microbiota however can be highly relevant because of the close interaction with the epithelial cells of the host, so that signals from these bacteria can be sensed by the host. (Opportunistic) Pathogens often found in the mucus layer are *Clostridium perfringens*, the causative agent of necrotic enteritis, and *Enterobacteriaceae*, such as *Escherichia coli*. Beneficial microbes that can be mucosa-associated are often anaerobic butyrate producers (Levine, Looft, Allen, and Stanton, 2013; Van den Abbeele *et al.*, 2013). Mucin degradation can therefore be problematic when it supports outgrowth of pathogens in the gut (such as *C. perfringens*). On the other hand, mucin usage by beneficial butyrogenic microbes (such as *Akkermansia mucinophila*) is an example of a symbiotic relation between microbes and the host, in which the host is providing substrates that are used by the bacteria to produce health-promoting end products, in close proximity to the epithelium (Van den Abbeele *et al.*, 2011). The ratio between beneficial and potentially harmful microbes can thus be important to maintain gut health.

Intestinal health and its definitions

Intestinal health is a term that is not easy to define in a simple way, because many different diseases and syndromes of various nature can affect the gut. The simplest definition of an optimal gut health is the absence of any gross or microscopical lesions, and the absence of pathogenic micro-organisms in the gut. While gross lesions are often correlated with the presence of pathogens (e.g. necrotic enteritis, coccidiosis), microscopical defects and inflammatory events are not always clearly linked with a particular pathogen. Macroscopical lesion scoring systems for necrotic enteritis and coccidiosis exist, and even macroscopical scoring systems for evaluating general intestinal health have been developed, based on different gut wall characteristics (Teirlynck *et al.*, 2011). Its use is, however, time-consuming and requires specific expertise. Macroscopical intestinal health scoring systems have been evaluated by correlating the scores with morphological measurements of the gut wall. Often, but not always, poor intestinal health is associated with shorter villi and the excessive presence of inflammatory cells in the gut wall. The epithelial barrier, and the appearance of breaches in this barrier, is considered to be of crucial importance in gut health, as increased intestinal epithelial permeability causes inflammation because of stimulation of TLRs by bacterial molecules and feed antigens that make contact with the basolateral surface of the epithelial cells and cells in the mucosa (including immune cells) (Peterson and Artis, 2014). Furthermore, increased intestinal permeability can cause micro-organisms to translocate (e.g. enterococci, *E. coli*), causing systemic disease (e.g. chondronecrosis with osteomyelitis). Poor intestinal health is often associated with an overgrowth by specific pathogens too. While this can be clear such as in the case of for example necrotic enteritis, in which huge numbers of pathogenic NetB-toxin producing *C. perfringens* are present, most of the time this is much less obvious. Indeed, the most important zoonotic agents (*Salmonella, Campylobacter*) cause no pathologies in the gut, but still one cannot say that the gut health of chickens carrying these organisms is optimal. Even more, opportunistic pathogens are always present in the chicken gut, but their involvement in disease requires the presence of predisposing triggers. It becomes more and more clear that the microbiota composition is of importance in maintenance of gut health, and that the 'one pathogen – one disease' concept is old-fashioned and is replaced by the concept of 'disease complex' that consists of a complex interaction between environment – host – microbiota, the so-called 'pathobiome' (Vayssier-Taussat *et al.*, 2014).

Microbial metabolites and intestinal health

Bacteria are able to ferment sugars, proteins and fat in the gut and in this way can produce a variety of metabolites that can affect other micro-organisms as well as the host. Depending on the substrate and the microbiota composition, a shift in the nature and the composition of these metabolites can occur, with consequential effects on intestinal health. To ferment complex substrates, bacteria use a network in which specific bacterial species catabolize specific steps in the substrate breakdown. One thus needs a high microbial diversity for this reason.

Fermentation of complex polysaccharides, such as cellulose, pectins, starch, and arabinoxylans will result in production of oligosaccharides of variable chain length. These oligosaccharides on their turn can be used by other bacterial groups to form short chain fatty acids (acetic, propionic and butyric acid), lactic acid and gases (CO_2, H_2), in addition to other metabolites. An important health-promoting metabolite is butyric acid. Butyric acid is anti-inflammatory, promotes proliferation of epithelial cells in the gut and strengthens the epithelial barrier (by increasing transepithelial resistance). In addition, it is a signal molecule that lowers virulence of pathogens such as *Salmonella*. High amounts of butyric acid are produced in the gut by bacteria from the families *Ruminococcaceae* and *Lachnospiraceae* from the Firmicutes phylum. A part of the butyric acid producers from the *Lachnospiraceae* consume lactic acid. This is a typical example of cross-feeding, in which a group of bacteria delivers substrates that can be used by another group of bacteria to produce end-metabolites. The breakdown of polysaccharides to oligosaccharides that can be used by other bacterial members is also a cross-feeding example. Acetic acid is important for the host in lipogenesis and propionic acid in gluconeogenesis. Apart from lactic and short-chain fatty-acids, also gases are produced during sugar fermentation. Hydrogen (H_2) is therefore important because it can be converted by hydrogenotrophic bacteria for the production of acetic acid, methane or hydrogen sulphide (H_2S). At high concentrations the latter is toxic for intestinal epithelial cells. Competition for hydrogen is thus important because this steers either or not production of a harmful compound, due to competition between hydrogen sulphide producers and acetogenic bacteria that consume hydrogen. Interestingly, several hydrogen sulphide producers are sulphate reducing bacteria that consume lactate and thus go in competition with lactate-utilizing butyrate producers, hereby again illustrating that competition for molecules between bacterial groups drives gut health (Scott *et al.*, 2013).

Proteins are ideally digested in the small intestine, and amino acids and peptides taken up by the host epithelium. They can however also reach the distal intestinal tract where they can be fermented by proteolytic bacteria. Amino acid fermentation can occur through deamination with the production of short chain and branched chain fatty acids (isobutyrate, isovalerate, 2-methylbutyrate) and ammonia (NH_3). Deamination of aromatic amino acids leads to phenolic molecules, such as p-cresol, indolpropionate and –acetate and phenylacetate. A second route of bacterial amino acid conversion is through decarboxylation, by which a variety of amines and polyamines are produced. Sulphur-containing amino acids (cysteïne, methionine) can cause H_2S formation. It is a common belief that excessive protein fermentation is harmful through the production of toxic molecules (e.g. NH_3, H_2S, phenolic derivatives) although strikingly, no information is available for many of the produced molecules. Protein fermentation is mainly associated with lactobacilli, *Clostridia* and genera such as *Escherichia, Proteus, Klebsiella, Desulfovibrio* and others (mostly *Proteobacteria*) (Windey *et al.*, 2012).

Microbial degradation of fat is much less investigated but would lead to reductions in short chain fatty acid production and thereby enhance inflammation.

Although not metabolites in the strict sense, many other microbial molecules can also be of importance for intestinal health. An example is LPS, being a highly inflammatory molecule because of TLR-4 activation.

Steering the microbial production of metabolites by nutrition

The gut microbiota composition can be altered by changing the feed composition (substrate changes) and providing substrates (prebiotics), bacteria (probiotics) or combinations. Also certain 'antibacterial' molecules such as organic acids, essential oils (aldehydes and phenols) and clearly also antibiotics change the microbiota composition. Feed enzymes can also alter the microbiota composition as they change substrate availability in the gut. While a lot of work has been done on investigating the effects of specific feed additives on animal performance, a good description of their mechanism of action is often lacking. Taking into account the potential effect of the above mentioned microbiota groups and microbial metabolites, it is clear that a microbial shift should be as such that the ratio between beneficial vs harmful microbial groups or the ratio between beneficial and harmful metabolites is optimal (or high). The major issue is that one needs much more information on the identity of the microbiota members and their effects on host health and even more the information on the effects on host health of the variety of potential microbial metabolites is very scarce. From what is known, some simple rules can be deduced. For example, it is known that butyrate is a key component that is essential for an optimal gut health (Hamer *et al.*, 2008). Microbial populations that are producing high levels of butyrate belong (among others) to the families of *Lachnospiraceae* and *Ruminococcaceae* (Pryde *et al.*, 2002). It is also clear that certain specific bacterial groups can be harmful. These contain (opportunistic) pathogens, such as *C. perfringens* and *E. coli*, but it could as well be expanded to for example the *Enterobacteraceae* family or sulphate reducing bacteria (SRB, producing hydrogen sulphide) (Ijssennagger *et al.*,

2016; Pedron and Sansonetti, 2008). Therefore feed additives could be used that specifically target these microbial groups. The simplest strategy could be to supplement poultry diets with butyrate producing strains. An example is the *Ruminococcaceae* strain *Butyricicoccus pullicaecorum*, which has been shown to improve feed conversion, and reduce the number of *Enterobacteraceae* and enterococci in the gut of broilers (Eeckhaut *et al.*, 2016). Even more, administration of this strain to broilers reduces small intestinal damage in a necrotic enteritis model. The concept that butyrate production in the distal gut improves small intestinal damage has been explained by activation of enteroendocrine cells that secrete small peptide hormones, such a GLP-2, in the circulation. GLP-2 is an epitheliotrophic and anti-inflammatory peptide. In theory high butyrate production can also be achieved by supplementing strains that take part in the cross-feeding cascade that promote butyrate production. This can be a reason why supplementing lactic acid bacteria can be of value, because, apart from their own specific effects (e.g. H_2O_2 and bacteriocin production), they can stimulate butryate production by lactic acid consuming *Lachnospiraceae* strains. One can stimulate butyrate production in the gut by using prebiotic compounds. It has been shown that XOS promote butyrate production by stimulating cross-feeding reactions leading to lactate and further butyrate (De Maesschalck *et al.*, 2015). As XOS is a degradation product of arabinoxylans, this implicates that feed enzymes can also work through release of prebiotic compounds that are converted to beneficial metabolites. It has indeed been shown that enzymatic treatment of wheat releases butyrogenic substrates that were considered to be short-chain arabinoxylans. Feedstuffs such as small particle size wheat bran are also proven to stimulate butyrate production and lower *Salmonella* colonization in broilers (Vermeulen *et al.*, 2017). There are thus quite some potential strategies to improve or maintain gut health.

Conclusions

The gut microbiota is complex, although its composition is more and more being identified, mainly because of fast (and cheaper) DNA sequencing tools that became available in the last decade. While this 'virtual' composition will be defined in detail for broilers too, there is a serious lack of information on the effect of individual strains and their specific metabolites on host health. Also a clear understanding of the breakdown cascade of feed substrates, the involvement of microbial groups and the potential outcomes in terms of production of metabolites is not fully unravelled. This has been done for polysaccharides to some extent but for proteins this is still a black box. Therefore, even though there is a lot of novel information on steering the microbiota composition to promote gut health and animal performance, the future should bring fundamental research on the network 'substrate-bacteria-metabolite-host' to further understand the ecology in the gut. It also needs to be emphasized that these future gut health promoting compounds are of no use on farms without good management practices that embraces biosecurity.

References

Arumugam, M., Raes, J., Pelletier, E., Le Paslier, D., Yamada, T., Mende, D.R. and Meta, H.I.T. Consortium. 2011. *Nature* 473: 174-180.
De Maesschalck, C., Eeckhaut, V., Maertens, L., De Lange, L., Marchal, L., Nezer, C. and Van Immerseel, F. 2015. *Applied and Environmental Microbiology* 81: 5880-5888.
Eeckhaut, V., Wang, J., Van Parys, A., Haesebrouck, F., Joossens, M., Falony, G. and Van Immerseel, F. 2016. The Probiotic Butyricicoccus pullicaecorum Reduces Feed Conversion and Protects from Potentially Harmful Intestinal Microorganisms and Necrotic Enteritis in Broilers. *Frontiers in Microbiology* 7.
Hamer, H.M., Jonkers, D., Venema, K., Vanhoutvin, S., Troost, F.J. and Brummer, R.J. 2008. *Aliment Pharmacol Ther* 27: 104-119.
Huttenhower, C., Gevers, D., Knight, R., Abubucker, S., Badger, J.H. and Chinwalla, A.T. and the Human Microbiome Project Consortium. 2012. *Nature* 486: 207-214.
Huyghebaert, G., Ducatelle, R. and Van Immerseel, F. 2011. *Veterinary Journal,* 187: 182-188.
Ijssennagger, N., van der Meer, R. and Van Mil, S.W.C. 2016. *Trends in Molecular Medicine* 22: 190-199.
Levine, U. Y., Looft, T., Allen, H.K. and Stanton, T.B. 2013. *Applied and Environmental Microbiology* 79: 3879-3881.

Lozupone, C.A., Stombaugh, J.I., Gordon, J.I., Jansson, J.K. and Knight, R. 2012. *Nature* 489: 220-230.

Niewold, T.A. 2007. *Poultry Science* 86: 605-609.

Pedron, T. and Sansonetti, P. 2008. *Cell Host and Microbe* 3: 344-347.

Pedroso, A.A., Menten, J.F.M., Lambais, M.R., Racanicci, A.M.C., Longo, F.A. and Sorbara, J.O.B. 2006. *Poultry Science* 85: 747-752.

Peterson, L.W. and Artis, D. 2014. *Nature Reviews Immunology* 14: 141-153.

Pryde, S.E., Duncan, S.H., Hold, G.L., Stewart, C.S. and Flint, H.J. 2002. *FEMS Microbiology Letters* 217: 133-139.

Qin, J.J., Li, R.Q., Raes, J., Arumugam, M., Burgdorf, K.S., Manichanh, C., Meta, H.I.T. Consortium. 2010. *Nature* 464: 59-65.

Scott, K.P., Gratz, S.W., Sheridan, P.O., Flint, H.J. and Duncan, S.H. 2013. *Pharmacological Research* 69: 52-60.

Teirlynck, E., Gussem, M.D.E., Dewulf, J., Haesebrouck, F., Ducatelle, R. and Van Immerseel, F. 2011. *Avian Pathology* 40: 139-144.

Van den Abbeele, P., Belzer, C., Goossens, M., Kleerebezem, M., De Vos, W.M., Thas, O. and Van de Wiele, T. 2013. *Isme Journal* 7: 949-961.

Van den Abbeele, P., Van de Wiele, T., Verstraete, W. and Possemiers, S. 2011. *FEMS Microbiol Rev.* 35: 681-704.

Vayssier-Taussat, M., Albina, E., Citti, C., Cosson, J.F., Jacques, M.A., Lebrun, M.H. and Candresse, T. 2014. *Frontiers in Cellular and Infection Microbiology* 4.

Vermeulen, K., Verspreet, J., Courtin, C.M., Haesebrouck, F., Ducatelle, R. and Van Immerseel, F. 2017. *Vet Microbiol* 198: 64-71.

Wei, S., Morrison, M. and Yu, Z. 2013. *Poultry Science* 92: 671-683.

Windey, K., De Preter, V. and Verbeke, K. 2012. *Mol Nutr Food Res* 56: 184-196.

Wise, M.G. and Siragusa, G.R. 2007. *Journal of Applied Microbiology* 102: 1138-1149.

Zoetendal, E.G., Rajilic-Stojanovic, M. and de Vos, W.M. 2008. *Gut* 57: 1605-1615.

Intestinal digestive function and the influence of diet and feeding management

B. Svihus
Norwegian University of Life Sciences, P.O. Box 5003, 1432 Aas, Norway; birger.svihus@nmbu.no

Summary

The small intestine is under tremendous stress due to the large quantities of feed that needs to be handled by modern poultry breeds. The feed must be digested and the nutrients must be absorbed during the very short retention time in the small intestine. However, there is increasing evidence for impaired feed efficiency – at least for some birds in a broiler flock – due to suboptimal small intestinal function. This indicates that changes to improve digestive function and/or feeding management is increasingly required, both from a welfare/digestive health perspective and from a feed efficiency point of view. It is now clear that the gizzard plays a pivotal in nutrient digestion and absorption in the small intestine, both through the digestion process that takes place here, and in regulating flow of material into the small intestine. Stimulation of gizzard function through coarse grinding or addition of coarse fibres are therefore warranted. However, intermittent feeding and/or changes from pelleted diets to coarse mash feed may also be required to maximize intestinal function and gut health, at least in broiler chickens.

Introduction

The intestine of the modern chicken has had to adapt to tremendous changes due to intensive breeding for number of eggs for layers and growth rate for broiler chickens. A 30 day old male broiler chicken, for example, consumes a quantity of feed equivalent to 10% of its live weight per day, and the digestive tract will thus have to handle slightly over seven gram of feed per hour. To put this in perspective, a 75 kg person would have to eat more than 450 gram per hour during the 16 awake hours to have an equal food intake relative to body weight; equivalent to one normal loaf of bread per awake hour.

It is logical to assume that this high production rate and thus high feed intake makes the intestine vulnerable to impaired functionality. The impaired functionality can be due to insufficient development of the digestive tract, or it can be due to external factors such as microflora or insufficiencies in the feed. In severe cases of impaired functionality it may be easy to observe this dysfunction, for example where *Clostridium perfringens* has resulted in necrosis of the digestive tract wall. As pointed out in a recent review, undigested nutrients fuels microflora proliferation (Moran, 2014). Another example is when lack of structural components has resulted in a dilated proventriculus and a non-functional gizzard where feed passively flows through into the intestine. However, in many cases a suboptimal functionality may take place without such conspicuous signs of malfunction. In fact, it has been indicated that modern fast-growing broilers may exhibit suboptimal ability to digest nutrients compared to slower-growing strains, indicating that the requirement of modern broiler chickens for consuming large quantities of feed results in that a fraction of nutrients are not digested (Carre *et al.*, 2008).

A number of studies have shown that chicks are rapidly adapting to increased digestion when fed at hatch, as indicated by high activity levels of disaccharidase (Mahagna and Nir, 1996) and α-amylase (Sklan and Noy, 2000) two days after hatch. Even when measured on material collected from the ileum of broiler chickens, starch digestibility, using this quantitatively most important nutrient as example, has often been observed to be above 0.95 even for pelleted diets (Svihus, 2001; Hetland *et al.*, 2002; 2003; Svihus *et al.*, 2004; Hetland *et al.*, 2007). This high digestive capacity of poultry is truly impressive, not the least in broiler chickens, where pelleted diets and a high appetite results in material passing through the digestive tract in less than 5 hours (Svihus *et al.*, 2010; Svihus *et al.*, 2002). Extent of gelatinization during pelleting is normally below 20%, so starch in poultry diets is to a large extent present in the form of complex semi-crystalline native starch granules, which in human- and pig-related experiments have been demonstrated to be slowly and incompletely digested even after several hours digestion time (Shrestha *et al.*, 2012; Willamil *et al.*, 2012; Stein and Bohlke, 2007).

The impressive capacity of broiler chickens to digest even unprocessed starchy ingredients is further illustrated by the fact that even when whole untreated cereals have been used in large quantities, starch digestibility has been observed to be very high. Svihus *et al.* (1997) found ileal starch digestibility to be 0.98 in a mash diet containing 70% untreated whole barley as the only cereal source, and Svihus and Hetland (2001) found that 50% of the birds exhibited an ileal starch digestibility above 0.94, even when given a pelleted diet with 38.5% whole wheat. Even more astonishing were the results by Hetland *et al.* (2002), who observed an ileal starch digestibility of 0.98 in diets with 44% whole wheat mixed with pelleted other ingredients. The starch digestion process in such cases cannot commence before grinding in the gizzard, and since digestibility was based on analyses of contents collected from the ileal segment between Meckel's diverticulum to the ileo-cecal junction, this means that the whole starch digestion and glucose absorption process has taken place during the short retention time in the duodenum and jejunum, estimated by Rougiere and Carre (2010) to be around one hour.

Despite the above indications of high starch digestion capacity, starch digestibility values below 0.9 (measured either on ileal and/or total tract level) have also been seen in a large number of experiments (Wiseman *et al.*, 2000; Maisonnier *et al.*, 2001; Marron *et al.*, 2001; Svihus, 2001; Svihus and Hetland, 2001; Weurding *et al.*, 2001; Carré *et al.*, 2002; Hetland *et al.*, 2002; Carré *et al.*, 2005; Zimonja and Svihus, 2009). In several of these experiments, the low digestibility for a group of birds has been characterized by a large individual variation and a very low digestibility of some individuals. Thus, some birds seems to fail to digest feed properly. This mini-review will explore this anomaly, under the hypothesis that suboptimal digestibility is a characteristic of at least a portion of birds in a flock of broilers. The focus will mainly be on starch. This review leans heavily on two previous reviews (Svihus 2014a,b).

Proventriculus/gizzard function

It is becoming increasingly clear that functionality of the gizzard is perhaps the most important cause for failure of digestion.

The proventriculus and gizzard are the true stomach compartments of birds, where hydrochloric acid and pepsinogen are secreted by the proventriculus and mixed with contents due to muscular movements in the gizzard. However, the gizzard has an important additional function in grinding feed material, since this is not done in the mouth. Thus, the gizzard contains strongly myolinated muscles and has a koilin layer which due to its sand-paper like surface will aid in the grinding process. Grinding activity and the regulation of this activity in the gizzard has been described in detail by Duke (1992).

It has been shown repeatedly that when structural components such as whole or coarsely ground cereals or fiber materials such as hulls or wood shavings are added, the pH of the gizzard content decreases by a magnitude of 0.2 to 1.2 units (Gabriel *et al.*, 2003; Engberg *et al.*, 2004; Bjerrum *et al.*, 2005; Huang *et al.*, 2006; Gonzales-Alvarado *et al.*, 2008; Jimenez-Moreno *et al.*, 2009; Senkoylu *et al.*, 2009; Sacranie *et al.*, 2012; Svihus *et al.*, 2013). The logical explanation for this is the increased gizzard volume and thus a longer retention time which allows for more hydrochloric acid secretion. Since feed usually has a pH close to neutral, high feed intake can be expected to result in an elevated gizzard pH unless gastric juice secretion is able to increase in accordance with intake. This is probably the main reason why gizzard pH is reported to be higher with pelleted diets when compared to mash diets (Engberg *et al.*, 2002; Huang *et al.*, 2006; Frikha *et al.*, 2009), although less structure due to the grinding effect of pelleting will also contribute to this effect (Engberg *et al.*, 2002; Svihus *et al.*, 2004). As reviewed extensively by Svihus (2011), the increase in size of the gizzard when the diet contains structural components in the form of coarse fibers or cereals improves digestive function both through an increased retention time, a lower pH and a better grinding. This, probably combined with a better synchronisation of feed flow, has been shown to improve nutrient utilization.

Yet unpublished experiments carried out at our lab have demonstrated that birds with a underdeveloped gizzard due to lack of structural components will allow for a very rapid flow of feed material into the small intestine when starved and then refed. As observed before in experiments with turkeys Jackson and Duke, 1995), refeeding of birds which have been starved to empty the digestive tract will result in that the entire length of the small intestine will be filled with contents within an hour. Although the flow is surprisingly fast independent of gizzard function, the amount was observed to be considerably larger for birds without a well-functioning gizzard. Even more conspicuously, the starch content in the ileal material was much higher for these birds. In fact, the starch content in ileal material one hour after refeeding was virtually nondigested feed material, with a starch content of between 30 and 40% of the dry matter. Although this may indicate a dramatic loss of nutrients in such a situation, assessment of excreta has shown that although the nutrient content is higher than normal in this situation, most of the nutrients have disappeared. Although the microflora, for example through passage into the ceca, may have contributed, it is perhaps unlikely that such large quantities of nutrients can be removed by fermentation within a couple of hours. Thus, digestion in the small intestine remains a more plausible mechanism. Since Kadhim et al (2011) has shown that the ileum has a considerable amylase activity and Ferrer *et al.* (1994) has demonstrated a considerable absorption capacity of the ileum, one possibility is that digestion and absorption continues through the whole small intestine. Another possibility is that reflux allows for digestion and absorption in more anterior parts of the small intestine.

Form of feed and digestive tract functionality

One important factor is the form of the feed, which to a large extent will determine feed intake. Pelleting of the diet will usually increase feed intake of broiler chickens by 10 to 20% (Engberg *et al.*, 2002; Svihus *et al.*, 2004), and thus will increase the demands on an already high-performing digestive system. An increase in digestibility when diets were given as mash compared to as pellets was observed by Svihus and Hetland (2001), and indicates that pelleting may cause an overload of the digestive system. Engberg *et al.* (2002) found significantly higher levels of digestive enzymes when diets were given as mash compared to pellets, and also showed that pelleted diets resulted in a much more poorly developed gizzard than when mash diets were given. Thus, since the gizzard probably has an important role as a feed-flow regulator (Svihus, 2011), it is possible that the combined effect of a high feed intake and a low gizzard-stimulating effect increases the risk of a too rapid passage of material through the digestive tract. This fits with conclusions made by Rougiere and Carré (2010), who based on passage studies concluded that retention time in the proventriculus/gizzard was a major limiting factor for digestion in broiler chickens. A high feed intake due to pelleting may therefore have particularly detrimental effects when there are no structural components in the diet and therefore a small and under-developed gizzard. Environmental conditions may be important in this context, since birds will to some extent compensate for lack of structural components in the diet by eating litter materials such as wood shavings if available (Hetland *et al.*, 2005; Hetland and Svihus, 2007). Since pelleted diets are used commercially for broiler chickens, this means that the use of mash diets under experimental conditions may not reflect the commercial reality in terms of digestibility and digestive function. Further, it is possible that limitations to feed intake of broiler chickens would result in improvements in digestive function. One possible way of restricting feed intake would be to restrict access to feed at periods during the day. An advantage of such a feeding regime would be to increase retention time in the crop, which has been shown to have advantageous effects on feed efficiency (Svihus *et al.*, 2010; 2013). However, it is uncertain whether such intermittent feeding would restrict feed intake, since it has been shown that birds rapidly learn to fill up their crop and to use this as a reservoir for up to 5 hours after feeding. Thus, an alternative/additional way to restrict feed intake could be to switch from pelleted diets to mash diets. In addition to savings in processing costs, and additional advantage of such a feeding system would be that it would be easier to include coarse ingredients, since the structure of those would not be destroyed by the pelleting process.

References

Bjerrum, L., K. Pedersen and R.M. Engberg. 2005. *Avian Diseases* 49: 9-15.

Carré, B., A. Idi, S. Maisonnier, J.P. Melcion, F.X. Oury, J. Gomez and P. Pluchard. 2002. *British Poultry Science* 43: 404-415.

Carré, B., S. Mignon-Grasteau and H. Juin. 2008. *World's Poultry Science Journal* 64: 377-390.

Carré, B., N. Muley, J. Gomez, F.X. Oury, E. Lafitte, D. Guillou and C. Signoret. 2005. *British Poultry Science* 46: 66-74.

Duke, G.E. 1992. *Poultry Science* 71: 1-8.

Engberg, R.M, M.S. Hedemann and B.B. Jensen. 2002. *British Poultry Science* 44: 569-579.

Engberg, R.M., M.S. Hedemann, S. Steenfeldt and B.B. Jensen. 2004. *Poultry Science* 83: 925-938.

Ferrer, R., Gil, M., Moret, M., Oliveras, M. and Planas J.M. 1994. *Pflügers Archive* 426:83-88.

Frikha, M., H.M. Safaa, M.P. Serrano, X. Arbe and G.G. Mateos. 2009. *Poultry Science* 88: 994-1002.

Gabriel, I., S. Mallet and M. Leconte. 2003. *British Poultry Science* 44: 283-290.

Gonzales-Alvarado, J.M., E. Jiménez-Moreno, D.G. Valencia, R. Lázaro and G.G. Mateos. 2008. *Poultry Science* 87: 1779-1795.

Hetland, H. and B. Svihus. 2007. *Journal of Applied Poultry Research* 16: 22-26.

Hetland, H., B. Svihus and M. Choct. 2005. *Journal of Applied Poultry Research* 14: 38-46.

Hetland, H., B. Svihus and Å. Krogdahl. 2003. *British Poultry Science* 44: 275-282.

Hetland, H., B. Svihus and V. Olaisen. 2002. *British Poultry Science* 43: 416-423.

Hetland, H., A.K. Uhlen, K.H.K. Viken, T. Krekling and B. Svihus. 2007. *British Poultry Science* 48: 12-20.

Huang, D.S., D.F. Li, J.J. Xing, Y.X. Ma, Z.J. Li and S.Q. Lv. 2006. *Poultry Science* 85: 831-836.

Jackson, S. and G.E. Duke. 1995. *Physiology Behaviour* 58: 1027-1034.

Jimenez-Moreno, E., J.M. González-Alvarado, R. Lázaro and G.G. Mateos. 2009. *Poultry Science* 88: 1925-1933.

Kadhim, K.K., A.B.Z. Zuki, M.M. Noordin, S.M.A. Babjee and M. Zamri-Saad. 2011. *Afr. Journal of Biotechnology* 10: 108-115.

Mahagna, M. and I. Nir. 1996. *British Poultry Science* 37: 359-371.

Maisonnier, S., J. Gómez and B. Carré. 2001. *British Poultry Science* 42: 102-110.

Marron, L., M.R. Bedford and K.J. McCracken. 2001. *British Poultry Science* 42: 493-500.

Moran, E.T. 2014. *Poultry Science* 93, 3028-3036.

Rougiere, N. and B. Carré. 2010. *Animal* 4: 1861-1872.

Sacranie, A., Svihus, B., Denstadli, V., Moen, B., Iji, P.A. and Choct, M. 2012. *Poultry Science* 91, 693-700.

Senkoylu, N., H.E. Samli, H. Akyurek, A.A. Okur and M. Kanter. 2009. *Ital. Journal of Animal Science* 8: 155-163.

Shrestha, A.K., J. Blazek, B.M. Flanagan, S. Dhital, O. Larroque, M.K. Morell, E.P. Gilbert and M.J. Gidleya. 2012. *Carbohydrate Polymers* 90: 23-33.

Sklan D. and Y. Noy. 2000. *Poultry Science* 79: 1306-1310.

Stein, H.H. and R.A. Bohlke. 2007. *Journal of Animal Science* 85: 1424-1431.

Svihus, B. 2014a. *Poultry Science* 93, 2394-2399.

Svihus, B. 2014b. *Journal of Applied Poultry Science* 23, 306-314.

Svihus, B. 2011. *Worlds Poultry Science Journal* 67: 207-223.

Svihus, B. 2001. *Animal Feed Science and Technology* 92: 45-49.

Svihus, B. and H. Hetland. 2001. *British Poultry Science* 42: 633-637.

Svihus, B., H. Hetland, M. Choct and F. Sundby. 2002. *British Poultry Science* 43: 662-668.

Svihus, B., V.B. Lund, B. Borjgen, M.R. Bedford and M. Bakken. 2013. *British Poultry Science* 54: 222-230.

Svihus, B., K.H. Kløvstad, V. Perez, O. Zimonja, S. Sahlström, R.B. Schüller, W.K. Jeksrud and P. Prestløkken. 2004. *Animal Feed Science and Technology* 117: 281-293.

Svihus, B., R.K. Newman and C.W. Newman. 1997. *British Poultry Science* 38: 390-396.

Svihus, B., A. Sacranie, V. Denstadli and M. Choct. 2010. *Poultry Science* 89: 2617-2625.

Plenary session 02. Feeding strategies and gastrointestinal health

Weurding R.E., A. Veldman, W.A.G. Veen, P.J. van der Aar and M.W.A. Verstegen. 2001. *Journal of Nutrition* 131: 2329-2335.
Willamil, J., I. Badiola, E. Devillard, P.A. Geraert and D. Torrallardona. 2012. *Journal of Animal Science* 90: 824-832.
Wiseman, J., N.T. Nicol and G. Norton. 2000. *World's Poultry Science Journal* 56: 305-331.
Zimonja, O. and B. Svihus. 2009. *Animal Feed Science and Technology* 149: 287-297.

The potential of perinatal nutrition: *in ovo* and prestarter feeding

P.R. Ferket
Prestage Department of Poultry Science, North Carolina State University, Raleigh, NC 27695-7608, USA; peter_ferket@ncsu.edu

Summary

Based on Darwin's theory of adaptive evolution and Mendel's fundamental laws of heritability, breeding and genetic selection for increased growth rate and meat yield has dramatically advanced the production efficiency of poultry during the last 50 years, and this trend is expected to continue well into the future. Now, the period of embryonic and neonatal development is approaching 50% of the productive life of modern broilers, turkeys, and ducks. Although genetic selection does dictate how maternal and paternal genes that are inherited by their progeny, we are becoming more aware that nutrition and management may influence how those inherited genes are expressed. Epigenetics is the rising science of programming gene expression during critical developmental periods, which subsequently allow an animal to metabolically or physiologically adapt to specific dietary or environmental conditions. In poultry, epigenetic programming can occur during two critical periods: during the period of gametogenesis when breeding stock are adolescents, and during egg formation when egg nutrients are consumed by the embryo *via* amniotic fluid prior to hatch and yolk through to the first few days after hatch. Adaptive conditioning can be advanced further by nutritional or physiological imprinting during the first days after hatch. This paper discusses implications of nutritional and physiological stress of breeders on the epigenetic response of progeny, and how it can be modified by perinatal nutrition, including amniotic fluid supplementation by *in ovo* feeding and early feeding technologies. Epigenetic and adaptive conditioning of neonatal nutrition will also be discussed in the context of a 'programmed nutrition' strategy to increase production efficiency and meat quality. As new molecular biology tools to measure gene expression become increasing affordable and robust, the study of epigenetic programming by perinatal nutrition will become an increasingly popular field of research. Moreover, emerging perinatal nutrition technologies for hatcheries will likely make programmed nutrition a commercial reality in the future as genetic selection for performance efficiency continues.

Introduction

Modern agriculture constantly strives to maximize biological performance of food production in an effort to optimize economic efficiency, profit potential, and sustainability. Commercial poultry production is among the most efficient and progressively successful of all food production sectors. What factors does it take for continued success in efficient poultry production? It takes the right genetics, combined with optimum health and management practices, and an optimized nutrition and feeding program. Efficiency and sustainability depends on the ability of a poultry production company to achieve competitive production indicators, including average daily gain, days to market weight, feed (caloric) conversion, livability, flock uniformity, and processing yields. However, profitability largely depends on how well a poultry production company meets consumer demand. Consumers want wholesome, safe, and affordable food. They want poultry products that look good, and are enjoyable to eat. Moreover, the most affluent consumers also want to buy their food from companies that excel in environmental stewardship and animal welfare. After proper management of commercial genetic stock, nutrition and feed is the most variable component of economic efficiency and profitability, as it represents 70 to 80% of live production costs.

Genetic selection is continually changing the 'playing field' of production potential for the poultry industry; but it is the expression of this genetic potential that drives growth performance, health, and ultimately the profitability of poultry production. Growth performance and meat yield has improved linearly by about 1% each year, and 85% of this improvement is attributed to genetic selection of broilers (Havenstein *et al.*, 2003) and turkeys (Havenstein *et al.*, 2007). One may argue that nutritional advancements have not kept pace with genetic selection as metabolic disorders and apparent nutritional deficiencies continue to arise, which mandate diet formulation constraints to be updated. However, the time has come to close this pace gap as we learn to harness the power of perinatal nutritional imprinting and adaptive conditioning to program the expression of genes associated with socioeconomically important traits.

The ancient Greek philosopher, Aristotle, theorized an individual's traits are acquired from their parents and contact with their environment. In simple terms, Aristotle's theory of developmental destiny means that all life on this planet is programmed to succeed in its given environment! Just a generation before Mendel's time, Jean Lamark was a proponent of the inheritance of acquired characteristics. The so-called Lamarckism theory emphasized the use and disuse of organs as the significant factor in determining the characteristics of an individual, and it postulates that any alterations in the individual could be transmitted to the offspring through the gametes. Despite many attempts, this inheritance of acquired characteristics has never been experimentally verified. Furthermore, many of Lamarck's examples, such as the long neck of the giraffe, can be more satisfactorily explained by means of natural selection.

The academic discipline of genetics has followed Mendel's basic concepts for over a century, and it was reinforced by the discovery and sequencing of DNA. Genetics describes the inheritance of information on the basis of DNA sequence. As the DNA sequence fragments were scientifically associated with certain biological traits, genomic scientists began to realize the importance of the gene expression. It was not until the recent introduction of molecular biological tools that the science of epigenetics and conditional imprinting has emerged. We can now study gene expression by mRNA up-regulation, proteomics and metabolomics. Now, many molecular geneticists agree that gene expression in response to environmental cues can be passed on to future generations. The old Greek philosophers and Jean Larmarck may have been right after all; they just did not have the scientific tools to prove it! There is now growing evidence that nutrition and environmental stimuli of parent stock and their progeny during the perinatal period may literally program how an animal's genes are expressed as an adaptive response to increase the chances of survival. This new science of 'gene expression programming' is Epigenetics; it is the inheritance of information on the basis of gene expression, or inherited adaptation.

What is epigenetic or adaptive conditioning?

I am the son of immigrants who came to Canada in 1956 to farm as my ancestors did before them. My parents left the Netherlands so their children would not experience what they had experienced when they were children during World War II. They occasionally spoke of how the Nazi soldiers took nearly all the food their family farm produced, leaving barely enough of the food they toiled to produce to eat for themselves. Towards the end of World War II there was a national famine that caused over 30,000 people to starve to death because of scarce food supplies from war-torn agricultural lands, and an unusually harsh winter. Detailed birth records collected during that 'Dutch Winter Famine' provided scientists with useful data for analyzing the long-term health effects of prenatal exposure to famine. The children of this famine to 3 generations have unusually high incidence of developmental and adult disorders, including low birth weight, short body height, diabetes, obesity, coronary heart disease, and cancer, (Pray, 2004). In another study, Kaati *et al.* (2002) correlated grandparent's prepubertal access to food with diabetes and heart disease. Remarkably, a pregnant mother's diet can affect the expression of her genes in such a way that not only her children, but her grandchildren and possibly great-grandchildren inherit the same health problems. Using data from a small Swedish community, Pembrey *et al.* (2006) observed that epigenetic effects are sex-linked. Grandfathers who had access to surplus food during their slow growth phase (9 to 12 years of age) begot more diabetic grandsons than grandfathers who did not have as much food available to them before they reached puberty. In contrast, the biggest effect of food supply in grandmothers occurred when she was a fetus and infant, and it affected the mortality rate of their granddaughters. These responses suggest that information is being captured at key stages of egg and sperm formation, and is passed on to the offspring, possibly as modifications to the epigenome.

Epigenetics literally means 'on genes', and refers to all modifications to genes other than changes in the DNA sequence itself. DNA within each cell is wrapped around proteins called histones. Both the DNA and histones are covered with chemical tags, to form what is called the epigenome. These chemical tags react to signals to the outside world, such as diet and stress. Some parts of the epigenome are wrapped and unreadable, and other parts are relaxed and readable for expression. A good instructional video that describes the basic concepts of epigenetics can be viewed at http://learn.genetics.utah.edu/content/epigenetics/intro.

Epigenetic imprinting of genes occurs most often by differential methylation of DNA at the promoter regions of specific genes that can permanently modulate an organism's adaptive response to adverse stimuli during critical periods of development. Particularly, early-life programming can turn on 'Thrifty' genes that permanently reprogram normal physiological responses to survive environmental stressors, including moderate nutrient deficiency, and thus increase the chances of passing on their genes to the next generation. Evidence for epigenetic programming is demonstrated by swarming locusts: the swarming phenotype is environmentally influenced by drought conditions and the trait is passed onto the next generation until the population finds better conditions.

Transgenerational epigenetic or adaptive conditioning may explain some of the blessings and curses observed as a result of our system of commercial poultry production. Consider how we manage the weight of broiler or turkey breeders before and during egg production: this is during the critical epigenetic period of gametogenesis. Broiler breeder nutrition and feeding management likely has an important epigenetic effect on progeny. Consider how we manage and incubate commercial hatching eggs: this is during the critical epigenetic period of *de novo* methylation of somatic cells in the embryo. Environmental conditions (*i.e.* temperature and oxygen concentration) in the incubator may program epigenetic responses that affect subsequent metabolism. Consider how we manage chicks during the first few days after hatch. Feeding behavior, nutrition, and brooding conditions can affect metabolism and the development of breast muscle, the skeleton, and immune system.

In ovo feeding jump-starts perinatal development

Phenotypic characteristics that are programmed or imprinted to succeed in it's given environment and diet happen most effectively when the animal is young, and it is the first few meals that usually make the difference. For example, all honeybees are genetically similar, but what predestines a bee to become a worker or a queen is what the larvae are fed. Likewise, poultry may be programmed to succeed with the desired phenotypic traits by nutritional modification during the perinatal period: the 3 days before hatch and the 3 days after hatch. The chick's first meal occurs when it imbibes the amnion prior to hatch, and so this is the first opportunity for nutritional programming. By *in ovo* feeding (Uni and Ferket, 2003; US Patent No. 6,5692,878), nutrient balance and key metabolic co-factors of the amnion meal can be modified and influence subsequent phenotypic traits of economic importance for the poultry industry.

The benefits of *in ovo* feeding on early growth and development of broilers and turkeys have been demonstrated by several experiments in our laboratory (Uni and Ferket, 2004). *In ovo* feeding has increased hatchling weights by 3% to 7% (*P*<.05) over controls, and this advantage is often sustained at least until 14 days post-hatch. The degree of response to *in ovo* feeding may depend upon genetics, breeder hen age, egg size, and incubation conditions (i.e. the epigenotype). Above all, IOF solution formulation has the most profound effect on the neonate. Positive effects have been observed with IOF solutions containing NaCl, sucrose, maltose, and dextrin (Uni and Ferket, 2004; Uni *et al.*, 2005), β-hydroxy-β-methyl butyrate, egg white protein, and carbohydrate (Foye *et al.*, 2006ab), Arginine (Foye *et al.*, 2007), zinc-methionine (Tako *et al.*, 2005), butyric acid

(Salmanzadeh *et al.*, 2015), IGF-1 (Liu *et al.*, 2012), and L-glutamine (Shafey *et al.*, 2013). In addition to the increased body weights typically observed at hatch, the positive effects of *in ovo* feeding may include increased hatchability (Uni and Ferket, 2004; Uni *et al.*, 2005); advanced morphometic development of the intestinal tract (Uni and Ferket, 2004; Tako *et al.*, 2004) and mucin barrier (Smirnov *et al.*, 2006); enhanced expression of genes for brush boarder enzymes (sucrase-isomaltase, leucine aminopeptidase) and their biological activities, along with enhanced expression of nutrient transporters, SGLT-1, PEPT-1, and NaK ATPase (Tako *et al.*, 2005; Foye *et al.*, 2007); increased liver glycogen status (Uni and Ferket, 2004; Uni *et al.*, 2005; Tako *et al.*, 2004; Foye *et al.*, 2006a); enhanced feed intake initiation behavior (de Oliveira, 2007); increased breast muscle size at hatch (Uni *et al.*, 2005; Foye *et al.*, 2006a), breast muscle growth and meat yield (Kornasio *et al.*, 2011), and improved skeletal development (Yair *et al.*, 2015). *In ovo* feeding clearly advances the digestive capacity, energy status, and development of critical tissues of the neonate by about 2 days at the time of hatch. Using scanning electron microscopy, Bohórquez *et al.* (2008) observed that *in ovo* feeding significantly increased functional maturity and mucus secretion of goblet cells of villi of ileum and ceca of turkey poults. Associated with these goblet cells was the colonization of lactobacilli. Therefore, *in ovo* feeding may help improve the colonization resistance of enteric pathogens of neonatal chicks and poults. Based on the rapidly growing number of peer-reviewed publications from around the world, *in ovo* feeding consistently shows promising benefits, especially if applications can be done without compromising hatchability.

In ovo feeding offers promise of sustaining the progress in production efficiency and welfare of commercial poultry. Although selection for fast growth rate and meat yield may favor the modern broiler to become a more altricial, proper early nutrition and *in ovo* feeding may help these birds adapt to a carbohydrate-based diet and metabolism typical of a precocial bird at hatch. Our original research on *in ovo* feeding has established a new science of neonatal nutrition that many other scientists are now pursuing. As a result, we are all gaining greater understanding of the developmental transition from embryo to a juvenile bird. Now more work on *in ovo* feeding application technology and hatchery logistics must be done before *in ovo* feeding can be widely adopted for commercial practice.

Potential of post-hatch nutrition on nutritional imprinting

The first few days post-hatch is the second part of the perinatal period that can imprint production traits by adaptive conditioning of gene expression. Chicks can be imprinted to enhance their tolerance to immunological, environmental, or oxidative stress. Nutritional programming during the perinatal period can also influence energy and mineral utilization or requirement, while other bioactive dietary components may 'program' enteric microflora colonization that affect gut health and food safety. For example, Yan *et al.* (2005) reported that conditioning broilers fed a diet low in calcium and phosphorus for 90 hours post-hatch improves intestinal calcium and phosphorus absorption at 32 days of age, and increases the expression of the gene for the mineral transporter protein throughout the life of the bird. Angel and Ashwell (2008) demonstrated that broilers fed a moderately deficient conditioning diet for the first 90 hour post-hatch were more tolerant to a P-deficient grower and finisher diet, but they were also heavier, had better feed conversion, and they had higher tibia ash and P retention. The work of Angel and Ashwell demonstrate that epigenetic imprinting and nutritional adaptation to low dietary Ca and P is indeed possible and likely for other minerals as well.

Based on the concepts of epigenetics, imprinting, and adaptive conditioning presented above, several experiments has been done to test various nutritional programming strategies at the Alltech-University of Kentucky Nutrition Research Alliance Coldstream Farm and Alltech's Center for Animal Nutrigenomics and Applied Animal Nutrition. By evaluating the expression patterns of key functional gene groups, dietary amounts of nutrients that affect homeostatic balance were discovered to depend on the form of the nutrient, levels of and interactions among nutrients, and the timing of administration. Feeding chicks a specifically-formulated diet during the first 72 hours post-hatch has been developed to 'condition' the gut for better nutrient utilization and program metabolism that ultimately affects production efficiency, carcass composition, and meat quality. Chicks that have been fed the appropriate conditioning diet, followed by a complementary growing and finishing diet, have improved growth performance and feed efficiency through to market age, and over 70% higher calcium and phosphorus digestion than controls. A programmed nutrition strategy can literally change the nutrient requirement and production efficiency, and may yield a response greater than any single feed additive on the market. Not only can programmed nutrition increase production efficiency that is so important to poultry producers, there is evidence that it improves the meat quality consumers demand, which yields greater potential profits from the poultry products produced. Broilers that have been raised on a programmed nutrition strategy have reduced carcass fat and produce breast meat that has more appealing color, less drip losses during storage, improved oxidative stability, and lower cooking losses.

Although feeding broilers a special nutritional conditioning diet for just 72 hours after hatch presents great opportunities, it is logistically difficult to accomplish in practice using current production systems. Moreover, variation in the time and stress exposure between hatch-pull and placement will affect the effectiveness of the 3-day nutritional conditioning period. However, recent hatch-brood technology (http://www.hatchbrood.nl/hatchbrood/product.php) offers a practical means to deliver specially formulated diets during the first 2 or 3 days post-hatch in the controlled environment of a hatchery. The hatchery of the future will be a place that will do much more than simply hatch and vaccinate chicks: it will also be the place where the chicks will be conditioned better tolerate the challenges of life, and be programmed for optimum nutrient efficiency. Nutritional science is no longer a matter of supplying minimally required nutrients in the ideal balance to achieve desired production and welfare goals. We now know that nutrition is a process that can be programmed to succeed by strategic perinatal diet manipulation by *in ovo* and post-hatch feeding.

References

Angel, R. and Ashwell, C.M. 2008. Dietary conditioning results in improved phosphorus utilization. *Proceedings of the XXIII World's Poultry Congress, Brisbane, Australia, June 30 - July 4, 2008.*

Bohórquez, D.V., Santos, Jr., A.A. and Ferket, P.R. 2008. *Poultry Science* 87(Suppl 1): 139.

Havenstein, G.B., Ferket, P.R., Grimes, J.L., Qureshi, M.A. and Nestor, K.E. 2007. *Poultry Science* 86: 232-240.

Havenstein, G.B., Ferket, P.R. and Qureshi, M.A. 2003. *Poultry Science* 82: 1509-1518.

De Oliveira, J.E., P.R. Ferket, C.M. Ashwell, Z. Uni and C. Heggen-Peay. 2007. *Poultry Science* 86(Suppl 1): 214.

Foye, O.T., Uni, Z., and Ferket, P.R. 2006a. *Poultry Science* 2006 85: 1185-1192.

Foye, O.T., Uni, Z., McMurtry, J.P. and Ferket, P.R. 2006b. *International Journal of Poultry Science* 5: 309-317.

Foye, O.T., Uni, Z. and Ferket, P.R. 2007. *Poultry Science* 86: 2343-2349.

Kaati, G., Bygren, L.O. and Edvinsson, S. 2002. *European Journal of Human Genetics* 10: 682-688.

Kornasio, R., O. Halevy, O. Kedar and Z. Uni. 2011. *Poultry Science* 90: 1467-1777.

Liu, H., Wang, J., Zhang, R, Chen, X., Yu, H, Jin, H, Li, L., Han, C., Xu, F., K, B., He, H. and Xu, H. 2012. *Journal Ce. Physiology* 227:1465-1475.

Pembrey, M.E., Bygren, L.O., Kaati, G., Edvinsson, S., Northstone, K., Sjostrom, M., Golding, J., and the ALSPAC Study Team. 2006. *European Journal of Human Genetics* 14: 159-166.

Pray, L. A. 2004. *The Scientists* 18: 13-14.

Salmanzadeh, M., H. Aghdam Shahryar and A. Lotfi. 2015. *Kalkas Univ Ved Fak Derg* 21:19-25.

Shafey, T.M., A.S. Sami and M.A. Abouheif. 2013. *Journal of Animal Veterinary Advances* 12: 135-139.

Smirnov, A., Tako, E., Ferket, P.R. and Uni, Z. 2006. *Poultry Science* 85: 669-673.

Tako, E., Ferket, P.R. and Uni, Z. 2004. *Poultry Science* 83: 2023-2028.

Tako, E., Ferket, P.R. and Uni, Z. 2005. *Journal of Nutritional Biochemistry*. 15: 339-346.

Uni, Z., Ferket, P.R., Tako, E. and Kedar, O. 2005. *Poultry Science* 84: 764-770.

Plenary session 03 and 04. Precision feeding

Uni, Z. and Ferket, P.R. 2003. Enhancement of development of oviparous species by *in ovo* feeding. United States Patent No. 6,592,878.

Uni, Z. and Ferket, P.R. 2004. *World's Poultry Science Journal* 60: 101-111.

Yair, R., S. Shahar and Z. Uni, 2015. *Poultry Science* 94: 2695-2707.

Yan, F., Angel, R., Ashwell, C., Mitchell, A. and Christman, M. 2005. *Poultry Science* 84: 1232-1241.

How to survive a 'challenging childhood': rearing the laying hen!

R.P. Kwakkel[1]*, M.L. Elling-Staats[1] and M.M. Van Krimpen[2]
[1]Wageningen University and Research, Department of Animal Sciences, Animal Nutrition Group, P.O. Box 338, 6700 AH Wageningen, the Netherlands; [2]Wageningen University and Research, Department of Animal Sciences, Livestock Research, P.O. Box 338, 6700 AH Wageningen, the Netherlands; rene.kwakkel@wur.nl

Summary

A one-day old chick that needs to develop into a pullet and young hen faces many different challenges during early life among which dehydration during the first few days and the many stressful vaccinations throughout rearing are major ones. An easy start could be enhanced by feeding wet-based diets to enhance starter feed intake capacity, basic body development and gut maturation. To avoid adult feather pecking behaviour and maintain uniformity during rearing, the use of fibrous and coarse diets and management seems crucial during rearing to stimulate a proper feeding behaviour, a well-developed gut and functioning microbiome. Finally, prior to onset of lay, the young hen need to efficiently use minerals such as Ca and P, to ensure optimal bone structure and a good egg shell quality for even a laying period where hens age up to 100 weeks.

Introduction

A one-day old chick that needs to develop into a pullet and young hen faces many different challenges during early life among which dehydration during the first few days and the numerous vaccinations throughout the rearing period play a major role. So a good start (early feeding concept) is an important factor for proper young hen management. Next to that, there is a large variation in vaccination policies throughout Europe. Nordic countries, such as Sweden and Finland, follow a quite restrictive system. Whereas in the Netherlands, with dense populations of flocks, a more intensive vaccination scheme is followed as standard procedure. Young rearing stock is being vaccinated against several diseases (such as Marek, Infectious Bronchitis, New Castle Disease, Gumboro, Avian Metapneumovirus TRT, Fowl Pox, Infectious Laryngotracheitis ILT, Salmonella; some vaccinated multiple times). During rearing, pullets and young hens are being vaccinated with different vaccines at days 1, 7, 14, 28, 42, 49, 56, 65, 84, 105 and 112 post hatch, and they show a reduced appetite (up to 10% of daily consumption) for a few days post vaccination, and as a consequence, a retarded growth at several moments during the rearing period (sources: GD Animal Health; Verbeek Hatchery Holland). On-going discussions at EU-level advocate a more restrictive policy over countries.

Uniformity seems to be of utmost importance. It seems that the main objective in immature layer nutrition focuses on a high uniformity, which is a challenge due to the fact that individual birds may respond differently to nutrient intake and conversion efficiency (*residual feed intake*), but moreover, they also react differently to the numerous vaccinations during rearing (Figure 1).

A very nice research tool (and in the future a potential commercial tool) is the Precision Feeding (PF) system, being developed at the University of Alberta (UA) by the group of Dr Zuidhof. The PF system is a *feed station* for individual birds to enter that not only monitors feed intake and BW, but also controls feed intake (according to the BW curve) of individual free run chickens. Up to now, most research has been carried out with broiler breeders. Uniform flocks can be produced with Coefficients of variation as low as 3% in broiler breeder pullet BW can be reached in this PF system (Zuidhof, 2015).

Until we reach a situation that birds in a flock can be individually fed in a PF system like dairy and sows, we still need to feed our flock *as a whole* an adequate amount of nutrients throughout the day in an almost ad-lib fed system, taking into account all types of variation.

Plenary session 03 and 04. Precision feeding

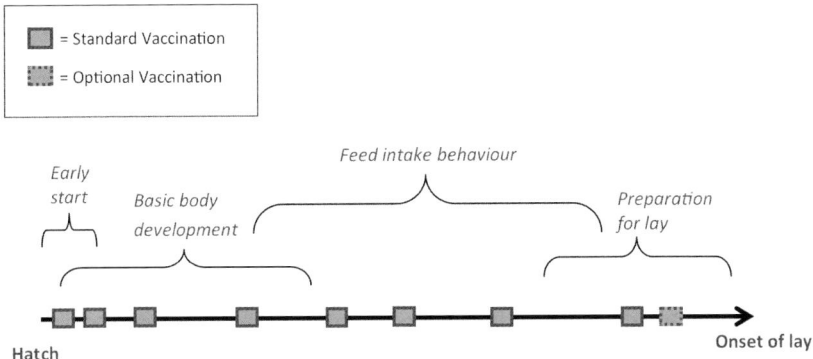

Figure 1. Time schedule (cascade of events) from hatch to 18 weeks of age.

One of the propositions in this paper is that 'feed intake drives performance' and to the authors' opinion, an adequate and stimulating feed intake should be the main focus of pullet rearing.

In this paper, we tried to discriminate nutritional factors influencing performance (growth and development), (feeding) behaviour, and gut health (ingredient selection). Whereas most factors are interrelated and as such researched and reported by colleagues in the field, it was quite a task to separate evidence based research in particular paragraphs. We hope the reader don't mind!

Well, as the reader already might think: there is a lot to discuss on rearing hen nutrition. Yes, there is, and therefore it seems awkward that in the past 20 years (1996-2016) very few papers have been published in scientific journals on layers/rearing hens relative to broilers. A quick search in publication databases such as Scopus, Web of Science and PubMed yielded a distribution of 15 vs 85 between rearing/laying vs broiler (breeder) studies. It's time to focus again on the rearing and laying hen. We hope this paper may be a stimulant to do so!

A challenging start: *being thirsty!*

Every year, about 0.5 million pullets in the Netherlands (up to 1.5%) die in their first week of life due to dehydration, causing a significant economic loss of 1.1 million Euro (source: KWIN, 2016-2017; Data Livestock Research WUR). Thus, dehydration, as a consequence of an impaired water and feed intake of young chickens, is a significant problem in (all segments of) the poultry industry. One-day old pullets, transported from the hatchery to the rearing farm, do not receive any feed or water, causing a time lag of up to 72 hours before having access to these nutrients (Willemsen *et al.*, 2010). On farm, pullets may have difficulty to find the nipples, and this results in high mortality rates during the first week of life, mainly attributed to dehydration. Hence, day-olds are at severe risk of dehydration in common practice and this is both a welfare and economic concern.

Welfare is compromised in several ways during this early post-hatch period: a novel environment (microflora), a change from yolk (liquid) to solid (crumble and mash) feed, and a low feed intake as such may cause variation in initial growth and development and impaired flock uniformity at a later stage. The first few days of life are of great importance for the development of the gastrointestinal tract (GIT) in young birds. In fact, the weight of the small intestine relative to the chick's body weight as well as major changes in the mucosal GIT peaks at 7 days of age (Iji *et al.*, 2001; in a broiler study). Moreover, several studies indicated that withholding feed for a short period after hatch (up to 3 days) impairs GIT development and immune defense against various challenges (a.o., Simon *et al.*, 2015).

Plenary session 03 and 04. Precision feeding

A nutritional intervention such as feeding a 'porridge'-type diet may be an efficient tool to enhance a quick start for young pullets. A liquid-based diet in the first few days may reduce the risk of dehydration, as the birds ingest moisture together with feed in porridge form. From broiler research we know that liquid-based or 'wet feeding' increases birds' feed intake and growth (Yasar and Forbes, 1999) and these effects seem to be more pronounced if such a diet is fed during early life (Mai, 2007).

In recent trials in our lab with day-old broilers, such liquid feeds (1:1 for water:feed) increased total air-dry feed intake compared to dry feeds in a choice-fed system, indicating its attractiveness to young chickens. A wet diet including whole wheat (WW) was most attractive to young chickens (Figure 2). When such porridge-type diets include a coarser dietary structure (larger particle sizes), birds may develop functional gizzards and this may improve protein digestibility and a healthy gut, ultimately stimulating a fast early development. Preliminary results showed that choice-fed chicks were 10% heavier at d 21 as compared to chicks fed an *ad lib* dry control diet (1.14 vs 1.03 kg). The choice-fed chickens could choose between a dry and wet diet; they choose in more than 90% for the wet diet.

The GIT is equipped with an intrinsic immune system where innate immune cells support intestinal development and homeostasis by protecting the intestine against pathogens (Rescigno, 2011). The expression of these innate immune cells is influenced by nutrients and metabolites from bacterial and dietary origin (Bostick *et al.*, 2015).

A project to be started aims to develop a liquid-based feeding strategy (with larger particle sizes) for day-olds (both laying-type pullets and broilers). This liquid-based feed will be used as a novel way to modulate the GIT within the first 10 days of life, which we consider the critical window for gut innate immune development and nutrient digestibility. This new liquid-based feed should predisposition the GIT towards a better developed GIT, an improved innate immune response, a balanced microflora, and hence, more bird welfare, and lower feed costs.

Optimal (early) body development: *an old model*

In the nineties of the past century, previously evidence based research in Israel on growth and development of broiler breeders (Soller *et al.*, 1984), in combination with the classical growth

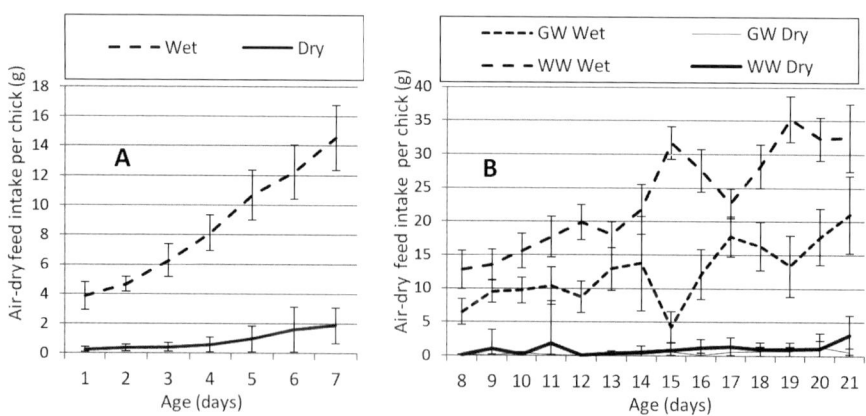

Figure 2. Preference test for broilers. Panel A: diets consisting of crumbled pellets were fed as dry and wet, provided in two troughs per pen (day 0-7; n=12 pens); Panel B: diets consisting of 4 mm pellets were fed as dry and wet, with either whole wheat (WW) or ground wheat (GW) included, provided in four troughs per pen (days 8-21; n=12 pens) (data not published).

modelling theory from the Hammond school (Hammond, 1932), was used as a model to start research on 'critical growth phases' in body development of laying-type pullets (Kwakkel *et al.*, 1993), with the aim to provide feeding strategies for an optimal start at onset of lay and to elucidate how important target weights at a certain age are. Body composition seemed more important than BW *per se*. It was revealed that an early basic body development (protein, water and ash) was the main determinant in initiating physiological processes that result in the development of the reproductive tract, as depicted in terms of a *pubertal growth spurt* (3rd phase in Figure 3). This growth spurt, as part of a multiphasic growth model, seems to be the major determinant of young layer egg performance.

Despite the ongoing focus on growth and development of (female) broiler breeders by Research groups in Belgium (Decuypere *et al.*, 2006), Scotland (Hocking, 2007), the Canadian group of Robinson (Renema *et al.*, 2007), and more recently, the Dutch group (Van Emous *et al.*, 2015), hardly any effort has been executed towards laying-type pullet growth.

Although the industry is aware of the importance of an early basic body development in pullets, only a *single paper* was dedicated to growth modelling and allometric relationships between body organs (Lieboldt *et al.*, 2016b). In this study, allometric relationships indicated that an early development of supply organs (heart, liver, gizzard and pancreas) was crucial for early BW gain. Some authors reported dose-responses of amino acids on early pullet growth for lysine and methionine (Halle, 2002), for threonine (Bonato *et al.*, 2015), and for L-arginine (Lieboldt *et al.*, 2016a, 2016b). A response to higher levels of protein during the last phase of rearing is mostly lacking, as illustrated by the digestible lysine study of Jardim *et al.* (2011).

Recently, a combined effort has started between the University of Alberta and Wageningen University to investigate the impact of uniformity on pullet growth and development. In these studies, the Precision Feeding system (individual feeding), as explained in the Introduction of this paper, will be used to test the interaction between different growth models for pullets and the provision of more or less density diets adjusted throughout the grower phase on uniformity and subsequent laying performance.

Bone development in loose systems: *fly like an eagle*

Bone development is an important aspect in relation to performance and welfare in rearing and laying hens. Bone fractures and bone related disorders are one of the most important welfare problems and also negatively affect performance of laying hen flocks. A large number of hens

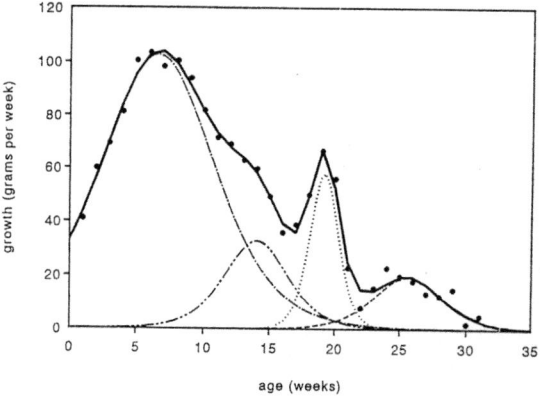

Figure 3. Multiphasic development in body components of a rearing and young mature hen (adapted from Kwakkel et al., *1993).*

have or have had keel bone fractures at the end of the laying period, likely caused by collisions with perches and other elements in the housing (Harlander-Matauschek *et al.*, 2015).

Although different factors may play a role in bone development, there are two factors (light and feed) that are of main interest, as there is little knowledge on the different aspects and underlying mechanisms. In rearing hens, light may affect bone strength through stimulating bird activity and reducing fearfulness of the birds and hence reduce the risk for bone fractures. Aspects of lights such as intensity or colour during rearing may influence the behaviour and preference of birds later in life (nest choice) and may help the hens to move adequately in a three dimensional space (Moinard *et al.*, 2004; Gunnarsson *et al.*, 2008). The main focus in this paper, however, will be on *nutrition and bone quality*.

The fatty acid composition of a diet affects bone development (Swiatkiewicz *et al.*, 2009; Toscano *et al.*, 2015). The most important nutritional factors related to bone formation, however, are Ca, P and vitamin D. Phosphorus is of particular interest, because the world-wide store of P is limited (Elser and Bennett, 2011). It is a challenge to increase the dietary P-efficacy of poultry, without negatively affecting performance and bone quality. Table values of P-requirement of rearing hens are (with updates in matrices) still based on the datasets from 1998 (Schutte *et al.*, 1998). At that time, however, hens were housed in cages, which might have affected bone development of the birds. Therefore, there is a need for validating the current P recommendations for rearing hens in loose housing systems or even under free range conditions.

A project at WUR has recently started where the aim is to understand how dietary interventions (such as NPP levels) in early life affect lifelong mineral utilisation for egg shell quality and bone development in rearing and laying birds. This information can be used to develop diets and feeding strategies to improve bone quality and make better use of limited resources of minerals. In a recent trial in our lab (unpublished data), WL-pullets from 16 weeks of age were fed mash diets with a low or moderate NPP level (1.6 or 2.9 g/kg, respectively), dry or wet (1:1 feed:water ratio) and with fine or coarse particle sizes in a 2×2×2 factorial design. It was hypothesized that coarse and wet diets would counteract the negative effects of a low NPP on performance and bone strength in the early laying period. As shown in Table 1, however, dietary NPP level did not affect performance or bone breaking strength in these young hens. The NPP requirement for young layers might, therefore, be overestimated in practice. Effects of this low NPP level later on in the laying period has not yet been determined. Interestingly, the reduction of tibia breaking strength from start of lay to 28 weeks of age tended to be lower for hens fed the coarse diets ($P=0.084$), indicating that dietary structure might play a greater role in bone mineralisation of young layers than NPP level. Moreover, ready-to-start young layers had a higher feed intake (g DM) on the wet diets, particularly during the period from 16-17 weeks (10%).

Evidence is available for neonatal programming of gene expression related to Ca and P efficiency in oviparous species (Ashwell and Angel, 2010). Because metabolic programming usually occurs during the first days of life (both embryonic and early post hatch), it can be hypothesised that nutrients supplied to the breeders, e.g. dietary Ca and P levels, already program the Ca and P absorption and utilisation of the offspring. A few examples: it has been shown that providing low protein diets to broiler breeders resulted in a faster post-hatch growth rate and in increased breast meat yield (Rao *et al.*, 2009). Providing omega-3 rich lipids to maternal diets improved development of the progeny by influencing organic matrix quality and mineralization in embryos (Liu *et al.*, 2003). Results of Favero *et al.* (2013) in broiler breeders indicated that supplementing additional organic Zn, Mn and Cu on top of a control diet increased tibia calcification and thickness of tibia of the progeny. Until now, the impact of interventions in layer breeder diets on the metabolic pathways that improve bone quality of the offspring has not been researched and is thus unknown.

In ovo feeding of minerals may be a valuable tool to study mechanisms of bone development during incubation (Yair *et al.*, 2015; in broilers). There is little knowledge on the effects and

Table 1. Preliminary data on performance parameters of laying hens fed diets with moderate or low NPP level, dry or wet diet moisture, and fine or coarse structure in the pre-layer (16-17 wk of age) and the early production period (23-27 wk of age) (data unpublished).

Effect	Pre layer (16-17 wk)		Early production (23-27 wk)					26-27 wk	22 wk	28 wk
	ADFI[1] (g)	BW gain (g/wk)	ADFI (g/d)	Rate of lay	FCR	Egg mass (g/hen/d)	Egg weight (g)	Egg-shell breaking strength (N)	Tibia breaking strength[4] (Nmm)	Tibia breaking strength[4] (Nmm)
NPP level[2]										
Moderate	83.9	89.1	114.9	91.04	2.27	50.66	55.76	48.8	145.54	132.35
Low	82.9	87.5	114.7	90.72	2.29	50.41	55.47	48.8	139.34	137.33
SEM	1.32	3.21	0.46	0.59	0.03	0.35	0.14	0.38	4.96	4.78
Diet moisture										
Dry	79.5	86.8	113.5	90.53	2.27	50.12	55.39	48.5	141.70	136.90
Wet	87.3	89.8	116.0	91.22	2.28	50.95	55.84	49.1	143.19	132.78
SEM	1.20	3.20	0.45	0.59	0.03	0.35	0.14	0.38	5.00	4.79
Structure[3]										
Fine	83.2	88.2	115.5	91.68	2.26	51.14	55.79	48.2	141.19	128.84
Coarse	83.7	88.4	114.1	90.08	2.29	49.92	55.43	49.3	143.70	140.83
SEM	1.33	3.21	0.46	0.59	0.03	0.35	0.14	0.38	5.00	4.65
Source of variation (P-values)										
NPP level	0.421	0.739	0.715	0.695	0.689	0.539	0.039	0.981	0.387	0.467
Diet moist.	<0.001	0.501	<0.001	0.402	0.716	0.080	0.003	0.192	0.835	0.547
Structure	0.720	0.955	0.012	0.051	0.435	0.007	0.010	0.028	0.726	0.084

[1] ADFI is based on air-dry feed intake; intake of wet diets was corrected.
[2] NPP level: 'moderate' was 2.6 and 2.9 g/kg and 'low' was 1.5 and 1.6 g/kg in the Pre-layer and Early production period, respectively.
[3] Structure: finely ground (Geometric Mean Diameter (GMD) =390) and coarsely ground (GMD=599) mash diets, particle sizes determined via wet-sieve analysis.
[4] Tibia breaking strength (Nmm) = Energy (compression × force) required to reach point of fraction.

at which stage of embryonic development minerals may affect bone development, and how it interacts with feeding programs in (layer) breeder.

Ontogeny of feather pecking behaviour: *no stress, please!*

Feather pecking (FP) in laying hens is another major welfare issue and results in a significant economic loss for the farmer (Nicol *et al.*, 2013; Rodenburg *et al.*, 2013). Severe feather pecking in a flock can progress into cannibalism, resulting in high levels of mortality (up to 30%). The EU-ban on the use of traditional battery cages since January 2012 and the growing concern with the acceptability of beak trimming asks for a mind shift in the layer industry. Currently, beak trimming is already banned in some EU countries, whereas it is no longer allowed in the Netherlands from September 2018 onwards. Such policies (loose housing and intact beaks) make flocks more vulnerable for FP behaviour and therefore a need for a proper layer management is evident.

Plenary session 03 and 04. Precision feeding

FP is a multifactorial problem, where both genetic and environmental factors play a major role in the aetiology of feather pecking. Large genetic differences exist between lines (Kjaer, 1995), but also housing conditions, such as a lack of appropriate litter material (Huber Eicher and Sebo, 2001) and large group sizes (Bilcík and Keeling, 2000) increase the incidence of FP as well.

Also nutritional conditions, such as eating and foraging play a role in the incidence of FP behaviour. Jungle fowl spend 60% of their time on eating and foraging (Dawkins, 1989). In modern laying hens, feeding related behaviours varies between 29 and 52% of the daily time budget, depending on diet structure and foraging material, (Aerni et al., 2000). The hypothesis that FP behaviour is redirected ground pecking towards feathers of conspecifics in the absence of adequate foraging incentives in modern poultry farming (Blokhuis and Van der Haar, 1992) was later confirmed in studies, showing that the incidence of FP was inversely related to the time spent eating and foraging (Van Krimpen, 2008). Once developed, FP behaviour seems to be persistent and difficult to extinguish (Bestman et al., 2009).

Nutritional interventions, such as low energy and high coarsely ground, insoluble fibre diets, increase eating time and show less FP and cannibalism (Van Krimpen et al., 2008). Highly insoluble fibre accumulates in the gizzard as a function of particle size distribution, and it is thought that this accumulation triggers satiety (Hetland et al., 2004; Van Krimpen et al., 2011), which is also assumed to reduce FP behaviour. While nutrition seem to play a crucial role in ontogeny and prevention of FP behaviour (Rodenburg et al., 2013), up to now the mechanism is still unclear.

Generally speaking, hens seem less prone to FP (*'less stressed'*) if they can better adapt to non-natural environmental conditions as present in modern husbandry systems. The ability to adapt successfully requires sufficient genetic potential, building up experience during early life and receiving the opportunity and resources to adapt (Star et al., 2006). Nutrition during early life, however, was believed to play a minor role in the ontogeny of feather pecking. Previous findings in our lab contradicted this statement, indicating that a more fibrous, less energy dense diet during rearing may reduce feather pecking up to 20% in later life (van Krimpen et al., 2009; Figure 4). It was hypothesized that pullets were increasingly 'imprinted' on feed as pecking substrate if dietary energy dilution level (by adding sand to the diet) in the mash rearing diet increased.

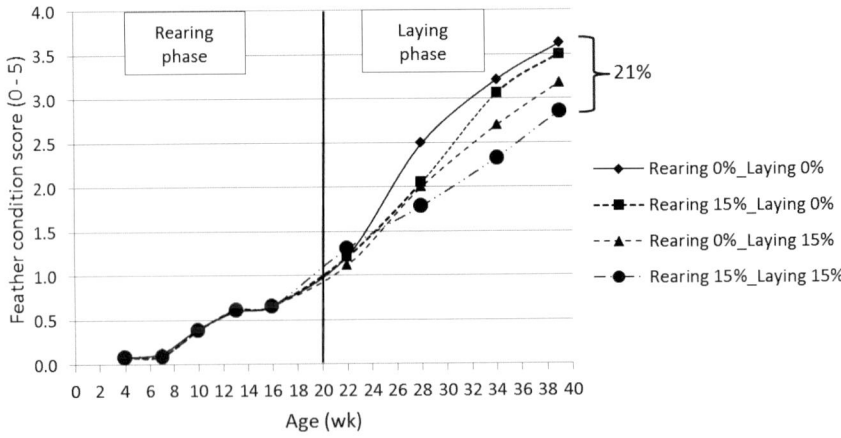

Figure 4. Interaction effects between dietary nutrient dilution level during rearing and laying phase on feather condition score, ranging from 0 (intact feathers, no injuries or scratches) to 5 (completely denuded area) (adapted from Van Krimpen et al., 2009).

These findings were confirmed by a another trial from our group, that concluded that dietary dilution affected time budgets of pullets, as shown by more feeding-related behaviour, resulting in less feather pecking behaviour (Qaisrani *et al.*, 2013; Figure 5).

Hence, it seems that FP behaviour in mature hens is influenced by early life experiences (Newberry *et al.*, 2006; Bestman *et al.*, 2009; Van Krimpen *et al.*, 2009). If pullets experience early in life (*is there a critical window?*) appropriate dietary substrates (such as fibrous diets) for feeding and/or adequate litter material for foraging, they seem to learn to direct their pecks to the correct materials and reduce the potential to direct their pecks to the feathers of conspecifics as a suitable substrate for foraging or feeding (Chow and Hogan, 2005). It has been proposed that a predisposition to develop feather pecking at young ages is associated with differences in serotonin turnover in the brain (Van Hierden *et al.*, 2004a).

A project has been initiated at our lab (March 2017), aiming to study the development of FP behaviour in several commercial lines of rearing hens (differing in potential feather pecking incidence). In this project, studies are *under construction* to investigate whether manipulation of feeding management (feeding frequency, feeding level, method of diet provision, diet composition) during the juvenile period alters behavioural (eating and foraging time budgets), neuroendocrine (e.g. serotonin and dopamine) and metabolic parameters (satiety related gut hormones, such as GLP-1, PYY, ghrelin, and CCK), and hence the sensitivity of pullets to develop FP later in life.

The project described above aims to unravel underlying physiological mechanisms, such as satiety and the activation of the gut-brain axis, as related to a combination of specific nutrient supply and nutrient density that alters eating and foraging behaviour, as well as the motivation for FP.

That does not mean that the industry currently lacks any 'preventing activity' as being concerned about the occurrence of FP. It is common practice nowadays to provide fibrous components to pullets and (organic) layers, either by supplementing roughages (e.g. alfalfa, carrots, silages; a.o. Steenfeldt *et al.*, 2007) in troughs (choice fed) or bales to the birds, or by providing fibrous diets (including oat hulls or oat bran). These supplements contribute to more natural behaviour of the birds (longer eating time and attractive barn elements), and at the same time, provide the gut with contraction stimulating material.

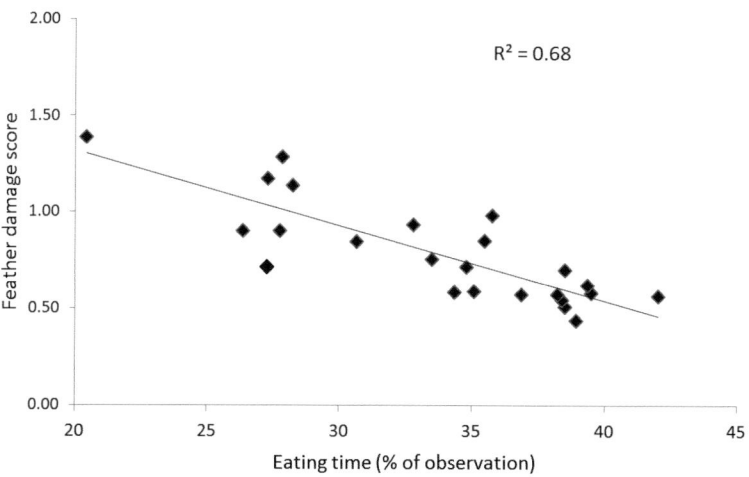

Figure 5. Relation between eating time and feather damage score in Lohmann Brown pullets during the rearing period. Feather damage score ranged from 0 (intact feathers, no injuries or scratches) to 5 (completely denuded area) (adapted from Qaisrani, 2013).

Plenary session 03 and 04. Precision feeding

Fibre, feed form, structure and microbiota: *a perfect gut-feeling*

Structural – *more coarse* – diets have a counter-intuitive effect on bird performance: large particles are likely to be more efficiently digested, because these larger particles stimulate gut contractions, slow-down passage rate (due to retrograde peristalsis), and improve gizzard development (Svihus, 2011; Kwakkel *et al.*, 2014; Qaisrani, 2014). A well-developed gizzard is associated with increased pancreatic enzymes secretion and improved grinding capacity (Engberg *et al.*, 2002). A wealth of data on gut development and its potential effects on gut health is available in literature, but all evidence is from broiler data.

To the authors' knowledge very limited research has been undertaken in this field with rearing and laying hens, except for a series of interesting studies conducted at Mateos' lab, studying the impact of different sources of fibrous materials (straw vs sunflower hulls vs sugar beet pulp), cereals (corn vs wheat) and feed form (crumble or pellet vs mash) (a.o., Frikha *et al.*, 2009; Saldana *et al.*, 2015; Guzman *et al.*, 2015). Some of these data have been summarized below (Table 2).

It was revealed from these studies that growth performance was mostly hampered if mash was fed instead of pellets or crumble, and if an insoluble fibre source such as straw was included in the rearing diets at too high levels. GIT parameters, such as relative gizzard weight and gizzard pH,

Table 2. Results from Spanish research on the effects of different feed forms (pellet, crumble and mash) and fibre content (2 and 4% straw) in brown egg laying pullets diets on Average Daily Feed Intake (ADFI), Average Daily Gain (ADG), and Feed Conversion Ratio (FCR) from 0-17 weeks of age; data on gizzard weights and pH are summarized for different ages during rearing (data adopted from Frikha et al., 2009; Saldana et al., 2015 and Guzman et al., 2015); particle sizes were determined by dry sieve analyses.

Treatment and reference	ADFI (g)	ADG (g)	FCR	Rel. full gizzard weight (% BW)		Gizzard pH
Frikha *et al.*, 2009[1]				45 d	120 d	120 d
Pellet	54.7[a]	12.4[a]	4.42[a]	2.7[a]	1.9[a]	3.99[a]
Mash	52.8[b]	12.0[b]	4.39[a]	4.2[b]	2.3[b]	3.46[b]
Saldana *et al.*, 2015[2]				70 d	120 d	120 d
Crumble	51.9[a]	12.3[a]	4.21[a]	3.1[a]	2.9[a]	3.88[a]
Mash	50.4[b]	11.5[b]	4.36[b]	5.0[b]	4.4[b]	3.17[b]
Guzman *et al.*, 2015[3]				70 d	120 d	70 d
Straw 0%	48.9[a]	12.0[a]	4.09[a]	4.2[a]	3.6[a]	2.94[a]
Straw 2%	49.3[a]	11.8[a]	4.16[b]	4.3[a]	3.8[a]	2.60[a]
Straw 4%	49.6[a]	11.8[a]	4.20[b]	4.7[b]	4.0[a]	2.43[a]

[a,b]: a common superscript within a column and reference means *P*>0.05.
[1] Pullets were fed for 45 days a pellet or mash diet; thereafter all pullets received a mash diet to 120 days; GMD varied between 837 and 899 for the mash diets (1 to 120 days of age) and between 597 and 634 for the pellet diets (1 to 45 days of age)
[2] Pullets were fed for 120 days a crumble or mash diet; GMD varied between 1,275 and 1,766 for the crumble diets and between 768 and 1,204 for the mash diets (1 to 120 days of age).
[3] Pullets were fed for 120 days a control (0% straw) and experimental diets (2 and 4% straw); GMD varied between 862 and 1,080 for the straw 0% diet, between 840 and 1,050 for the straw 2% diet and between 924 and 995 for straw 4% diet (GMD data shown are from 1-70 days of age)

being indicators of gut health, however, were improved if pullets had been fed mash or increased levels of insoluble fibre throughout rearing. As shown by Braz *et al.* (2011), fibre content of grower diets for pullets may aid in gut development, it might, however, also hamper BW gain up to target ages, if inclusion levels become too high and dilute the intake of other essential nutrients. So, although smaller, similar effects as shown in many broiler studies do account for pullets as well. The unique challenge is to find the balance between a quick early start from hatch (pellet or crumble feeding) without losing the promising effects of a fibrous (coarse) diet on early gizzard development and, as demonstrated by Yokhana et al (2016) a potential increase of digestive enzymes.

Only very few studies investigated the change in microbiota modulated by dietary changes in layer pullets. A recent paper of Chalvatzi *et al.* (2016) indicates that the use of a clay mineral may have beneficial effects, both on uniformity as well as on the presence of beneficial microbiota.

Studies that are undertaken to investigate performance in poultry related to gut health and structure of a diet should consider to use a wet sieve analysis (Goelema *et al.*, 1999) to obtain a particle size distribution that mimics the particles that flow down from the crop towards the gizzard.

Pullet nutrition for a 100 week production period: *preparation for later*

In the Netherlands, a focus on sustainable egg production for layers has resulted in a project entitled 'A*ctive2-be-fit*', where the main objective is to prolong the laying cycle (up to 100 weeks of age or more) under conditions of a still good egg shell quality, reduced mortality, good persistency in the 2nd half of the laying cycle, while using hens with intact beaks. In this project the young immature hen during rearing and transition receives a lot of attention. Some of the aspects of a good start have been discussed already in previous paragraphs in this paper. A topic that still needs attention is the Ca and P provision in the diet (degradability of phytate; potential reduction of NPP; use of phytase). Knowledge on Ca/P dynamics seems of utmost importance to hypothesize the relationship between mineral intake and the physico/chemical conditions in the GIT (De Vries *et al.*, 2010).

It is therefore important to investigate feeding strategies that allow a more efficient use of P. Such strategies might involve changing diet structure (coarseness) and moisture (wet feeds), as these factors positively affect feed intake capacity, gastro-intestinal morphology (larger villi), and nutrient digestion (reduced viscosity of gut contents and increased retention time in proventriculus and gizzard). It has been demonstrated that feeding broilers coarse corn increases bone ash and P and Ca retention as well, potentially due to a decreased digesta pH (Kilburn and Edwards, 2001).

An interesting study in this respect was conducted by Abdollahi *et al.* (2016) in broilers, where an improved P-degradability of phytate was observed if a Ca-source was provided separately from the diet (that including phytase). The hypothesis is that if Ca is fed separately from the diet, it cannot occupy phytate binding sites and form complexes that hamper phytate degradation by phytase. This type of feeding management could be a starting point to investigate for late rearing hens as well.

Concluding remarks

Young pullets need to develop into a well-performing hen, and this development is challenged by life-cycle events such as potential dehydration during the first few days post-hatch, the multiple vaccinations throughout the rearing period, with, as a consequence, a reduced feed intake capacity and thus insufficient intake of essential amino acids and minerals prior to lay. All events that increase the risk of a *stressed* bird and a flock with a decreased uniformity at onset of lay.

Plenary session 03 and 04. Precision feeding

Some ideas '*how to survive such a challenging childhood*' have been presented in this paper, such as early wet feeding to stimulate feed intake and appetite, thereby provoking a quick maturation of the gut and the immune system; feeding insoluble fibre rich and/or coarsely ground diets in the second phase of rearing to avoid FP behaviour and stimulate a well-developed gut and functioning microbiome and finally assuring an adequate Ca and P supply to ease transition into lay.

Nowadays, our demands in terms of a profitable young hen producing eggs for an even extended laying period means that the pullet deserves a '*healthy and joyful childhood*'.

Acknowledgments

In the process of writing this paper several people have helped the authors in providing valuable information. In alphabetical order: Dr Naomi de Bruijn and Dr Janine Wiegel (GD Animal Health, NL) Albert Dijkslag (For Farmers, NL), Jan van Harn and Izak Vermeij (WUR, Livestock Research, NL), Dr Martin Zuidhof (University of Alberta, Canada). To all: thank you!

References

Addollahi, M.R., Duangnumsawang, Y., Kwakkel, R.P., Steenfeldt, S., Bootwalla, S.M. and Ravindran, V. 2016. *Animal Feed Science and Technology* 219: 48-58.

Aerni, V., El Lethey, H. and Wechsler, B. 2000. *British Poultry Science* 41: 16-21.

Ashwell, C.M. and Angel. R. 2010. *Rev. Bras. Zootec.* 39 (Suppl 1): 268-278.

Bestman, M., Koene, P. and Wagenaar, P. 2009. *Applied Animal Behaviour Science* 121: 120-125

Bilcík, B. and Keeling, L.J. 2000. *Applied Animal Behaviour Science* 68: 55-66.

Blokhuis, H.J. and Van Der Haar, J.W. 1992. *British Poultry Science* 33: 17-24.

Bonato, M.A., Sakomura, N.K., Gous, R.M., Dourado, L.R.B., Rafael, J.M., Fernandes, J.B.K. 2015. *British Poultry Science* 56: 361-369.

Bostick, J.W. and Zhou, L. 2016. *Cellular and Molecular Life Sciences* 73: 237-252.

Braz, N.M., Freitas, E.R, Bezerra, R.M., Cruz, C.E.B., Farias, N.N.P., Silva, N.M., Sá, N.L. and Xavier, R.P.S. 2011. *Revista Brasileira de Zootecnia-Brazilian Journal of Animal Science* 40: 2744-2753.

Chalvatzi, S., Kalamaki, M.S., Arsenos, G. and Fortomaris, P. 2016. *Journal of Applied Microbiology* 120: 1033-1040.

Chow, A. and Hogan, J.A. 2005. *Applied Animal Behaviour Science* 93: 283-294.

Dawkins, M.S. 1989. *Applied Animal Behaviour Science* 24: 77-80.

De Vries, S., Kwakkel, R.P. and Dijkstra, J. 2010. Dynamics of Calcium and Phosphorus Metabolism in Laying Hens. In: *Phosphorus and Calcium Utilization and Requirements in farm Animals*. Ed: Vitti, D.M.S.S. and Kebreab, E. CAD International, ISBN: 978 1 84593 626 6.

Decuypere, E., Hocking, P.M., Tona, K., Onagbesan, O., Bruggeman, V., Jones, E.K.M., Cassy, S., Rideau, N., Metayer, S., Jego, Y., Putterflam, J., Tesseraud, S., Collin, A., Duclos, M., Trevidy, J.J. and Williams, J. 2006. *World's Poultry Science Journal* 62:443-453.

Elser, J. and Bennett, E. 2011. N*ature* 478: 29-31.

Favero, A., Vieira, S.L., Angel, C.R., Bos-Mikich, A., Lothhammer, N., Taschetto, D., Cruz, R.F.A. and Ward, T.L. 2013. *Poultry Science* 92: 402-411.

Frikha, M., Safaa, H.M., Serrano, M.P., Arbe, X. and Mateos, G.G. 2009. *Poultry Science* 88: 994-1002.

Goelema, J.O., Smits, A., Vaessen, L.M. and Wemmers, A. 1999. *Animal Feed Science Technology* 78: 109-126.

Gunnarsson, S., Heikkilä, M. and Valros, A. 2008. *Animal Science* 58: 93-99.

Guzmán, P., Saldaña, B., Kimiaeitalab, M.V., García, J. and Mateos, G.G. 2015. *Poultry Science* 94: 2722-2733.

Halle, I. 2002. *Archiv fur Geflugelkunde* 66: 66-74.

Hammond, J. 1932. *Journal of the Royal Agricultural Society England* 93: 131-145.

Handboek Kwantitatieve Informatie Veehouderij. 2016. Wageningen Livestock Research; KWIN-V 2016-2017.

Harlander-Matauschek, A., Rodenburg, T.B., Sandilands, V., Tobalske, B.W. and Toscano, M.J. 2015. *World's Poultry Science Journal* 71: 461-472.

Plenary session 03 and 04. Precision feeding

Hetland, H., Choct, M., and Svihus, B. 2004. *World's Poultry Science Journal* 60: 415-422.

Hetland, H., Svihus, B. and Choct, M. 2005. *Journal of Applied Poultry Research* 14: 38-46.

Hocking, P.M. 2007. Optimum feed composition of broiler breeder diets to maximise progeny performance. Proceedings 16th European Symposium on Poultry Nutrition (ESPN), 26-30 August 2007, Strasbourg, France.

Huber Eicher, B. and Sebo, F. 2001. *Applied Animal Behaviour Science* 74: 223-231.

Iji, P.A., Saki, A. and Tivey D.R. 2001. *British Poultry Science* 42: 505-513.

Jardim, R.D., Stringhini, J.H., Cafe, M.B., Leandro, N.S.M., Andrade, M.A. and de Carvalho, F.B. 2011. *Revista Brasileira De Zootecnia-Brazilian Journal of Animal Science* 40: 1947-1954.

Kilburn, J. and Edwards, H.M. 2001. *British Poultry Science* 42: 484-492.

Kjaer, J. B. 1995. *Applied Animal Behaviour Science* 44: 257-281.

Kwakkel, R.P., Ducro, B.J. and Koops, W.J. 1993. *Poultry Science* 72: 1421-1432

Kwakkel, R.P., Wartena, F.C. And Moquet, P.C.A. 2014. Coarse diets for poultry: effects on N-efficiency and gut health. In Proceedings of the 5th International Broiler Nutritionists' Conference 'Poultry Beyond 2020', 13-17 April 2014, Queenstown, New Zealand.

Lieboldt, Ma., Halle, I., Frahm, J., Schrader, L., Weigend, S., Preisinger, R., Breves, G. and Dänicke, S. 2016a. *Journal of Poultry Science* 53: 8-21.

Lieboldt, Ma., Halle, I., Frahm, J., Schrader, L., Weigend, S., Preisinger, R., Breves, G. and Dänicke, S. 2016b. *Journal of Poultry Science* 53: 136-148.

Liu, D., Veit, H.P. Wilson, J.H. and Denbow. D.M. 2003. *Poultry Science* 82: 463-473.

Mai, A.K. 2007. Wet and coarse diets in broiler nutrition: Development of GI tract and performance. PhD Thesis Wageningen University, 141 pp. ISBN: 978 90 8504 682 0.

Moinard, C., Statham, P., Haskel, M.J., Mccorquodale, C., Jones, R.B. and Green, P.R. 2004. *Applied Animal Behaviour Science* 85: 77-92.

Newberry, R.C., Keeling, L.J., Estevez, I. and Bilcik, B. 2007. *Applied Animal Behaviour Science* 10: 262-274.

Nicol, C.J., Bestman, M., Gilani, A.M., De Haas, E.N., De Jong, I.C., Lambton, S., Wagenaar, J.P., Weeks, C.A. and Rodenburg, T.B. 2013. *World's Poultry Science Journal* 69: 775-788.

Qaisrani, S.N. 2014. Improving performance of broilers fed lower digestible protein diets. PhD Thesis Wageningen University, 182 pp, ISBN: 978 94 6257 031 3 (WUR library http://edepot.wur.nl/313161).

Qaisrani, S.N., Van Krimpen, M.M. and Kwakkel, R.P. 2013. *Poultry Science* 92: 591-602.

Rao, K., Xiw, J., Yang, X. and Chen, L. 2009. *British Journal of Nutrition* 102: 848-857

Renema, R.A., Rustad, M.E., Robinson, F.E. 2007. *World's Poultry Science Journal* 63:457-472

Rescigno, M. 2011. *Trends in Immunology* 32: 256-264.

Rodenburg, T.B., Van Krimpen, M.M., De Jong, I.C., De Haas, E.N., Kops, M.S., Riedstra, B.J., Nordquist, R.E., Wagenaar, J.P., Bestman, M. and Nicol C. J. 2013. *World's Poultry Science Journal* 69: 361-373.

Saldana, B., Guzman, P., Safaa, H.M., Harzalli, R. and Mateos, G.G. 2015. *Poultry Science* 94: 2650-2661.

Schutte, J.B., Kwakkel, R.P. and Langhout, D.J. 1998. Research on optimal P-levels in rearing diets in relation to performance and bone quality during lay (in Dutch). Internal report ILOB-TNO/LUW-WIAS no I98-31095, 35 pp.

Simon, K., De Vries Reilingh, G., Bolhuis, J.E.,Kemp, B. and Lammers, A. 2015. *Poultry Science* 94: 2041-2048.

Soller, M., Eitan, Y. and Brody, T. 1984. *Poultry Science* 63: 1255-1261.

Star, L., Ellen, E., Uitdehaag, K. and Brom F. 2006. Robustness of laying hens: An ethical approach. Ethics and the politics of food. From: Ethics and the politics of food. Ed: KAISER, M. and LIEN, M.E. Wageningen Academic Publishers, ISBN: 978-90-8686-008-1

Steenfeldt, S., Kjaer, J.B. and Engberg, R.M. 2007. *British Poultry Science* 48: 454-468.

Svihus, B. 2011. *World's Poultry Science Journal* 63: 457-472

Swiatkiewicz, S., Koreleski, J., and Arczewska, A. 2009. *Acta Veterinaria Brno* 79: 185-193.

Toscano, M.J., Booth, F., Wilkes, L.J., Avery, N.C., Brown, S.B., Richards, G. and Tarlton, J.F. 2015. *Poultry Science* 94: 823-835.

Van Emous, R.A., Kwakkel, R.P., Van Krimpen, M.M. and Hendriks, W.H. 2015. *Applied Animal Behaviour Science* 168: 45-55.

Van Hierden, Y.M., De Boer, S.F., Koolhaas, J.M. and Korte, S.M. 2004. *Behavioral Neuroscience* 118: 575-583.

Plenary session 03 and 04. Precision feeding

Van Krimpen, M.M. 2008. Impact of nutritional factors on eating behaviour and feather damage of laying hens. PhD Thesis Wageningen University, 225 pp. ISBN: 978 90 8504 948 7.

Van Krimpen, M.M., Kwakkel, R.P., Van Der Peet-Schwering, C.M.C., Hartog, L.A.D. and Verstegen, M.W.A. 2008. *Poultry Science* 87: 485-496.

Van Krimpen, M.M., Kwakkel, R.P., Van Der Peet-Schwering, C.M.C., Den Hartog, L.A. and Verstegen, M.W.A. 2009. *Poultry Science* 88: 759-773.

Van Krimpen, M.M., Kwakkel, R.P., Van Der Peet-Schwering, C.M.C., Den Hartog, L.A. and Verstegen, M.W.A. 2011. *British Poultry Science* 52: 730-741.

Willemsen, H., Debonne, M., Swennen, Q., Everaert N., Careghi, C., Han, H., Bruggeman, V., Tona, K. and Decuypere, E. 2010. *World's Poultry Science Journal* 66: 177-188.

Yair, R., Shahar, R., and Uni, Z. 2015. *Poultry Science* 94: 2695-2707.

Yasar, S. and Forbes, J.M. 1999. *British Poultry Science* 40: 65-76.

Yokhana, J.S., Parkinson, G. and Frankel, T.L. 2016. Poultry Science 95: 550-559.

Zuidhof, M.J. 2015. Improving Broiler Performance through Breeder Nutritional Management. Proceedings 20th European Symposium on Poultry Nutrition (ESPN), 24-27 August 2015, Prague, Czech Republic.

Sequential and choice feeding in laying hens, how much does it help to adapt to the egg cycle?

C. Hamelin[1], A. Molnar[2,3] and Y. Nys[4]*
[1]CCPA, 35150 Janzé, France, [2]ILVO, Scheldeweg 68, 9090 Melle, Belgium, [3]Experimental Poultry Center, Poiel 77, 2440 Geel, Belgium, [4]INRA, UR83, 37380 Nouzilly, France; yves.nys@inra.fr

Summary

Conventional feeding system for laying hens relies on a complete mash feed available *ad libitum*. Under such complete diet system, bird consumption is mainly controlled by energy requirement but in case of high heterogeneity in level of egg production between hens or when the environmental temperatures varies, the birds might either over or under-consume protein and Ca. Sequential, loose-mix and choice feeding have been proposed as alternative feeding systems. These give the possibility to hens to make a self-selection between energy, protein sources including vitamins and minerals and large particle size calcium sources, to adapt to individual requirements and to hourly change in hen requirement. In addition, use of local whole cereals spares the energy needed for transport, milling and mixing the cereals. This review explores the physiological basis for sequential, loose-mix and choice feeding and evaluates the impact of these systems on egg production and quality.

Introduction

In conventional feeding system, feed intake is mainly controlled by energy intake but in choice feeding, it relies on dietary energy concentration, size of particles, specific appetite for Ca (Calcium) (Mongin and Sauveur, 1974) and possibly on ability to regulate nutrient intake based on daily egg formation cycle (Chah and Moran, 1985; Leeson and Summers, 1997).

Three choice feeding systems have been studied as alternatives to complete diet: the free choice system (1) offers two or three diets in different containers simultaneously. Hens are able to select between whole grains, a protein concentrate and Ca particles to meet their requirement (Robinson, 1985; Forbes and Covasa, 1995). In the loose-mix feeding system (2) only one diet is offered throughout the day in a single trough, containing the three nutrients with different particle sizes mixed together. This system allows simultaneous distribution in a single diet and exploits hen's ability to sort out the whole cereal, the Ca particles and the protein concentrate (Picard *et al.*, 1997; Umar-Faruk *et al.*, 2010a,b). In sequential feeding system (3), diets (most frequently two) with different composition in energy, protein and Ca are distributed successively in the same trough throughout the day to supply the specific nutrients required during sequential phases of the daily egg formation cycle (Umar-Faruk *et al.*, 2010, 2011; Traineau *et al.*, 2013). This review explores the physiological basis in favour of hourly changes in hen nutrient requirement and evaluates the impact of these feeding systems on egg production and quality.

Kinetic of egg component synthesis and secretion

Birds are oviparous and produce a daily egg containing all the essential nutrients for the development of an embryo. Egg is composed of three main compartments, separated by membranes, each showing a particular composition: emulsion of triglyceride-rich lipoproteins and proteins in yolk, salty solution of proteins in albumen and Ca carbonate in eggshell (Nys and Guyot, 2011). Egg components are produced sequentially by two different anatomical structures, the ovary and the oviduct. The liver produces the egg yolk components. In hens, hepatic lipogenesis and lipemia are increased, respectively, 15- and 20-fold at sexual maturity. Amongst the 10,000 oocytes present on the ovary at hatch, 2,000 oocytes slightly develop for about 60 days throughout the hen production cycles. Amongst them, a limited number (5 to 8) simultaneously grow for 8-10 days forming the particular hierarchy of follicles (Nys and Guyot, 2011). The synthesis of yolk is therefore a continuous process in the liver of mature hens throughout the laying period and is regulated by sex steroid hormones and additional ovarian factors. The precursors of the yolk are transported via the bloodstream and deposited in the ovary during the last phase of rapid growth of the yolk for about 8 days. At ovulation, the largest ovarian follicle releases a mature egg yolk in the oviduct. Then specialized segments of the oviduct synthesize and sequentially secrete around the yolk the constituents of the outer vitelline membrane, the egg white (albumen), the shell membranes then the eggshell in about 24 h. This temporal and spatial sequence is precisely controlled by the hormonal secretion of the ovary and hypothalamus. All albumen proteins are continuously synthesized by the magnum. Protein synthesis occurs regardless of the presence or absence of an egg in the oviduct but might be accelerated during the passage of the egg in the magnum according to Muramatsu *et al.* (1991). These authors estimated that the protein amount accumulated in granules at the apical region of secretory cells corresponded to that of two eggs. In contrast, albumen proteins are secreted only for a short period of about 3.5 h. Secretion occurs by pinocytosis and is induced by the dilatation of the magnum by the forming egg. The isthmus is the site where the two shell membranes are secreted for 1 hrs. Then the egg enters the uterus 6 hrs after ovulation and stays in this compartment for nearly 20 hrs. The egg is firstly hydrated (plumping), gets its ovoid shape and when in close contact with the uterine mucosa, the process of mineralisation is initiated. From 10 to 22 hrs after ovulation, a large quantity of Ca carbonate is quickly deposited, together with a small proportion of proteins composing the organic matrix of the shell (Nys and Guyot, 2011). This linear deposition stops about 2 hrs before egg expulsion (oviposition) then the shell is covered with the cuticle. The transfer of eggshell precursors is an hourly process. No Ca storage occurs in the shell gland before shell formation. The amount needed for forming a shell (2 g of Ca) is very high, therefore the pool of blood ionic Ca is renewed every 12 min during the 12 hrs of rapid shell deposition. Ca is provided directly by the diet or indirectly by bone resorption. The shell formation occurs mainly during the night and any factor correcting the desynchronization of dietary Ca supply and shell formation improves the eggshell strength (Nys, 2017). Eggshell carbonates are directly issued from blood CO_2 and the pH reduction in the uterine fluid or plasma during shell formation underlines the short term supply of carbonates to the eggshell. The hourly regulation of uterine ionic transport is demonstrated by the stimulation of gene expression of some ionic transporters during the process of Ca carbonate secretion (Brionne *et al.*, 2014). The turnover of isotopic deposition of carbon in egg compartments (^{13}C) reflecting rate of protein synthesis can be explored by substituting wheat by corn in the laying hen diet because these cereals show different isotopic signatures (Segalen *et al.*, 2013). The first isotopic change in the egg compartment and the plateau (reflecting duration of synthesis) are respectively reached after two and six days for albumen, four and eleven days for yolk but only one and two days for the eggshell. In conclusion, the secretion of egg precursors follows a short temporal sequence which is different from the long period of synthesis of yolk and albumen precursors.

Regulation of feed intake in hens

Feed intake mainly relies on energy concentration of the diet, both being linearly correlated (Bouvarel *et al.*, 2011). However, the hourly pattern of feed intake is related to the nycthemeral rhythm and to the egg formation cycle. Two peaks of feed intake are observed, the first one (40%) between 4 to 6 hrs and the second one (60%) at 14 to 16 hrs after light-on (Kesavarz, 1998a). The second peak of consumption is clearly associated to the shell formation (Mongin and Sauveur, 1974). In early literature, hens were proven to eat slightly more when albumen proteins were secreted compared to those not forming albumen (28 vs 19 g; Morris and Taylor, 1967). Some authors therefore questioned the capacity of hens to select between protein and energy sources during the day to meet any specific requirement. They observed in hens separately fed a cereal (energy source) and a protein concentrate, this aptitude of hens at least when considering the whole day requirement (Cunnings 1994; Forbes and Covasa, 1995). There was, however, no evidence for any hourly regulation of feed consumption with the exception of Ca. In contrast, the choice feeding for Ca when supplied as particles is clearly demonstrated.

Particle size of dietary components influences feed intake. Birds are reluctant to pick up fine particles containing supplemented amino acids, vitamins or minerals and might under-consume these premixes (Bennett and Classen, 2003). Distribution of a fine rather than a more coarse meal (31% vs 9% particles <0.5 mm) reduced feed consumption (-4 g), laying rate (-3.4%) and egg weight (-0.9 g) (Joly, 2008). Feed intake increased when coarser particle size are slightly enlarged (13-16 vs 21-22% particles <0.63 mm), without any effect on hen productivity (Safaa *et al.* 2009). In 30 weeks old hens, feed intake was increased when corn particle size was 1.1 mm compared to 0.64 or 0.87 mm, egg production and egg weight however were similar (Amornthewaphat *et al.*, 2007). Portella *et al.* (1988) also observed that 24 weeks old hens had preference for larger particles in a mash feed. Crumbled or pelleted diets stimulate feed consumption in hens, avoid under-consumption of fine particles but reduce the duration of feed consumption. Ileal digestibility of starch was slightly improved by feeding hens a coarse feed compared to a fine feed (Ruhnke *et al.*, 2015). Whole grains favoured the muscular development of the gizzard to more efficiently grind the grain once egg production begins (Amerah *et al.*, 2007). Therefore in practice, meals with larger particles of cereals are more frequently used than crumbles and breeder company guidelines recommends to have 75-80% of the particles between 0.5-3.2 mm. The preference of hens for large particles can reinforce competition within a flock and impair flock uniformity (Forbes and Cosava, 1995).

Training of hens for alternative feeding systems

Hens are sensitive to any change in the feeding system. Training of the birds at an early age to accustom them to whole cereals was proven to be an important issue either in broilers or in laying hens irrespective of the feeding system. It is recommended to introduce whole cereals at 15-18 weeks to pullets to avoid any failure to select feed at later ages (Forbes and Covasa, 1995). The use of collective cage or groups of hens in free range systems favours social interaction which facilitates learning to select between different diets and to meet nutrient requirements compared to individual caged birds (Meunier-Salaün and Faure, 1984). Process of learning can also be accelerated by using experienced birds (Mastika and Cummming, 1987). In conclusion, an adjustment period at about 15 weeks of age for commercial layers allows the birds to learn how to choice-feed before they are exposed to the nutritional demands of egg production and build up the Ca reserves.

Free choice feeding and its effect on hen performance and egg quality

Free-choice feeding studies report either similar or improved performance or egg quality compared to a conventional diet (Table 1). Chah and Moran (1985) fed hens separately an energy mash containing ground corn as the main cereal (2,800 ME/kg, 8% CP; 0.75% Ca), protein pellets (50% CP, 2,500 kcal ME/kg, 0.2% Ca) and oyster shell flakes and found similar feed intake, egg production, and egg weight compared to a conventional diet (2,800 kcal ME/kg, 18% CP, 3.75% Ca). Haugh unit of eggs did not differ, whereas shell resistance to deformation increased with free choice feeding. In a similar study, Olver and Malan (2000) offered whole corn instead of ground, with pelleted protein concentrate and limestone for pullets already from 7 weeks of age and followed the birds' performance until to 80 weeks. During the laying period there was no difference in feed intake or egg production, but higher egg weight and shell thickness and lower FCR in the free-choice fed group. Higher egg weight and lower FCR were also reported by Karunajeeva (1978) but a lower feed intake was observed when offering a protein concentrate and whole wheat to hens. Similarly, Leeson and Summers (1979) reported lower feed intake and larger egg weight when hens were simultaneously offered a high energy, protein (3,065 kcal/kg; 19.2%), low Ca (0.47%) diet and a low energy, protein (1,740 kcal/kg; 10.7%), high Ca (13.1%) diet. Egg production was similar to the conventional diet (2,794 kcal ME/kg, 17.1% CP, 3% Ca). However, the effects of free choice feeding were reported to be variable across birds contributing to increased heterogeneity in the flock (Forbe and Cosava, 1995). Lower egg quality, impaired plumage, and increased feed intake have also been reported in flocks fed in choice feeding (Tauson and Elwinger, 1986; Albustany and Elwinger, 1988).

Table 1. Effect of free-choice feeding compared to conventional feeding

Reference	Performance traits[1]							
	Egg production	Egg weight	Egg mass	Feed intake	Energy intake	Protein intake	Ca intake	FCR
Karunajeewa, 1978		↑		↓				↓
Leeson and Summers, 1979	=	↑		↓	↓	↓	↑	
Farrell et al. 1981	=	=		↓	↓	=		↓
Tausson and Elwinger, 1986	=	↑	↑					
Chah and Moran, 1985	=	=		=	↓	↓	↓	
Olver and Malan, 2000	=	↑		=	↓	↓	↑	↓

[1] ↓: represents a significant decrease ↑: represents a significant increase; =: no significant difference, left empty: parameter is not reported.

Alternatively, Farrell *et al.* (1981) improved FCR relative to conventional feeding when offering three free-choice diets based on energy sources: maize in 100%, maize in 60% mixed with cassava in 40% and maize (94%) mixed with palm oil (6%). The egg production and egg weight were similar but feed intake was higher with the commercial diet and lower with the maize and cassava mixture compared to other treatments.

Loose-mix feeding and its effect on hen performance and egg quality

Whole grains and protein concentrate

Loose-mix feeding studies based on whole grains and a protein concentrate lead to contradictory results in laying hens (Table 2). Incorporating 45% whole wheat in a complete laying diet resulted in unchanged egg production and feed efficiency as reported by Ouart *et al.* (1986). Umar-Faruk *et al.* (2010a) mixed whole cereal and protein concentrate in a ratio 50:50 (121 g daily). This loose-mix feeding resulted in comparable feed intake, egg production, egg mass and feed efficiency similar to the conventional system. An increased whole cereal proportion to 60% as whole barley, reduced egg production, feed efficiency, egg specific gravity but increased feed intake, egg weight compared to a mash diet with same amount of ground barley (Bennett

Table 2. Effect of loose-mix feeding compared to conventional feeding.

Reference	Whole cereals	Performance traits[1]							
		Egg prod.	Egg weight	Egg mass	Feed intake	Energy intake	Protein intake	Ca intake	FCR
Ouart et al. 1986	37%	=	=	=	=				=
Umar-Faruk et al. 2010a	50%	=	=	=	=	=	=		=
Benett and Classen, 2003	60%	↓	↑		↑				↑
Henuk et al. 2000	60%	=	=		↓	=	=	↓	↓
Blair et al. 1973	70%	=			↓	=	↓	↓	↓
Robinson, 1985	75%	↓	↑		↑				↑
Umar-Faruk et al. 2010b	75%	=	=		↓	=	↓		

[1] ↓: represents a significant decrease ↑: represents a significant increase; =: no significant difference, left empty: parameter is not reported.

and Classen, 2003). However, when 60% of whole wheat was mixed with 40% of a protein concentrate in pellet form, similar egg production and egg weight, but lower feed intake and better feed efficiency were found compared to a complete mash diet (Henuk *et al.*, 2000). Increasing the amount of whole grain to 75% (wheat and oats) in the diet mixed with granulated limestone and decreasing the protein concentrate to 25% lead to decreased egg production, feed efficiency and higher mortality (Robinson, 1985). With the same ratio of wheat and balancer diet (75:25) Umar-Faruk *et al.* (2010b) did not report any significant reduction in egg production or increase in FCR, however, they did found reduced feed intake and egg mass compared to a conventional diet. In a similar loose-mix system where different whole cereals (wheat, barley and maize) were mixed with a protein concentrate and oyster shell grit in a ratio 70:23:7, Blair *et al.* (1973) also reported similar egg production, but lower feed intake and better feed conversion in the loose-mix system compared to a conventional mash.

These contrasting results highlight that the relative proportion of whole cereals (50 to 70%) and protein concentrate (mash or pellet) can be a crucial factor when preparing loose-mix diets.

Calcium and other micro ingredients

It is well-recognized that dietary supply of large Ca particles allows hens to express their specific appetite for Ca and improves eggshell quality and bone mineralisation. According to Guinotte and Nys, 1991, eggshell-breaking strength was higher when birds were fed particulate seashells or limestone (in general 2-4 mm particle size) than those fed ground limestone (<0.5 mm). Tibia breaking strength and ash percentage were greatly increased with the use of particulate Ca sources. Some recent works confirmed the benefits of coarse Ca sources on egg shell quality. Coarse limestone (0.8-2 mm) increased egg shell breaking strength compared to the addition of fine limestone (<0.5 mm) in brown hens aged 50 weeks (Skřivan *et al.*, 2016). Coarse Ca improves relative shell density (Safaa *et al.*, 2008; Pizzolante *et al.*, 2011) or egg shell breaking strength (Skřivan *et al.*, 2016; Lichovnikova, 2007; Koreleski andSwiatkiewicz, 2004; Pizzolante *et al.*, 2011). Nevertheless some works reported no beneficial effects (Pelicia *et al.*, 2009, Cufadar *et al.*, 2011). This suggests that the source of coarse Ca or experimental conditions (age of hens, dietary Ca level) might have an influence on shell quality. Attention should be paid to the proportion of the coarser particles as Koreleski and Swiatkietwicz, (2004) found that coarse particles refusal percentage can increase with 100% inclusion levels of coarse limestone or oyster shells in the feed with brown hens from 25 to 72 weeks of age.

Ahmad and Balander (2004) substituted 50% of fine limestone with oyster shells from 28 to 64 weeks of age in the afternoon diet of white hens along with a reduction of the available P from 0.45% to 0.32%. The plasmatic level of P was decreased by 25% with oyster shells but only by 6% in hens fed lower available P. The effect of oyster shells was prominent because of decreased bone mobilisation during the night. Therefore the particle size of Ca sources greatly influences the requirement of P. Other authors underline the interest to supply different levels of P throughout the day, in addition to changes in dietary Ca to favour bone accretion just after oviposition.

New research at Euronutrition showed that using 100% coarse carbonate and coarse phosphate in place of fine sources can be both beneficial to the performance and shell quality of brown hens (46-62 weeks). As the coarse carbonate level in the diet increased, particle size distribution of the mash diet changed: the amount of particles <0.2 mm decreased close to zero, whereas the amount of greater particles (2-4 mm) increased. Premix particle size can also influence performance and egg quality as shown in a recent CCPA trial: including a premix with a coarse presentation (>65% of particles was 1-2 mm) in the diet of brown hens (18-26 weeks) resulted in a higher egg mass (+1.7 g/d) and a decreased feed conversion (-0.08) compared to hens receiving a standard premix (>65% of particles was 0.1-0.5 mm). The premixes had the same composition (vitamins and trace elements) except the particle size distribution. These results highlight that a premix with coarse particle size can secure the intake of micro nutrients (vitamins and trace elements), and helps to better cover the birds' needs. In addition, both trials confirmed the preference of laying hens for >1 mm and <4 mm particles.

Sequential feeding and its effect on hen performance and egg quality

Sequential feeding of a protein rich diet in the first half of the day followed by a low protein diet in the second half of the day resulted in similar egg production compared to feeding a diet with constant protein level during the day (Penz and Jensen, 1991; Keshavarz, 1998a). However, it was shown that protein supply at different times of the day can affect albumen weight, indicating that more protein might be needed after oviposition to meet the protein requirements for the synthesis of the albumen (Penz and Jensen, 1991). When not only protein level, but also Ca level was adjusted offering a high energy, high protein and low Ca diet in the morning (3,040 ME, 18% CP, 0.5% Ca) and a low energy, low protein and high Ca diet in the afternoon (2,320 ME, 13.7% CP, 9.5% Ca) feed efficiency was improved but with similar egg production and lower egg weight compared to a conventional diet (Lee and Ohh, 2002). De los Mozos *et al.* (2012) also reported similar performance (egg production, egg weight, egg mass), but slightly lower feed intake and therefore improved feed efficiency when decreasing the energy and protein level of the afternoon diet (2650 ME, 15.5% CP) compared to a single diet (2750 ME, 17% CP). A lower Ca supply in the afternoon reduced shell quality, but the lower Ca intake in the morning has no effect (Keshavarz, 1998a,b). This is also observed when cereals with low Ca content are provided in the morning. The balancer diet of the afternoon with high Ca ensures normal shell formation and quality.

The use of whole cereals as the energy source in the morning combined with a protein concentrate fed in the afternoon lead in most studies to improved feed efficiency in laying hens. Higher feed intake due to higher pellet consumption was only found when cereals (wheat, barley and corn) and concentrated pellets were fed *ad libitum* in sequential system (Blair *et al.*, 1973), without affecting egg production compared to a conventional mash diet. However, when restricted feeding was used providing cereals and protein concentrate in a ratio 50:50 (121 g/d), intake was lower and feed efficiency largely improved in sequential feeding compared to both conventional or loose-mix system (Table 3; Umar-Faruk *et al.*, 2010a). The egg production, egg weight and egg mass were similar in all feeding systems. The improvement in feed efficiency in the sequentially fed group was due to lower wheat intake. In addition, feeding whole wheat favours feed efficiency by developing the gizzard and the diet digestibility. When increasing the amount of whole wheat to 75% fed sequentially with a balancer mash diet, egg production, egg weight and feed intake were similar to those of hens fed 50% whole wheat sequentially (Umar-Faruk *et al.*, 2010b). Both treatments were comparable to conventional feeding in terms of egg production and feed efficiency. Indeed, wheat intake and egg weight are lower when ground wheat is fed instead of

Table 3. Effect of sequential feeding of whole cereals compared to conventional feeding.

Reference	Performance traits[1]						
	Egg production	Egg weight	Egg mass	Feed intake	Energy intake	Protein intake	Ca intake	FCR
Blair *et al.* 1973	=	=		↑	=	↑	↑	
Umar-Faruk *et al.* 2010a	=	=	=	↓	↓	↓		↓
Umar-Faruk *et al.* 2010b	=	=	=	=				=
Umar-Faruk *et al.* 2011	=	=	=	↓	↓			↓
Traineau *et al.* 2013	=	↑	=	↓	↓			↓

[1] ↓: represents a significant decrease ↑: represents a significant increase; =: no significant difference, left empty: parameter is not reported.

whole wheat and egg production and egg mass remains unaffected. The use of ground wheat or corn either at 100% or mixed with whole wheat in a ratio 80:20, lowered egg weight and egg mass, without affecting egg production compared to whole wheat in sequential feeding (Traineau *et al.*, 2013). Feeding ground cereals was less efficient than whole wheat in sequential feeding but was still superior in terms of feed efficiency to conventional mash.

Conclusions

Experimental studies on choice feeding systems have been conducted with individual or small groups of laying hens in cages for relatively short period. Therefore, production scale studies are needed in practice and should include alternative housing systems (aviary with or without free-range). Group size, activity of birds, genotype (white vs brown hens) might affect performance and feeding behaviour in choice feeding system. In addition, offering daily different diets requires good management to avoid competition between birds and over-consumption of the balanced diet. The difficulty of providing separate feeders limits the use of free choice feeding in practical conditions. The loose-mix system is easier to settle practically but depends on the aptitude of hens to sort heterogeneous particle size nutrients. Clear benefits of coarse Ca particles are now well recognized: better mineral status and shell quality. A further decrease in proportion of fine particles using particulate phosphate or premix looks promising. The use of whole grains should be carefully considered because it can increase competition, heterogeneity in the flock and even feather pecking. Contrasting results within loose mix feeding studies highlight the need for more technical knowledge from the farmers but there is potential in using on farm grown cereals. Some development of sequential feeding at farm level have been tested using two silos with morning diet (high energy, high protein, P, low Ca) and evening diet (low energy, low protein, High Ca, low P) or by adding a third silo dedicated to particulate Ca distributed in the evening. Main expectation is to reduce FCR and to improve shell quality. Sequential feeding of P and Ca can strongly reduce bone mobilization and the excretion of P and might positively impact both welfare and environment. Potential feed cost savings could be done compensating the higher cost for more storage capacity.

References

Ahmad, H.A., Balander, R.J. 2004. *International Journal Poultry Science* 3: 100-111.
Al Bustany, Z. and Elwinger, K. 1988. *Swedish Journal of Agricultural Research* 18: 31-40.
Amerah, A.M., Ravindran, V., Lentle, R.G., Thomas, D.G. 2007. *World's Poultry Science Journal* 63: 439-455.
Amornthewaphat, N., Attamangkune, S., Songserm, O., Ruangpanit, Y, Thomawong, P. 2007. Proc. of 16th Eur. Sym. on Poultry Nutrition: 479-482.
Bennett, C.D. and Classen, H.L. 2003. *Poultry Science* 82: 147-149.
Bhatti, B.M. and Morris, T.R. 1978. *British Poultry Science* 19: 125-128.
Blair, R., Dewar, W.A., Downie, J.N. 1973. *British Poultry Science* 14: 373-377.
Bouvarel, I., Nys, Y. and Lescoat, P. 2011. In: Improving the Safety and Quality of Eggs and Egg Products, Vol. 1 Cambridge: Woodhead Publishing, pp. 261-99.
Brionne, A., Nys, Y., Hennequet-Antier, C. and Gautron, J. 2014. *BMC Genomics* 15:220.
Cufadar, Y., Olgun, O. and Yildiz, A.Ö. 2011. *British Poultry Science* 52: 761-768.
Chah, C. and Moran, E.T. 1985. *Poultry Science* 64: 1696-1712.
Cumming, R.B. 1994. In '9th European Poultry Conference', Vol. II, pp. 219-222. WPSA, Glasgow.
De los Mozos, J., Gutierrez del Alamo, A., van Gerwe, T. and Sacranie, A. 2012. Proc. of World's Poultry Conference: 283-286.
Farrell, D.J., Ramlah, H., Hutagalung, R.I. 1981. *Tropical Animal Production* 6: 1.
Forbes, J.M. and Covasa, M., 1995. *Worlds Poultry Science Journal* 51: 149-165.
Guinotte, F. and Nys, Y. 1991. *Poultry Science* 70: 583-592.
Henuk, Y.L., Thwaites, C.J., Hill, M.K. and Dingle, J.G. 2000. Proceedings Australian Poultry Symposium: 117-120.
Joly, P. 2008. Nutrition Management Institut de Selection Avicole. Saint Brieuc, France.

Plenary session 03 and 04. Precision feeding

Karunajeewa, H. 1978. *British Poultry Science* 19: 699-708.

Keshavarz, J. 1998a. *Poultry Science* 77: 1320-32.

Keshavarz, K, 1998b. *Poultry Science* 77: 1333-46.

Koreleski, J. and Swiatkiewicz, S. 2004. *Journal of Animal and Feed Science* 13: 635-645.

Lee, K.H., Ohh, Y.S. 2002. *Korean Journal Poultry Science* 29:195-204.

Leeson, S. and Summers, J.D. 1979. *Poultry Science* 58: 646-651.

Leeson, S. and Summers, J.D. 1997. Commercial poultry nutrition. 2 Ed. University Books Guelph, On. Canada.

Lichovnikova, M. 2007. *British Poultry Science* 48: 71-75.

Muramatsu, T., Hiramoto, K. and Okumura, J. 1991. *Comparative Biochemistry Physiology* B 99: 141-146.

Mastika, M. and Cumming, R.B. 1987. In: D.J. Farrell (Ed.) *Recent Advances in Animal Nutrition in Australia*. University of New England, Armidale, Australia, 260-282.

Meunier-Salaün, M.C and Faure, J.M. 1984. *Applied Animal Behaviour Science* 12: 129-141.

Mongin, P. and Sauveur, B. 1974. *British Poultry Science* 15: 349-359

Morris, B.A. and Taylor, T.G. 1967. *British Poultry Science* 8: 251-257.

Nys, Y. 2017. In: Roberts J (Ed.) Achieving sustainable production of eggs-Vol 2. Chap1. Burleigh Dodds Science publishing.

Nys, Y. and Guyot, N. 2011. In: Y. Nys, Y., M. Bain, M. & and F. Van Immerseel, F. (Eds.), Improving the Safety and Quality of Eggs and Egg Products. Vol1, Cambridge: Woodhead Publishing. pp. 83-132.

Olver, M.D. and Malan, D.D. 2000. *S. Afr. J. Anim. Sci.* 30: 110-114.

Ouart, M.D., Marion, J.E. and Harms, R.H. 1986. *Poultry Science* 65: 1015-1017.

Pelicia, K., Garcia, E.A., Faitarone, A.B.G., Silva, A.P., Berto, D.A., Molino, A.B. and Vercese, F. 2009. *Brazilian Journal of Poultry Science* 11: 87-94.

Penz, A.M. and Jensen, L.S. 1991. *Poultry Science* 70: 2460-2466.

Picard, M., Melcion, J.P., Bouchot, C. and Faure, J.M. 1997. *INRA Productions Animales* 10: 403-414.

Pizzolante, C.C., Saldanha, E.S.P.B., Laganá, C., Kakimoto, S.K., Togashi, C.K. 2011. *Brazilian Journal of Poultry Science* 13: 103-111.

Portella, F.J., Caston, L.J. and Leeson, S. 1988. *Canadian Journal of Animal Science* 68: 915-922.

Robinson, D. 1985. *British Poultry Science* 26: 299-399.

Ruhnke, I., Röhe, I., Krämer, C., Boroojeni, F.G., Knorr, F., Mader, A., Schulze, E., Hafeez, A., Neumann, K., Löwe, R. and Zentek, J. 2015. *Poultry Science* 94: 692-699.

Safaa, H.M., Serrano, M.P., Valencia, D.G., Frikha, M., Jiménez-Moreno, E. and Mateos, G.G., 2008. *Poultry Science* 87: 2043-51.

Safaa, H.M., Jiménez-Moreno, E., Valencia, D.G., Frikha, M., Serrano, M.P. and Mateos G.G. 2009. *Poultry Science* 88: 608-614.

Segalen, L., Mills, M., Sebilo, M., Labourdette, N., Vaury, V. and Nys, Y. 2013. In: World's Poultry Science Journal 69 supplement. 15 European Symposium on the Quality of Eggs and Egg Products. Bergamo, ITA.

Skřivan, M., Englmaierová, M., Marounek, M., Skřivanová, V., Taubner, T., Vít, T. 2016. *Czech Journal Animal Science* 61: 473-480.

Tauson, R. and Elwinger, K. 1986. *Acta Agriculture Scandinavian* 36: 129-146.

Traineau, M., Bouvarel, H., Mulsant, C., Roffidal, L., Launay, C. and Lescoat, P. 2013. *Poultry Science* 92: 2475-2486.

Umar-Faruk, M., Bouvarel, I., Même, N., Rideau, N., Roffidal, L., Tukur, G.M., Bastianelli, D., Nys, Y. and Lescoat, P. 2010a. *Poultry Science* 89: 785-796.

Umar-Faruk, M., Bouvarel, I., Même, N., Roffida, L., Tukur, H.M., Nys, Y. and Lescoat, P. 2010b. *British Poultry Science* 51:811-820.

Umar-Faruk, M., Bouvarel, H., Mallet, S., Ali, M.N., Tukur, H.M., Nys, Y. and Lescoat, P. 2011. *Animal* 5: 230-238.

Avinesp model: predicting growth and nutritional requirements of broilers

N.K. Sakomura[*], L. Hauschild, M.P. Reis and N.T. Ferreira
Animal Science Department, UNESP – Univ. Estadual Paulista, Jaboticabal, Brazil;
sakomura@fcav.unesp.com.br

Summary

Growth models can be used as tools to help nutritionists to define feeding programs aiming sustainability and profitability. The objective is to present the Avinesp model and how it could be used in poultry industry to develop nutritional programs. To simulate the growth of broilers in this model, it's necessary to insert the bird's genotype, environmental conditions and feed composition. A feed formulator was included in Avinesp, allowing the user to formulate the feed as well. The concept applied in this model was based on previous studies published by Emmans and collaborators. The model provides daily values for energy and amino acid requirements, performance and body composition. In addition, a calibration procedure may be used to adjust some model parameters and simulate performance matching as close as possible the observed performance. To demonstrate the capacity of Avinesp for the poultry industry, simulations were performed using data from a Brazilian integration. Based on these simulations, the industries can establish their nutritional program for specific conditions, allowing to reduce feed costs and decreasing ambient pollution by applying the precise nutrition concept.

Introduction

In general, nutritional programs for poultry are based on either requirement tables or breeder's guidelines. However, the dietary levels change according to the feed intake, which is mainly affected by the genetic potential and environmental conditions. The growth models that consider these factors to estimate the feed intake and the genetic potential to determine the requirements could be used as a tool to define nutritional programs in the poultry industry. In this context, growth models are used to estimate nutritional requirements and simulate growth for different feeding scenarios applied in poultry production.

In the poultry industry, nutritionists need a tool to make quickly decision that aim both sustainability and profitability in different production scenarios, such as the changes in the genetic potential, environmental conditions and ingredient prices. The previous knowledge about the environmental effects on the birds' response is extremely necessary to formulate the best feed and improve feed efficiency. In this case, the simulation models are able to simulate the response of the birds and allows nutritionists to anticipate and predict the needs of the birds, consequently, nutritionists can make a better decision that will improve feed efficiency, decrease feed costs and optimize performance, achieving a precise nutrition.

Considering those needs for poultry nutritionists and the use for academic purposes, the Avinesp, a Brazilian software, was developed by the Nutrition and Modelling research group of UNESP-Jaboticabal. The Avinesp model considers the growth potential of the bird, diet, and environmental conditions to perform an accurate prediction of the birds' growth and nutritional requirements.

The objective of this paper is to present the Avinesp Model and to demonstrate how it can be used as a tool for poultry industry, allowing nutritionists to apply the precise nutrition concept.

Model description

The model proposed here is based on the concepts developed in the past studies (Emmans, 1974; Emmans, 1981; Emmans, 1987; Emmans, 1994, 1999) and currently is described with more details by Hauschild *et al.* (2015).

A first concept of the model is to describe the genetic potential of the bird. The potential for protein deposition in the body (free of feathers) and feathers is described using the Gompertz equation. The potential for lipid deposition is estimated based on the ratio between lipid and protein deposition as proposed by Emmans (1981). The deposition of ash and water is described in the model by allometric ratio to body protein, since ash and water components does not vary much among sexes or genotypes (Gous *et al.*, 1999). The growth parameters that describe the broiler strains (Cobb and Ross) and the layer strains (Hy Line and Hisex pullets) were obtained from previous studies conducted at the Poultry Science Laboratory (Lavinesp) at UNESP-Jaboticabal.

The model estimates the current status of the bird, i.e. the empty body weight gain (carcass with feathers and no gut fill) on a given day. This gain is added to empty body weight of the previous day to determine the current body weight. Similarly, each body component is determined. The model estimates the desired feed intake and actual feed intake. The physical capacity of the digestive tract is taken into account to estimate feed intake.

The external effects on bird's response, such as the environmental temperature and heat production are considered, affecting the energy requirement for maintenance and feed intake, consequently, the growth rate. Feed intake is not affected if heat production is in-between minimum and maximum heat loss, however, if heat production is below or above the limits of heat loss, the birds will attempt to adjust the heat production by modifying the feed intake and keeping heat loss between minimum and maximum. Therefore, temperature is considered in the model and may affect heat loss, consequently, feed intake.

Applicability of Avinesp software

To demonstrate the applicability of the Avinesp model in the poultry industry as a tool in decision making, we performed simulations using data from a Brazilian integration.

Twenty flocks with approximately 50,000 broilers each were used in the simulation. The feed program consisted of four diets according to the feeding program for growing birds (1 to 7, 8 to 21, 22 to 33, and 34 to 42 days old). For each flock, the growth curve of Cobb males was calibrated based on observed body weight from database. Body weight data were obtained in the same integration of Brazil, but the flocks were raised in different poultry houses.

Figure 1A shows the average body weight of each broiler flock in the period from 1 to 42 days old. Although the flocks have the same genetic line and the nutritional program, a variation on average body weights of each flock was observed. This variation on body weight reflects in different lysine requirements (Figure 1B).

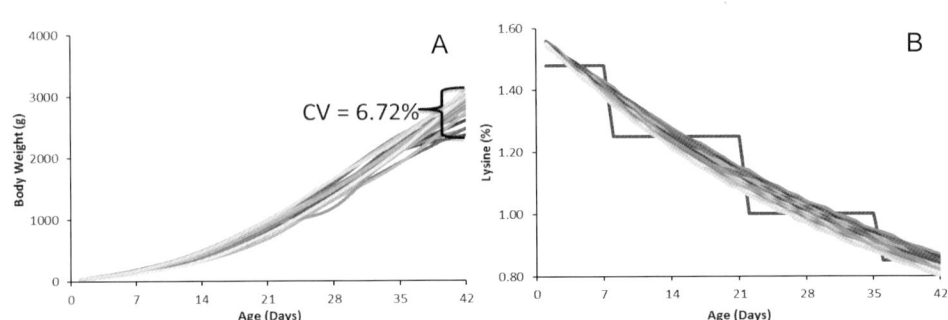

Figure 1. Observed average body weight (A), the lysine concentration in the diet (darkest straight line) and the estimated daily lysine requirement (B) of 20 flocks of 50,000 male Cobb chickens each. CV = coefficient of variation.

Plenary session 03 and 04. Precision feeding

Because feed intake increases as the broilers grow, the concentration of dietary protein required is reduced. However, it is not practical to reduce the dietary concentration every day, hence the use of a phase-feeding programme. In commercial broiler production, feeding programmes are usually divided into phases, using three to five diets decreasing in protein throughout the production cycle. Assuming that the birds have *ad libitum* access to the feed, they are fed with sub-optimal levels of protein (amino acids) at the beginning of each phase and excesses towards the end of each phase as illustrated in Figure 1B.

At the start of each phase, the birds will attempt to consume sufficient amount of the limiting amino acid(s) by overconsuming energy, which is deposited as body lipid and, thus, decreasing feed efficiency during this period. As each phase progresses the under-supply of amino acids becomes less and then at a point will become excessive in relation to the requirement. At this point the bird is able to draw on the excessive body lipid supplies to obtain much of the energy required, and feed efficiency is improved.

In order to elucidate the effect of under- and over-feeding protein on body composition of broilers, it was performed simulations of different nutritional programmes.

Six feeding programmes were simulated, divided into three phases: Starter (1 to 21 d), Grower (22 to 35 d) and Finisher (35 to 42 d), two energy levels with three levels of protein for each phase (Table 1).

Diets with lower protein levels (program 3–HE/LP and 6–LE/LP) result in a higher feed and ME intakes, consequently, higher lipid deposition in the body (Figure 2B). In addition, in the feeds with lower protein, the bird cannot consume sufficient lysine to meet its requirement, resulting in a reduction in body protein deposition (Figure 2A). Gous *et al.* (1990) reported that birds fed unbalanced diets, where energy is not the first limiting nutrient, retain lipid in excess. Programs 4–LE/HP and 5–LE/MP (lower ME and higher and medium CP) provided ME intakes close to the requirement and lipid deposition according to genetic potential. The program 5–LE/MP (lower ME and medium CP) provided ME and lysine according to the requirements, consequently, higher protein deposition and lipid deposition according to genetic potential (Figure 2). These results showed that in suitable ME to CP ratio, the lipid deposition in the body depend of the genetic potential.

The effect of under- and overfeeding on each feeding phase may affect body composition, as observed in program 1–HE/HP, 2–HE/MP and 5–LE/MP. There is an increase in energy intake, especially when changing the diet phase, consequently, there is a greater deposition of lipid in the body (Figure 2). Emmans (1981) reported that body lipid reserves are labile, changing on account of feed level, previous feed offered, and environmental conditions. In addition, growing broilers with *ad libitum* access to feed with lower ratio between the first-limiting amino acid and

Table 1. Feeding programmes with diets (1 to 6) consisting in the combination of high (HE) or low metabolizable energy (LE) and high (HP), medium (MP) or low crude protein (CP) level (LP) during three rearing phases.

Programmes	Starter (1 to 21 d)	Grower (22 to 35 d)	Finisher (35 to 42 d)
1 – HE/HP	3,250 kcal / 26% CP	3,350 kcal /24% CP	3,400 kcal / 22% CP
2 – HE/MP	3,250 kcal / 22% CP	3,350 kcal / 20% CP	3,400 kcal / 18% CP
3 – HE/LP	3,250 kcal / 18% CP	3,350 kcal / 16% CP	3,400 kcal / 14% CP
4 – LE/HP	3,050 kcal / 26% CP	3,150 kcal / 24% CP	3,200 kcal / 22% CP
5 – LE/MP	3,050 kcal / 22% CP	3,150 kcal / 20% CP	3,200 kcal / 18% CP
6 – LE/LP	3,050 kcal / 18% CP	3,150 kcal / 16% CP	3,200 kcal / 14% CP

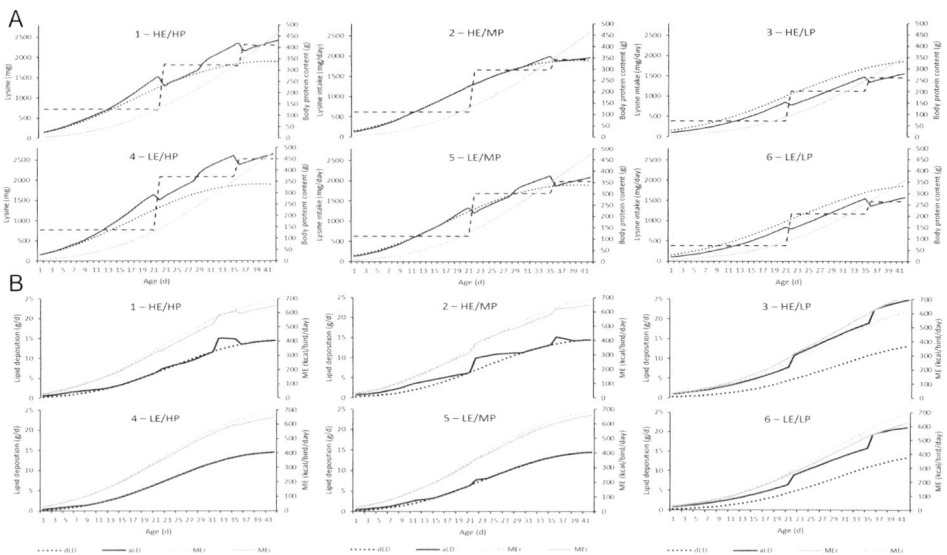

Figure 2. (A) Lysine requirement estimated (dotted line), dietary lysine concentration by phase (dashed line), lysine intake corrected by the actual feed intake (solid black line) and body protein content (solid grey line). (B) Actual lipid deposition (solid black line), desired lipid deposition (dotted black line), metabolizable energy requirement (dotted grey line) and metabolizable energy intake (grey line) for each feed program as a function of age. HE=high energy, LE=low energy, HP=high protein, MP=medium protein, and LP=low protein.

energy, take more time and consume more energy to reach a given weight with more fat in the body (Jackson *et al.*, 1982; Gous *et al.*, 1990).

The feed intake is dependent, among other factors, on the state of the animal at the time and this feed intake will define the amount of protein and lipid that will be deposited each day (Gous *et al.*, 2012). In Figure 2, there is a small variation in the deposition of protein in the body among the 1–HE/HP, 2–HE/MP, 4–LE/HP and 5–LE/MP programs, except for programs 3–HE/LP and 6–LE/LP, because the concentration of protein in the diet was lower at all stages, resulting in lower deposition of protein in the body. In addition, program 3–HE/LP showed the highest lipid deposition, since the birds increased feed intake to meet their lysine requirement. Programs 1–HE/HP, 4–LE/HP and 5–LE/MP had the lowest lipid deposition, since the diet provided sufficient lysine and energy to meet the requirements. Based in this simulation the programme 5–LE/MP provided the best results of body composition. The ratio between energy and the first-limiting amino acid affects body composition dynamically, thus, the nutritionist can make decisions about nutritional levels for broilers aimed at maximising the quality of the carcass. This diagnostic allows adjusting diet formulation and making decision to improve performance, reduce nutrient excretion and improve quality of carcass.

Conclusions

Calibrating the growth potential of the birds according to real condition enables to make simulations for growth and requirements of broilers at any condition, providing a diagnostic considering many aspects of poultry production.

Avinesp software could be used as a diagnostic tool for decision-making, since enable to simulate how changes in nutritional programs affect the performance and body composition, also estimate nutritional requirements for different genetic potential, environment condition and feed. These assumptions allow to establish a more adequate nutrition, adjusted to specific situation, promoting a precise nutrition.

References

Emmans, G.C. 1974. The effect of temperature on performance of laying hens. In: Energy requirements of Poultry, pp. 79-90.

Emmans, G.C. 1981. A model of the growth and feed intake of *ad libitum* fed animals, particularly poultry. In: Computers in Animal Production. Animal production Occasional, pp. 103-110.

Emmans, G.C. 1987. *World Poultry Science* 3: 208-227.

Emmans, G.C. 1994. *British Journal of Nutrition* 71: 801-821.

Emmans, G.C. 1999. Energy Flows. In: A Quantitative Biology of the Pig, pp. 363-377.

Gous, R.M., Moran, E.T., Jr., Stilborn, H.R., Bradford, G.D. and Emmans, G.C. 1999. *Poultry Science* 78: 812-821.

Gous, R.M., Emmans, G.C. and Fisher, C. 2012. *South African Journal of Animal Science* 42: 63-73.

Gous, R.M., Emmans, G.C., Broadbent, L. A. and Fisher, C. 1990. *British Poultry Science* 31: 495-505.

Hauschild, L. Sakomura, N.K. and Silva, E.P. 2015. AvinespModel: Predicting poultry growth, energy and amino acid requirements. In: Nutritional modeling for pigs and poultry 188-207.

Jackson, S., Summers, J.D. and Leeson, S. 1982. *Poultry Science* 61: 2224-2228.

Lopez, G., de Lange, K. and Leeson, S. 2007. *Poultry Science* 86: 2162-2171.

Regulation of feeding behaviour after feeding disproportionate amounts of amino acids

N. Tous and E. Esteve-Garcia[*]

IRTA, Animal Nutrition and Welfare Program, Mas de Bover, Ctra. Reus-El Morell, Km 3.8, 43120 Constantí, Spain; enric.esteve@irta.cat

Summary

Amino acid disproportions in the diet have been classified as: toxicities, antagonisms, imbalances and deficiencies. Briefly, toxicity refers to the detrimental effects caused by the excess of a single amino acid; antagonism is caused by a structurally similar amino acid and affects other amino acids which are not necessarily first limiting; imbalances refer to the effect of surpluses amino acids other than the one that is limiting for growth or maintenance and are prevented by the addition of the limiting amino acid; and deficiency refers to the lack of one or more essential amino acids being a simple form of an imbalance. In all cases, the first sign is the depression in feed intake. In the case of imbalances, given a choice, animals select a balanced diet avoiding unbalanced diets and they prefer a protein-free to an imbalanced diet although the former will not support animal growth. This fact indicates that animals are able to reject non-balanced diets and to select adequate diets in order to maintain homeostasis. This response is quick and several regulating mechanisms are described. After feeding an imbalanced diet the level of the limiting amino acid in blood is reduced. This reduction is also observed at brain level, particularly in the anterior piriform cortex (APC). When the levels of an amino acid decrease enough, its transfer RNA will be uncharged, blocking the initiation of protein synthesis. Thereafter, secondary responses are evoked to other brain areas involved in the regulation of feed intake.

Amino acid interactions

The appropriate selection of diets containing adequate amounts of amino acids is essential for survival. In farm practice, diets are formulated to meet nutrient requirements. Since more and more synthetic amino acids are available it is possible to formulate diets with lower protein levels which are less costly and have less impact on the environment. However, this could result in the balance of amino acids to be altered and a new amino acid becoming limiting because there is less information on amino acid requirements of those that are less limiting. In this review, the consequences of feeding disproportionate amounts of amino acids will be discussed, in particular those affecting feed intake, growth, the feeding behaviour and the possible mechanisms involved.

H. H. Mitchell and H. M. Scott in the late 1950s and early 1960s defined the term Ideal Chick Protein as a blend of indispensable amino acids that exactly meets an animal's requirement for protein accretion and maintenance, with no deficiencies and no excesses. For diets not meeting this criteria Harper (1956) and Harper *et al.* (1970) defined four groups of amino acid interactions: deficiencies, imbalances, antagonisms, and toxicities. All of them have in common the fact that one of the earliest and most apparent signs is a depression in food intake.

The term deficiency refers to the lack of one or more essential amino acid in the diet. All diets deficient in one or more amino acids can be considered imbalanced. Thus, a deficiency is a simple form of an imbalance (Baker, 1974).

Harper (1956) suggested the term antagonism when the addition of one amino acid causes an adverse effect which is corrected by the addition of a structurally similar amino acid which need not be the limiting amino acid in the original diet. Antagonism is distinguished from an imbalance because the supplementation of the limiting amino acid does not prevent or alleviate the growth retardation (Harper and Rogers, 1965). Two examples are the lysine-arginine antagonism in which excess of Lys increases the requirement for Arg (Jones, 1964; O'Dell and Savage, 1966; Allen *et al.*, 1972) and the branched-chain amino acid antagonism involving Leu, Ile and Val in which an excess of one or two branched-chain amino acids increases the requirement for the other (Spolter and Harper, 1961; Allen and Baker, 1972). Excess of Leu has been reported to increase the activity if muscle branched-chain amino acid aminotransferase in chicks (Smith and Austic, 1978), hepatic branched-chain α-ketoacid dehydrogenase activity in rats (Wohlhueter and Harper, 1970), the oxidation of Ile and Val in chicks (Smith and Austic, 1978; Calvert et al., 1982) and the oxidation of Ile in rats (Phansalkar et al., 1970). If significant amounts of Ile and Val are lost via catabolic processes, this may explain the failure of force feeding to prevent adverse effects on the antagonism. The metabolic consequences of the lysine-arginine antagonism in chicks include depressed creatinine synthesis (Jones et al., 1967; Austic and Scott, 1975), increased activity of kidney arginase and urea excretion (Jones et al., 1967; Austic and Nesheim, 1970), and, when Lys levels are highly excessive, increased urinary excretion of Arg (Nesheim, 1968; Boorman, 1971).

Toxicity is defined as a condition in which retardation of growth and other detrimental effects occur as a consequence of feeding excessive amounts (2 to 10 times the requirement) of single amino acids, and is not prevented by the addition of another amino acid to the diet. Apart from growth depression, specific gross or histopathological lesions may occur, depending on the amino acid present in excess. Some studies in rats reported liver, kidney, skin, pancreas and testis damage when fed diets with toxic levels of Tyr, Cys, Met and Ser (Harper et al., 1970). Methionine (Katz and Baker, 1975), Phe (Elkin and Rogler, 1983) and Cys (Baker, 2006) provide relevant examples of amino acid toxicity in chickens. It appears that those amino acids which serve as precursors for a number of other compounds and which cannot be channelled in great quantity into major catabolic pathways, are capable of producing the greatest toxic effects. In contrast, amino acids that are readily oxidized and involved in relatively fewer metabolic appear to be less toxic. Indeed, animals can tolerate an excess of an individual amino acid better when the protein content of the diet is high (Leung and Rogers, 1975).

Harper (1956) restricts the term imbalance to cases that a growth depression is observed in diets containing a suboptimal level of amino acids and this growth-depressing effect is prevented by small supplement of the amino acid that is most limiting. There are indicators that imbalances induced by high-protein diets formulated from intact ingredients may reduce the efficiency of utilization of the first-limiting amino acid in poultry (D'Mello, 2003). Amino acid imbalances can be induced by adding either a relatively small amount of an indispensable amino acid or a relatively large quantity of a mixture of amino acids lacking the limiting one into low protein diets (Park, 2006). Severity of the depression in food intake and growth increased with each increase in the quantity of the unbalancing amino acid mixture added to the low protein basal diet (Leung et al., 1968a). Initially, it was believed that imbalances only occurred in low protein diets deficient in one amino acid. However, the supplementation of amino acid to diets containing adequate dietary protein also appeared to show the typical consequences of amino acid imbalances (Davis and Austic, 1994; Park and Austic, 1998; Park and Austic, 2000). It was also believed that only indispensable amino acids, not dispensable amino acids might induce imbalance. However, imbalances also occurred in case of dispensable amino acids although much larger amounts were required to depress growth than needed with indispensable amino acids (Peng and Harper, 1970; Tews et al., 1979; Tews et al., 1980). Only histidine, Ile, Lys, Met, threonine, and tryptophan have been used in the studies of amino acid imbalances (Park, 2006).

Taking into account the effects of feeding disproportionate amounts of amino acids on feed intake, the balance of all amino acids is critical to optimise performance.

Feeding behaviour when feeding imbalanced diets

Rats given a choice between an unbalanced and a protein-free diet consistently rejected the unbalanced diet and ate mainly the protein-free which did not support any growth at all (Sanahuja and Harper, 1962; Leung and Rogers, 1986). Pant et al. (1972) suggested that selection was not induced by the nutritional quality of the diet, but it was rather a question of which diet would satisfy the metabolic need and restore the homeostasis of the animals. Experiments by Leung et al.

Plenary session 03 and 04. Precision feeding

(1986) showed that rats selected protein free or corrected diets containing quinine-HCl (a negative taste cue) over imbalanced diets containing Na-saccharin (a positive taste cue), indicating that their aversion to the imbalanced diet still occurred when selection of the alternative diet was against the taste preferences shown with balanced diets. Furthermore, force-feeding imbalanced diets in order to prevent the usual depressions of food intake resulted in normal growth rate and body composition (Benevenga et al., 1968; Leung et al., 1968c; Harper et al., 1970). In addition, hormone administration such as cortisol or thyroxine injections (Kumta and Harper, 1961; Leung et al., 1968b) and cold (Harper and Rogers, 1966) also prevented the depression of food intake by increasing the protein degradation in some tissues and resulting in increased levels of the blood limiting amino acid.

Imbalances occur so rapidly that the retarded growth may be attributed to depressed food intake but not to digestion or absorption (Sugahara and Kubo, 1992), smell or taste (Leung and Rogers, 1969; Leung et al., 1972); palatability (Benevenga et al., 1968), stomach emptying, urinary or faeces excretion or oxidation of amino acids (Yoshida et al., 1966; Pant et al., 1974). Studies showed that rats reject a deficient diet from within 2-3 hours after exposure (Leung and Rogers, 1986). However, some studies in rats showed an adaptation to amino acid imbalance. After few days feeding rats an imbalanced diet restored feed intake and growth (Davis and Austic, 1994). In addition, the reduction of the limiting amino acid in blood was no longer significant after 5 days of supplying the imbalanced diet (Tews et al., 1980). Contrary to rats, it appears unlikely that chickens adapt to imbalanced diet (Park, 2006).

A consistent finding that parallels the depression of food intake is that the plasma concentration of the amino acid that is limiting in the diet falls markedly within a few hours (Gray et al., 1960; Alam et al., 2014). There are two possible mechanisms involved in the alteration of plasma amino acid pattern: an increase in the rate of catabolism of the first-limiting amino acid, which causes a reduction in its plasma concentration or that amino acids in excess in the unbalanced diets induce more protein synthesis or suppresses breakdown in the liver and, as a consequence the limiting amino acid becomes even more limiting in the liver and its plasma concentration decreases (Rogers, 1996). This fact is supported by the stimulation of the hepatic ribosome aggregation and a higher incorporation of radiolabeled amino acid into liver in rats fed imbalanced diets (Yoshida et al., 1966; Yasukawa and Yoshida, 1980); suggesting that protein synthesis in the liver is increased while the supply of the limiting amino acid for peripheral tissues is reduced but not to a level that will depress protein synthesis in peripheral tissues (Benevenga et al., 1968). The homoeostatic capacity of the body to restore the free amino acid pattern of the blood, muscle, and possibly those of some other organs or tissues, to normal is exceeded and, in response food intake is depressed, and hence growth is retarded (Harper and Rogers, 1965).

More recent studies evaluated enzyme activities related with the degradation of each limiting amino acid. The most studied imbalances are the ones of Ile, threonine (Thr), and Phe followed by Leu and Lys. In all cases, the enzymes catalysing the limiting amino acid increase suggesting that these enzymes may contribute to the increased amino acid requirement associated with the imbalance and the decrease in plasma amino acid concentration may be also a consequence of the increased catabolism. Park and Austic (1998) showed that Ile imbalance increased the basal and total activities of hepatic branched-chain keto acid dehydrogenase (BCKAD). Similarly occurred in case of Thr, activities of Thr dehydrogenase (TDH) in chicks fed the Thr imbalanced diet were higher than chicks fed the basal diet. Indeed, the highest TDH activities in both basal and imbalanced group occurred 1.5 h after the diets were fed (Davis and Austic, 1994; Yuan et al., 2000), although previous studies had not found an increase in Thr aldolase and TDH activities (enzymes that catalyse the first step in the three degradation pathways of Thr in chicken), suggesting that net Thr catabolism was not markedly increased (Davis and Austic, 1982a,b). In case of Phe, catabolism is initiated by the activities of 2 enzymes: Phe hydroxylase (PAH), which catalyses the conversion of Phe to Tyr, and Phe-pyruvate aminotransferase (PAT), which catalyses the formation of phenylpyruvic acid. In chicks fed the Phe imbalance, PAT activity was higher in liver and kidneys. However, Phe imbalance did not alter PAH activity. This suggests that Phe imbalance increased the degradation of Phe via PAT (Lartey and Austic, 2008, 2009; Lu and Austic, 2009).

Several studies in rats found that dietary disproportions of amino acids can alter the flux of specific amino acids across the blood-brain barrier. Depressions in influx along with associated changes in brain amino acid concentration caused by a deficiency of an essential amino acid in the diet appear to be ultimately accompanied by depressions of food intake and growth (Tews *et al.*, 1988; Beverly *et al.*, 1991; Beverly *et al.*, 1993). Amino acids are transported across the BBB by carriers which are depending on the size and charge of amino acids (Oldendorf and Szabo, 1976). Competition for uptake into brain occurs between individual amino acids within specific groups (neutral, basic and acidic) because their Km values for transport are similar to or lower than normal plasma concentrations (Tews *et al.*, 1980); hence, alterations in plasma amino acid concentrations can alter flux of amino acids into brain (Lutz *et al.*, 1975).

Feed intake regulation at brain level

Involvement of the brain was suggested by the finding that the feeding response was reversed by infusion of the limiting indispensable amino acid (IAA) and that a much lower quantity was needed when infused through the carotid artery than through the jugular vein (Leung and Rogers, 1969). Also, levels of the limiting IAA decreased as rapidly in brain tissue as they did in plasma (Peng *et al.*, 1972). Koehnle *et al.* (2004) found that levels of the limiting IAA are decreased in anterior prepyriform cortex (APC) tissue by 56% at 21 min. Hence, the prepyriform cortex, located in the anterior ventrolateral forebrain, appears to be the area responsible for detecting imbalances. Lesions in this area caused rats to eat more of the imbalanced or deficient diet when compared to intake controls and prefer the imbalanced diets over a protein-free diet (Leung and Rogers, 1971). Rogers and Leung (1973) concluded from these observations that the anterior prepyriform cortex may contain areas sensitive to the concentration of the growth – limiting amino acid, sending inhibitory signals to the lateral hypothalamic feeding area to curtail food intake. Intravenous injections of large amounts of a single amino acid can lower the cerebral concentrations of other amino acids by inhibiting their uptake and stimulating their utilization in cerebral tissue (Roberts, 1968). Indeed, injecting dietary limiting essential amino acid (DLAA) within the PPC is necessary to increase the intake of imbalanced diets. This increase is not observed when the amino acid injected either into a more posterior area of the pyriform or the medial amygdala, suggesting that the decline in DLAA concentration in the PPC is the sole signal for the reduction in intake of imbalanced diets. Alternatively, the differences in timing may reflect the time necessary for diffusion of the DLAA throughout the PPC or to a specific site(s) within the PPC. The 3- to 4-h delay in restoring (or preventing) the food intake depression may also indicate that some aspect of metabolism, possibly protein synthesis and/or transport, are necessary before the feeding response is evident (Beverly *et al.*, 1990b, a). The subsequent microinjection and histochemical studies into the anterior piriform cortex (APC) have confirmed this finding (Beverly *et al.*, 1990b, a; Gietzen *et al.*, 1998). Parallel work in chicks indicated that the analogous brain area in birds serves the same function (Firman and Kuenzel, 1988). On the other hand, it has been suggested that the hypothalamus (Tome, 2004), including the lateral hypothalamic area (Blevins *et al.*, 2000) and the dorsomedial hypothalamic nucleus (Bellinger *et al.*, 1999), should also be considered in the food-intake responses to IAA depletion or high protein levels. The most likely reconciliation of these divergent views is that the responses in the APC appear first, and that subsequently the projections from the APC (Gietzen *et al.*, 1998) affect other brain areas for the behavioural responses. Different circuits may mediate the initial recognition and secondary conditioned responses to imbalanced diets. There are several studies indicating that different neuroreceptors such as norepinephrine (Gietzen *et al.*, 1986), serotonin 5-HT$_3$ (Hammer *et al.*, 1990; Terry-Nathan *et al.*, 1995) and somatostatin (Nakahara *et al.*, 2012) participate in the regulatory process. Further studies are needed to understand the role of these mechanisms on feeding regulation.

Plenary session 03 and 04. Precision feeding

In the earliest steps leading to the initiation of translation in protein synthesis, AAs are acylated to tRNA by their cognate amino-acyl tRNA synthetase enzymes. Clearly, a continuous supply of all AA must be available for protein synthesis to proceed. If levels of an IAA decrease enough (AA deficiency), its tRNA will be deacylated (i.e. 'uncharged'). Therefore, the accumulation of uncharged tRNA in the APC starts which blocks the initiation of protein synthesis via the non-depressing kinase 2 (GCN2) system followed by loss of the GABAergic inhibitory control in the APC circuit and increased glutamatergic activity. Glutamatergic axons reach many brain areas associated with the behavioural responses. The authors attributed this pathway as the primary sensor of the IAA deficiency (Gietzen and Aja, 2012). The GCN2 pathway is a pathway conserved from yeast (Wek *et al.*, 2006). Hence, the discovery of such an ancient pathway in the sensing of IAA deficiency *in vivo* underscores the importance of protein homeostasis (Gietzen and Rogers, 2006). Other mechanisms downstream from the IAA-sensor include several genes (Kilberg *et al.*, 2005) sodium cotransport (Blais *et al.*, 2003) and intracellular calcium (Sharp *et al.*, 2004).

Conclusions

Dietary selection for appropriate amino acid balance is essential for omnivores due to the inability to synthetize some amino acids. Animals are able to recognise unbalanced diets. The identification of an imbalance occurs in the first hours after exposure, resulting in reduction of feed intake and growth. Several mechanisms are described in rats which are presumably similar to those in chicks. After feeding, the level of the limiting amino acid is reduced in both blood and brain, particularly in the anterior piriform cortex (APC). The primary signal appears to be the presence of uncharged tRNA in APC which then evokes secondary responses to other brain areas regulating feed intake.

References

Alam M.R., *et al.* 2014. *British Poultry Science* 55: 605-609.
Allen N.K., Baker D.H. 1972. *Poultry Science* 51: 1292-1298.
Allen N.K., *et al.* 1972. *Journal of Nutrition* 102: 171-180.
Austic R.E., Nesheim M.C. 1970. *Journal of Nutrition* 100: 855-867.
Austic R.E., Scott R.L. 1975. *Journal of Nutrition* 105: 1122-1131.
Baker D.H. 1974. *FEEDSTUFS* 15: 21-22.
Baker D.H. 2006. *Journal of Nutrition* 136: 1670S-1675S.
Bellinger L.L., *et al.* 1999. *American Journal of Physiology – Reg. I.* 277: R250-R262.
Benevenga N.J., *et al.* 1968. *Journal of Nutrition* 95: 434-444.
Beverly J.L., *et al.* 1990a. *American Journal of Physiology – Reg. I.* 259: R709.
Beverly J.L., *et al.* 1990b. *American Journal of Physiology – Reg. I.* 259: R716.
Beverly J.L., *et al.* 1991. *Journal of Nutrition* 121: 1287-1292.
Beverly J.L., *et al.* 1993. *Physiology Behaviour* 53: 899-903.
Blais A., *et al.* 2003. *Journal of Nutrition* 133: 2156-2164.
Blevins J.E., *et al.* 2000. *Brain Research* 879: 65-72.
Boorman K.N. 1971. *Comparative Biochemistry and Physiology – Part A: Physiology* 39: 29-38.
Calvert C.C., *et al.* 1982. *Journal of Nutrition* 112: 627-635.
D'Mello J.P.F. 2003. *Amino Acids in Animal Nutrition*, 2nd ed., edited by J.P.F. D'Mello. Wallingford, Oxon, UK: CABI Publishing.
Davis A.J., Austic R.E. 1994. *Journal of Nutrition* 124: 1667-1677.
Davis A.T., Austic R.E. 1982a. *Journal of Nutrition* 112: 2170-2176.
Davis A.T., Austic R.E. 1982b. *Journal of Nutrition* 112: 2177-2186.
Elkin R.G., Rogler J.C. 1983. *Poultry Science* 62: 647-658.
Firman J.D., Kuenzel W.J. 1988. *Brain Reserach Bulletin* 21: 637-642.
Gietzen D.W., Rogers Q.R. 2006. *Trends Neuroscience* 29: 91-99.
Gietzen D.W., Aja S.M. 2012. *Molecular Neurobiology.* 46: 332-348.
Gietzen D.W., *et al.* 1986. *Physiology Behaviour* 36: 1071-1080.
Gietzen D.W., *et al.* 1998. *Journal of Nutrition* 128: 771-781.
Gray J.A., *et al.* 1960. *Canadian Journal of Biochemistry & Physiology* 38: 435-441.

Plenary session 03 and 04. Precision feeding

Hammer V.A., *et al.* 1990. *American Journal of Physiology – Reg. I.* 259: R627.
Harper A.E. 1956. *Nutrition Reviews* 14: 225-227.
Harper A.E. and Rogers Q.R. 1965. *Proceedings of the Nutrition Society* 24: 173-190.
Harper A.E. and Rogers Q.R. 1966. *American Journal of Physiology* 210: 1234.
Harper A.E., *et al.* 1970. *Physiological Reviews* 50: 428-558.
Jones J.D. 1964. *Journal of Nutrition* 84: 313-321.
Jones J.D., *et al.* 1967. *Journal of Nutrition* 93: 103-116.
Katz R.S. and Baker D.H. 1975. *Journal of Animal Science* 41: 1355-1361.
Kilberg M.S., *et al.* 2005. *Annual Review of Nutrition* 25: 59-85.
Koehnle T.J., *et al.* 2004. *Journal of Nutrition* 134: 2365-2371.
Kumta U.S. and Harper A.E. 1961. *Journal of Nutrition* 74: 139-&.
Lartey F.M. and Austic R.E. 2008. *Poultry Science* 87: 291-297.
Lartey F.M. and Austic R.E. 2009. *Poultry Science* 88: 774-783.
Leung P.F. and Rogers Q.R. 1971. *American Journal of Physiology* 221: 929-935.
Leung P.M.B., *et al.* 1968a. *Journal of Nutrition* 95: 483-492.
Leung P.M.B. and Rogers Q.R. 1969. *Life Sciences* 8: 1-9.
Leung P.M.B. and Rogers Q.R. 1975. *Total parenteral nutrition*, edited by H. Ghadimi. New York: John Wiley and Sons.
Leung P.M.B. and Rogers Q.R. 1986. *Physiology Behaviour* 37: 747-758.
Leung P.M.B., *et al.* 1968b. *Journal of Nutrition* 96: 139-151.
Leung P.M.B., *et al.* 1968c. *Journal of Nutrition* 95: 474-482.
Leung P.M.B., *et al.* 1972. *Physiology Behaviour* 9: 553-557.
Leung P.M.B., *et al.* 1986. *Physiology Behaviour* 38: 255-264.
Lu J. and Austic R.E. 2009. *Poultry Science* 88: 2375-2381.
Lutz J., *et al.* 1975. *American Journal of Physiology* 229: 229.
Nakahara K., *et al.* 2012. *Amino Acids* 42: 1397-1404.
Nesheim M.C. 1968. *Journal of Nutrition* 95: 79-87.
O'Dell B.L. and Savage J.E. 1966. *Journal of Nutrition* 90: 364-370.
Oldendorf W.H. and Szabo J. 1976. *American Journal of Physiology* 230: 94-98.
Pant K.C., *et al.* 1972. *Journal of Nutrition* 102: 131-142.
Pant K.C., *et al.* 1974. *Journal of Nutrition* 104: 1584-1596.
Park B.C. 2006. *Asian Australas. Journal of Animal Science* 19: 1361-1368.
Park B.C. and Austic R.E. 1998. *Journal of Nutrition Biochemistry* 9: 687-696.
Park B.C. and Austic R.E. 2000. *Poultry Science* 79: 1782-1789.
Peng Y. and Harper A.E. 1970. *Journal of Nutrition* 100: 429-437.
Peng Y., *et al.* 1972. *American Journal of Physiology* 222: 314-321.
Phansalkar S.V., *et al.* 1970. *Proceedings of the Society for Experimental Biology and Medicine* 134: 262-263.
Roberts S. 1968. *Progress in Brain Research* 29: 235-243.
Rogers Q.R. 1996. *Protein metabolism and nutrition*, edited by Cole et. al., pp. 279-301.
Rogers Q.R. and Leung P.M. 1973. *Federation Proceedings Journal* 32: 1709-1719.
Sanahuja J.C. and Harper A.E. 1962. *American Journal of Physiology* 202: 165-170.
Sharp J.W., *et al.* 2004. *Neuroscience* 126: 1053-1062.
Smith T.K. and Austic R.E. 1978. *Journal of Nutrition* 108: 1180-1191.
Spolter P.D. and Harper A.E. 1961. *American Journal of Physiology* 200: 513-518.
Sugahara K. and Kubo T. 1992. *British Poultry Science* 33: 805-814.
Terry-Nathan V.R., *et al.* 1995. *American Journal of Physiology – Reg. I.* 268: R1203.
Tews J.K., *et al.* 1979. *Journal of Nutrition* 109: 304-315.
Tews J.K., *et al.* 1980. *Journal of Nutrition* 110: 394-408.
Tews J.K., *et al.* 1988. *Journal of Nutrition* 118: 756-763.
Tome D. 2004. *Br. Journal of Nutrition* 92: S27-S30.
Wek R.C., *et al.* 2006. *Biochemical Society Transactions* 34: 7-11.
Wohlhueter R.M. and Harper A.E. 1970. *Journal of Biological Chemistry* 245: 2391-2401.
Yasukawa T. and Yoshida A. 1980. *Journal of Nutrition Science Vitaminology (Tokyo)* 26: 461-473.
Yoshida A., *et al.* 1966. *Journal of Nutrition* 89: 80-90.
Yuan J.-H., *et al.* 2000. *Journal of Nutrition* 130: 2746-2752.

Nutritional requirements of broilers with different growth capacity

M.A. Grashorn
WG Poultry Science, Inst. of Animal Science (460), University of Hohenheim, Stuttgart, Germany;
michael.grashorn@uni-hohenheim.de

Summary

For a long time, broiler breeding focused on improving growth capacity. This resulted also in increased nutrient requirements of birds. Accordingly, recommended contents of amino acids in diets have been adapted continuously, both in nutrition tables and in breeder guidelines. Meanwhile, extensive broiler meat production is gaining more interest and a higher market share demanding adapted nutrition recommendations for broilers with a lower growth capacity. Currently, there do not exist any proven recommendations for supplying slower growing broiler genotypes with amino acids. Thus, the present paper aimed to derive some preliminary estimates. In a first step, broiler genotypes were assigned to three categories according to their growth capacity. Available information on amino acids supply from literature was used to derive supply recommendations. The categories were slow, medium and fast growth, with the fast growing broilers having the highest requirements. Recommendations for medium and slow growing genotypes were fixed as 10 and 20% lower than for the fast growing genotype, respectively. Recommendations for layer-type genotypes were arranged between medium and slow growing genotypes.

Introduction

In the beginning of the last century, the breeding of specific meat-type chickens emanated from purebred lines, and White Rocks have been one of the first broiler genotypes. In the middle of the last century, population genetics was introduced into commercial broiler breeding. Since then, performance improved substantially by mainly applying reciprocal recurrent selection (Table 1). As the life market weight remained rather constant, intensive selection resulted in a halving of fattening duration and in an improvement of feed conversion ratio by nearly 1 kg.

Distinct improvements in husbandry conditions (climatic control, hygienic status, etc.) and mainly in feeding supported this process. Anyway, nutrient requirements of broiler breeds in the 1950ies have been distinctly lower than today. First recommendations for supplying broilers with nutrients have been set up by NRC in the 1960ies and have been adapted several times, the 1994 edition of NRC being the latest one (NRC, 1994). The development of growth capacity and nutrient supply was investigated in detail by Gerald Havenstein, comparing the 1957 situation firstly with the situation in 1991 (Havenstein *et al.*, 1994) and secondly in 2001 (Havenstein *et al.*, 2003). From the 2003 publication is it clearly visible, that actually used contents of nutrients in compound feeds increased continuously. In this 44 years period, contents of energy, crude protein

Table 1. Development of fattening performance of broilers from the 1950ies until today. Results of station testing of broilers in Germany.

Year	Fattening duration (d)	Market life weight (g)	Feed conversion ratio (g/g)
1950	63	1,300	2.45
1970	56	1,450	2.25
1990	42	1,700	1.92
2010	32	1,600	1.60
2015	30	1,600	1.55

and essential amino acids like methionine + cysteine and lysine increased by 10.7, 8.0, 21.3 and 5.9%, respectively, in the starter diets (Havenstein *et al.*, 2003). The content of methionine increased most distinctly as this essential amino acid is the most important one in broilers, being involved in growth and maintenance. As the methionine content of most feeding stuffs is rather low and cannot meet requirements of broilers, diets are supplemented with free methionine. The development of the technology to produce huge amounts of free amino acids was one of the basic requirements to further improving growth performance of broiler breeds. But, within the last 20 years customers got more and more interested in extensively produced broiler meat as they believe that intensive broiler meat production with fast growing genotypes is in contradiction to animal welfare and results in impaired meat quality. Already in the 1970ies the French Label Rouge programme was launched which prescribed the use of 'slow growing' genotypes. This is also the main requirement of fattening organic broilers to allow for a prolongation of the fattening period to 81 days, at least. Meanwhile, in broiler meat production a shift towards label programs based on the 'extensive barn system' (EC 543/2008) is visible. Here, also slow growing genotypes have to be used. Up to now, neither a sound definition of growth capacity nor clear information on nutrient requirements for slow growing broiler genotypes is available.

Definition of growth

Information on the performance capacity of the different broiler genotypes is necessary for deriving nutrient requirements. Table 2 summarizes fattening results from different purebred, slow and fast growing broiler breeds. The different growth capacities indicate their different requirements, which are highest for the fast growing broilers and lowest for the purebreds. Results of a growth curve analysis including a bigger number of different broiler genotypes (Table 3) underline these differences. According to this analysis, it is not possible to distinguish just between fast and slow growth. It seems more appropriate to have three categories: slow – <35, medium – 35-50 and fast – >50 g average daily gain (ADG). This also indicates the necessity

Table 2. Fattening performance of different chicken purebreds and hybrids (Hörning et al., 2010).

Breed	Age at slaughter (d)	Market life weight (g)	Feed conversion ratio (g/g)
Cochin	2,242	119	2,97
Brahma	2,564	112	3.63
Kabir	2,633	74	2.58
Sasso	2,856	70	2.46
Olandia	2,465	67	2.44
Isa JA 757	2,656	57	2.14
Ross 308	2,794	43	1.64

Table 3. Growth characteristics of fast, medium and slow growing broiler breeds derived by Gompertz growth curve analysis Grashorn et al., 2012).

	Final asymp. body weight (g)	Age at max. growth (wks)	Body weight at max. growth (g)	Average daily gain at max. growth (g)
Fast	6,700	6.8	2,450	79
Medium	4,200	6.2	1,500	52
Slow	3,450	8.4	1,270	36

to have differentiated information on nutrient requirements of fast, medium and slow growing broiler genotypes, respectively.

Methods to deriving nutrient requirements

This paper focusses on methionine as it is first limiting amino acid in broilers. In monogastric animals requirements for methionine can only be derived in combination with the total supply of crude protein and energy. This is done by dose-response experiments. Here the amino acid of interest is supplemented to a defined basic diet, which should cover the primary nutrient requirements, in equal steps. The amount of the supplemented amino acid is plotted against the response (e.g. daily weight gain) and is evaluated by regression analysis. The peak of the curve indicates the maximum response and can be defined as the recommended supplementation level to reaching maximum performance. As this procedure is rather laborious, requirements are mainly determined for the four first-limiting amino acids – methionine, lysine, threonine, and tryptophan. Another approach is the factorial deduction of the amino acids requirement. Here the nutrient composition of the animal body is used as a basis (GfE, 1999). But, data on the nutrient composition of the broiler body are rare. Peter *et al.* (1998) found an average protein content in the fresh matter of the whole animal body of 186 g/kg. GfE (1999) report contents of methionine, lysine, threonine and tryptophan in the animal body without feathers (and feathers only) of 2.5 (0.6), 7.0 (1.8), 3.9 (4.8) and 1.0 (0.7) g/16 gN, respectively (GfE, 1999). For calculating the real requirements the requirement for maintenance (38, 41, 65 and 10 mg/kg life body mass) and the total conversion (0.70, 0.68, 0.65, 0.66) has to be considered. Further points of interest are the relation between amino acids, i.e. ideal protein, and the amino acids digestibility, which can be determined best in broilers as ileal digestibility (Rodehutscord *et al.*, 2004).

Recommended amino acids supply of fast growing genotypes

Based on the information given above, the requirements for amino acids can be calculated. During early life (1-21 days after hatching) this amounts to roughly 16.9, 6.2, 11.4 and 2.7 mg methionine, lysine, threonine and tryptophan/ g daily weight gain, respectively, without considering the gender (GfE, 1999). The amounts of methionine, threonine and tryptophan in relation to lysine should be 0.35, 0.67 and 0.16, respectively. The values reported by GfE (1999) for methionine, threonine and lysine are lower than the values given by NRC (1994) in the first weeks of life (Table 4).

Calculating the contents of methionine, lysine, threonine and tryptophan in diets by using the GfE values results in 11.0, 4.1, 5.1 and 1.2 g/kg, respectively. This is distinctly lower than the values given by the breeders. The recommended contents for methionine and lysine by GfE amount to about 75 and 85% of the contents recommended by Ross Breeders for the periods 1-10 and 11-24 days of life, respectively. The corresponding ratios for threonine and tryptophan are 53 and 58%, respectively. The higher values of Ross Breeders also consider the digestibility

Table 4. Recommendation for dietary contents of essential amino acids in broiler starter diets (1-21 days of life; final diets ME=13.0 MJ/kg) (NRC, 1994; GfE, 1999, Ross 308 Nutrition Specifications, 2014).

	NRC (1994) (g/MJ ME)	GfE (1999) (g/MJ ME)	Final diet (g/kg)	Ross 308	
				Day 1-10[a] (g/kg)	Day 11-24 (g/kg)
Lysine	0.82	0.85	11.0	14.4	12.9
Methionine	0.37	0.31	4.1	5.6	5.1
Threonine	0.60	0.57	5.1	9.7	8.8
Tryptophan	0.15	0.13	1.2	2.3	2.1

[a] ME=12.6 MJ/kg.

of amino acids, which was assumed to be about 90%. The corresponding contents of digestible lysine, methionine, threonine and tryptophan in the period 1-10 days of life are 12.8, 5.1, 8.6 and 2.0 g/kg diet, respectively. This underlines the assumption that amino acids requirements reported by NRC and GfE fit more to genotypes with medium or low growth capacity than for high performing genotypes.

Recommended amino acids supply of slower growing genotypes

In the previous paragraph, a distinct difference between the recommended amino acids supply of broilers within the first 21 days of life between NRC (1994), GfE (1999) and Ross Breeders was shown. Fatufe *et al.* (2004) used diets for cocks of a fast growing broiler genotype and of a layer breed (Lohmann Brown) with 3.8 to 16.8 g lysine/kg and found the best growth rate (days 8 to 21 of life) for the broilers with 12.5 and for the layers with 10.4 g lysine/kg. In a comparable study on threonine, Rosa *et al.* (2001) achieved the best body weight gain in broilers with 7.7 g threonine/kg diet, whereas, the highest weight gain in layers was already reached with 7.0 g. But, layers also grew well with only 6.3 g threonine/kg diet. Carrasco *et al.* (2014) achieved a sufficient growth rate of Isa J 257 broilers with 9.4 g lysine, 4.0 g methionine, 8.9 g threonine and 2.1 g tryptophan/kg diet (ME=11.9 MJ/kg). Ritteser (2016) used diets (ME=12.0 MJ/kg) with 10.6 g lysine, 4.0 g methionine, 6.9 g threonine and 1.6 g tryptophan for feeding Isa JA 957 broilers during the starter period (1-21 day of life) to determine ileal amino acids digestibility of organic feeding stuffs and observed the same body weight gain as for diets with higher contents of these amino acids.

Currently, no basic study exists where amino acids requirements have been determined for slow growing broilers. Therefore, requirements of broiler genotypes with lower growth capacity have to be derived from existing nutrition tables and breeder recommendations (Table 5). Here it is clearly visible that table values define the minimum supply with amino acids. According to Fatufe *et al.* (2004) rearing of layer pullets is possible with these rather low contents of amino acids. In a similar way, both Carrasco *et al.* (2014) and Ritteser (2016) achieved a good growth rate with also rather low amino acids contents with slower growing JA 257 and JA 757 broilers, respectively,

Table 5. Comparison of recommended amino acids contents in diets from nutrition tables, chicken breeders and of used contents in literature (starter period, 1-21 d of life).

	ME (MJ/kg)	Lysine (g/kg)	Methionine (g/kg)	Threonine (g/kg)	Tryptophan (g/kg)
Tables					
NRC (1994)	13.0	10.7	4.8	7.8	2.0
GfE (1999)	13.0	11.0	4.1	5.1	1.2
Fast growing broilers					
Aviagen Ross 308 Nutrition Specifications (2014)	12.6	14.4	5.6	9.7	2.3
Slow growing broilers					
Aviagen Rowan Ranger Management (2016)	12.6	14.0	5.4	9.5	2.2
Isa JA 757 (Ritteser, 2016)	12.0	10.6	4.0	6.9	1.6
Carrasco *et al.* (2014)	11.9	9.4	4.0	8.9	2.1
Layer breeds					
Lohmann Management Guide Alternative Systems (2016)	12.0	12.0	4.8	8.0	2.3
Fatufe *et al.* (2004)	13.7	10.4			
Rosa *et al.* (2010)	13.4			6.3	

although these broilers have to be assigned to the medium growth category. Here it is surprising to see that layer breeders recommend higher amino acids contents. Maybe, the higher safety margins are necessary to better control feather pecking. Another interesting point in the table is the high recommended amino acids contents for the slower growing genotype Rowan Ranger.

According to the management guide this breed cannot be considered as really slow growing as it reaches its market weight within 50 to 56 days. Thus, the conclusion of this table is, that breeders of fast and slower growing broilers as well as breeders of laying hens recommend a rather high amino acids supply, probably exceeding the real requirements, even when considering the fact that these values do not consider the real digestibility of amino acids. Depending on the choice and combination of feeding stuffs, digestibility of amino acids may be really variable. Furthermore, the energy content of the diet is also of importance as high energy diets also have to contain higher amounts of amino acids. Anyway, it can be expected that digestibility of all first limiting amino acids exceeds 80% in commercial compound feed. Based on the provided information it can be concluded that the knowledge on the amino acids requirements of broiler genotypes with a slower growth rate is still insufficient. This holds also for layer breeds, although the situation is there more complex due to the feather pecking problem.

Conclusions

Based on the current (limited) knowledge adapted amino acids contents can be suggested for the different genotypes as shown in Table 6. The recommended contents of Ross Breeders for fast growing broilers can be assumed as adequate, whereas, the contents for medium and slow growing genotypes have to be reduced by 10% each. These contents should be sufficient for an appropriate growth and reduce the waste of amino acids. In contrast, despite lower amino acids contents should be claimed for layer breeds the contents recommended by the breeder have been taken over to preventing problems with feather pecking. This deduction of recommended contents of amino acids for different broiler genotypes is only a very rough approach, leaving the necessity to conduct experiment to close this gap.

Table 6. Estimated adapted amino acids contents in diets for fast, medium and slow growing broiler genotypes as well as for layer hybrids (g/kg).

	ME (MJ/kg)	Lysine	Methionine	Threonine	Tryptophan
Broiler					
Fast growth	12.5	14.0-14.5	5.5-6.0	9.0-9.5	2.0-2.3
Medium growth	12.5	12.6-13.1	5.0-5.4	8.1-8.6	1.8-2.1
Slow growth	12.0	11.3-11.8	4.5-4.9	7.2-7.7	1.6-1.8
Layer					
Adjusted	12.0	12.0	5.0	8.0	2.0

Plenary session 03 and 04. Precision feeding

References

Carrasco, S., Bellof, G. and Schmidt, E. 2014. *Livestock Science* 161: 114-122.

Fatufe, A.A., Timmler, R. and Rodehutscord, M. 2004. *Poultry Science* 83: 1314-1324.

GfE, 1999. Empfehlungen zur Energie- und Nährstoffversorgung der Legehennen und Masthühner (Broiler). DLG-Verlag, Frankfurt a.M., Germany.

Grashorn, M.A., Lorenz, C. and Schmidt, E. 2012. An attempt to defining slow growth in broiler chicken. 24th World's Poultry Congress, Salvador de Bahia, Brazil, 5-9 August.

Havenstein, G.B., Ferket, P.R., Scheideler, S.E. and Larson, B.T. 1994. *Poultry Science* 73: 1785-1794.

Havenstein, G.B., Ferket, P.R. and Qureshi, M.A. 2003. *Poultry Science* 82: 1500-1508.

Hörning, B., Trei, G., Ludwig, A. and Rolle, E. 2010. Eignung unterschiedlicher Herkünfte für die ökologische Haltung von Masthähnchen. BÖLN 06OE217, Germany, Project Report.

NRC. 1994. Nutrient requirements of poultry. 9th edition, National Academy Press, Washington D.C., USA.

Peter, W., Dänicke, S. and Jeroch, H., 1998. *Archiv Geflügelkunde* 62: 132-140.

Ritteser, C., 2016. Bestimmung der präcecalen Verdaulichkeistkoeffizienten für heimische Energie- und Proteinfuttermittel für die Bio-Hühnermast. Doctoral Thesis Sc. Agr. University of Hohenheim, Stuttgart, Germany.

Rodehutscord, M., Kapocius, M., Timmler, R. and Diekmann, A. 2004. *British Poultry Science* 45: 85-92.

Rosa, A.P., Pesti, G.M., Edwards Jr., H.M. and Bakalli, R.I. 2001. *Poultry Science* 80: 1710-1717.

Plenary session 05. Optimized use of feed ingredients

The next big steps for feed enzymes

M. Choct
University of New England, Armidale, NSW 2351, Australia; mchoct@une.edu.au

Summary

This brief review aims to bring about discussion regarding the opportunities and challenges facing xylanase application in poultry diets. These include but are not necessarily limited to: (1) relating enzyme activity to bird performance; (2) elucidating the roles of xylanase hydrolysis by-products in gut health and nutrition; and (3) manipulating the anatomical and physiological processes of birds to enhance the efficacy of existing enzymes. In addition, some speculations are made about finding novel uses of proteases such as deactivation of anti-nutrients in feed. More research on substrate characterisation will underline future advances in enzyme application.

Introduction

Enzymes have become an essential additive in monogastric animal feed over the past three decades. Commonly used enzymes are phytase and glycanases. In recent years, the use of proteases has also increased. Phytase application is widespread with its mechanism of action well understood and practical benefits widely accepted. Thus, this paper will focus on novel ways of applying xylanases and proteases in poultry diets.

Xylanases and their substrates

The substrates for xylanases are xylans, which is the second most abundant group of polysaccharides in nature. In a feed context, it accounts for 50-70% of the non-starch polysaccharides (NSP) present in cereal grains.

Xylans refer to a large number of polysaccharides that have a backbone of β-1,4 linked xylose residues. In cereal grains, they are often known as arabinoxylans or pentosans, which have varying proportions of the xylan backbone substituted with α-L-arabinofuranose at O-2 and O-3 positions. In contrast, xylans from woods have their β-D-xylopyronose backbone units substituted at C-2 positions with 1,2-linked 4-O-methyl-β-D-glucuronic acid residues.

There is a wide range of variation in xylan structures and functions. The simple differences are: (1) molecular weight and arabinose to xylose ratio, with the latter indicating the variation in the degree of substitution by arabinose side chains; and (2) solubility and viscosity, where a xylan must be soluble to be viscous, but not necessarily all soluble xylans are viscous. To expand on this last point, arabinoxylans from wheat and rye are viscous (Bedford and Classen, 1992), but those from rice are not (Smits and Annison, 1996).

There are many complex variations among xylans. They may have different side chains with respect to sugar composition and linkage types, they can be partially methylated or acetylated or esterified, and they can be crossed linked with other cell wall constituents. Voragen *et al.* (1992) classified xylans into the following four families, on the basis that they:
1. only have side chains of single terminal units of α-L-arabinofuranosyl substituents (as described above for arabinoxylans in cereal grains);
2. only have α-D-glucuronic acid or its 4-O-methyl ether derivative as substituents;
3. have both α-D-glucuronic (and 4-O-methyl-α-Dglucuronic) acid and α-L-arabinose as substitutes, and
4. have terminal β-D-galactopyranosyl residues on complex oligosaccharide side chains. Such xylans are typically found in perennial plants.

These variations mean that they require different xylanases to unlock them. Xylanases are classified into glycoside hydrolases (GH) families based on the similarities of their amino acid sequences (Coutinho and Henrissat, 1999). As of January 2017, there are at least 141 GH families registered at the CAZY site (https://www.cazypedia.org/index.php/Glycoside_Hydrolase_Families). Commercial exploration for xylanases has focussed primarily on the GH10 and 11 endoxylanases (β-1,4-endoxylanase; EC 3.2.1.8), although strong xylanase activity has also been characterised in GH5, 7, 8 and 43 (Collins *et al.*, 2005). The both GH 10 and 11 xylanases act on the glycosidic bonds in the middle of the xylan backbone. Most commercial xylanases belong to this group and they reduce viscosity, or break the cell wall architecture, or both. However, not all endo-xylanases act the same way. The majority of GH 11 xylanases only attack the xylan backbone at the uninterrupted sequences (Gruppen *et al.*, 1993) whereas GH 10 xylanases are more versatile and can cleave the links adjacent to side chains (Biely *et al.*, 1997). Furthermore, there are endoxylanases that have affinity for either soluble or insoluble xylans (Moers *et al.* 2005). There are other xylan-degrading enzymes, namely β-1,4-xylosidases (EC 3.2.1.37) and exoxylanases, which remove successive D-xylose residues from the non-reducing termini of xylans or of xylo-oligomers. This group of xylanases is not commonly used in the feed industry.

Indeed, the multiplicity and heterogeneity of both the enzyme (xylanase) and the substrate (xylan) pose the biggest challenge for the next breakthrough for the feed enzyme industry.

The current challenges for xylanases

Tremendous advances in the application of xylanases have occurred over the past thirty years, including yield, stability and activity. However, there remain significant opportunities and challenges for the feed xylanase industry. The following are a few thoughts that reiterate areas of recent focus in xylanase research.

Relationship between xylanase activity and bird performance

Rightly or wrongly some researchers and end-users place significant importance on enzyme activity. This is understandable because the first step in knowing whether or not a product will be beneficial to bird performance is to know if the enzyme is active. There are two primary methods for measuring xylanase activity, the reducing sugar method and the dye-release assay. The 'reducing sugar method' actually refers to two assays, the 3,5 dinitorsalicylic acid (DNS) reducing sugar method (Sumner, 1925; Miller, 1959), and the Nelson-Somogyi (NS) method (Nelson, 1944; Somogyi, 1952). The dye-release assay is the chromogenic tablet technique (McCleary, 1992).

There has been much validation of, and debate about, these methods (Bailey *et al.*, 1992; McCleary and McGeough, 2015). But none of these methods would mean much if anyone wanted to use enzyme activity as a measure of nutrient release or bird performance (Ravindran, 2013) because it is not the case that the higher the enzyme activity the better the bird performance. The issues are related to the difficulty in: (1) understanding the substrate and its effect on bird performance; (2) quantifying the key mechanism whereby the substrate affects bird performance; and, not least, (3) testing the effect of enzymes under standard dietary and husbandry conditions. Thus, a practically relevant enzyme activity assay will have to await more work on substrate characterisation.

Fine-tuning of xylanase hydrolysis products

The modes of action of xylanases in terms of substrate specificity, substrate affinity, and the types of hydrolysis end-products they release *in situ* are not fully understood. However, it is known that different xylanases release different amounts and types of carbohydrate moieties, including monomers, oligomers and other low-molecular weight xylans. The roles of the hydrolysis products, xylo-oligosaccharides (XOS) or arabnoxylo-oligosaccharides (AXOS), in humans (Childs *et al.*, 2014; Lin *et al.*, 2016) and in poultry (Pourabedin *et al.*, 2015) have become an interesting area of development. Indeed, XOS seem to be unique prebiotics. Mäkeläinen *et al.* (2010a,b) reported that XOS were fermented with high specificity by strains of Bifidobacteria. Furthermore, the prebiotic effects of XOS include optimizing colon function, increasing or changing composition of short chain fatty acids (SCFAs), increasing mineral absorption, immune stimulation, and increased ileal villus length (Kim *et al.*, 2011). The use of XOS in poultry diets appears to be particularly beneficial for butyrate-producing gut microflora (De Maesschalck *et al.*, 2015). These authors demonstrated improved bird performance, which may have resulted from the availability of more fuels (butyrate) for epithelial cells and hence intestinal epithelial integrity. De Maesschalck *et al.* (2015) used a test product that contained 35% XOS, with chain lengths ranging from 2-7, and 65% of maltodextrin. It will be important to conduct studies that examine individual XOS molecules in a more pure form in the future.

Most commercial xylanases can produce a range of low-molecular weight xylan fragments, including xylobiose, xylotriose, xylotetrose and numerous other moieties (Morgan *et al.*, 2017). The types of XOS produced depend on the substrate, the xylanase, and the gut environment of the animal. Thus, there is an impetus to examine various xylanases from: (1) affinity and specificity of soluble and insoluble substrates and the amounts and types of low-molecular weight xylans produced; (2) side-chain cleaving characteristics; and (3) the most effective site of XOS release in the gastrointestinal tract (GIT) of animals. Using insoluble xylans as raw materials for *in situ* production of XOS will require a major shift in thinking when it comes to xylanase production. This is because, traditionally, xylanases are provided primarily to depolymerise soluble xylans to reduce the negative effects of digesta viscosity on nutrient digestion and absorption. There has been much debate about the 'nutrient encapsulation' effect of cell wall NSP and the benefit of breaking down cell wall architecture using enzymes. However, the mechanisms for finding the types of xylanases required to do this, how they work in concert with other enzymes, and to what extent the xylans are broken down are not yet fully elucidated. With the roles of XOS becoming more prominent in the beneficial action of xylanases, a greater attention should perhaps be given to the production of XOS when using xylanases in monogastric animal diets.

Enhancing xylanase efficacy by changing feeding practices

In intensive poultry production, well-processed feed is offered to birds on a continuous basis. This practice does not promote a long retention time of food in the crop or the gizzard (Classen *et al.*, 2016). However, crop and gizzard holding can be manipulated through management and feeding practices (Moen *et al.*, 2012; Classen *et al.*, 2016). Rodrigues *et al.* (2017) conducted an experiment where broilers were fed diets intermittently with or without enzymes. Intermittent feeding resulted in more digesta being retained in the crop and gizzard for a substantial period of time. With this increased preconditioning and enhanced mechanical grinding in the foregut, the enzymes were more effective in improving bird performance. The implications of this initial study are that there is scope to change feeding practices to add value to existing enzymes. In the meantime, it also brings up new questions regarding how enzymes are selected in terms of pH optima and affinity for the secondary substrates released earlier in the GIT because the pH in the chicken gut ranges from high acidic in the foregut to near neutral in the hindgut.

Proteases

This review will only touch upon a couple of questions regarding the use of proteases in feed. Proteases have a widespread industrial application, including their use as a feed additive. Unlike carbohydrases, proteases have a common bond, i.e. a peptide bond, to cleave. However, like other enzymes, proteases have a degree of specificity. For instance, while aminopeptidases liberate one amino acid or a small peptide at a time by acting on the N termini of proteins, carboxypeptidases do the same thing by acting on the C termini; likewise, endopeptidases cleave peptide bonds in the middle of a protein molecule, but serine proteases will hydrolyse a peptide bond that has either tyrosine, phenylalanine or leucine. Thus, there is a great deal of scope to further explore the use of proteases in feed. A couple of thoughts are outlined below for expanding use of proteases in feed:

Plenary session 05. Optimized use of feed ingredients

1. More characterisation of substrates: It is essential to identify and characterise feed proteins to match proteases that will have real impact.
2. Deactivation of anti-nutrients: many anti-nutrients present in feed ingredients are proteins. Proteases targeting some anti-nutrients may be explored, making it possible to find a more natural or cost-effective solution to address the problems associated with anti-nutrients.

Conclusions

Great advances have been made in both technology and application of enzymes over the past thirty years. Now there is an increased demand for substrate characterisation, for rapid and accurate measurements of functionality and efficacy of enzymes, and for new application of existing enzymes.

References

Bailey, M.J., Biely, P. and Poutanen, K. 1992. *Journal of Biotechnology* 23: 257-270.
Bedford, M.R. and Classen, H.L. 1992. *Journal of Nutrition* 122: 560-569.
Biely, P., Vrsanska, M., Tenkanen, M. and Kluepfel, D. 1997. *Journal of Biotechnology* 57: 151-166.
Childs, C.E., Roytio, H., Alhoniemi, E., Fekete, A.A., Forssten, S.D., Hudjec, N., Lim, Y.N., Steger, C.J., Yaqoob, P., Tuohy, K.M., Rastall, R.A., Ouwehand, A.C. and Gibson, G.R. 2014. *British Journal of Nutrition* 111: 1945-1956.
Classen, H.L., Apajalahti, J., Svihus, B. and Choct, M. 2016. *Worlds Poultry Science Journal* 72: 459-472.
Collins, T., Gerday, C. and Feller, G. 2005. *FEMS Microbiology Reviews* 29: 3-23.
Coutinho, P.M. and Henrissat, B. 1999. Carbohydrate-active enzymes: an integrated database approach. In: Gilber HJ, Davies GJ, Henrissat B, Svensson B (Eds), Recent Advances in Carbohydrate Bioengineering, The Royal Society, Cambridge, pp 3-12.
De Maesschalck, C., Eeckhaut, V., Maertens, L., De Lange, L., Marchal, L., Nezer, C., De Baere, S., Croubels, S., Daube, G., Dewulf, J., Haesebrouck, F., Ducatelle, R., Taminau, B. and Van Immerseel, F. 2015. *Applied Environmental Microbiology* 81: 5880-5888.
Gruppen, H., Kormelink, F.J.M., Voragen, A.G.J. 1993. In: Wenk C, Boessinger M (Eds), Proceedings of the 1st Symposium on Enzymes in Animal Nutrition, Kartause Ittingen, Switzerland, pp 276-280.
Kim, G.B, Seo, Y.M., Kim, C.H. and Paik, I.K. 2011. *Poultry Science* 90: 75-82.
Lin, S.H., Chou, L.M., Chien, Y.W., Chang, J.S. and Lin C.I. 2016. Gastroenterology Research Practice. doi:10.1155/2016/5789232
Mäkeläinen, H., Forssten, S., Saarinen, M., Stowell, J., Rautonen, N. and Ouwehand, A.C. 2010a. *Beneficial Microbes* 1: 81-91.
Mäkeläinen, H., Saarinen, M., Stowell, J., Rautonen, N. and Ouwehand, A.C. 2010b. *Beneficial Microbes* 1: 139-148.
McCleary, B.V. 1992. Measurement of endo-1,4-β-xylanase. In: Visser J, Beldman G, Kusters-van Someren MA, Voragen AGJ (Eds), Xylans and xylanases. Pp 161-169. Amsterdam: Elsevier Science.
McCleary, B.V. and McGeough, P. 2015. *Applied Biochemestry and Biotechnology* 177: 1152-1163.
Miller, G.L. 1959. *Analytical Chemistry* 31: 426-428.
Moen, B., Rudi, K., Svihus, B. and Skanseng B. 2012. *Journal of Applied Microbiology* 113: 1176-1183.
Moers, K., Celus, I., Brijs, K., Courtin, C.M. and Delcour, J.A. 2005. *Carbohydrate Research* 340: 1319-1327.
Morgan, N.K., Bedford, M.R. and Choct, M. 2017. *Carbohydrate Polymers* (in press)
Nelson, N. 1944. *Journal of Biological Chemistry* 153: 375-380.
Pourabedin, M., Guan, L. and Zhao, X. 2015. *Microbiome* 3: 15.
Ravindran, V. 2013. *Journal of Applied Poultry Research* 22: 628-636.
Rodrigues, I., Toghyani, M., Svihus, B., Bedford, M.R., Gous, R.M. and Choct, M. 2017. In: European Symposium on Poultry Nutrition, Salou, Spain.
Smits, C.H. and Annison, G. 1996. *World's Poultry Science Journal* 52: 203-221.
Somogyi, M.J. 1952. *Journal of Biolological Chemistry* 195: 19-23.
Sumner, J.B. 1925. *Journal of Biological Chemistry* 65: 393-395.

Voragen, A.G.J., Gruppen, H., Verbruggen, M.A. and Vietor, R.J. 1992. Characterization of cereals arabinoxylans. In: Xylan and Xylanases. J Visser, G Beldman, MA Kuster-van Someren and AGJ Voragen (Eds). Elsevier, Amsterdam. In: Visser J, Beldman G, Kusters-van Someren MA, Voragen AGJ (Eds), Xylans and xylanases, pp 51-67. Amsterdam: Elsevier Science.

Using science and business for a better soybean meal in poultry nutrition

M. Sifri
Sifri Solutions LLC, P.O. Box 5291, Quincy, IL 62305, USA; mamduhsifri@gmail.com

Summary

Understanding the science and business about soybean meal is critical to achieve the goal of better soybean meal in poultry nutrition. It requires knowing the facts and acting upon them. These include the use of all pertinent compositional and biological attributes in assessing soybean meal.

Introduction

To achieve a better soybean meal for poultry, it is essential to understand the science and business associated with it (Bajjalieh, 2012). This write-up is an outline of the most critical attributes of soybean meal and how to quantitatively assess its relative value for poultry nutrition.

Definitions

The National Oilseed Processors Association (http://www.nopa.org) has provided the official specifications for soybean meal. There are primarily two soybean meal specifications; 44% or 47.5-49.0% protein products. These meals are produced by cracking, heating and flaking dehulled soybeans and reducing the oil content of the conditioned flakes using hexane or homologous hydrocarbon solvents. The extracted flakes are cooked and sold as such or as a ground soybean meal.

The business of soybean meal

The estimated world production of soybeans in2016 is 333.41 million metric tons. Parallel to that production is the estimated world output of 225.50 million metric tons of soybean meal (www.soymeal.org/infosource/august16/index.html). The major producers of soybeans are USA, Argentina and Brazil; however, the major users of soybean meal are USA, China, European Union and Brazil. Depending on the exchange rate, that translates into an economic value of $ 120 billion, annually.

The science of soybean meal

There is a plethora of scientific knowledge and application of soybean meal in poultry nutrition. The references and websites listed in this article represent a fraction of the available literature.

Historical background

The soybean meal industry has effectively promoted the product domestically and internationally. Concurrent with all marketing and promotional efforts, there is still a definite need for further improvements in evaluating soybean meal for its use in poultry feed. It is not surprising that such a goal is not realized due to the emphasis on price.

Country of origin of soybean meal

Country of origin of soybean meal is of great interest for business and science since it has an impact on its composition. The reasons for differences in soybean meal may include the meal origin within a country (Thakur and Hurburgh, 2007), the seed (Wilcox and Shibles, 2001), the environmental temperatures (Wolf *et al.*, 1982) and the countries of origin such as Brazil, China and USA (Grieshop and Fahey, 2001). It is typical that country of origin has been promoted to help in differentiating the pertinent attributes. Thus, many international surveys were conducted to elucidate the differences by de Costa-Sinova *et al.* (2008), Frikha *et al.* (2012) and Garcia-Rebollar *et al.* (2016). The surveys demonstrated that nutrient composition and digestibility may vary with soybean meal origin and were consistent with higher protein content and lower trypsin inhibitor activity (TIA) as reported by de Costa-Sinova *et al.*, (2008). Such efforts were extended to assess the relationship between digestibility, crude protein content, KOH solubility, reactive lysine, neutral detergent fiber (NDF) and oligosaccharide (Frikha *et al.*, 2012). From a business point of view, these extensive studies led to the judgement that soybean meals from USA and Brazil were processed under better conditions than that from Argentina. A more comprehensive assessment of soybean meal origin as it relates to the multitude of nutrients was reported by Garcia-Rebollar *et al.* (2016). This clearly illustrates the importance soybean meal country of origin in any purchasing decision.

Soybean meal quality

The subject of soybean meal quality is intertwined with most of the aspects covered in this manuscript. During the last century, tremendous progress has been made by the soybean meal industry, research institutions, universities and individual scientists, worldwide. Those achievements led to the establishment of quantitative techniques and measurements to help the soybean meal industry and the poultry industry. It is evident that using only compositional attribute might not lead to a conclusive assessment; however, utilizing a few of them or all of them may lead to a quantitative decision. It is prudent to list the major soybean meal attributes and comment on them briefly. Some of the measurements are more closely associated with the country of origin. Historically, the soybean meal industry and the poultry industry considered; primarily, dry matter (DM), ash, crude protein (CP), ether extract (EE) for fat content and crude fiber (CF) as the basis for the soybean meal price. It was also recognized that measurements of sucrose, stachyose, raffinose, amino acids, minerals, and neutral detergent fiber improved the assessment of soybean meal. Eventhough these measurements were chemically quantitative, they were not totally reflective for the meal biological and performance values. Other measurements gained tremendous attention because of their practical applications. Such measurements included urease activity (UA), protein dispersibility index (PDI), potassium hydroxide protein solubility (KOH Protein Solubility), trypsin inhibitor activity (TIA) and heat damage indicator. All these parameters were addressed comprehensively by Garcia-Rebollar (2016). Some of the pioneering efforts that paved the way for better understanding of these measurements include Araba and Dale (1990), Balloun (1980), Choct *et al.* (2010), Evonik (2010), Lee *et al.* (1991), Lee *et al.* (2004), Parsons *et al.* (1991) and Van Eys (2012).

Impact of genetics on soybean meal

Due to the direct and indirect effect of genetics on soybean meal including the use of genetically modified organisms (GMO), it is pertinent to consider any changes that might affect the soybean meal value. Even though the genetic component of soybean meal is of great importance, it is not within the scope of this manuscript to address it in detail.

Impact of other modifiers on soybean meal

The impact of components that are not part of the inherent composition of the soybean meal should not be ignored. They include alternative feed additives (modifiers) such as enzymes, prebiotics, probiotics, short and medium chain fatty acids, plant extracts, yeast cell components or antibodies. Their impact on the soybean meal value for poultry nutrition can be critical. Managing the impact of such additives on soybean meal brings tremendous challenges. Using this approach mirrors what is happening in the industry where such components are used at varying degrees and combinations in the poultry industry. Recently, Sifri (2016a,b,c) outlined the impact of such modifiers and their interaction with nutrients which may result in changes of soybean meal value. Valuable reviews have been published to address this issue (Adeola and Cowieson, 2011; Bedford and Partridge, 2001; Choct, 1997; Ravindran, 2013).

Plenary session 05. Optimized use of feed ingredients

It is unfortunate that the use of modifiers in poultry nutrition is often associated with numerous limitations. Some of those limitations come from unrealistic and unsubstantiated claims for the contributions of such products. Even though there are numerous publications about the impact of feed enzymes on soybean meal nutrition, there is still a serious void in assessing their values properly. The exception to these concerns is the use of phytase in poultry nutrition. The impact of phytase on the release of phytate phosphorus and other nutrients in soybean meal is well recognized. The presence and availability of many sources of phytase to the poultry industry continues to be a great opportunity; however, it continues to bring new confusions related to their contributions. It is incumbent upon the enzyme industry to provide credible documentations about the specificity and contributions of their enzymes. The use of non-starch polysaccharide (NSP) enzymes in poultry feeds containing soybean meal face more challenges. This is primarily because there are many different enzymes and different substrates. As it pertains to the use of feed enzymes, it is of interest to recognize the potential role of protease enzymes on the value of soybean meal. The judgement on protease is mixed and the application appears to be limited to specific situations (Douglas *et al.*, 2000; Fritas *et al.*, 2011; Ghazi *et al.*, 2002; Simbaya *et al.*, 1996). Regardless of the outcome, feed enzymes have the potential to help in improving soybean meal nutritional value for poultry.

Impact of other processes on soybean meal

Physical manipulations of soybean meal may hold great potential in improving the nutritional value for poultry nutrition. The resulting changes in soybean meal must be factored in when assessing its nutritional value.

Enzyme hydrolysis

Examples of these processes include enzymatically hydrolyzing raffinose and stachyose in soybean meal (Graham *et al.*, 2002) or by using varieties that are inherently low in oligosaccharides (Baker *et al.*, 2011) which resulted in better chick and broiler performance.

Particle size

Varying the particle size of soybean meal has also gained popularity where larger particle size was better utilized than smaller particles in broilers (Kilburn and Edwards, 2004). However, Pacheco *et al.*, 2013 concluded that soybean meal particle size over 1,300 micrometers depressed body weight but improved protein digestibility in broilers.

Feed form

Studies on feed form by Serrano *et al.*, 2012 documented that the feed form and soybean meal source are related where crumbling and pelleting of higher soybean meal protein from the USA lead to better broiler performance.

Elusieve process

A more in-depth study was conducted by Srinivasan *et al.*, 2013 using the Elusieve process which is a combination of sieving and eutriation (air classification technique). This classification has been successful in separating fiber from soybean meal and other products such as ground corn and distillers dried grains with solubles (DDGS). The study demonstrated that the Elusieve process of soybean meal lead to improved broiler performance.

Plenary session 05. Optimized use of feed ingredients

Processed full fat soybean

Comparisons between full fat soybeans with soybean meal lead to the conclusion that both full fat processed soybeans and soybean meal may be used successfully when formulated based on their respective nutritional contributions (Hamilton and Niven, 2000).

Soy protein concentrate and isolate

Comparisons between soybean meal, soy protein concentrate and soy protein isolates (Jankowski *et al.*, 2009) demonstrated that caecal fermentation in young turkeys and body weight were reduced; however, feed utilization was improved in association with the gradual reduction of oligosaccharides in soy protein concentrate, soy protein isolate.

Conclusions

To achieve a better soybean meal in poultry nutrition, it requires the use of science in the evaluation.

Acknowledgements

Great appreciation goes to Dr. G.G. Mateos and all his colleagues whose names are listed in the references of this manuscript. Their contributions were invaluable in preparing this manuscript.

References

Adeola, O. and A. J. Cowieson. 2011. *Journal of Animal Science* 89: 3189-3218.

Bajjalieh, N. 2012. *Feedstuffs* 84 (22), May 28.

Baker, K.M., Utterback, P.L., Parsons, C.M. and Stein, H.H. 2011. *Poultry* Science 90: 390-395.

Balloun, S.L. 1980. Soybean Meal in Poultry Nutrition. Ovid Bell Press, Fulton, MO.

Bedford, M. R. and A. J. Cowieson. 2012. *Animal Feed Science and Technology* 73: 76-85.

Bedford, M. R. and G. G. Partridge. 2001. Enzymes in Farm animal nutrition. CABI Publishing, CAB International, Wallingford, Oxon, OX10 8DE, UK.

Choct, M. 1997. Feed Milling International. June Issue, pp 13-26.

De Coca-Sinova, A., Valencia, D.G., Jiménez-Moreno, E., Lázaro, R. and Mateos, G.G. 2008. *Poultry* Science 87:2613-2623.

Douglas, M. W., Parsons C. M. and M. R. Bedford. 2000. *Journal of Applied Poultry Research* 9: 74-80.

Evonik, 2010. Special Edition. Analytic: AminoRed. AminoNews. Evonik-Degussa GmbH, Hanau-Wolfgang, Germany.

Freitas, D.M., Vieira, S.L., Angel, C.R., Favero, A. and Maiorka, A. 2011. *Journal of Applied Poultry Research* 20: 322-334.

Frikha, M., Serrano, M.P., Valencia, D.G., Rebollar, P.G., Fickler, J. and Mateos, G.G. 2012. *Animal Feed Science and Technology* 178: 103-114.

Garcia-Rebollar, P., Camara, L., Lazaro R.P., Dapoza, C., Perez-Maldonado, R. and Mateos, G.G. 2016. *Animal Feed Science and Technology* 221: 245-261

Ghazi, S., Rooke J. A., Galbraith H. and Bedford M. R. 2002. *British Poultry* Science 43: 70-77.

Graham, K.K., Kerley, M.S. and Firman, G.L. 2002. *Poultry Science* 81: 1014-1019.

Grieshop, C.M., Kadzere, C.T., Clapper, G.M., Flickinger, E.A., Bauer, L.L., Frazier, R.L. and Fahey, G.C., 2003. *Journal of Agricultural Food Chemistry* 51: 7684-7691.

Hamilton, R.M.G. and Niven, M.A. 2000. *Canadian Journal of Animal Science* 80: 483-488.

Jankowski, J., Juskiewicz, J., Gulewicz, K., Lecewicz, A., Slominski, B.A. and Zdunczyk, Z. 2009. *Poultry* Science 88: 2132-2140.

Kilburn, J. and Edwards, H.M. 2004. *Poultry* Science 83: 428-432.

Pacheco, W.J., Stark, C.R., Ferket, P.R. and Brake, J. 2013. *Poultry* Science 92:2914-2922.

Parsons, C.M., Hashimoto, K., Wedekind, K.J. and Baker, D.H., 1991. *Journal of Animal Science* 69: 2918-2924.

Plenary session 05. Optimized use of feed ingredients

Ravindran, V. 2013. *Journal of Applied Poultry Research* 22: 628-636.

Serrano, M.P., Valencia, J., Mendez J. and Mateos G.G. 2012. *Poultry* Science 91: 2838-2844.

Sifri, M. 2016a. Measurements, proofs for modifier's economic efficacy in poultry feeds. Poultry Service Industry Workshop (PSIW), Banff, Alberta, Canada.

Sifri, M. 2016b. The science and application of non-starch polysaccharides (NSP) enzymes for poultry and pig nutrition 'what are the facts'. 77th Minnesota Nutrition Conference, Prior Lake, Minnesota, USA.

Sifri, M. 2016c. Interactions of nutrients and feed additives and their impact on gut health. 5th Mediterranean Poultry Summit (MPS 5) of WPSA, Italy, Spain, France.

Simbaya, J., Slomminski, B. A. Guenter W., Morgan A. and Campbell L. D. 1996. *Animal Feed Science and Technology* 61: 219-234.

Srinivasan, R., Lumpkins, B., Kim, E., Fuller, L. and Jordan, J. 2013. *Journal of Applied Poultry Research* 22: 177-189.

Thakur, M. and Hurburgh, C. 2007. *Journal of American Oil Chemists Society* 84: 835-843.

Van Eys, J.E., 2012. Manual of Quality Analyses for Soybean Products in the Feed Industry, 2nd ed. USSEC, Chesterfield, MO.

Wilcox, J.R. and Shibles, R.M. 2001. *Crop Science* 41: 11-14.

Wolf, R., Cavins, J., Kleiman, R. and Black, L. 1982. *Journal of American Oil Chemists Society* 59: 230-232.

Selected resources and websites for soybean meal

www.feedipedia.org
www.soymeal.org
www.nopa.org
www.ussec.org
https://unitedsoybean.org
www.soyconnection.com
https://soygrowers.com

Plenary session 05. Optimized use of feed ingredients

Quality control and nutritional value of fats

R. Codony[1]*, F. Guardiola[1], A. Tres[1] and A.C. Barroeta[2]
[1]Nutrition, Food Science and Gastronomy Department, INSA, Universitat de Barcelona, Faculty of Pharmacy, Av. Joan XXIII s/n, 08028 Barcelona, Spain; [2]Animal Nutrition and Welfare Service (SNiBA) Universitat Autònoma de Barcelona, Veterinary School, 08193 Bellaterra, Spain; rafaelcodony@gmail.com

Summary

During the last four decades, chemical control of oil and fat quality has largely evolved with powerful methodologies available for the assessment of different parameters. However, this fact poses the question of which ones are really significant for a basic routine control and which are only complementary. Different aspects should be taken into account in order to approximate the energetic value of fats and oils. During the last two decades, several prediction equations have been proposed to calculate this, but they do not fit well with *metabolizable energy* observed in broilers when non-conventional fats or blends are included in the diet. Moreover, control of fat repercussions in the sensory properties and in meat fatty acid composition must also be considered. A first gap in the analytical control of feed fats and oils is the need of a better definition of the significance of the values corresponding to the main parameters currently used. And the second would be to propose a new methodology for a more practical, quick and simple assessment of quality parameters (i.e. NIR spectrophotometry for a quality multiparametric determination). This paper presents the current knowledge of quality control of dietary oils and fats, which are the main questions to solve, and what could be the future for more reliable and simplified analytical control schemes. Our aim is to follow a practical point of view, but giving the best scientific support for the different analytical applications proposed.

Fat sources for feed uses and main quality control parameters

Fat sources for broiler and laying-hen diets and their chemical composition

Several types of fat materials can be used as feed ingredients in order to enhance the energetic value and provide some specific and essential nutrients for poultry. Several types of fat materials can be used for feed manufacturing, with the most frequently used at this moment being vegetable oils (mainly palm, sunflower, rapeseed and soybean), but also animal fats (mainly lard and avian fat), and *acid oils from chemical refining/fatty acid distillates from physical refining* are also used (Mateos *et al.* 1996; Nuchi *et al.* 2009; Ravindran *et al.* 2016). For the purpose of this text, we use the generic term 'feed fats' for any of these materials. They are mainly selected according to different criteria, such as nutritional value for the animal (at a particular age), cost, market availability, chemical composition, stability and physical properties. Major attention is paid to composition differences among these types of fat materials, mainly in fatty acid (FA) compostion. But other particular aspects, such as % free fatty acids (FFA or Acidity); peroxide value (PV); secondary oxidation (TBA or *p*-anisidine (PAV) values); and impurities+ moisture+ non-saponifiable (MIU) must also be considered, since they can affect certain quality aspects. Fats and oils are mainly constituted by triacylglycerols (TAG), which are esters of fatty acids and glycerol. Other glyceridic components can be present, but in much lower proportions (phospholipids and glycolipids). The non-saponifiable fraction (around 2%-5% in conventional oils and fats) can contain more than 100 different components in variable amounts, being sterols the major portion. Another important group of components of the unsaponifiable matter is known as vitamin E, including α, β, γ, and δ-tocopherols and -tocotrienols, which have relevant nutritive and antioxidant properties. On the other hand, the typical composition of *acid oils from chemical refining* and *FA distillates from physical refining* is quite different than that of fats and oils, with less % TAG, relevant % of mono- and di-acylglycerols (MAG, DAG), 40%-90% FFA, and higher values of the %non-saponifiable fraction. Feed fats can undergo different chemical degradations during processing, handling and storage, such as hydrolysis, oxidation, polymerization and isomerization, whose products can decrease fat nutritional value and give rise to non-desirable compounds (Nuchi *et al.* 2009). The main process negatively affecting fat and feed quality is lipid oxidation, which yields both primary and secondary oxidation products, whose structures and amounts depend on the oxidation conditions to which they are subjected (time, temperature, oxygen concentration, and presence of substances with pro-oxidant and antioxidant effects).

Repercussions of dietary fats' composition on nutritional value

The nutritional value of fats is mainly related to their high energetic value in comparison with other feed ingredients. This depends on several characteristics of fat composition, and also on the physiological status of the animal (Ravindran *et al.* 2016). However, this value is variable according to two basic factors. First, their proportion of FA (supplying the major part of fat energy) and second, their digestibility. A lot of studies have reported results on animal trials in order to establish the influence of the chemical composition of feed fat on digestibility in chickens. In monogastric animals, and according to Mateos *et al.* (1996), four main factors can be defined as affecting digestibility and the energetic value of fats: (1) the content of *gross energy*; (2) the FFA vs FA ratio in the TAG form; (3) the level of FA unsaturation, and (4) the FA chain length. However, these factors need some remarks, for instance, in relation to Point 2, not only the % FFA must be taken into account but also the proportion of MAG and DAG accompanying these FFA, because both type of glycerides can have an emulsifier action, improving global fat digestibility. Regarding Point 3, it could be said that not all unsaturated FA has the same positive effect; and with respect to Point 4, it can be said that chain length will only have a significant effect when particular oils and fats are used, such as lauric fats rich in medium-/short-chain FA, or fish oils rich in very-long-chain FA. For the rest, the major FA proportion always corresponds to the C16 and C18 length. The clearest effect observed is the decrease in the energetic value when fats contain more FFA and more saturated FA, a fact that is more relevant for young animals. In line with this, several prediction equations have been proposed in order to calculate metabolizable energy for broilers according to the U/S values (unsaturated/saturated FA ratio) and FFA proportion (Wiseman *et al.* 1990; Wiseman *et al.* 1998; Jorgensen *et al.* 2000). However, some other fat characteristics and metabolism-dependent factors can influence digestibility; for example, the distribution of FA into the TAG moieties. Since FA located in the sn-2 position are more easily absorbed than are those located in sn-1 and -3 positions, this can particularly affect the absorption of long-chain saturated FA (Karupaiah y Sundram 2007; Vilarrasa *et al.* 2015). Moreover, this FA distribution in TAG affects the TAG melting point, which is directly related to the emulsification process in the gut (Bracco 1994). Emulsification can also be improved by the presence of particular proportions of some fat compounds, such as phospolipids, MAG and DAG, and unsaturated FA, which improves the micelle formation (Vilarrasa *et al.* 2015). This leads to a favoured FA absorption and to an improved FA digestibility, particularly of the long-chain saturated FA. Thus, when relevant changes are introduced in the fat added to feeds, such as fat blends, fat by-products, etc., prediction equations can have low accuracy, and thus, specific animal trials should be carried out (Ravindran *et al.* 2016). As a conclusion, it could be said that the calculation of the 'energetic value' of a fat material is a quite complex task, and prediction equations should take into account not only acidity and the U/S ratio, but also other composition values with great repercussion on the energetic value, particularly for non-conventional fat materials (acid oils, etc.). Additionally, it could be necessary to know and quantify the effect of blending fat materials, which can provide synergistic effects on the whole digestibility value.

Quality control parameters and their significance

Many analytical parameters can be determined in order to define the nutritive value or other quality aspects of a 'feed fat': Insoluble impurities; % Unsaponifiable; % Non-eluted material; Peroxide value; Secondary oxidation (TBA value, *p*-anisidine value, Conjugated dienes …); % TAG dimers and polymers; Total fatty acid content; Iodine value (IV); Saponification value (SV); FA composition (including mainly % linoleic, % omega-6 and omega-3 FA, % saturated FA, % unsaturated FA, % *trans* FA); TAG composition; % Phospholipids; Distribution of FA at TAG sn-2 position; Acidity (or %FFA); % MAG, DAG and TAG; tocopherol and tocotrienol contents; content of other specific components of the unsaponifiable fraction; other specific compounds (i.e residues and contaminants). In order to simplify the approach of quality control, we could classify these parameters into three groups. First, those that can be considered as basic quality parameters, and which should always be determined in order to establish the global fat quality (according to Regulations of the EU or the corresponding country). Then, we can find a second group not having a so universal value, and which should only be performed when the fat shows particular characteristics that must be checked; and third, we find a series of quality parameters indicating very specific characteristics, which are only interesting in particular cases (certain products of degradation or contamination; markers for checking technological processes and conditions, etc.).

Figure 1 shows the correspondance between the main analytical parameters usually performed for the feed fat-quality assessment, and the fat composition or degradation compounds evaluated. Obviously, among these parameters we find a group that determine big or complex lipid fractions, whose global values are key for the calculation of the energetic value. Then, we could find other parameters giving more specific information related to concrete aspects of nutritive value. This figure already includes a reduction in the number of parameters, avoiding redundances and those less useful, but still some discussion can be held regarding the different applications that can be found for parameters providing similar information. Thus, we can talk about some parameters that are frequently performed although they do not show real significance for the quality assessment of several fats and oils. As a summary of the relationships between this group of quality parameters and their practical significance, we can give some examples. Obviously, before testing which parameter could give us the most significant information, we have to take into account which is the aim of the control and the type of sample. A typical case for discussion is the usefulness of PV as a key parameter to assess the level of degradation in fats and oils. The use of the standard iodometric method gives us information about the peroxide groups present in the FA chains but not about other oxidated forms that could also be present. For a complete evaluation of global oxidation, we cannot perform only PV, and we should choose other complementary parameters according to our objective (secondary oxidation index, polymers, etc.). Moreover, in a whole assessment of the non-digestible FA, we must also take into account that other parameters, such

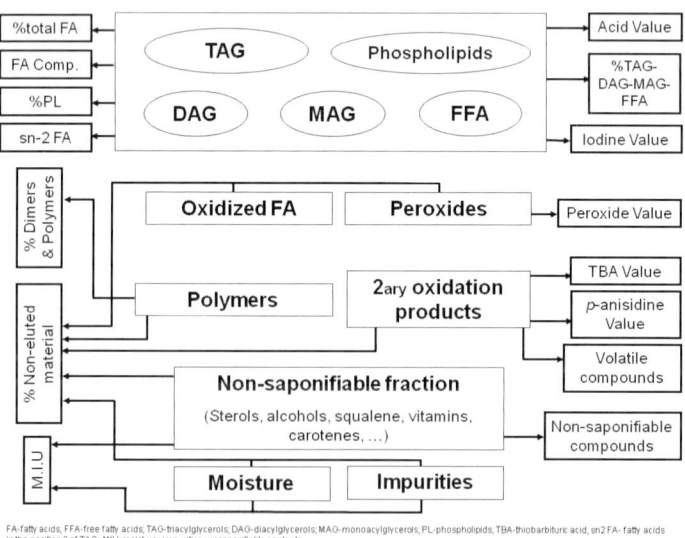

FA-fatty acids; FFA-free fatty acids; TAG-triacylglycerols; DAG-diacylglycerols; MAG-monoacylglycerols; PL-phospholipids; TBA-thiobarbituric acid; sn2 FA- fatty acids in the position 2 of TAG; MIU-moisture+impurities+unsaponifiable contents.

Figure 1. Relationships between fat materials' components and the analytical parameters for their corresponding determination.

as '% Impurities' or '% Non-eluted material', include certain oxidated derivatives of FA. Another interesting case is the assessment of hydrolytic degradation. Usually, acidity is the classical, and the only determination performed related to TAG hydrolysis. This value gives an idea of the quality of the fat and can also be used to estimate how digestible fat can be affected, since FFA are worst absorbed than 2-MAG. But it is also known that MAG and DAG have emulsifying properties and an increased % of them could improve FFA absorption. Thus, according to our objective, the determination of the acidity could not be enough, and determination in the fat sample of all lipid classes (FFA, MAG, DAG and TAG) could be needed. Finally, the case of Iodine Value can be discussed. In fact, it only gives a global idea of unsaturation of the fat determining its melting properties in the gut, which are directly related to fat digestibility. But if we determine the FA composition by gas chromatography (basic determination in all control laboratories), we have more detailed information about the % of each FA (obviously including their chain lenght and number of unsaturations). From this information, we can also extrapolate IV or the melting point. Additionally, we can accurately know the % of some FA that are essential nutrients for the animal (linoleic, linolenic, arachidonic). This is only a very brief discussion, but it serves to illustrate that in some cases our objectives will demand getting information about other specific components, for which we will search specific analytical methods for their determination. A more extended discussion of methodology can be found in Codony *et al.* 2010.

Application of vibrational spectroscopy in fat quality assessment

Vibrational spectroscopy (Raman, IR) is nowadays largely applied in the field of the quantitive evaluation of purity and authentication (Nunes 2014; Gasperini *et al.* 2007; Abbas *et al.* 2009; Mba *et al.* 2014), and determination of major components in many food and feed materials and meats (Cozzolino 2014; Prieto *et al.* 2009). Particularly, NIR has become a basic analytical tool in control laboratories. But in the last few years, a lot of successful applications of NIR have been reported in the literature. Thus, it is not only used for major feed components (water, protein, fat, fibre, etc.) or for authentication purposes, but also determination of relevant fat and oil quality parameters such as acidity, PV, FA composition, IV, *trans*-FA, cloud point, etc. has been achieved (Nunes 2014). We focus our comment particularly on NIR or vis-NIR, since this is a technique already highly introduced in control laboratories for feed composition and it could be interesting for them to extend its application to the assessment of different quality parameters. Its advantages include low cost, short measurement time, no need of solvent use, and the generation of multi-component analysis in a single run. In all cases, the use of chemometrics is unavoidable, and the different types of statistical analysis to be applied should be individually studied and calibrated by the analyst. Several multivariate statistical models exist that are commonly applied, and the analyst has to choose the most suitable one for each analytical problem. The most frequent statistical approach observed in the literature is the use of PLS regression, although for certain cases PCR has also been reported as the most useful one. In order to establish a reliable NIR determination of any fat property, the analyst should carefully take into account several aspects: select the most suitable spectrum region(s) related to the corresponding functional group(s) to be measured; select the most adequate measurement mode (absorbance, transmittance, reflectance); selection of the calibration mode and the validation set; and for quantitative determinations, definition of the most adequate linear-regression equations. Some NIR equipments include 'ready to use' prediction equations prepared by the NIR supplier. However, it must be taken into account that these equation should be better developed on NIR spectra obtained from a samples similar to the real samples to get accurate prediction values. It is recommendable to follow some existing Guidelines (AOCS 2009a). Table 1 gives a compilation of the main studies published during the last two decades. As we can see, very different analytical control objectives have been covered: characterization, authentication or adulteration detection; addition of by-products; FA composition; primary and secondary oxidation levels; acidity, IV and other traditional chemical indexes; etc. Almost all of these papers compare the values obtained by IR application with those obtained by the corresponding official or recommended method and, in some cases, they also compare NIR with the application of other IR or Raman techniques. An AOCS Official Method for IV determination (AOCS 2009b) and a DGF *German Standard Method* for the determination of

Table 1. Main studies reporting the use of IR techniques for the determination of components and quality parameters in feed fats and oils.

Reference	Aims	Type of measure
Siemens & Daun, 2005	NIR determination of FA composition of vegetable oils	Reflectance spectra between 400 and 2,500 nm
Endo et al. 2005	NIR determination of IV and Saponification Value in fish oils	Spectra between 9,000 and 7,560 cm^{-1}
Foca et al. 2016	FT-NIR determination of IV and FA composition in pig fat	Reflectance spectra between 12,500 and 3,800 cm^{-1}
Li et al. 1999	FT-NIR determination of SV, IV and cis-trans content in edible oils	Transmittance spectra between 10,000 and 4,000 cm^{-1}
Cox et al. 2000	FT-NIR International collaborative study for determination of IV	Transmittance spectra between 9,100 and 7,560 cm^{-1}
Li et al. 2000b	FT-NIR for discrimination of edible oils and determination of IV	Transmittance spectra between 12,000 and 4,500 cm^{-1}
Hendl et al. 2001	FT-IR determination of IV in vegetable oils.	Absorbance derivative between 4,000 and 400 cm^{-1}
Yang et al. 2005	FT-IR, FT-NIR and FT-Raman discriminant analysis of edible oils	Absorbance spectra between 3,500 and 500 cm^{-1} for FT-IR, and between 8,000 and 4,000 cm^{-1} for FT-NIR
Prieto et al. 2014	NIR determination of FA composition and IV in pig fat	Absorbance spectra between 400 and 2,500 nm
Adewale et al. 2014	FT-NIR determination of FA and IV in waste animal fat blends	Absorbance spectra between 800 and 2,500 nm
AOCS, 2009b	AOCS Official Method Cd 1e-01. FT-NIR determination of IV	Transmittance spectra between 9,100 and 7,560 cm^{-1}
Mba et al. 2014	FT-NIR characterization of palm and canola oil blends	Absorbance spectra between 12,000 and 4,000 cm^{-1}
Gasperini et al. 2007	FT-IR for classification of oil co- and by-products	Transmittance spectra between 4,000 and 400 cm^{-1}
Abbas et al. 2009	FT-Raman for discimination of animal fats and oil by-products	Raman spectra between 3,600 and 200 cm^{-1}
Wojcicki et al. 2015	NIR and MIR for assessing oxidation in edible oils	Reflectance spectra between 9,000 and 4,500 cm^{-1} for NIR, and between 3,500 and 500 cm^{-1} for MIR
Lagardere et al. 2004	FT-NIR determination of acidity and PV in virgin and refined olive and sunflower oils	Absorbance spectra between 5,100 and 4,520 cm^{-1} for Acidity, and between 7,500 and 6,300 cm^{-1} for PV
Li et al. 2000a	FT-NIR determination of PV in oils	Differential spectra between 4,800 and 4,520 cm^{-1}
Yildiz et al. 2001	NIR determination of oxidation in vegetable oils	Transmittance spectra between 400 and 2,500 nm
Yildiz et al. 2002	NIR determination of PV in vegetable oils	Transmittance spectra between 400 and 2,500 nm
Cayuela et al. 2013	Vis./NIR determination of olive oil oxidation levels and of oxidative stability	Transmittance spectra between 350 and 2,500 nm
Ng et al. 2011	NIR determination of degradation in used vegetable frying oils	Transmittance spectra between 400 and 2,500 nm
Talpur et al. 2014	FT-MIR determination of quality parameters in used frying cottonseed oil	Reflectance spectra between 4,000 and 650 cm^{-1}
Gertz et al. 2013	DGF Standard Method. FT-NIR screening of used frying fats, and determination of acid value, p-AV, %TPM, %TG-polymers	Absorbance spectra between 11,500 and 4,000 cm^{-1}
Engelsen, 1997	Vis./NIR determination of frying oil deterioration	Transmittance spectra between 400 and 2,500 nm
Graham et al. 2012	NIR and Raman detection of transformer and mineral oils in feed oils	Transmittance spectra between 9,000 and 4,500 cm^{-1} for NIR spectra
Moreira et al. 2015	NIR determination of critical properties of vegetable oils for biodiesel production	Absorbance spectra between 10,000 and 4,000 cm^{-1}, and with some selected bands
Chen & Zhao, 2014	Vis./NIR determination of water content in biodiesel.	Absorbance at 8 selected sensitive wavelengths

acidity, *p*-AV, %TAG-polymers and %Total Polar Materials (Gertz *et al.* 2013) even exist. These studies, as a whole, give a clear idea that the application of IR methods, combined with a suitable statistical analysis of the spectra (or certain concrete regions) is currently a quite developed technology that could simplify the quality control, and for sure new official or recommended methods could be approved in the near future for several parameters, not only for fats and oils but also for complex feeds and for meat, eggs, meat products, etc.

References

Abbas, O., Fernández Pierna, J.A. *et al.* 2009. *Journal Molecular Structure* 924-926: 294-300.

Adewale, P., Mba, O. *et al.* 2014. *Vibrational Spectroscopy* 72: 72-78.

AOCS. Analytical Guidelines Am 1a-09. 2009a: Near Infrared Spectroscopy Instrument Management and Prediction Model Development. American Oil Chemists' Society, Champaign, IL (USA).

AOCS. 2009b. Official Method Cd 1e-01, Determination of Iodine Value by Pre-calibrated FT-NIR with Disposable Vials. American Oil Chemists' Society, Champaign, IL (USA).

Bezerra da Costa, G., Sousa Fernandes, D.D. *et al.* 2016. *Food Chemistry* 196: 539-543.

Bracco, U. 1994. *American Journal of Clinical Nutrition* 60 (Suppl): 1002S-1009S.

Cayuela, J.A., Moreda, W. *et al.* 2013. *Journal Agricultural and Food Chemistry* 61: 8056-8062.

Chen, L. and Zhao, Y. 2014. *Transactions of Chinese Society of Agricultural Engineering* 30: 168-173.

Codony, R., Guardiola, F. *et al.* 2010. XXVI Curso de Especialización FEDNA, Madrid-Chapter 7. (http://fundacionfedna.org/sites/default/files/10CAP_VII.pdf).

Cox, R., Lebrasseur, J. *et al.* 2000. *Journal American Oil Chemists' Association* 77: 1229-1234.

Cozzolino, D. 2014. *Food Research International* 60: 262-265.

Endo, Y., Tagiri, M. *et al.* 2005. *Journal of Food Science* 70: 127-131.

Engelsen, S.B. 1997. *Journal American Oil Chemists' Association* 74: 1495-1508.

Foca, G., Ferrari, C. *et al.* 2016. *Food Analytical Methods* 9: 2791-2806.

Gasperini, G., Fusari, E. *et al.* 2007. *European Journal Lipid Science and Technology* 109: 673-681.

Gertz, Ch., Fiebig, H.J. *et al.* 2013. *European Journal Lipid Science and Technology* 115: 1193-1197.

Graham, S.F., Haughey, S.A. *et al.* 2012. *Food Chemistry* 132: 1614-1619.

Hendl. O., Howell, J.A. *et al.* 2001. *Analytica Chimica Acta* 427: 75-81.

Jorgensen. H., Gabert, V.M. *et al.* 2000. *Journal of Nutrition* 130: 852-857.

Karupaiah, T., Sundram, K. 2007. *Nutrition and Metabolism* 4: 16-33.

Lagardere, L., Lechat. H. *et al.* 2004. *Oleagineux, Corps Gras, Lipides* 11: 70-75.

Li, H., van de Voort. F.R. *et al.* 2000. *Journal American Oil Chemists' Association* 77: 137-142.

Li, H., van de Voort. F.R. *et al.* 2000. *Journal American Oil Chemists' Association* 77: 29-36.

Li, H., van de Voort. F.R. *et al.* 1999. *Journal American Oil Chemists' Association* 76: 491-497.

Mateos, G.G., Rebollar, P,G. *et al.* 1996. XII Curso de especialización FEDNA, Madrid-Chapter 1. (http://fundacionfedna.org/sites/default/files/96CAP_I.pdf).

Mba, O., Adewale, P. *et al.* 2014. *Industrial Crops and Products* 61: 472-478.

Moreira, S.A., Sarraguça, J. *et al.* Fuel 150: 697-704.

Ng, L.C., Wehling, R.L. *et al.* 2011. *Journal of Agricultural and Food Chemistry* 59: 12286-12290.

Nuchi, C., Guardiola, F. *et al.* 2009. *Journal of Agricultural and Food Chemistry* 57: 1952-1959.

Nunes, C.A. 2014. *Food Research International* 60: 255-261.

Prieto, N., Roehe, R. *et al.* 2009. *Meat Science* 83: 175-186.

Prieto, N., Uttaro, B. *et al.* 2014. *Meat Science* 98: 585-590.

Ravindran, V., Tancharoenrat, P. *et al.* 2016. *Animal Feed Science and Technology* 213: 1-21.

Siemens, B.J., Daun, J.K. 2005. *Journal American Oil Chemists' Association* 82: 153-157.

Talpur, M.Y., Kara, H. *et al.* 2014. *Talanta* 129: 473-480.

Vilarrasa, E., Codony, R. *et al.* 2015. *Poultry Science* 94: 1527-1538.

Wiseman, J., Powles, J. *et al.* 1998. *Animal Feed Science Technology* 71: 1-9.

Wiseman, J., Salvador, F. *et al.* 1990. *Poultry Science* 70: 1527-1533.

Wójcicki, K., Khmelinskii, I. *et al.* 2015. *Food Chemistry* 187: 416-423.

Yang, H., Irudayaraj, J. *et al.* 2005. *Food Chemistry* 93: 25-32.

Yildiz, G., Wehling, R.L. *et al.* 2001. *Journal American Oil Chemists' Association* 78: 495-502.

Yildiz, G., Wehling, R.L. *et al.* 2002. *Journal American Oil Chemists' Association* 79: 1085-1089.

Updating the available P requirements of broilers

V. Khaksar, B. Meda and A. Narcy[*]
URA, INRA, 37380, Nouzilly, France; agnes.narcy@inra.fr

Summary

Optimization of P supply in broiler feed appears crucial and remains a major environmental and economic challenge for poultry production. The aim of this work was to update the available P (aP) requirements of broilers, using a factorial approach. The nutrient requirements were evaluated by considering the needs for 'maintenance' plus that for 'growth' which were estimated from determination of endogenous P losses and of body phosphorus content, respectively. Literature on endogenous P losses in broilers is very scarce and the findings are very variable between and within studies. The regression of endogenous P losses (g/d) against body weight (kg) revealed a quadratic relationship ($-0.0021\,BW^2 + 0.0185\,BW + 0.0034$, with a mean value of endogenous phosphorus losses of 0.2 g/kg dry matter intake). In the factorial approach, body P deposition (BPD) is the main parameter used to calculate the aP requirements for growth and the estimation is strongly influenced by dietary P and Ca levels. A literature database was used to estimate the regression between optimal cumulative BPD (g) and cumulative BWG (kg). The relationship was linear (4.68 BWG). The overall model underlined the predominant contribution of the growth requirement (>90%) to the overall aP requirements compared to that for maintenance. The current model values presented quite the same aP requirements up to 30 days of age and afterward decreased. This may be due to the priority of bone mineralisation in the first days and then to the higher deposition rate of soft tissues in upper ages in modern broilers.

Introduction

Phosphorus (P) is one of the essential minerals for the birds, due to its critical functions on metabolic processes (cellular metabolism and regulatory mechanisms). It is also required to attain birds' optimum genetic potential in growth (mainly muscle), feed efficiency and skeletal integrity (bone mineralization) (Park *et al.*, 2009). It was demonstrated in broilers that for achieving maximum bone mineralization, a higher P supply is required compared to the P requirement for maximal growth and feed efficiency (Létourneau *et al.*, 2010): bone is made of hydroxyapatite and contains 85% of the body's total P. A marginal dietary P supply reduces bone development and broiler performance, while a high P content of diet increased environmental pollution (e.g. eutrophication) and diet cost (Selle and Ravindran, 2007). So, P supply in broiler feed needs to be optimized and remains a major environmental and economic challenge for poultry production. The precise knowledge of the P requirements of the birds and of its availability allow to minimize the dietary P supply.

Modelling phosphorus requirements

It is clearly demonstrated that non-phytate phosphorus (NPP) is highly available in contrast to phytate P but with variability depending on the sources. Thus it is necessary for optimizing the P requirements to precisely evaluate the available P (aP) for broilers (WPSA Working Group report, 2013). Available P (aP) is the part of dietary total P that, at marginal level of P supply, can be utilised to cover the P requirements of the animal. Poultry nutritional requirements can be evaluated by 2 systems: empirical or factorial. The empirical approach determines the P requirement as the onset of the plateau of response in dose-response trials. These data vary widely firstly because of the methods used to analyse the data and secondly because of differences in animal factors (age, gender and strain), diet composition and environmental conditions. It should be also considered that running these trials are time consuming and costly. Alternatively, the requirements can be determined using the factorial approach: the nutrient requirements are calculated by evaluating the needs for 'maintenance' plus that for 'growth' (soft tissues development and bone mineralization). In this approach, aP requirement for maintenance is determined by estimating endogenous P losses (EPL) and that for growth by measuring the body phosphorus content (BPC).

Requirements for maintenance

Literature on endogenous P losses in broilers is very scarce. The estimations and values of EPL are very variable between and within studies (Table 1). This variability in EPL values may result from differences in age, gender or strain of broilers, type of diet (conventional or semipurified diets: P-free diets) or approaches used to estimate the EPL. Al-Masri (1995) determined the excreta EPL in male Lohmann broilers at a level of 30 to 135 mg/b/d (297 to 1,352 mg/kg DM intake) using an isotope-dilution technique. These birds were reared from 14 to 29 and were fed conventional diets with different dietary Ca:total P ratios ranging between 2.5:1 and 1:1 (Ca was changed but total P remains constant). A substantial amount of P is excreted through urine, especially when the levels of Ca is low, and is recovered in excreta. In two other studies, in which the birds were fed semi-purified diets, pre-ceacal EPL values were 65 and 45.5 mg/b/d (446 and 272 mg/kg DM intake) in 28d male and 35d mixed Ross broilers, respectively (Rutherfurd et al., 2002 and 2004). Excreta EPL was 115 mg/b/d (983 mg/kg DM intake) in 26d female Ross broilers fed semi-purified diets (Cowieson et al., 2004).

Dilger and Adeola (2006) run an experiment on broilers fed conventional diets (conventional soybean meal (SBM) or low-phytate SBM) with different Ca:total P ratios ranging between 0.56:1 and 1:1 (Ca and total P were changed). Excreta EPL was estimated to be 190.50 and 395.8 mg/kg DM intake (21 and 43.5 mg/b/d) for conventional and low-phytate SBM, respectively, in 22d male Ross 308 broilers. EPL estimation by Dilger and Adeola (2006) was based on the regression method. The mean excreta EPL value of Dilger and Adeola (2006) was 293.2 mg/kg DM intake (32 mg/b/d).

The latest excreta EPL estimation (Liu et al., 2012) was 20 to 57 mg/b/d (148 to 420 mg/kg DM intake) in 26d male Arbor Acres broilers fed semi-purified diets and underwent different metabolic conditions. The birds were exposed to different times of feeding, fasting and excreta collection during the experiment. Mean excreta EPL value was 263 mg/kg DM intake (35.5 mg/b/d). The comparison of EPL value of Dilger and Adeola (2006) and Liu et al. (2012) showed quite

Table 1. Summary of the literature on endogenous P losses (EPL) in broilers.

References	Al-Masri (1995)	Rutherfurd et al. (2002 & 2004)	Cowieson et al. (2004)	Dilger and Adeola (2006)	Liu et al. (2012)
EPL*, mg/b/d	**30-135[#]**	65 & 45.5	**115[#]**	21 & 43.5	**20-57[#]**
EPL*, mg/kg DMI	297-1352	**446 & 272[#]**	983	**190.5 & 395.8[#]**	148-420
Strain	Lohmann	Ross	Ross	Ross	Arbor.[1]
Gender	Male	Male & Mix.	Female	Male	Male
Age, day	14-29	28 & 35	26	22	26
Type of diet[2]	Conv.	P-free	P-free	Conv.	P-free
P collection[3]	Excr.	Prec.	Excr.	Excr.	Excr.
P determination[4]	Isotope.	Spect.	Spect.	Regr.	Spect.

* Original EPL values (measurements) were expressed either in mg/b/d or mg/kg DM intake (DMI) by different authors (original data in bold font). To ease the comparison, we calculated the alternative unit using the Performance Objectives of the respective strain.
[1] Arbor: Arbor Acres.
[2] Conv: conventional diet.
[3] Excr.: excreta; Prec.: prececa.
[4] Spect.: spectrometrically; Isotope.: isotope-dilution technique; Regr.: regression method.

similar mean excreta EPL (293.2 vs 263 mg/kg DM intake). In animals fed conventional diets, endogenous losses result from a large variety of factors including feed intake, the nature of the material fed and the presence of anti-nutrients (Ferraz de Oliveira, 1998), while, semi-purified diets alleviate some of these effects, especially those associated with variation in feed intake.

According to Figure 1, when regressing endogenous P losses (EPL, g/d) against body weight (kg), a quadratic relationship was obtained with EPL (g/d) = –0.0021 BW^2 + 0.0185 BW + 0.0034.

Requirements for growth

Body P content is the main parameter used for evaluating aP requirements needed for the factorial approach. However, there are few direct data (experimental measure of P in the whole carcass). Most frequently, the P composition was estimated by measuring apparently retained phosphorus from balance trials. We used a database including 41 publications (109 experiments and 427 treatments) to estimate values of body P content (BPC) and retained phosphorus (ARP) in broilers of different ages, genders and strains. This provides a global overview of the variability of these parameters, especially those induced by the composition of the diet. The analysed dietary total P and Ca were considered from the analysed nutrients in each experimental diet which were respectively 83 and 82% of the whole data, otherwise, the calculated ones were considered.

Body P deposition (BPC and/or ARP) was mainly affected by dietary Ca and P levels (Figure 2). Phosphorus deposition in soft tissues is fixed over a wide range of P and Ca levels (Narcy et al, 2015). Consequently, its deposition in the skeleton drives the overall body P deposition.

The optimum for body P deposition was evaluated by considering only treatments with the following ranges for dietary Ca and total P levels. For Ca, the range was set to: ±15% of dietary Ca level recommended for Ross 308 (version 2014) and Cobb 500 (version 2015), i.e. ranges between 7.14-11.29 and 6.46-9.29 g Ca/kg diet for grower and finisher phases, respectively. Total P ranges were calculated to obtain on average a Ca:total P ratio of 1.4:1 (WPSA Working Group Report, 2013) taking into account the grower and finisher dietary Ca levels (ranges): 5.10-8.06 and 4.61-6.64 g total P/kg diet for grower and finisher phases, respectively. In accordance with this procedure, 112 optimal data of BPD have been extracted. The regression of cumulative BPD (g) against cumulative BWG (kg) revealed a linear relationship:

BPD (g) = 4.68 BWG (R^2=0.97).

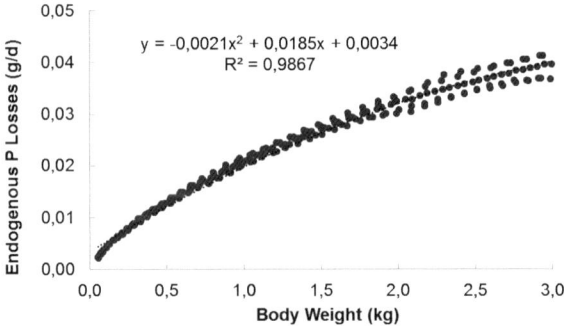

Figure 1. Estimation of the relationship between body weight (kg) and phosphorus endogenous losses (g/d) using performance objectives of two conventional strains of broilers (Male and female Ross 308 and Cobb 500) and an acceptable value of endogenous phosphorus losses of 0.2 g/kg dry matter intake.

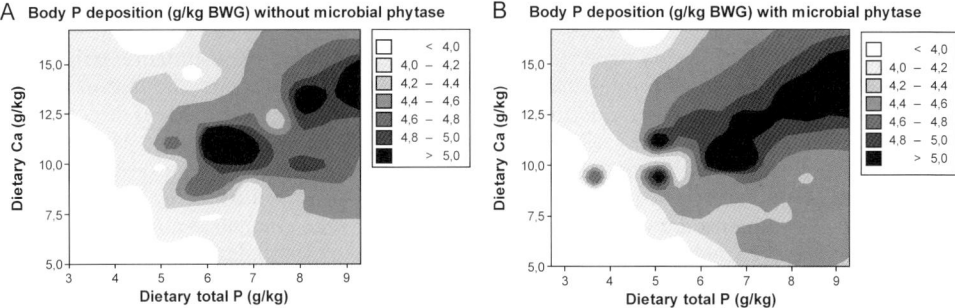

Figure 2. Contour plot of body phosphorus deposition against dietary total P (axis X) and Ca (axis Y) levels in broilers fed diets without (A) or with (B) microbial phytase. The database included 41 publications (109 experiments and 427 treatments): Al-Masri (1995), Johnston and Sothern (2000), Ravindran et al. (2000), Rutherfurd et al. (2002), Viveros et al. (2002), Cowieson et al. (2004), Dilger et al. (2004), Rutherfurd et al. (2004), Applegate (2005), Dilger and Adeola (2006), Mondal et al. (2007), Leytem et al. (2008a and 2008b), Liem et al. (2008), Manangi and Coon (2008), Olukosi et al. (2008), Olukosi and Adeola (2008), Plumstead et al. (2008), Zhou et al. (2008), Ziaei et al. (2008), Han et al. (2009), Manangi et al. (2009), Selle et al. (2009), Thomas and Ravindran (2010), Deepa et al. (2011), Abudabos (2012), Delezie et al. (2012), Demirel et al. (2012), Liu et al. (2012), Rousseau et al. (2012), Shastak (2012a), Shastak et al. (2012b), Chung et al. (2013), Liu et al. (2013), Narcy et al. (2013), Rousseau et al. (2013), Van Krimpen et al. (2013), Mutucumarana et al. (2014), Hamdi et al. (2015), Shang et al. (2015) and Belloir (2016).

The validity of this model was confirmed by using an external database including 90 optimal individual data of body P content from 8 experiments carried out between 2012 and 2015 (Narcy *et al.*, unpublished results). By plotting estimated BPD against measured BPD, we observed a low and non-significant intercept, a slope close to 1 and a high R^2 of 0.95, and confirmed our predictive equation.

Total requirements

The evolution of total requirements (i.e. including both maintenance and growth ones) for as-hatched Ross 308 is presented in Figure 3. The total requirements range between 0.07 and 0.48 g/d, but the contribution of maintenance in the total requirements is below 10%.

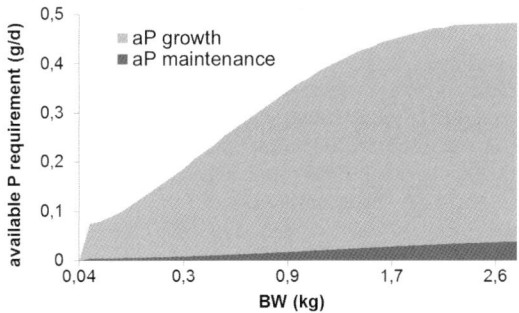

Figure 3. Example of the partitioning of available phosphorus requirements between maintenance and growth for a commercial strain of broiler (As-hatched Ross 308).

Conclusions

The approach described in this paper provides a practical tool for the estimation of available phosphorus requirements of broilers. Even though the effect of age, gender and/or strain is not directly considered in this modelling approach, these effects have been indirectly taken into account by considering the variability of existing data from the literature. However, this approach allows its use in different production contexts (i.e. age, gender, strain) by considering two variables sensitive to these parameters (i.e. BW and BW gain). It confirms that phosphorus maintenance requirements are very low compared to growth requirements. Values obtained with the current model presented quite the same aP requirements up to 30 days of age and afterward decreased which may be respectively due to the priority of bone mineralisation in the first days and then higher deposition rate of soft tissues in upper ages in modern broilers.

Acknowledgement

The authors would like to acknowledge members of the WPSA Phosphorus Working Group for their expertise and BASF for financial support.

References

Abudabos, A.M. 2012. *AJAVA* 7: 288-298.
Al-Masri, M.R. 1995. *British Journal of Nutrition* 74: 407-415.
Applegate, T.J. 2005. *Poultry Science* 84: 742-747.
Belloir P., 2016. PhD Thesis.
Chung, T.K., S.M. Rutherfurd, D.V. Thomas and P.J. Moughan. 2013. *British Poultry Science* 54: 362-373.
Cowieson, A.J., T. Acamovic and M.R. Bedford. 2004. *British Poultry Science* 45: 101-108.
Deepa, C, G.P. Jeyanthi and D. Chandrasekaran. 2011. *Asian J. Poultry Science* 5: 28-34.
Delezie, E., L. Maertens and G. Huyghebaert. 2012. *Poultry Science* 91: 2523-2531.
Demirel, G., A.Y. Pekel, M. Alp and N. Kocabağlı. 2012. *Journal of Applied Poultry Research* 21: 335-347.
Dilger, R.N. and O. Adeola. 2006. *Poultry Science* 85: 661-668.
Dilger, R.N., E.M. Onyango, J.S. Sands and O. Adeola. 2004. *Poultry Science* 83: 962-970.
Fan, M.Z., T. Archbold, W.C. Sauer, D. Lackeyram, T. Rideout, Y. Gao, C.F. de Lange and R.R. Hacker. 2001. *Journal of Nutrition* 131: 2388-2396.
Ferraz de Oliveira, M.I. 1998. Ph.D. Thesis, University of Aberdeen, Aberdeen.
Hamdi, M., S. Lopez-Verge, E.G. Manzanilla, A.C. Barroeta and J.F. Perez. 2015. *Poultry Science* 94: 2144-2151.
Han, J.C., X.D. Yang, H.X. Qu, M. Xu, T. Zhang, W.L. Li, J.H. Yao, Y. R. Liu, B.J. Shi, Z.F. Zhou and X.Y. Feng. 2009. *Journal of Applied Poultry Research* 18: 707-715.
Johnston, S.L. and L.L. Southern. 2000. *Poultry Science* 79: 1485-1490.
Létourneau-Montminy, M.P., A. Narcy, P. Lescoat, J.F. Bernier, M. Magnin, C. Pomar, Y. Nys, D. Sauvant and C. Jondreville. 2010. *Animal* 4: 1844-1853.
Leytem, A.B., G.P. Widyaratne and P.A. Thacker. 2008a. *Poultry Science* 87: 2466-2476.
Leytem, A.B., P. Kwanyuen and P. Thacker. 2008b. *Poultry Science* 87: 2505-2511.
Liem, A., G.M. Pesti and H.M. Edwards Jr. 2008. *Poultry Science* 87: 689-693.
Liu, J.B., D.W. Chen and O. Adeola. 2013. *Poultry Science* 92: 1572-1578.
Liu, S.B., S.F. Li, L. Lu, J.J. Xie, L.Y. Zhang and X.G. Luo. 2012. *Poultry Science* 91: 1879-1885.
Manangi, M.K. and C.N. Coon. 2008. *Poultry Science* 87: 1577-1586.
Manangi, M.K., J.S. Sands and C.N. Coon. 2009. *International Journal of Poultry Science* 8, 10: 919-928.
Mondal, M.K., S. Panda and P. Biswas. 2007. *International Journal of Poultry Science* 6, 3: 201-206.
Mutucumarana, R.K., V. Ravindran, G. Ravindran and A.J. Cowieson. 2014. *Poultry Science* 93: 412-419.
Narcy, A., X. Rousseau, M. Magnin and M.P. Letourneau-Montminy. 2013. Optimisation des apports phosphocalciques chez le poulet de chair. Presented at Eastern Nutrition Conference, Québec, CAN.
Narcy, A., X. Rousseau, N. Même, M. Magnin and Y. Nys. 2015. ESPN, Prague. Olukosi, O.A., A.J. Cowieson, and O. Adeola. 2008. *British Poultry Science* 49, 4: 436-445.
Olukosi, O.A. and O. Adeola. 2008. J. *Poultry Science* 45: 192-198.

Park, K.W., A.R. Rhee, J.S. Um, and I.K. Paik. 2009. *Journal of Applied Poultry Research* 18, 3: 598-604.

Plumstead, P.W., A.B. Leytem, R.O. Maguire, J.W. Spears, P. Kwanyuen and J. Brake. 2008. *Poultry Science* 87: 449-458.

Ravindran, V., S. Cabahug, G. Ravindran, P.H. Selle and W.L. Bryden. 2000. *British Poultry Science* 41: 193-200.

Rousseau, X., M.P. Letourneau-Montminy, N. Meme, M. Magnin, Y. Nys and A. Narcy. 2012. *Poultry Science* 91: 2829-2837.

Rousseau, X. 2013. PhD Thesis. François-Rabelais University, Tours.

Rutherfurd, S.M., T.K. Chung, P.C.H. Morel and P.J. Moughan. 2004. *Poultry Science* 83: 61-68.

Rutherfurd, S.M., T.K. Chung and P.J. Moughan. 2002. *British Poultry Science* 43: 598-606.

Selle, P.H. and V. Ravindran. 2007. *Animal Feed Science and Technology* 135: 1-41.

Selle, P.H., V. Ravindran and G.G. Partridge. 2009. *Animal Feed Science and Technology* 153: 303-313.

Shang, Y., A. Rogiewicz, R. Patterson, B.A. Slominski and W.K. Kim. 2015. *Poultry Science* 94: 955-964.

Shastak, Y. 2012a. PhD Thesis (Chapter 5). Institute of Animal Nutrition. University of Hohenheim.

Shastak, Y., M. Witzig, K. Hartung and M. Rodehutscord. 2012b. *Poultry Science* 91: 2201-2209.

Thomas, D.V. and V. Ravindran. 2010. Asian-Australian Journal of Animal Science 23: 68-73.

Van Krimpen, M.M., J.Th.M. Van Diepen, P.G. Van Wikselaar, P. Bikker and A.W. Jongbloed. 2013. Wageningen UR Livestock Research. Report 670.

Viveros, A., A. Brenes, I. Arija and C. Centeno. 2002. *Poultry Science* 81: 1172-1183.

WPSA Working Group No 2: *World's Poultry Science* 69: 687-698.

Zhou, J.P., Z.B. Yang, W.R. Yang, X.Y. Wang, S.Z. Jiang and G.G. Zhang. 2008. *Journal of Applied Poultry Research* 17: 331-339.

Ziaei, N. J.H. Guy, S.A. Edwards, P.J. Blanchard, J. Ward and D. Feuerstein. 2008. *British Poultry Science* 49: 195-201.

The significance of mycotoxins in future poultry production

G. Antonissen[1,2*] and S. Croubels[1]

[1]Department of Pharmacology, Toxicology and Biochemistry, Faculty of Veterinary Medicine, Ghent University, Salisburylaan 133, 9820 Merelbeke, Belgium; [2]Department of Pathology, Bacteriology and Avian Diseases, Faculty of Veterinary Medicine, Ghent University, Salisburylaan 133, 9820 Merelbeke, Belgium; gunther.antonissen@ugent.be

Summary

Poultry and feed industry together are facing one of the major challenges of the future, namely producing safe and healthy animal protein in a sustainable way. Only by supplying qualitative, safe and healthy feed livestock industry will be able to improve animal health, welfare and production levels. The presence of mycotoxins in feed represents a severe threat for animal health and welfare, and poses relevant research challenges in the field of feed toxicology. Climate changes may affect the global distribution of mycotoxigenic fungi and their mycotoxins. At present, acute mycotoxicosis caused by high doses is rare. Ingestion of low to moderate amounts of mycotoxins is more common and generally does not result in obvious intoxication. However, these low to moderate contamination levels may impair intestinal health, immune function and/or animal susceptibility to infectious diseases. Besides, mycotoxin co-occurrence is a common feature. The toxic effect of the mycotoxin mixture can be more severe than when mycotoxins alone occur. In the past, research has focused mainly on legislatively regulated mycotoxins such as aflatoxins, ochratoxin A, deoxynivalenol, T-2 toxin, fumonisins, and zearalenone. In contrast, toxicological data about the frequently occurring modified mycotoxins and emerging mycotoxins is rather limited.

Introduction

Among the most important safety risks for the future feed industry and security of the feed supply chain are mycotoxins. Mycotoxins are secondary metabolites produced by toxigenic fungal species present on a multitude of crops (Dellafiora and Dall'Asta, 2017; Pinotti et al., 2016). Kovalsky et al. (2016) reported that the yearly median mycotoxin concentrations for the period 2014 and 2015 were globally increased for DON, fumonisins (FBs), and zearalenone (ZEN), with doubled median concentrations compared to 2012 and 2013. Moreover, the Fusarium mycotoxins ZEN, DON, and FBs are globally the most frequently occurring mycotoxins contaminating 88%, 79%, and 67% of 1113 tested feed samples in the period 2012-2015 (Kovalsky et al., 2016). Although sometimes high contamination levels are present, the majority of the feed samples was found to comply with European Union regulations and recommendations on the maximal tolerable concentrations in feed (Kovalsky et al., 2016; European Commission, 2006).

Climate change and mycotoxin occurrence

Since mycotoxin production is highly connected to environmental factors such as temperature and humidity, mycotoxicology is likely to be affected by climate change (Paterson and Lima, 2010). Available information suggests that slightly elevated CO_2 concentrations and interactions with temperature and water availability may stimulate growth of some mycotoxigenic species, especially under water stress (Magan et al., 2011). Furthermore, Elsgaard et al. (2012) predict a shift in the cropping patterns of corn and wheat in Europe due to climate changes. The production of corn is suggested to be characterized by generally increased cropping shares and a northwards expansion. Wheat is defined by divergent changes in cropping patterns, with increase in the northern parts and decrease in the southern parts of Europe. Climate change will induce a shift in composition of Fusarium species affecting cereal crops. Among Fusarium species associated with Fusarium Head Blight disease (FHB) in wheat, the prevalence of F. graminearum, the main DON producer, has already increased in Central Europe and is likely to increase in the North due to expected future weather conditions (Paterson and Lima, 2010; Madgwick et al., 2011;

Miraglia *et al.*, 2009; Parikka *et al.*, 2012). *F. verticillioides*, a producer of FBs, is currently the most common species on corn in Southern Europe. Since FBs have been associated with both dry weather during grain filling and late season rains, the production of these toxins will probably increase in Central and Northern Europe due to climate change (Miraglia *et al.*, 2009; Parikka *et al.*, 2012).

Cumulative health risk assessment of co-occurring mycotoxins

Mycotoxigenic fungi are usually capable of producing more than one mycotoxin, and feed raw materials are commonly infected with various fungal species at a time. Compound feed is particularly vulnerable to multiple mycotoxin contamination as it typically contains a mixture of several raw materials (Kovalsky *et al.*, 2016; Grenier and Oswald, 2011; Streit *et al.*, 2012). Although mycotoxin co-occurrence is the rule rather than the exception, current legislation (European Commission, 2006) does not address co-contamination and associated risks. For many years, toxicological investigation on mycotoxins was mostly carried out in single-molecule studies, which allow understanding of the standalone and system-dependent potency of the various mycotoxins. The response of animals to exposure to more than one mycotoxin can be the same as the response to each toxin individually (additive effect), more than the predicted sum of the responses to each individual mycotoxin (synergistic) and, more rarely, less than the predicted response to each toxin individually (antagonistic) (Grenier and Oswald, 2011; Streit *et al.*, 2012). However, *in vivo* data reporting the effects of mycotoxin co-occurrence is very limited (Grenier and Oswald, 2011). For example, an additive interaction of aflatoxins (AFs) and ochratoxin A (OTA) was observed on egg production and feed efficiency in laying hens (Verma *et al.*, 2007). Tessari *et al.* (2006) showed a synergistic decrease of the antibodies titres against Newcastle disease following simultaneous exposure to AFs and FBs.

Modified mycotoxins

Occasionally, manifested clinical symptoms due to intake of mycotoxin-contaminated feed are significantly greater than what would be expected based on the feed contamination level. This has led to the discovery of modified or masked mycotoxins, which initially owed their name to the 'ability' to escape detection by routine analytical methods (Gareis *et al.*, 1990; Rychlik *et al.*, 2014). At the level of food and feed, mycotoxins can be subjected to biological modification through conjugation by plants and fungi, or through chemical modification, either thermally or non-thermally, e.g. by food processing. The term 'masked mycotoxins' is kept for the fraction of biologically modified mycotoxins that are conjugated by plants, such as DON-3-glucoside (Rychlik *et al.*, 2014; Berthiller *et al.*, 2012). These modified forms might endanger animal and human health as they are possibly hydrolysed into their free toxins in the digestive tract, and may consequently contribute to an unexpected high toxicity. As modified toxins represent an emerging issue, it is not a surprise that toxicological data are scarce to non-existent (Berthiller *et al.*, 2012). Broekaert *et al.* (2015) demonstrated that for broiler chickens, the orally absorbed fractions of 3-acetyl-DON (18%) and especially 15-acetyl-DON (42%) are higher than or at least equal to that of DON (11%). Furthermore, 3-acetyl-DON, which is reported to be less toxic than DON, is completely hydrolyzed presystemically to DON, whereas 15-acetyl-DON is reported to be more toxic and not fully hydrolyzed. This results in a 'worst case scenario' for broilers, where each mole of 15-acetyl-DON could be as toxic as 4 moles of DON, and where the less toxic 3-acetyl-DON can be regarded as equally toxic as DON because it is completely hydrolyzed presystemically. Furthermore, hydrolysis of modified mycotoxins can also be species dependent, for example broiler chickens do not hydrolyse DON-3-glucoside to DON, in contrast to the complete presystemic hydrolysis in pigs (Broekaert *et al.*, 2016).

Emerging mycotoxins

The group of 'emerging' mycotoxins are neither routinely determined nor legislatively regulated, and also received much less scientific attention in the past compared to AFs, FBs, DON, T-2 toxin

(T-2), ZEN, and OTA. Common members of this group are aflatoxin precursors, ergot alkaloids, enniatins, beauvericin, moniliformin, and *Alternaria* toxins (Jestoi, 2008; Streit *et al.*, 2013). Recent surveys (Kovalsky *et al.,* 2016; Streit *et al.,* 2013) showed a high occurrence for enniatins (e.g. beauvericin (83%-98%), enniatin B (ENNB) (71-92%) and enniatin B1 (ENN B1) (69-92%), moniliformin (76-79%), alternariol (80%), alternariol methyl ether (82%), and tenuazonic acid (65%). In contrast, most aflatoxin precursors and ergot alkaloids, occur less frequently and particulary in low contamination levels, e.g. ergocristine (4-11%) and sterigmatocystin (2.3-8%). Although livestock is frequently exposed, little is known about the toxicity and toxicokinetics, including absorption, distribution, metabolism and excretion (ADME processes) of emerging mycotoxins in humans and animals. This information is essential to assess the health risk in affected humans and animals and also to estimate the possible carry-over of these mycotoxins into animal-derived tissues and products. Recently, Fraeyman *et al.* (2016) demonstrated that both ENN B1 and ENN B are poorly absorbed in broiler chickens after oral administration, in sharp contrast with ENN B1 in pigs. Absolute oral bioavailabilities of only 5% and 11% for ENNs B1 and B, respectively, were noted, resulting in a low systemic exposure to these mycotoxins. Consequently, systemic adverse effects due to the consumption of contaminated feed are probably limited. Nevertheless, ENN concentrations in the gastrointestinal tract can be high, and *in vitro* studies demonstrated the cytotoxicity of ENNs in different cell types (Jestoi, 2008; EFSA, 2014; Vejdovszky *et al.,* 2016). ENN B was more cytotoxic compared to nivalenol, DON, and ZEN in Caco-2 cells (Vejdovszky *et al.,* 2016). Besides, the impact of ENNs on the gastrointestinal microbiota should be investigated. Therefore, further research is imperative to examine the local adverse health effects of ENNs in the intestine of broiler chickens.

Sub-clinical levels of mycotoxins negatively affect gut health

At present, acute mycotoxicosis caused by high doses is rare. Ingestion of low to moderate amounts of mycotoxins is more common and generally does not result in obvious intoxication. However, these low to moderate contamination levels may impair intestinal health, immune function and/or animal susceptibility to infectious diseases (Antonissen *et al.*, 2014a).

Mycotoxins negatively impact on the intrinsic component of the intestinal barrier through the modulation of intestinal epithelial integrity and epithelial cell renewal and repair. By measuring the transepithelial electrical resistance (TEER), several *in vitro* and *ex vivo* studies indicate that mycotoxins such as DON and FB_1 are able to increase the permeability of the intestinal epithelial layer of human, porcine and avian origin (Antonissen *et al.*, 2014b; Osselaere *et al.*, 2013; Maresca *et al.*, 2002; Sergent *et al.*, 2006; Pinton *et al.*, 2009). Also the viability and proliferation of animal and human intestinal epithelial cells can be negatively affected by mycotoxins (Antonissen *et al.,* 2014b, 2015a). The extrinsic barrier consists of secretions and other substances that are not physically part of the epithelium, but which affect the barrier function of the epithelial cells, such as the mucus layer, secretion of antimicrobial peptides, chemokines and cytokines, and reactive oxygen species (ROS). It was observed that feeding a DON and/or FBs contaminated diet, at contamination levels close to the EU maximum guidance levels, to broiler chickens affected the duodenal mucus layer by decreasing the *intestinal mucin 2* (*MUC2*) gene expression and altering the duodenal mucin monosaccharides composition, i.e. N-acetyl-neuraminic acid, N-acetyl-galactosamine, and galactose (Antonissen *et al.,* 2015b). Several mycotoxins are also able to modulate the production of cytokines *in vitro* and *in vivo,* and induce oxidative stress (Osselaere *et al.,* 2013; Antonissen *et al.,* 2015b; Bondy and Pestka, 2000). Feeding a FBs contaminated diet to broiler chickens for 15 days resulted in a reduced diversity of the ileal microbiota compared to the control group. The abundance of the immunomodulating bacterium *Candidatus* Savagella was decreased in chickens fed a FBs contaminated diet. FBs also modulated the presence of *Lactobacillaceae* in the ileum (Antonissen *et al.,* 2015a).

Plenary session 06. Hot Topics: sustainability on poultry feeding

Since the intestinal tract is also a major portal of entry to many enteric pathogens and their toxins, mycotoxin exposure could increase the animal susceptibility to these pathogens. Girgis *et al.* (2008, 2010) showed a delayed cell-mediated immune response against coccidiosis in broiler chickens fed a diet naturally contaminated with *Fusarium* mycotoxins. Besides, Grenier *et al.* (2016) suggested a stronger intestinal inflammation induced by coccidial challenge in birds fed a DON contaminated diet, and thereby these birds required more regulatory T cells to control inflammation. A more pronounced occurrence and severity of *Eimeria maxima* and *E. tenella* lesions were observed and oocyst yield was higher in chickens fed a DON and/ or FBs contaminated diet.

Recently, it was also demonstrated that feeding a DON or FBs contaminated diet is a predisposing factor for the development of *Clostridium perfringens* induced necrotic enteritis in broiler chickens (Antonissen *et al.,* 2014b, 2015a). This coincides with negative effects on selected components of the intrinsic and extrinsic intestinal barrier of the chicken host, i.e. villus height, tight junctions, mucus, oxidative stress, and microbiota homeostasis.

Besides, in pigs, it was demonstrated that the trichothecenes DON and T-2 alter a *Salmonella* Typhimurium infection by promoting *Salmonella* invasion, trans-intestinal epithelial passage, and the uptake by alveolar macrophages (Vandenbroucke *et al.*, 2011; Verbrugghe *et al.*, 2012). In T-2-challenged broiler chickens, an increased level of *Salmonella* Typhimurium-related organ lesions or mortality was seen (Ziprin and Elissalde, 1990). Additionally, moniliformin and FB_1 delayed systemic *Escherichia coli* (avian pathogenic *E. coli*, APEC) clearance in broilers and turkeys after intravenous administration (Li *et al.,* 2000a,b).

Treatment of infectious diseases under pressure by mycotoxin exposure?

Mycotoxins may affect poultry susceptibility to infectious diseases (Antonissen *et al.,* 2014a). However, exposure to mycotoxins may also have an impact on poultry pharmacotherapy, such as an impact on the disposition and pharmacokinetics of orally administered drugs. As a drug is actively or passively absorbed from the intestinal lumen to the systemic compartment, it must cross two major barriers: the mucus layer and intestinal mucosa. *in vitro*, porcine intestinal epithelial barrier disruption by DON and T-2 promoted transepithelial passage of the antibiotics doxycycline and paromomycin (Goossens *et al.*, 2000). *in vivo,* an increased oral bioavailability of chlortetracycline was observed in pigs after feeding a diet contaminated with T-2 (Goossens *et al.,* 2013).

Besides, intestinal drug metabolizing enzymes (a.o. cytochrome P450 (CYP450) isoenzymes) and ABC drug transporter proteins (a.o. P-glycoprotein (P-gp) or MDR1 and MRP2) can also be influenced by a prolonged exposure to mycotoxins. Recent findings suggest that concurrent administration of drugs with FBs contaminated feed might alter the pharmacokinetic characteristics of CYP1A4 and MDR1 substrate drugs, such as enrofloxacin, in broiler chickens (Antonissen *et al.,* 2017).

Conclusions

Mycotoxins will have an important role in future poultry productions. In the near future, there is reason to believe that climate change trends may result in higher levels of certain mycotoxins in crops. The negative effects of mycotoxins on poultry health will manifest mainly on the sub-clinical level, by negatively affecting the immune system and intestinal health. Further research is necessary to investigate the importance of mycotoxin co-occurrence, modified and emerging mycotoxins, role of mycotoxins in infectious diseases, vaccine and therapy failure.

Plenary session 06. Hot Topics: sustainability on poultry feeding

References

Antonissen, G., Martel, A., Pasmans, F., Ducatelle, R., Verbrugghe, E. *et al.* 2014. *Toxins* 6: 430-452.

Antonissen, G., Van Immerseel, F., Pasmans, F., Ducatelle, R., Haesebrouck, F. *et al.* 2014. *PloS One* 9:e108775.

Antonissen, G., Croubels, S., Pasmans, F., Ducatelle, R., Eeckhaut, V. *et al.* 2015. *Veterinary Research* 46: 98.

Antonissen, G., Van Immerseel, F., Pasmans, F., Ducatelle, R., Janssens, G.P.J. *et al.* 2015. *Journal of Agricultural and Food Chemistry* 63: 10846-10855.

Antonissen, G., Devreese, M., De Baere, S., Martel, A., Van Immerseel, F. *et al.* 2017. *Food and Chemical Toxicology* 101: 75-83.

Berthiller, F., Crews, C., Dall'Asta, C., Saeger, S.D., Haesaert, G. *et al.* 2012. Masked mycotoxins: a review. *Molecular Nutrition & Food Research* 57: 165-186.

Bondy, G.S. and Pestka, J.J. 2000. *Journal of Toxicology and Environmental Health Part B* 3: 109-143.

Broekaert, N., Devreese, M., De Mil, T., Fraeyman, S., Antonissen, G. *et al.* 2015. *Journal of Agricultural and Food Chemistry* 63: 8734-8742.

Broekaert, N., Devreese, M., van Bergen, T., Schauvliege, S., De Boevre, M. *et al.* 2016. Archives of Toxicology, DOI 10.1007/s00204-016-1710-2.

Dellafiora, L. and Dall'Asta. C. 2017. *Toxins* 9: 18.

European Food Safety Authority. 2014. Scientific Opinion on the risks to human and animal health related to the presence of beauvericin and enniatins in food and feed. *EFSA Journal* 12 (8), 3802.

Elsgaard, L., Børgesen, C.D., Olesen, J.E., Siebert, S., Ewert, F. *et al.* 2012. *Food Additives and Contamintans part A* 29: 1514-1526.

European Commission Recommendation 576/2006/EC of 17 August 2006 on the presence of deoxynivalenol, zearalenone, ochratoxin A, T-2 and HT-2 and fumonisins in products intended for animal feeding. *Official Journal of the European Union* L 229:7.

Fraeyman, S., Devreese, M., Antonissen, G., De Baere, S., Rychlik, M. *et al.* 2016. *Journal of Agricultural and Food Chemistry* 64: 7259-7264.

Gareis, M., Bauer, J., Thiem, J., Plank, G., Grabley, S. *et al.* 1990. *Journal of Veterinary Medicine Series B* 37: 236-240.

Girgis, G.N., Sharif, S., Barta, J.R., Boermans, H.J. and Smith, T.K. 2008. *Experimental Biology and Medicine* 233: 1411-1420.

Girgis, G.N., Barta, J.R., Girish, C.K., Karrow, N.A., Boermans, H.J. *et al.* 2010. *Veterinary Immunology and Immunopathology* 138: 218-223.

Goossens, J., Pasmans, F., Verbrugghe, E., Vandenbroucke, V., De Baere, S. *et al.* 2000. *BMC Veterinary Research* 8: 245.

Goossens, J., Devreese, M., Pasmans, F., Osselaere, A., De Baere, S. *et al.* 2013. *Journal of Veterinary Pharmacology and Therapeutics* 36: 621-624.

Grenier, B. and Oswald, I. 2011. *World Mycotoxin Journal* 4: 285-313.

Grenier, G., Dohnal, I., Shanmugasundaram, R., Eicher. S.D., Selveraj, R.K., *et al.* 2016. *Toxins* 8: 231.

Jestoi, M. 2008. *Critical Reviews in Food Science and Nutrition* 48: 21-49.

Kovalsky, P., Kos, G., Nährer, K., Schwab, C., Jenkins, T. *et al.* 2016. *Toxins* 8: 363.

Li, Y., Ledoux, D., Bermudez, A., Fritsche, K. and Rottinghaust, G. 2000. *Poultry Science* 79: 26-32.

Li, Y., Ledoux, D., Bermudez, A., Fritsche, K. and Rottinghaus, G. 2000. *Poultry Science* 79: 871-878.

Madgwick, J.W., West, J.S., White, R.P., Semenov, M.A., Townsend, J.A. *et al.* 2011. *European Journal of Plant Patholol ogy* 130: 117-131.

Magan, N., Medina, A. and Aldred, D. 2011. *Plant Pathology* 60: 150-163.

Maresca, M., Mahfoud, R., Garmy, N. and Fantini, J. 2002. *The Journal of Nutrition* 132: 2723-2731.

Miraglia, M., Marvin, H.J., Kleter, G.A., Battilani, P., Brera, C. *et al.* 2009. *Food and Chemical Toxicology* 47: 1009-1021.

Osselaere, A., Santos, R., Hautekiet, V., De Backer, P., Chiers, K. *et al.* 2013. *PloS One* 8: e69014.

Parikka, P., Hakala, K. and Tiilikkala, K. 2012. *Additives & Contaminants. Part A* 29: 1543-1555.

Paterson, R.R.M. and Lima, N. 2010. *Food Research International* 43: 1902-1914.

Pinotti, L., Ottoboni, M., Giormini, C., Dell'Orto, V. and Cheli, F. 2016. *Toxins* 8: 45.

Plenary session 06. Hot Topics: sustainability on poultry feeding

Pinton, P., Nougayrède, J.P., Del Rio, J.C., Moreno, C., Marin, D.E. *et al.* 2009. *Toxicology and Applied Pharmacology* 237: 41-48.

Rychlik, M., Humpf, H.U., Marko, D., Danicke, S., Mally, A. *et al.* 2014. *Mycotoxin Research* 30: 197-205.

Sergent, T., Parys, M., Garsou, S., Pussemier, L., Schneider, Y.J. *et al.* 2006. *Toxicology Letters* 164: 167-176.

Streit, E., Schatzmayr, G., Tassis, P., Tzika, E., Marin, D. *et al.* 2012. *Toxins* 4: 788-809.

Streit, E., Schwab, C., Sulyok, M., Naehrer, K., Krska, R. *et al.* 2013. *Toxins* 5: 504-523.

Tessari, E.N.C., Olivaira, C.A.F., Cardoso, A., Ledoux, D.R. and Rottinghaus, G.E. 2006. *British Poultry Science* 47: 357-364.

Vandenbroucke, V., Croubels, S., Martel, A., Verbrugghe, E., Goossens, J. *et al.* 2011. PloS One 6:e23871.

Vejdovszky, K., Warth, B., Sulyok, M. and Marko, D. 2016. *Toxicology Letters* 241: 1–8.

Verbrugghe, E., Vandenbroucke, V., Dhaenens, M., Shearer, N., Goossens, J. *et al.* 2012. *Veterinary Research* 43: 1-18.

Verma, J., Johri, T.S. and Swain, B.K. 2007. *Journal of the Science of Food and Agriculture* 87: 760-764.

Ziprin, R. and Elissalde, M. 1990. *American Journal of Veterinary Research* 51: 1869-1872.

Some trends of poultry production beyond 2017: Influence of consumer perception and sustainability.

G. Santomà[1,2]

[1]Working Group no. 2, European Branch W.P.S.A., [2]Trouw Nutrition Spain, Technical Manager. Gran Vía Carles III 84, Baixos S-3, 08028 Barcelona, Spain; g.santoma@trouwnutrition.com

Summary

This paper intends to give a global view to our job, not only as poultry nutrition specialists but also as 'prosumers' (pro-active consumers) to try to make us more aware of our important role in both, the food chain and the sustainability of our planet. For such a purpose the following subjects are covered: (1) poultry meat and eggs demand according to both, the world population growth and evolution of the main drivers of consumer's demand; (2) influence of food production on world sustainability, specifically of both, poultry meat and egg productions on climate change, and ways to decrease its impact through changing human eating patterns and decreasing food wastage. As a component of sustainability, antimicrobial resistance is also covered. The challenges that present these 2 broad subjects can only be faced through people consciousness, social change, and technology. In this way, finally this paper summarizes the main achievements, new developments and challenges of poultry nutrition to partner these 2 key areas. The design of feeding programs in a more holistic way is proposed.

Introduction

The main target of this paper is to give a global view to our job as poultry nutrition specialists in the different areas, to try to make us more aware of our important role in the food chain and sustainability of our planet. The sustainable development goals adopted by the General Assembly of the UN (2015), the Global Action Plan on Antibiotic Resistance proposed by the WHO (2015), or the agreement reached among the nations of the world in Paris December 2015, reflect the world consciousness to be responsible to feed properly the population and preserve the planet for future generations. To face the challenges that arise from those goals, a holistic approach and a better knowledge of all the disciplines involved is required.

Poultry production forecast

According to the UN (2015b) the current world population of 7.3 billion is projected to reach 9.7 billion in 2050 with about 70% living in urban areas, while incomes could increase by 2% a year (Mottet and Tempio, 2016). In a context of 9,15 billion people in 2050, Alexandratos and Bruisma (2012) projected that the demand for animal source food could grow by 70% between 2005 and 2050, from which the global demand for eggs would increase by 65% and poultry meat is expected to have the highest growth, with 121%, benefiting from its various competitive advantages (e.g. affordability, convenience, absence of religious restrictions, healthy image, limited GHG emissions, lower production costs, short rearing time and lower required investments). Backyard systems contribute to 8% of global eggs production and 2% of the poultry meat production (Mottet and Tempio, 2016). Small-scale poultry is a valuable asset to the local human population in many countries located in the tropical and sub-tropical environments (e.g. Lan Phuong et al., 2015; Fasina et al., 2016). Therefore poultry sector encompasses a contrast between dominant global large corporations and small-scale producers especially in the developing countries (Vaarst et al., 2015).

Plenary session 06. Hot Topics: sustainability on poultry feeding

Consumer perception to poultry products

Main drivers of food demand

Following Deloitte (2016), the key drivers of the consumer's choice are: Price, Convenience and Emotional Rewards. The two first are considered the Traditional key drivers, and among the emotional rewards, taste is also considered a traditional one, meanwhile the others are in continuous evolution. They are considered Evolving key drivers which include: Health & Wellness (nutritional content, organic production, natural ingredients, non-GMO, all natural and antibiotic free, etc.), the very related Safety (absence of allergens, company attributes, etc.), Social Impact (local sourcing, sustainability, animal welfare, treatment of employees, etc.), Experience (retail store layout, channel innovation, brand interaction, etc.), and Transparency (labelling, company website, etc.). Traditional value drivers continue to be the most important ones of consumer behaviour, and likely will be for the foreseeable future. However, evolving factors are becoming increasingly important for a large and diverse group of consumers around the world, especially in developed economies. Preference toward 'Evolving drivers' is highly correlated to two other market disruptors: social media and digital channel use have democratized information in such a way to empower consumers to become hyperconnected, proactive and demanding 'prosumers'. However, food retailers can have a large impact on the meat market. In the U.S. and Europe, a handful of companies control the largest part of the market (e.g. Sloyan, 2017).

Main evolving value drivers

Due to the socio-demographic changes that the world is experiencing, as household composition, aging people, new lifestyles and new eating habits, specificities of different generations (baby-boomers, X, Millenials), evolving drivers become more important. Several published papers analyze the purchasing patterns, perceptions and decision factors for poultry in different EU countries (e.g. Vucasovic, 2014; Walley *et al.*, 2015), where the main decision drivers were safety, welfare, organic, use by date, taste, tenderness and local production with different priorities depending on the country. It is relevant the growing demand for alternative poultry especially in north-west of Europe through segmenting the market with animal-welfare products (Van Horne and Bondt, 2014). Slow-growing broilers are consolidated in France and growing its market very fast in other countries (e.g. The Netherlands; Toudic, 2016). Deloitte (2016) shows health and wellness as the most significant evolving driver to USA consumers in food purchase, with antibiotic-free, cage-free eggs and slow-growing broilers as the most important increasing demand of poultry products (Thornton, 2016). Especially relevant is the case of the demand of cage-free eggs, where Widowski (2016) states that the public concern for the welfare of laying hens is rapidly changing the ways that eggs are being produced and marketed around the world.

Poultry production and sustainability

Main challenges of world sustainability

WWF report (2016) synthesizes the great amount of evidence showing the Earth system is under increasing threat: climate, biodiversity, ocean health, deforestation, the water cycle, the nitrogen cycle, the carbon cycle. Humanity currently needs the regenerative capacity of 1.6 Earths to provide the goods and services we use each. Many researchers suggest that we are entering a new era in which humans rather than natural forces are the primary drivers of planetary change: the Anthropocene. Lorek and Vergragt (2015) assert that research has hardly begun to investigate a possible and necessary societal transition to new and sustainable production and consumption systems, because it needs to cover the various dimensions (material, technological, economic, cultural, psychological, historical and political) in an integrated way. This transition requires fundamental changes in two global systems: energy and food (WWF, 2016).

Plenary session 06. Hot Topics: sustainability on poultry feeding

Animal-source foods and sustainability

According to De Laurentiis *et al.* (2014) the most recurrent pathways to sustainably manage and mitigate for the contributions that the food system makes to climate change are:
- producing more with less (it implies technology; section 5) and at a lower environmental cost;
- changing diets;
- reducing waste.

Animal-source foods and climate change

In order to keep the increase in global temperature below the crucial ceiling of 2 °C, emissions will have to be reduced by as much as 70% by 2050 (FAO, 2016). Tis target can only be achieved with the contribution of the agriculture sectors which now account for at least one-fifth of total emissions. Based on a Life Cycle Assessment approach (GLEAM 2.0, 2016), it is estimated that poultry supply chains emits about 836 million tons of CO_2 equivalent, about 11% of the total GHG emissions from livestock supply chains (7,1 $GtCO_2$), from which feed is the biggest contributor both, in broiler meat (78%) and in egg production (69%) (MacLeod *et al.*, 2013). In the EU, according to OECD-FA (2011) estimations from 2004 data, beef, sheep and pig production emit 22, 20, 7.5 kg CO_2-eq/kg of product respectively, meanwhile poultry meat and eggs have the lowest figures (5 and 3 respectively), except for cow milk (1.4).

Diet change

Meat consumption and Health. Several recent papers (e.g Virtanen *et al.*, 2015; Stangierski and Lesnierowski 2015; Mottet and Tempio, 2016) describe that animal-source foods, especially poultry meat and eggs, are important to nutrition and health, especially for children, pregnant women and for the elderly, mainly due to providing a wide range of nutrients (Ca, Fe, Zn, vitamins A, B2, B-12, etc.) difficult to obtain in adequate quantities from plants source alone. Beside the risk of food-borne diseases, there are reports and public consciousness that high levels of meat consumption considerably increase the risk of lifestyle diseases such as coronary heart disease, obesity, diabetes and cancer, especially from WHO (2015b) report on red and processed meat consumption. Most of the maximum consumption recommendations concern precisely to these types of meat with levels between 40 and 90 g/d, meanwhile poultry meat maximum recommendations are not proposed, although recommendations for total daily meat intake reach from 60 to 160 g and from 10 to 50 g for eggs (WHO, 2000; Wellesley *et al.*, 2015) from those institutions that have proposed a level. The average meat consumption globally is 115 g per day (UNEP, 2012), with enormous variation from region to region (e.g. 322 g in the USA and 19 in Africa), and large differences within regions (e.g. in Asia from 12 g in India to 160 in China). Europeans consume around 200 g.

Health risks of poultry rearing. The emergence and re-emergency diseases will remain a continuous major challenge given the intensity and frequency of contacts between humans and birds (e.g. recent Avian Influenza H5N8 rapidly spreading). Of the known animal diseases, 61% are zoonotic (IFAH, 2012).

Reduction of Meat consumption and Sustainability. *Greenhouse gas emissions.* The IPCC (2014) identifies changing diets as a significant though undeveloped area for action for GHG mitigation. If population and food consumption patterns continue, as mentioned in section 2, food production alone will reach (80% increase) the global targets for total greenhouse gas emissions in 2050 (Bajželj *et al.*, 2014). As previously said, reducing ruminant meat and dairy products has the highest impact on GHG emissions compared to other foods. According to Tilman and Clark (2014) a vegetarian diet could reduce both, emissions from food production by 55% per capita and 600 million Ha for cropland compared to the projected diet patterns in 2050.

Plenary session 06. Hot Topics: sustainability on poultry feeding

Land and water use. On a global scale, livestock takes up about 70% of all agricultural land (Steinfeld *et al.*, 2006). Despite poultry protein is more efficiently produced than that from pork or ruminants, poultry production is the sub-sector that requires the most land for cereal production with an estimated 93 million ha in 2010 (44% of the total cereal area required by the global livestock sector; Mottet *et al.*, 2016). Besides, according to Mottet and Tempio (2016) the estimated composition of the global feed ration of poultry includes 64% of human edible feed material. In total, 65% of the expansion in land use between 1960 and 2011 is due to increased production of animal products (Alexander *et al.*, 2015), driving to loss of biodiversity and releasing more greenhouse gases. On the other hand, small-scale agriculture and extensive husbandry on pastures is generally less harmful or even beneficial for biodiversity (FAO 2006). A further effect of increased land use is the threat of food security through land grabbing for the expansion of cropland for feed and pastures (Lovera 2015). Mekonnen and Hoekstra (2012) estimated that poultry needed an average of 4325 m3 of water per ton of meat and 3265 m^3 per ton of eggs, which account for 11% and 7% of the total water footprint of animal production. Those figures are the lowest among animal products.

Changing food consumption patterns. Based on a meta-analysis of 155 studies Stoll-Kleemann and Schmidt (2016) review the barriers, opportunities and steps in order to encourage the consumption of less meat. Verain *et al.* (2015) propose 'Flexeterianism' (People who don't abstain from meat as a matter of principle but eat less of it), to promote less meat intake. New dietary guidelines of the Chinese Ministry of Health are aiming to reduce by 2030 meat consumption by 50% of the current level of 50-55 kg/meat/person to reduce both gas emissions and country´s nutritional problems of obesity and diabetes (UECBV, 2016). Kottmeyer (2016) foresees China encouraging less meat and eggs goes 'global'.

Indicators of this trend are the tax that the Danish ethics council has called for on red meat in 2016, and the growing trend of vegetarians and vegans that nowadays reach about a billion people worldwide, in part due to cultural and religious factors (Leahy *et al.*, 2010).

New developments. Plant based protein foods as alternative to meat are gaining market and several big companies are investing in this area (Hughes, 2016). Other more long term projects include meat recreation in a laboratory. The first commercial laboratory hamburger is announced for about five years (Ghosh, 2015).

Food wastage

FAO (2011) has estimated that every year roughly one-third of the edible parts of food produced for human consumption is lost. This food loss accounts for about 8% of global GHG emissions as well as being a misuse of resources (FAO, 2016). Stenmark *et al.* (2016) estimates a food waste in the EU-28 of 88 million tonnes (173 kilograms of food waste per person; 20% of the total food produced). The sectors contributing the most to food waste are households (47 million tonnes) and processing (17 million tonnes). Apart from the contribution of several private companies and institutions (e.g. Food Bank), the feed industry is a key stakeholder in decreasing food wastage. According to zu Ermgassen (2016) Japan and South Korea recycle around 40% of their food waste as feed after a properly regulated heat treatment. According to zu Ermgassen *et al.* (2016), if the European Union lifted the ban imposed to food waste, if properly treated for feed, around 1.8 million hectares of land could be saved.

Human health and antimicrobial resistance (AMR)

The World Health Organization (WHO, 2014) predicts that if no additional measures are taken, the annual death toll attributable to AMR may rise to 10 million and exceed other causes such as cancer in 2050. UN General Assembly Declaration in September 2016 refers to AMR as the current biggest challenge to public health. In order to increase livestock's positive contribution to human health, and reduce their negative impact on AMR, animal health should be made a priority in public policies as a One Health approach (FAO-AGAL, 2016).

Poultry nutrition achievements and challenges

According to the FAO (2003), 70% of the increased production needed to meet the growing global demand for food and protein must come from the use of technology. This concept refers to the IPAT formula (Ehrlich and Holden 1971; Thøgersen 2014) where the impact (I) on the environment is determined by population (P), affluence (A) and technology (T). In the previous sections the P and A elements of the equation have been reviewed, and in this one the main T achievements and challenges concerning poultry nutrition will be overviewed.

Achievements

Thanks to a better efficiency, between 1960 and 2010, global average GHG emission per kg of product fell by an estimated 38% for milk, 45% for pork, 76% for chicken meat and 57% for eggs (Smith *et al.*, 2014). According to Fancher (2014) looking forward 10 years, strategic use of genetic resources available should provide plenty of variation for continued progress in broiler chicken productivity, welfare and environmental impact. As far as poultry nutrition achievements, Elwinger *et al.* (2016) have recently published a review in commemoration of the over 100 years since the founding of WPSA. As the authors say, this review gives a flavor of the advances in nutritional science and practice in poultry.

Challenges

After the discussion of this paper and following the thoughts of den Hartog *et al.* (2016) and Choct (2016), the importance of using a holistic approach to enable successful conversion of feed into high quality poultry protein in a sustainable way is evident. Apart from changing to a less meat diet and reducing food wastage, as far as poultry nutrition is concerned two strategies come to mind. Firstly, a strategy to increase productivity through precision nutrition methods and tools and big data management to economically optimize the feeding programs and reduce emissions into the environment. Targeted feed additive strategies applied to support gut health will contribute to establish a responsible, prudent use of antibiotics and will also fit in strategies to reduce the prevalence of food pathogens. Recent advances in science also highlight the importance of early life nutrition for later life performance, health and product quality. On the other hand poultry nutrition must give response to both, the evolving drivers of consumer´s meat and egg demand and to the small-scale production systems. Secondly, to save land, a strategy to explore novel sources of feed as worms, insects, larvae, aquatic plants and microbial proteins as presented several times at this Symposium and in our Journal (e.g. Khusro *et al.*, 2012; Świątkiewicz *et al.*, 2015) and elsewhere, especially Dutch researchers (e.g. Veldkamp *et al.*, 2013; van Krimpen *et al.*, 2013) and also Danish ones with the Starpro project (marine proteins including starfish) among other initiatives.

FAO proposal: sustainable animal diets (StAnD)

FAO (2014) has developed a concept of Sustainable Animal Diets (StAnD), integrating the importance of protecting the environment, efficient use of natural resources, socio-cultural benefits, and ethical integrity and sensitivity, in addition to currently recognized nutrition-based criteria in producing safe and economically viable feed. It is based on the Three-P dimensions of sustainability (Planet, People and Profit), complemented by the ethics of using a particular feed. Vaarst *et al.* (2015) take into consideration these aspects when they conclude that a system cannot be considered sustainable in itself if it forms part of a larger system that involves unsustainable practices or structures. To consider the environmental impact of feed ingredients, Leinonen *et al.* (2013) propose to quantify their Global Warming Potential, Eutrophication Potential and Acidification Potential by using the Life Cycle Assessment. This proposal represents an advance to this holistic approach in feed formulation and as Penz (2016) asserts, in the future, nutritionists will need to participate in sustainability initiatives.

Conclusions

As poultry nutrition specialists, when designing feeding programs, we have to take into account, not only the minimum cost but also both, the evolving key drivers of consumers´ demand and especially the sustainability, social, animal welfare, safety, human and animal health consequences of our decisions. This focus requires a holistic approach to poultry nutrition which implies a deeper knowledge and collaboration among the wide variety of disciplines involved.

References

Alexander P., Rounsevell M.D.A., Dislich C., Dodson J.R., Engström K. and Moran D. 2015. *Global Environmental Change* 35: 138-147.

Alexandratos N. and Bruinsma J. 2012. World agriculture towards 2030/2050. Rome.

Bajželj B., Richards K.S., Allwood J.M., Smith, P., Dennis J.S., Curmi E. and Gilligan C.A. 2014. *Nature Climate Change* 4: 1-6.

Choct M. 2016. Proceedings of XXV World's Poultry Congress. Beijing. L23: 122-127.

De Laurentiis V., Hunt D.V.L. and Rogers C.D.F. 2014. Proceedings World Sustainability Forum. 1-30 November. DOI: 10.3390/wsf-4-g003.

Deloitte. 2016. Capitalizing on the shifting consumer food value equation. https://www2.deloitte.com/content/dam/Deloitte/us/Documents/consumer-business/us-fmi-gma-report.pdf

Den Hartog L.A., Garcia Ruiz A.I., Smits C.H.M. and Scott T. 2016. Proceedings of XXV World's Poultry Congress. Beijing. L60: 285-288.

Ehrlich P. and Holden J. 1971. *Science* 171: 1212-1217.

Elwinger E., Fisher C., Jeroch H., Sauveur B., Tiller H. and Whitehead C.C. 2016. *World's Poultry Science Journal* 72: 701-720.

Fancher B.I. 2014. What is the upper limit to commercially relevant body weight in modern broilers? Aviagen. Huntsville. USA.

FAO (Food and Agriculture Organization) 2003. World Agriculture: towards 2015/2030: A FAO Perspective. Earthscan Publications Ltd. London.

FAO. 2006. Livestock's long shadow. Environmental issues and options. Rome.

FAO. 2011. Global food losses and food waste: extent, causes and prevention. Rome.

FAO. 2014. Towards a concept of sustainable animal diets. Rome.

FAO. 2016 The state of food and agriculture. Climate change, agriculture and food security. Rome.

FAO-AGAL, 2016. Synthesis Livestock and the Sustainable Development Goals. Rome.

Fasina F.O., Ali A.M., Yilma J.M., Thieme O. and Ankers P. 2016. *World's Poultry Science Journal* 72: 178-188.

GLEAM 2, 2016. Global Livestock Environmental Assessment Model. FAO. Rome.

Ghosh, P. 2015. BBC News. http://www.bbc.com/news/science-environment-34540193.

Hughes, D. 2016. Biomin's World Nutrition Forum. Vancouver.

IFAH (International Federation for Animal Health-Europe) 2012. Annual Report. Brussels.

IPCC (Intergovernmental Panel of Climate Change) 2014. Mitigation of climate change. Contribution to the Fifth Assessment Report of the IPCC. Geneva.

Kottmeyer R. 2016. Chicken Marketing Summit. http://www.wattagnet.com/articles/27597-bold-predictions-for-the-poultry-industrys-future.

Khusro M., Andrew N.R. and Nicholas A. 2012. *World's Poultry Science Journal* 68: 435-446.

Lan Phuong T.N., Dong Xuan K.D.T and Szalay I. 2015. *World's Poultry Science Journal* 71: 385-396.

Leahy E., Lyons S. and Tol R.S.J. 2010. ESRI working paper no. 340. The Economic and Social Research Institute. Dublin.

Leinonen I., Williams A.G., Waller A.H. and Kyriazakis I. 2013. *Agricultural Systems* 121: 33-42.

Lorek S. and Vergragt P.J. 2015. Handbook of Research on Sustainable Consumption. Eds.: Reisch L.A. and Thøgersen J. pp. 19-32. Edward Elgar Publishing. Cheltenham. UK.

Lovera M. 2015. Global Forest Coalition and Brighter Green, Asunción.

Plenary session 06. Hot Topics: sustainability on poultry feeding

MacLeod M., Gerber P., Mottet A., Tempio G., Falcucci A., Opio C., Vellinga T., Henderson B. and Steinfeld H. 2013. Greenhouse gas emissions from pig and chicken supply chains – A global life cycle assessment. FAO. Rome.

Mekonnen M.M. and Hoekstra A.Y. 2012. *Ecosystems* 15: 401-415.

Mottet A. and Tempio G. 2016. Proceedings of XXV World's Poultry Congress. Beijing. L1: 1-8.

Mottet A., de Hann C., Falcucci A., Tempio G. and Gerber, P. 2016. Contribution to the feed/food debate. Global Food Security, under review.

OECD-FAO. 2011. OECD-FAO agricultural outlook 2011-2020. Paris.

Penz A.M. 2016. Proceedings of XXV World's Poultry Congress. Beijing. L21: 111-115.

Sloyan M. 2017. http://www.wattagnet.com/articles/29436-questions-about-the-future-of-animal-protein-production.

Smith P., Bustamante M., Ahammad H., Clark H., Dong H., Elsiddig E.A., Haberl H., Harper R., House J., Jafari M., Masera O., Mbo, C., Ravindranath N. H., Rice C.W., Robledo Abad C., Romanovskaya A., Sperling F. and Tubiello F. 2014. Contribution of Working Group III to the Fifth Assessment Report of the IPCC. Cambridge University Press.

Stangierski J. and Lesnierowski G. 2015. *World's Poultry Science Journal* 71: 71-82.

Steinfeld H., Gerber P., Wassenaar T., Castel V., Rosales M. and De Haan C. 2006. Livestock's long shadow – environmental issues and options. FAO, Rome, Italy, pp. 390.

Stenmarck Å., Jensen C., Quested T. and Moates G. 2016. FUSIONS (Food Use for Social Innovation by Optimising Waste Prevention Strategies). IVL Swedish Environmental Research Institute. Stockholm.

Stoll-Kleemann S. and Schmidt U.J. 2015. Regional Environmental Change 16. Springer.

Świątkiewicz S., Arczewska-Wloseki A. and Jozefiak, D. 2015. *World's Poultry Science Journal* 71: 663-672.

Thøgersen, J. 2014. *European Psychologist* 19: 84-95

Thornton G. (2016). http://www.wattagnet.com/blogs/6-all-things-poultry/post/27519-is-your-poultry-production-business-new-normal.

Tilman D. and Clark M. 2014. *Nature* 515:518-522.

Toudic, C. 2016. National Chicken Council Chicken. Marketing Summit. 10-12 July. Hilton Head Island.

UECBV (European Union of Livestock and Meat Trade). 2016. Chinese trends in meat consumption in function of the environmental policy. 1040 Bruxelles. Ref: 9969.

UN (United Nations). 2015. Transforming our world: the 2030 Agenda for Sustainable Development. Resolution adopted by the General Assembly on 25 September 2015.

UN Dpt. of Economic and Social Affairs. 2015b. World Population Prospects: The 2015 Revision. Working Paper No. ESA/P/WP.241.

UNEP (Global Environmental Alert Service). 2012. http://www.unep.org/pdf/unep-geas_oct_2012.pdf.

Vaarst M., Steenfeldt S. and Hosrsted K. 2015. *World's Poultry Science Journal* 71: 609-620

Van Horne P.L.M and Bondt N. 2014. LEI Wageningen UR. Report LEI 2014-038.

Van Krimpen M.M., Bikker P., van der Meer I.M., van der Peet-Schwering C.M.C. and Vereijken J.M. 2013. Report 662. Livestock Research. Wageningen UR.

Veldkamp T., Van Duinkerken G., Van Huis A., Lakemond C.M.M., Ottevanger E., Bosch G. and Van Boekel M.A.J.S. 2012. Report 638. Livestock Research. Wageningen UR.

Verain M.C.D., Dagevos H. and Antonides G. 2015. In: Handbook of Research on Sustainable Consumption / Reisch L.A., Thogersen J., Edward Elgar. pp. 209-223.

Virtanen J. K., Mursu J., Tuomainen T-P., Virtanen H. and Voutilainen S. 2015. *Journal of Clinical Nutrition*, doi: 10.3945/%u200Bajcn.114.104109.

Vucasovic T. 2014. *World's Poultry Science Journal* 70: 289-301.

Walley K., Parrott P., Custance P., Meledo-Abraham P. and Bourdin A. 2015. *World's Poultry Science Journal* 71: 5-13.

Wellesley L., Happer C. and Froggatt A. 2015. Chatham House Report. The Royal Institute of International Affairs Chatham House. London.

WHO (World Health Organization) European Region 2000. CINDI dietary guide. http://www.euro.who.int/Document/E70041.pdf.

WHO 2014. Antimicrobial resistance: Global report on surveillance. http://apps.who.int/iris/bitstream/10665/112642/1/9789241564748_eng.pdf.

WHO. 2015. Global Action Plan on Antibiotic Resistance. Geneva. Switzerland.

WHO. 2015b. IARC. International Agency for Research on Cancer. Monographs evaluate consumption of red meat and processed meat. Press release no. 240. Lyon.

Widowski T.M. 2016. Proceedings of XXV World's Poultry Congress. Beijing. L15: 74-78.

WWF (World Wildlife Fund). 2016. Living Planet Report 2016. Gland. Switzerland.

Zu Ermgassen, E. K.H.J. 2016. http://www.allaboutfeed.net/Raw-Materials/Articles/2016/4/Regulate-the-use-of-food-waste-as-feed-2778480W.

Zu Ermgassen, E.K.H.J, Phalan, B., Green, R.E. and Balmford, A. 2016. *Food Policy* 58: 35-48.

Feeding broilers of the future

A.M. Penz, Junior
Cargill Animal Nutrition, Minneapolis, USA; mario_penz@cargill.com

Summary

The future of broiler feeding means a paradigm shift for nutritionists. They will continue been challenged to come up with low-cost feed formulations that provide best-performance results. However, they will need to interface more with other poultry production areas (management, pathology, environment, etc.) because the society is demanding more transparency on how production animals are being fed and treated. The trend toward environmental accountability and sustainability innovation will continue to be strong market drivers. The nutritionists will need to give more attention to ingredient quality; work on phase and sex feeding differentiation; concentrate on particle size and pellet quality, use additives more efficiently, paying special attention to enzymes. The consumers are requiring more AGP-free growing broilers and just started to challenge the poultry producers if the broilers would 'prefer' being produced in a slower speed of growing. So, new methods of feeding will need to be developed to meet these demands.

Introduction

In the last 50 years, the most important scientific advances were focused on poultry production knowledge; on quick adaptation of backyard production to industrialized business and on low cost of broiler meat and hens' eggs. Data from the Food and Agriculture Organization, of the United Nations (FAO) (2015) confirms broiler consumption is growing relative to other proteins. However, cost of production will remain a central issue and, at the same time, producers must be profitable enough to continue working with full efficiency.

Also, biosecurity is an issue that nutritionists will need to be involved. With the onset of avian influenza, there is increased attention on broiler losses and farm biosecurity. In addition, many consumers are confused about how avian pathology (*Salmonella, Campylobacter, Newcastle disease*) can affect human health. To avoid compromising farm biosecurity, governments, industry professionals and producers must work together to establish and to follow best practices, such as controlling the presence of backyard/wild chickens living close to production farms. In addition, farms must be as clean as possible, and employees must implement sanitation programs, established according to veterinary global recommendations.

In the last 20 years, consumers have been more active in expressing concerns on why production animals grow faster than ever, on food safety, on animal welfare, and there is an emphasis on sustainable production.

So, how can broiler nutrition and feeding innovate, respecting society concerns on sustainability? There are important issues to be addressed. Animals feed conversion must improve (use less feed), reducing excreta (pollutant) production. Nutrition must be managed, looking on animal water consumption. To achieve a more sustainable production, the only way is improving the understanding of precision nutrition so that producers can feed animals with low or no margin for error. This article will offer some suggestion for nutritionists who need to revisit concepts and innovate for the future of poultry production.

Aspects to be considered on future broiler nutrition

Broiler production free of antibiotic growth promoters (AGP)

The first proposal on producing broilers AGP free came from Sweden in 1986 (Cogliani *et al.*, 2011). Initial reactions to the proposal focused on a loss of broiler production efficiency, wherein the cost of production would increase. After many years of research, however, these assumptions are no longer accepted in the industry. Societal concern in Europe promoted the development of research on non-antibiotic additives (probiotics, prebiotics, essential oils, organic acids, antioxidants, etc.) and a more efficient use of enzymes, which preserves gut health, with minimal or no reduction on broiler performance. Simultaneously, technical broiler advisers reinforced the implementation of management best practices, feeding and biosecurity care, all improving infection prevention and minimizing production inefficiency.

While the production of AGP-free animals is now an option in many countries, the speed of implementation varies. In 2015, the U.S. finally joined the European movement. The move to AGP-free started in 2010, when the Food and Drug Administration (FDA) called for a strategy to phase out production use of medically important antimicrobial products and to bring the remaining therapeutic uses under the oversight of a veterinarian. This had an important consequence. Food chain and supermarket enterprises accepted the challenge and began publicizing that the ingredients used in their products were AGP-free. Consumers picked up on the news, and the AGP-free movement kept growing. Today, reducing antibiotics in production remains an important innovation priority for nutritionists.

Ingredient analysis

Knowing ingredient composition, with the help of qualified laboratory support, is mandatory if the nutritionists want to formulate feed with a lean safety margin. Historical data shows that many nutritionists underestimate ingredient nutrient values, which does not guarantee better broiler performance. Instead, it means an incremental nutrient loss and a possible increase on pollution (nitrogen and phosphorus). With the progress of NIR technology, mills have no excuse for not having constant nutrient evaluation on their ingredients (Black *et al.*, 2014). So, ingredients used in the diets are not just 'commodities' when used in feed formulation, but important nutrient drivers of production. New technologies like 'in-line feed formulation' are pushing the bounds of precision nutrition. The technology allows a NIR to be installed in the ingredients transportation line, before the mixer, which allows an immediate reading and consequent formulation and dosing, according to the nutrient composition of the ingredients that were read. This and other advancements are helping feed mills to deliver feed with exacting specification to produce better poultry, more sustainably.

Digestibility and phase feeding

The digestibility of the nutrients and energy vary in different production phases. For example, younger chicks are less efficient than older chicks (Noy and Sklan, 1995; Batal and Parsons, 2002a,b; Thomas *et al.*, 2008). In addition, composition and digestibility of ingredients can differ in various years and global regions. Knowing these digestibility differences and correctly calculating the required ingredient composition for the various phases will be an important skill for nutritionists in years to come. Getting digestibility calculations, before feed formulation, means a lessened diet cost, better host digestive tract health and increased environment sustainability.

In addition, to improving digestibility calculations, increasing the number of phases during production ensures less nutrient waste. For example, using five feeding phases, compared to three phases in broiler feeding, increases lysine delivery precision, and as a consequence, the other essential amino acids (ideal protein). Brewer *et al.* (2012b) working on phase feeding, identified that it did not affect weight gain but improved feed efficiency on two of three strains evaluated. However, phase feeding significantly reduced protein and amino acid consumption, making the cost of feeding cheaper than the conventional system. Similar results were seen by Brewer *et al.* (2012a). These findings confirmed what was observed by Angel *et al.* (2006). Moving from four to six broiler phases (0 to 42 days of age) and supplementing the six diets with amino acids (lysine, methionine, threonine, isoleucine, valine, arginine and tryptophan), the nitrogen excretion was reduced by 40% when compared to the control proposal (4 phases).

Plenary session 06. Hot Topics: sustainability on poultry feeding

Sometimes phase feeding is limited because of the feed mill structure. This new understanding of phase importance must be considered when a new feed mill project is developed and logistics are defined, so that the mill does not become the bottle neck of production.

The scientists are concern with the best use of nutrients and energy but, in the future, more research will be done using ingredient net energy instead of conventional metabolized energy, another method that will favour environmental sustainability.

Sex feeding

Generally, sex feeding is a controversial subject among nutritionists, hatcheries, feed mills, broiler production and slaughter plant managers. But, in the future, an increase focus on sustainability will require that all areas of production contribute logistically to improvement. The differences between male and female growth speed, body composition, nutrition requirement and behaviour are sufficient to justify separate sex production. This strategy promotes reduction of feed cost and, most importantly, reduction of slaughter weight variability. In mixed-sex production, males will need to stay in the barns longer to push the weight up to a specific average body weight that cannot be achieved by females. That procedure can double body weight variability, making male feed efficiency worse and increasing mortality, because of the extra days in the barn. The Aviagen manual for broiler 308 (2014) and the Cobb manual for broiler 500 (2015) do not offer information on sex nutrient requirement differences. However, Aviagen (2014) suggests that in the case of using separate sex feeding, modifications of nutrients and energy requirements should be considered. Both manuals show differences in the speed of weight gain and feed efficiency of different sexes and they differ on nutrients and energy recommendations. Therefore, broilers of each of these strains require different feed and nutrients. These considerations were confirmed by Faridy *et al.* (2015) when they evaluated data research from Cobb and Ross strains. Using meta-analysis, they concluded that males require more lysine than females and there is a difference of lysine requirement according to the strain. Also, lysine requirement increases with the increase of crude protein in the diet.

Particle size and pelleting

The nutritionists and feed mill managers must look more to particle size and feed pelleting. Feed with larger particle size promotes better gizzard development, gastric motility and gastro duodenal reflux in poultry. It improves digestion and reduces the entrance of pathogens in the intestine (Amerah *et al.*, 2008; Gabriel *et al.*, 2008, Svihus, 2011). Large particles require less energy during feed apprehension, as birds require fewer pecks to ingest the same amount of feed (Amerah *et al.*, 2007) and feed mills save energy from reduced grinding (Reece *et al.*, 1986).

Broilers fed pelleted diets showed better performance than those fed mash diets, with improvement directly related to pellet quality (McKinney and Teeter, 2004). Poultry fed pellets have higher dietary density; higher feed intake; reduction of energy for consumption; better starch and protein digestibility and reduction of feed waste (Amerah *et al.*, 2007; Dozier *et al.*, 2010). Zang *et al.* (2009) added that pellet improves intestinal function, as shown by increase in villi height and in villi height to crypt depth ratio. Most of these earlier findings were confirmed lately by Naderinejad *et al.* (2016), who reinforced that coarse corn particle size and pellet diets improve gizzard development and function and improve the use of nutrients and energy that promote better broiler performance.

Additives – enzymes

Sometimes enzymes are wrongly called additives. They are proteins. They can improve digestion of nutrients and energy but also offer nutrients and energy to the host. New enzyme technology has been growing very quickly in the last 15-20 years and it will continue to do so in the future. In 1996, Cowan *et al.* said that this technology improves ingredient digestibility and nutrient absorption. Also, Penz Jr. and Bruno (2010) reinforced that enzyme technology reduces pollutant excretion in animal waste. There are many enzymes available to help the digestion of phosphorus, calcium, carbohydrates, proteins and lipids of diets offered to the animals, especially poultry and swine. However, enzymes will not always work with 100% efficiency because they are exogenous sources. Their efficient use will depend on correct technical decisions, correct mixing and substrate availability.

There is a vast wealth of information available describing different modes of action and different enzyme products and inclusions, and nutritionists should look at enzyme use as another way to make poultry production more environmentally sustainable. Improvement on enzyme-promoted nutrients and energy digestibility might reduce waste disposal and nitrogen and phosphorus pollution. In addition, with the appeal of producing AGP-free broilers increasing, gut health is more important than ever. It is well known that around 70% of poultry immune response is provided by digestive tract cell stimulation, and well-used enzymes have a unique, positive effect on broiler immune response from the gut.

Conclusion

The future on poultry nutrition will follow the need to achieve more precise feed formulation. Safety margins on feed specifications will not be accepted. Precision nutrition will not compromise the animal efficiency and the cost of production and it will bring answers to the society that is willing to get more information on how the animals are fed and treated.

References

Amerah, A.M. *et al.* 2007. World's Poultry Science Journal 63: 439-455.

Amerah, A.M. *et al.* 2008. Poultry Science 87: 2320-2328.

Angel, R. *et al.* 2006. Workshop on Agricultural Air Quality: State of Science, pp. 460-463.

Aviagen 308. 2014. Nutrition specifications, pp 8.

Batal, A.B. and Parsons, C.M. 2002a. *Poultry Science* 81: 400-407.

Batal, A.B. and Parsons, C.M. 2002b. *Poultry Science* 81: 853-859.

Black, J.L. *et al.* 2014. Australian Poultry Science Symposium, pp. 23-30.

Brewer, V.B. *et al.* 2012a. *Poultry Science* 91: 1256-1261.

Brewer, V.B. *et al.* 2012b. *Poultry Science* 91: 1262-1268.

Cobb 500. 2015. Broiler performance and nutrient specification, pp.14.

Cogliani, C. *et al.* 2011. *Microbe* 6: 274-279.

Cowan W.D. *et al.* 1996. *Animal Feed Science and Technology* 60: 311-319.

Dozier, W.A. *et al.* 2010. *Journal of Applied Poultry Research* 19: 219-226.

FAO, 2015. World Agriculture: Towards 2015/2030. An FAO Perspective. www.fao.org/docrep/005/y4252e/y4252e05b.htm.

Faridi, A. *et al.* 2015. *Livestock Science* 181: 77-84.

FDA (Food and Drug Administration). 2015. *Animal Feed Science and Technology* 142: 144-162.

McKinney, L.J and Teeter, R.G. 2004. *Poultry Science* 83: 1165-1174.

Naderinejad, F., *et al.* 2016. *Animal Feed Science and Technology* 215: 92-104.

Noy, Y. and Sklan, D. 1995. *Poultry Science* 74: 366-373.

Penz, Jr, A.M. and D.G., Bruno. 2010. Proceedings, Conferência Facta de Ciência e Tecnologia Avícolas 28: 17-34.

Reece, *et al.* 1986. *Poultry Science* 65: 636-641.

Svihus, B. 2011. *World's Poultry Science Journal* 67: 207-224.

Thomas, D.V. *et al.* 2008. *British Poultry Science* 49: 429-435.

Zang, J.J. *et al.* 2009. *Asian-Australian Journal of Animal Science* 22: 107-112.

Abstracts

Effect of various fibrous ingredients on performance, gizzard weight and gut microflora of broilers

K. Bébin[1], D. Gardan-Salmon[1], C. Jacquot[2], M. Urdaci[2], M. Arturo-Schaan[1] and M. Panhéleux-Lebastard[1]
[1]Groupe CCPA, ZA du Teillay, 35150 Janzé, France, [2]Bordeaux Sciences Agro-Université de Bordeaux, Laboratoire de Microbiologie, UMR 5248, 1, Cours du Général de Gaulle, 33170 Gradignan, France; kbebin@groupe-ccpa.com

The value of fiber in poultry diets has been studied in a many research centres, highlighting differences between fibrous ingredients. In this study, we looked at the effect of different fiber sources on performance, gizzard weight and gut microflora of chickens at 35 days of age. Two dietary fibrous ingredients, beet pulp (BP) and oat hulls (OH) were compared to a control with no fiber supplementation (C). Each diet was distributed to 8 cages of 4 ROSS chickens until 35 days. The weight at 35 days was significantly higher ($P<0.05$) with the two fibrous diets BP and OH (2,330 and 2,314 g respectively) compared to the control (2,134 g) and the feed efficiency was improved with BP (1.62) compared to C and OH (1.66 and 1.64 respectively). In addition, the gizzard weight (% of body weight) was significantly higher ($P<0.01$) with diet OH (0.84%) compared to C and BP (0.75 and 0.69% respectively). In the small intestine, the populations of enterococci and coliforms were significantly decreased with the BP and OH diets. In the caeca, bifidobacteria increased with fibrous diets, mainly with the BP diet. In conclusion, the contribution of a specific type of fibrous ingredients in the diet helped to improve performance and digestive balance in broiler chickens.

Whole wheat choice feeding for heat stress mitigation in free range broilers

M. Singh[1], K.M. Prescilla[1] and A.J. Cowieson[2]
[1]University of Sydney, Cobbitty, NSW 2570, Australia, [2]DSM Nutritional Products, Kaiseraugst, 4303, Switzerland; mini.singh@sydney.edu.au

Conventional climate control systems are largely ineffective in a free range setting especially in summer months when opening of pop-holes can exceed the thermo-neutral zone of the bird causing heat stress. The aim is to test if provision of whole-wheat, a feed source high in startch and low in protein, and thus a lower heat increment, can mitigate heat stress. Spatial separation of the pelleted diet and whole grain may result in nutrient intake regulation in broilers that offer additional feeding strategies to manage feed intake under heat stress. A total of 800 day-old male Cobb 500 broilers were randomly allocated to two tretment groups each consisting of eight pens of 50 birds. Birds were fed either complete pelleted diet (CP: 18.82, AME: 13.18) or complete diet plus whole wheat (CP: 10.33, AME: 12.83) from d 14. Pop holes were opened from d 21, with the average daytime temperature of shed reaching 29.6 °C. Whole-wheat consumed was low, ranging from 1.8-14.19% of total feed intake, although it increased significantly ($P<0.001$) with increase in age of birds. However, whole-wheat intake(measured between 11 AM and 3 PM daily) had a strong negative correlation with increase in temperature ($R^2=0.55$). Blood samples showed significantly elevated levels as compared to standard ($P<0.05$) for K, Cl, pH, HCO_3^-, TCO_2 and haematocrit in birds on both diets, indicating the bird's inability to maintain homeostasis and cope with heat stress and respiratory alkalosis under free range conditions. However, birds fed whole-wheat showed significantly lower glucose ($P<0.01$), TCO_2 ($P<0.05$) and hemoglobin ($P<0.05$) levels. While whole-wheat feeding did not decrease susceptibility to heat stress, it may have provided a source high in energy to offset the thermoregulation requirements, resulting in better performance. Replacing grain in pellets with whole-wheat on the side would be a better choice feeding strategy to encourage higher levels of whole-wheat intake.

Effect of feed processing and structural components on starch digestion dynamics in broilers

K. Itani and B. Svihus

Norwegian University of Life Sciences, Department of Animal and Aquacultural Sciences, P.O. Box 5003, 1432 Aas, Norway; khaled.itani@nmbu.no

The hypothesis of this experiment was that stimulating gizzard function through structural components would affect digesta flow and increase starch digestion rate, and that this effect would be larger with pelleted (P) as compared to extruded (E) diet. Treatments were arranged in a 2×2 factorial design with 6 replicate birds per treatment, consisting of fine or coarse particle sizes and pelleting or extrusion as processing methods. 96 male broilers were divided into two groups and fed a P wheat-based diet with cellulose (C) or oat hulls (OH) at 50 g/kg. On day 20, these groups were subdivided into groups of 24 birds subjected to either 3 or 8 h-starvation. Prior to this, birds were given a starch- and marker-free diet for 10 h to ensure the digestive tract did not contain either one. After feed deprivation, birds were allowed 30 min access to P or E wheat-based diet with C. Excreta were collected quantitatively 90 min after feed access, and 4 times every 45 min thereafter. On day 21, all birds were treated as for day 20, but were starved 5 h. 24 were killed each time at 1, 2, 3 and 4 h after feed access. OH feeding increased gizzard size and holding capacity ($P<0.0001$). Digesta passed into the jejunum more rapidly at 1 h ($P=0.001$) and 2 h ($P=0.0184$) for birds fed diet with C. At 1 and 2 h, jejunal starch digestibility was significantly lower for P diet with C, with tendency ($P=0.0537$ and $P=0.0863$) for an interaction effect between fiber and processing. Extrusion resulted in higher ($P<0.05$) starch digestibility and tended ($P=0.0915$) to alleviate the negative effect of lack of OH on ileal digestibility. Birds killed at 4 h showed few such effects. Although no particularly high level of starch was found in the excreta, 8 h starved birds fed P diet exhibited larger ($P<0.05$) starch excretion when given C during the first 270 min. Some starch may be lost in the excreta when diets are pelleted, but higher starch gelatinization through extrusion or improved gizzard function can alleviate this problem.

Dietary free fatty acids and saturation degree modify lipid absorption dynamics in broiler chickens

R. Rodriguez-Sanchez[1], A. Tres[2] and A.C. Barroeta[1]

[1]Animal Nutrition and Welfare Research Group (SNIBA), Department of Animal and Food Science, Veterinary Faculty, Campus UAB, Travessera dels Turons-Edifici V, 08193 Bellaterra, Spain, [2]LiBiFOOD, Department of Nutrition and Food Science, Farmacy Faculty, University of Barcelona, Av Joan XXIII s/n, 08028 Barcelona, Spain; raquel.rodriguez@uab.cat

The aim of the present experiment was to study the influence of the level of free fatty acids (FFA) in dietary fat with different degree of saturation on the fatty acid (FA) absorption dynamics along the gastrointestinal tract (GIT) in broiler chickens. A total of 528 one-day-old broiler chickens were randomly distributed in 8 treatments (6 replicates/treatment). The 8 treatments resulted from the addition of a 6% of a saturated (P, palm) or an unsaturated fat source (S, soybean), and 4 increasing levels of FFA (0, 15, 40 and 60%) for each one of the fat sources. At 14 d samples of excreta and the digestive content from jejunum and ileum were collected. FAs were analysed using GC-FID and the apparent digestibility coefficients were calculated using the titanium marker ratio in the diet and digestive content or excreta. S diets had higher total fatty acid (TFA) digestibility coefficients than P diets in jejunum, ileum and excreta ($P<0.01$). FFA level had an effect on saturated fatty acid digestibility in ileum and excreta ($P<0.01$). TFA digestibility coefficients calculated in excreta were S0: 0.87a, S15: 0.86ab, S40: 0.79b, S60: 0.79ab, P0: 0.69c, P15: 0.69c, P40: 0.72bc and P60: 0.66c. Low levels of FFA resulted in similar digestibility coefficients than the diets with a conventional fat (S0 and P0). TFA absorption, calculated as a proportion of total digestibility in excreta, mainly took place at jejunum level (65-75%). TFA absorption rate decreased as the fat saturation degree and FFA level increased. Low levels of FFA could be used in broiler chicken diets with no detriments in FA digestibility.

In vitro fermentation of cassava leaf meal and palm kernel cake mixture with *Bacillus amyloliquefaciens*

Y. Rizal, A. Yuniza, T.D. Nova, W.A. Angga and A. Annisa
University of Andalas, Animal Science, Kampus Limau Manis, 25163 Padang, Indonesia; yosemini@yahoo.com

An experiment was conducted to determine the appropriate inoculum dose of *Bacillus amyloliquefaciens* and fermentation length in fermenting 80:20% cassava leaf meal (CLM) to palm kernel cake (PKC) mixture ratio based on alterations of its dry matter (DM) and nutrient contents. This experiment was performed in a 4×3 factorial arrangement of treatments in a completely randomized design with 3 replicates. The first factor was inoculum dose (4.8×10^{10}, 9.6×10^{10}, 14.4×10^{10} and 19.2×10^{10} cfu/g), and the second was fermentation length (4, 6 and 8 days). Measured variables were DM, crude fiber (CF) and crude lipid (CL) reduction percentage, and crude protein (CP) augmenting percentage. Results of experiment indicated that both inoculum dose and fermentation length affected ($P<0.01$) DM reduction. The inoculum dose and fermentation length tended to be interacted ($P<0.10$) in DM reduction. The lowest DM reduction (4.5%) was at 14.4×10^{10} inoculum dose and 6 days fermentation length. The CF reduction was affected ($P<0.01$) by inoculum doses, as well as by fermentation length. There was a significant interaction ($P<0.05$) between inoculum dose and fermentation length. The highest CF reduction (55.6%) was at 19.2×10^{10} inoculum dose and 8 days fermentation length. Inoculum dose did not influence ($P>0.05$) CL reduction, whereas fermentation length did ($P<0.01$). The interaction between inoculum dose and fermentation length was not detected. The highest CL reduction (25.7%) was at 6 days fermentation length. The CP augmenting was not affected ($P>0.05$) by inoculum dose, but it was by fermentation length ($P<0.01$). The highest CP augmenting (14.5%) was at 6 days fermentation length. Inoculum dose did not interact ($P>0.05$) with fermentation length. Thus, the appropriate inoculum dose of *B. amyloliquefaciens* and fermentation length in fermenting 80:20% CLM to PKC mixture ratio was 19.2×10^{10} for 8 days based on the highest CF reduction percentage.

Variability of amino acid digestibility in different field bean cultivars for broilers

J. Abdulla, S.P. Rose, A.M. Mackenzie and V. Pirgozliev
Harper Adams University, Edgmond, Newport, TF10 8NB, United Kingdom; vpirgozliev@harper-adams.ac.uk

The variability of amino acid digestibility (AAD) of ten UK field bean cultivar samples in chicken broilers was studied. Male Ross 308 broilers were reared in 96 floor pens (5 birds in a pen) from 7 to 21 d age. Birds were fed one of twelve mash diets. A wheat-soybean based control diet was formulated to contain 231 g/kg CP and 13.71 MJ/kg metabolisable energy (ME). Ten diets were then produced including 200 g/kg of one of the ten different field bean cultivars and 800 g/kg of the control feed. Additional diet was formulated that contained 100 g/kg of one of the bean samples and 900 g/kg of the balancer feed to allow testing of linear response to AAD. Each diet was fed to 8 pens following randomisation. Ileal digesta were collected at the end of the study, and AA digestibility of the bean samples was determined using a regression approach. Data were statistically compared using a randomized block ANOVA. Tuckey's range test was used to determine significant differences between field bean treatment groups. The mean AAD of the field bean cultivar samples was 0.817 for indispensable, 0.807 for dispensable, and 0.811 for total AA. The lowest mean AAD was observed for proline (0.623) and the highest for lysine (0.861). The mean digestibility of methionine and threonine was 0.821 and 0.810, respectively. The highest coefficient of variation was determined for methionine (12.6%) and the lowest was for lysine (4.2%). There were differences in AAD between the field beans. The lowest AAD was observed for proline in cultivar Sultan (0.466) followed by Clipper (0.585) and then Arthur (0.593) and the highest digestibility was for lysine in cultivar Wizard (0.883) followed by Honey (0.882). The lowest and the highest lysine, threonine and cystine digestibility coefficients were for cultivars Buzz (0.810) and Wizard (0.883), Sultan (0.703) and Fury (0.853), Sultan (0.680) and Honey (0.840), respectively. The AAD of field beans is generally high but varies between cultivars.

Meta-analysis of performance in broilers fed a starter diet with alternative sources to SBM

C. Pedersen and C. Broekner
Hamlet Protein, R&D, Saturnvej 51, 8700 Horsens, Denmark; cap@hamletprotein.dk

When balancing diets, different protein ingredients are used based on chemical compositions and nutritional values in starter broiler diets. Alternatives to SBM as only protein source are interesting in order to reduce the concentration of potassium in the diet, increase protein digestibility and reduce the content of antinutritional factors for full growth potential to be exploited. The aim of the current meta-analysis was to evaluate the findings of 19 performance trials, where different protein sources in the starter diet had been used. Nineteen performance trials conducted at universities and commercial settings in 14 countries, resulting in a total of 73 different starter diets were compared. In 18 of the trials at least 1 diet had SBM as only protein source and was considered the control (n=24). In 19 trials, 1 starter diet included HP AviStart (enzyme treated soy product) (n=40). In 5 trials the starter diets included different levels of HP AviStart (up to 10%). In 5 trials, the starter diets included 1 or more vegetable or animal based protein source (n=9). All 19 trials had 2 to 8 replicates. Majority of the trials finished between day 35 and 42 and the starter period was between 7 and 14 days. Diets were formulated according to local recommendations for energy, amino acids, minerals and vitamins. Response parameters were feed intake, slaughter weight and FCR. The effect of different starter diets on response parameters were evaluated statistically by a two-way ANOVA, using trial as random parameter in SAS, 9.4. No significant ($P=0.938$) effect of the starter diets on feed intake was found. A significant ($P=0.006$) higher slaughter weight was found for birds fed HP AviStart (2,149 g) vs control (2,104 g) and others being intermediate (2,133 g). The FCR was significantly ($P=0.003$) lower for HP AviStart (1.84) vs control (1.88) and others being intermediate (1.87). It can be concluded that using alternative protein sources to SBM in the starter diet can improved the overall performance of broilers.

Effect of replacement of soybean by rape seed plus peas on laying performance of two hen strains

I. Halle
Institute of Animal Nutrition (FLI), Bundesallee 50, 38116 Braunschweig, Germany; ingrid.halle@fli.de

Soybean meal is the most common protein source in diets of poultry species. The majority of this soybean meal is imported from non-European countries and often derived from genetically modified varieties. Legumes like peas could be used as an alternative high protein source in poultry diets. Objectives of this study were to test the effect of a total replacement of soybean meal with rape seed meal plus peas in hens' feed on laying performance and egg quality parameters of two strains of hens with different laying performance. 230 Lohmann Brown (LB) and 230 Lohmann Dual (LD) hens were randomly divided into 4 groups. The hens were kept in pens (23 hens per pen) with 5 pens per group. The study commenced when the hens were 22 weeks old and continued until the 6th laying month (168 days). In the treatment, soybean meal feed (SBM; 21.6%) was totally replaced by 12% rapeseed meal (REM; 8.2 mmol glucosinolate/kg) plus 35% pea 'James' (235 g crude protein/kg). Nutrient content of SBM/REM+peas diets was per kg feed: 11.25/11.3 MJ ME, 0.85/0.9% Lys, 0.7/0.7% Met+Cys, 5.3/7.2% crude fat and 3.1/3.3% linoleic acid. Data were analyzed via ANOVA (SAS) and the Student-Newman-Keuls-test ($P<0.05$). Daily feed intake, laying intensity and egg weight of LB hens was significantly higher compared with LD hens. While no effect of substituting SBM through REM/peas in the diet was seen on laying intensity, the egg weight of LB and LD hens was decreased. The daily egg mass production was lower in the LB REM/peas group compared with LB SBM hens and not significantly different between the two LD groups. During the complete trial period eggs of LD hens showed a significantly higher percentage of egg yolk and as a result a reduced part of egg white. The trial's results allow the conclusion that a total replacement of soybean meal with rape seed meal plus peas in feed of LB hens first of all reduces the egg weight and secondly increases the feed conversion of LD hens.

The use of processed soya in the starter diet improves final performance

H.V. Masey O'neill[1], H. Hall[1] and S. De Vos[2]
[1]AB Agri Ltd., Innovation Way, Peterborough PE2 6FL, United Kingdom, [2]Inve Belgium, 9200, Baasrode, Belgium; helen.maseyoneill@abagri.com

Early chick development and growth is known to be influential on later stage growth and can be influenced by high quality nutrition in the first week. The objective of this study was to test the value of a processed soya product (AlphaSoy 530, AgroKorn, Denmark),in starter diets for broilers, on final growth performance. AlphaSoy 530 undergoes a four step extrusion based process with the involvement of enzymic processing aids and was included at 0, 2.5% or 5% in the starter diet in place of wheat gluten and soya bean meal. The basal diet was based on wheat, maize, soya bean meal and wheat gluten (1.5%). All treatment diets were formulated to be iso-caloric and iso- nitrogenous with approximately equivalent levels of fat and fibre. Six hundred and thirty Ross 308 male broilers were offered one of three dietary starter formulations during d 0-10. Thereafter, all broilers were fed a common commercial diet. There were 6 cage replicates per treatment with 35 broilers per cage. At day 11, there was a tendency ($P<0.1$) for the 5% treatment to have higher BW and improved FCR relative to the control. At day 36, the BW of the 5% treatment was significantly higher than the control ($P<0.05$). The cumulative FCR at day 36 tended to be improved with the 5% treatment, relative to the 2.5% treatment. When FCR was corrected to a constant BW of 2,500 g, the 5% treatment was significantly better than all other treatments ($P<0.05$). The addition of AlphaSoy AS530 to starter broiler diets at 5% as a replacement for wheat gluten and soya bean meal, improves 36 d weight and FCR. The AS530 production process causes compositional changes to the soya, an improved energy value and reduced ANF levels relative to soya bean meal. This is likely to improve digestibility of nutrients and improved energy availability which may improve gut development and improve later growth.

Processed soya improves performance of broilers from 0 to 2 wk of age

H.V. Masey O'neill[1], D. Currie[2], A. Knox[2] and H. Hall[1]
[1]AB Agri Ltd, Innovation Way, Peterborough PE2 6FL, United Kingdom, [2]Roslin Nutrition Ltd, Gosford Estate, East Lothian EH32 0PX, United Kingdom; helen.maseyoneill@abagri.com

Early chick growth is influenced by high quality nutrition in the first week. The objective of this study was to test the value of a processed soya product (AS: AlphaSoy 530, AgroKorn, Denmark; 53% CP), at three dose rates (0, 7.5% or 15%) compared to an alternative soya product (ASP: 7.5% inclusion; 56% CP) and a wheat-maize soya control, on d14 growth development. AS undergoes a four-step extrusion based process with the involvement of enzymic processing aids and ASP undergoes a similar but proprietary process. The AS production process causes compositional changes to the soya, an improved energy value and reduced ANF levels relative to soya bean meal. This improves product quality and therefore feed efficiency and growth. All test ingredients were included in place of soya bean meal and one diet was fed throughout. The basal diet was based on wheat and soya bean meal and included phytase and xylanase (AB Vista, UK) at 100 g/tonne each. All treatment diets were formulated to be iso-caloric and iso- nitrogenous using the supplier's matrices for each test product. There were four treatments (control, ASP and 2× AS level) with 10 replicate pens of 35 Ross 308 male broilers per pen. At d7, the AS treatments were significantly heavier than the control and the ASP was intermediate ($P<0.01$). The FCR of the AS treatments were significantly better than the control and the ASP treatment was not different from the control or AS 7.5% treatment ($P<0.05$). At d14, the 15% AS treatment was signifcantly heavier than the control and ASP treatment and the 7.5% AS was intermediate ($P<0.05$). The 15% AS530 treatment had better FCR than the control and ASP treatments and the 7.5% AS was intermediate ($P<0.01$). The addition of AS to starter broiler diets at 15% as a replacement for wheat and soya bean meal, improves d7 and d14 BW and FCR. Using AS530 at 7.5% improves d7 FCR relative to the control.

Extrusion increases the use of pea seeds in broiler chicken nutrition

M. Hejdysz, S.A. Kaczmarek, A. Zaworska and A. Rutkowski
Poznań University of Life Sciences, Wojska Polskiego 28, 60-637 Poznan, Poland; marhej@up.poznan.pl

This study was conducted with broiler chickens to investigate the effect of different levels of pea seeds in raw and extruded form on performance, nutrient (crude protein, crude fat, starch) digestibility, AME_N value as well as excretion of total and free sialic acids. In total, 960 one-day-old male broiler chicks of the Ross 308 strain were used in the experiment. The study consisted of a completely randomized experimental design with a 6×2 factorial arrangement of treatments, and six levels of pea addition (100, 200, 300, 400, 500 g/kg diet) in raw or extruded form. In Exp., statistical analyses were performed with the two-way analysis of variance. Within each pea form, orthogonal polynomial contrasts were used to determine the effects of pea level on tested parameters. Extrusion had a positive impact, leading to a decrease in phytic P and resistant starch content in pea seeds. Birds receiving diets with different levels of pea in extruded form were characterized by better performance than broilers fed diets with different levels of pea in raw form. These differences were confirmed by ANCOVA ($P \leq 0.05$). Increasing the level of pea seeds in extruded form did not lead to a deterioration in nutrient utilization or AME_N value, which had been confirmed for raw seeds (Figure 1). Extrusion did not affect the excretion of total and free sialic acids, but the changes were confirmed for different levels of pea in diets. In conclusion, extrusion is one process which can increase the use of pea seeds in broiler chicken nutrition. Figure 1. Relationship between level of pea in raw or extruded form and AME_N value.

Influence of blue sweet lupines (*Lupinus angustifolius*) in the feed on the growth of broilers

I. Halle
Institute of Animal Nutrition (FLI), Bundesallee 50, 38116 Braunschweig, Germany; ingrid.halle@fli.de

Soybean meal is the most common protein source in diets of broiler chickens. The majority of this soybean meal is imported from non-European countries and often derived from genetically modified varieties. Blue sweet lupines are suited as ingredient of poultry feed due to their high protein content and low alkaloids level. Processing treatments such as heat processing are carried out with the aim to improve the digestibility of nutrients. Objectives of this study were to test the effect of a replacement of the protein source soybean meal (SBM) with legumes as lupines as seeds or after toasting in broilers' feed on performance. One-day old male 400 cockerels (ROSS) were randomly allocated into 5 treatment groups (8 pens/group) over a study period of 35 days. Feed and water were provided for *ad libitum* consumption. Live weight was recorded for each broiler individually whereas feed was weighed back weekly on a pen-basis. Soybean meal (32%) was gradually replaced by lupines (5/15%) as seeds or after toasting in the diets. The low-alkaloid lupine variety 'Borlu' contained 297/317 g/kg crude protein (seed/after toasting). All diets contained a balanced concentration of 21% crude protein and the essential amino acids. A N-balance trial was caried out with 9 replicates per group 1, 3, 5 and broilers from the same hatch at the age of 3 weeks. Data were analyzed via two-way ANOVA (SAS). The results of this study indicate that an inclusion rate up to a level of 15% blue sweet lupines or 15% toasted blue sweet lupines in broilers' diets was without negative effects on feed intake (87.2/85.2 g/broiler/day), growth performance (2,160/2,100 g), percentage of carcass (n=8 per group) and protein utility (n=9 per group) ($P>0.05$). The trial's results allow the conclusion that a replacement of soybean meal with blue sweet lupines up to a level of 15% does not affect the growing performance of broilers and toasting of lupines seed does not result in a better chicken performance.

Graded inclusion of white lupin meal depress laying hens performance and nutrients digestibility

S.A. Kaczmarek, M. Hejdysz and A. Rutkowski

Poznan University of Life Science, Department of Animal Feeding and Feed Management, Wolynska 33, 60-637 Poznan, Poland; sebak1@up.poznan.pl

The aim of the study was to determine the usefulness of white lupine (WL) in laying hen diets and its influence on birds' performance and egg weight. The experiment was conducted with 360 layer hens Hy Line Brown located in cages (3 birds/cage). The birds were randomly assigned to six treatments, each with 60 hens and during the period of 17 weeks, they were fed diets with increasing lupine meal content – 0, 6, 12, 18, 24 and 30%. The environmental conditions were managed according to the standard requirements for Hy Line Brown layer hens. All diets were formulated to be isonitrogenous and isocaloric. The body weight, laying rate, egg weight, feed intake and feed conversion were registered. There was no negative effect of WL inclusion on feed intake during the whole trial. The mean value of feed intake for 17 weeks amounted to 115 g per hen/day. In the experiment, a decrease in laying rate was recorded in treatments where 24 and 30% of WL was used. The mean value of laying rate for 17 weeks amounted to 95.4 (0-18% of WL), 90.7 (24% of WL) and 87.9% (30% of WL) ($P<0.01$). The egg weight was diversified already after 7 weeks of egg production and was, on average: 58.5 (0, 6, 12% of WL), 56.2 (18, 24 and 30% of WL) g ($P<0.01$). High concentration of WL had a negative effect on FCR. The mean value of FCR for 17 weeks amounted to 2.14 (0 and 6% of WL), 2.25 (18 and 24% of WL) and 2.37 (30% WL) kg/kg egg weight ($P<0.01$). Apparent ileal digestibility of dry matter, ether extract, crude protein and starch, linearly decreased ($P<0.05$) as WL increased from 0 to 300 g/kg. There was a quadratic effect ($P<0.05$) of WL dose on sialic acid excretion. It could be concluded that 18% of white lupine in layer hens diet could be used as a valuable protein source, without negative effect on laying rate and feed intake.

The energy value of rapeseed meals produced from UK-grown cultivars, for broilers

E.S. Watts[1], S.P. Rose[1], A.M. Mackenzie[1], L. Barnard[2] and V. Pirgozliev[1]

[1]NIPH, Harper Adams University, Shropshire, TF10 8NB, United Kingdom, [2]Danisco Animal Nutrition, Wiltshire, SN8 1XN, United Kingdom; ewatts@harper-adams.ac.uk

Data on the apparent metabolizable energy (AME) content of rapeseed meal (RSM) produced from different UK cultivars is limited. This study determined the AME content of 10 different RSM batches produced from 10 different rapeseed cultivars, for broilers. Seeds from each cultivar were cold-pressed and solvent extracted followed by air desolventization and micronizing. Batches were standardised to 3% oil to exclude the impact of differences in fat content. In total 11 diets were prepared, a balancer and 10 that contained 750 g/kg balancer and 250 g/kg RSM. During the trial (13-21 d age) male Ross 308 broilers were assigned to pens (5 birds each) and diets were fed to 6 pens following randomisation. Excreta were collected between 17-21 d age and AME was determined by regression. Balancer and RSM batches were analysed for gross energy, dry matter and total, soluble and insoluble non-starch polysaccharides (NSP). Data were compared by ANOVA and multiple regression was employed to assess the relationship between AME (MJ/kg DM) and the NSP content of the meals (g/100 g). The mean AME of the RSM samples was 10.90, values ranged from 10.37-11.51 (CV%=3.74) and the mean total NSP content was 20.46 (19.77-21.57 CV%=3.19). The insoluble NSP fraction greatly exceeded the soluble fraction (17.66 vs 2.90) with values ranging from 16.45-19.68 (CV%=5.91) and 1.65-4.07 (CV%=23.38) respectively. No statistical differences ($P<0.05$) were detected in the AME of the RSM samples and multiple regression indicated no relationship ($P<0.05$) between the observed numerical differences in AME and the NSP composition of the RSM batches. The results show that there is relatively little variation in the AME and NSP content of RSM compared to other alternative protein sources. Knowledge of this type provides the British feed industry with the necessary information to inform the formulation of poultry feeds using locally sourced RSM with confidence.

The effect of whole rapeseed on chickens performance gizzard weight and nutrient digestibility

S.A. Kaczmarek, M. Kubis and A. Rutkowski
Poznan University of Life Science, Department of Animal Feeding and Feed Management, Wolynska 33, 60-637 Poznan, Poland; sebak1@up.poznan.pl

It has been demonstrated that its feeding value could be affected by incomplete rupture of the seed structure during feed processing but coarse material improve gizzard functioning. The aim of the study was to investigate the effect of whole rapeseed on gizzard weight and nutrients digestibility. The experiment was conducted with 160 broiler male chickens, divided into two dietary treatments (finely ground rapeseed (FG) or intact seeds (IS)), 14 replications with eight birds in each. The fine material was achieved by grinding the rapeseed using Skiold disc mill (Skiold A/S, Denmark). Content of FG or IS in diet amounted; 5% (8-21 d), 10% (22-28 d) and 15% (28-35 d). Irrespective of incorporated form of rapeseeds (EG or IS) birds were characterized by similar body weight gain ($P<0.36$), feed intake ($P<0.19$) and FCR ($P<0.7$). Gizzard weight at 35 d of experiment, increased when IS were used ($P<0.0013$), (FG; 1.784 vs IS; 1.981% of BW). There were no differences in apparent ileal crude fat digestibility. Use of intact seeds improved ileal crude protein digestibility (0.677 vs 0.751, $P<0.0063$) and tended to increase ileal digestible energy content (12.3 vs 12.85 MJ/kg of diet, $P<0.0508$). The present trial demonstrates that whole rapeseed in diet had a greater effect on gizzard weight than finely ground seeds.

Effects of dietary rapeseed meal supplementation on cecal microbiota in laying hens

J. Wang, C. Long, G.H. Qi and S.G. Wu
Feed Research Institute, Chinese Academy of Agricultural Sciences, 12 Zhongguancun Nandajie, Haidian, 100081, Beijing, China, P.R.; qiguanghai@caas.cn

This experiment was conducted to investigate the effects of dietary rapeseed meal supplementation on cecal microbiota in laying hens. A total of 288 45-wk-old Jing Brown hens were randomly allocated into 4 groups that were fed a basal soybean diet supplemented with 0%, 7%, 14% and 21% rapeseed meal, respectively. All diets were formulated at a fixed level of crude protein of 16.02% and energy of 2,601 kcal of ME/kg, and varied crude fiber levels (2.50, 2.93, 3.33 and 3.74%). Each dietary treatment consisted of 8 replicates with 9 birds each. The trial lasted for 6 wk. At d 42 of the feeding trial, one bird were selected from each replicate for cecal chyme sampling. Dietary rapeseed meal supplementation increased bacterial abundance and diversity. Weighted UniFrac, Nonmetric Multidimensional Scaling, and analysis of similarity indicated distinct clustering was dependent on dietary rapeseed meal supplementation. Rapeseed meal linearly decreased the proportion of Bacteroides ($P<0.001$), but linearly increased the proportion of Proteobacteria ($P<0.001$), Synergistetes ($P<0.01$), Spirochaetes ($P<0.01$), Chloroflexi ($P<0.01$), Tenericutes ($P<0.05$) and other low-abundance phyla ($P<0.05$). Compared with the soybean meal diets, the diets supplemented with 7% rapeseed meal increased the ratio of Firmicutes to Bacteroides ($P<0.05$). The effects of rapeseed meal on cecal microbiota may not only due to the crude fiber but also other distinctive compounds, such as glucosinolates, polyphenols, and tannins. The abundance of 38 genera from 30 family varied with dietary types, and some of which were positively correlated with plasma glucose, lipids and total biliary acid ($P<0.05$). In conclusion, gut bacterial community composition of laying hens was altered by dietary rapeseed meal supplementation, which could ultimately influence gut metabolism of dietary components and host (layers) metabolism.

Sunflower cake based concentrate as a substitute for fishmeal in diets for broilers

T.N. Lenkova, T.A. Egorova and E.N. Andrianova
All-Russian Research and Technological Poultry Institute, Dept. of Nutrition, Ptitsegradskaya Str., 10, 141311 Sergiev Posad, Moscow Region, Russian Federation; andrianova@vnitip.ru

The effects of sunflower cake based concentrate (SCBC; protein 81.81%, fiber 0.04%, fat 0.21%) as a substitute for fishmeal in diets for Cobb-500 broilers from 1 to 38 days of age (35 birds per treatment) were studied. Control treatment was fed diet with 4% of fishmeal throughout the entire experiment; in experimental treatments 25, 50, 75 and 100% of fishmeal was replaced with SCBC. All diets were formulated as iso-energetic and iso-nitrous (on the basis of digestible amino acids). The substitutions were found to maintain live bodyweight at the level of control. FCR at 25% substitution was better by 0.6% compared to control, at 50% was at the level of control, at 75 and 100% substitution was worse than in control due to the respective increases (by 1.7 and 2.3%) in feed consumption. Digestibility of dietary nutrients at 25-75% substitution was at the level of control and only the highest level of SCBC inclusion led to certain decrease in these parameters (digestibility of DM lowered by 0.9% compared to control, protein by 0.8%, fat by 1.1%, nitrogen by 1.3%, lysine by 1.4%, methionine by 0.8% compared to control, $P>0.05$). All levels of substitution didn't influence physiological status of internal organs (liver, gizzard, heart), eviscerated carcass yield, and breast meat yield. Accumulation of vitamins A, E, and B2 in liver was higher in treatments fed SCBC while there were no effects on chemical composition of breast and leg muscles. Substitution of SCBC for fishmeal led to decrease in diet costs by 1.3-4.0% for starter phase and by 1.4-5.8% for finisher phase compared to control.

Rye – can commonly be considered a feed ingredient in poultry rations?

D. Boros, A. Fraś and K. Gołębiewska
Institute of Plant Breeding and Acclimatization, National Research Institute, Laboratory of Quality Evaluation of Plant Materials, Radzików, 05-870 Błonie, Poland; d.boros@ihar.edu.pl

Rye does have some limitations as component of poultry feed because of high content of water extractable arabinoxylans (WE-AX), that create a viscous environment in the small intestine. Cultivars with low content of WE-AX (1.6%) had markedly improved all productive parameters in broilers. Heritability of WE-AX and related viscosity are high in rye (0.7%), showing the possibility of genetic improvement. The current low price of rye (30% lower than wheat), attributes of modern rye cultivars and development of new generation of feed enzymes to optimize diets on total nutrient availability, imply its possible use in poultry diets. In order to check above assumption, a set of 18 cultivars of winter rye were characterized from three contrasting agro-climatic conditions in Poland. Grain was analyzed for protein, minerals, lipids, starch, alkylresorcinols, viscosity of water extract (WEV) and dietary fibre complex (TDF), including NSP with its soluble and insoluble arabinoxylans and lignin. Analysis of nutrients performed using standard procedures, while TDF with the Uppsala method. The two-way ANOVA (cultivar × location) analyses curried out for all qualitative parameters. Significant cultivar differences were found in the nutrient and antinutrient content of rye, additionally the environment have a large impact on it. The variation in nutrient and TDF was low among cultivars (below 6%), while higher in viscous properties of WE-AX (11%). The cultivars differed in WEV from 8.2 to13.6 mPa.s, while between locations 6.5 to 16.9 mP.s. The average content of protein was 10.4%, starch 58.8%, minerals 1.8% and lipids 2.1%. Concluding, a big variation in WE-AX and WEV indicates that rye cultivars with the lowest values of these two qualitative parameters could be considered in poultry feeding. Further breeding efforts are needed to make rye more attractive in poultry feeding.

Triticale varieties in broiler chickens diets

S. Alijosius, S. Bliznikas, R. Gruzauskas, V. Sasyte, A. Raceviciute-Stupeliene, V. Viliene and A. Dauksiene
Lithuanian University of Health Sciences, Institute of Animal Rearing Technologies, Tilzes str. 18, 47181, Kaunas, Lithuania; romas.gruzauskas@lsmuni.lt

This experiment was undertaken to evaluate the nutritional potential of triticale varieties grown in Lithuania. In experiment 400 Ross 308 broiler chickens were distributed in an entirely randomized design, with two treatments with four replications per group and 50 broilers in each group. Broilers were fed for 5 wk. a pelleted wheat–soybean meal based diets (control group C, 13.08 MJ/kg ME, 21% CP) in treatment group 15% of wheat (pentosans 6.48% DM, beta glucans 0.40% DM) were replaced by triticale of the variety 'SU Agendus' (experimental group T), having the lowest quantity of pentosans (4.92%) and beta-glucans (0.45% DM). Diet was formulated to meet the nutrient and energy requirement for broiler chickens. The development of GIT's organs and blood biochemical parameters were analysed. In diet of broilers the replacement of wheat by triticale 'SU Agendus' the FCR increased 2.6% and BW were decreased 0.4% ($P>0.05$). In addition of 15% triticale in the compound feed of broiler chickens, the blood biochemical parameters – cholesterol and high-density lipoprotein concentration decreased by 7.4 and 11.9%, respectively ($P<0.05$). The internal organs weight in group T were increased compared to the group C ($P<0.05$). The results of the trial confirmed that triticale 15% with a lower pentosans and beta-glucans contents can be used for broiler chickens diets.

Nutritional evaluation of barley and wheat for energy for broiler chickens

O. Adeola
Purdue University, Animal Sciences, 915 W. State Street, 47907 West Lafayette, IN, USA; ladeola@purdue.edu

Five diets were used to determine the energy value of barley and wheat for broiler chickens from day 15 to 22 post hatching. Three-hundred and twenty birds were grouped by weight and assigned to eight replicate cages of eight birds per cage and five cages in each replicate of five diets. The five diets consisted of a corn-soybean meal reference diet and four test diets in which barley or wheat partly replaced reference diet at 100 or 200 g/kg diet. Dry matter (DM) of barley and wheat were 890, and 899 g/kg, respectively; the gross energy were 4567 and 4,456 kcal/kg DM, respectively. Barley addition to the reference diet linearly reduced ($P<0.01$) ileal digestibility of DM, nitrogen, and energy and Ileal digestible energy, whereas wheat addition did not affect any of the ileal digestibility measurements for the test diets. Neither of the cereal grains had any significant effect on nitrogen utilization in diets. Energy metabolizability of the test diets decreased ($P<0.05$) with barley addition. Both metabolizable energy and nitrogen-corrected metabolizable energy of the test diets decreased ($P<0.05$) with barley addition but wheat addition did not affect these response criteria. Ileal digestible energy, metabolizable energy and nitrogen-corrected metabolizable energy of barley and wheat were derived using the regression method. Ileal digestible energy value for the barley sample evaluated was 2,364 kcal/kg DM and the respective metabolizable energy and nitrogen-corrected metabolizable energy of barley were 2,894 kcal/kg DM and 2,841 kcal/kg DM. For wheat, the respective ileal digestible energy, metabolizable energy, and nitrogen-corrected metabolizable energy for broiler chickens were 3,413, 3,713, and 3,372 kcal/kg DM.

Diet preference tests for a wet versus dry feed, wheat structure and pellet size in broiler chickens

J. Van Loon, M.M.P. Raaijmakers, M.L. Elling-Staats, A.F.B. Van Der Poel and R.P. Kwakkel
Wageningen University, Animal Nutrition Group, P.O. Box 338, 6700 AH Wageningen, the Netherlands;
miranda.elling-staats@wur.nl

Diet preferences of broiler chicks were studied during 3 different choice-fed periods (P1, P2, P3). In P1 (1 to 7 days), chickens received a crumble starter diet with 250 g/kg ground wheat (GW) in two separate feed troughs with dry and wet (1:1 feed to water). During P2 (8 to 21 days), chickens received diets in 4 separate feed troughs with whole wheat (WW) and GW in a 4 mm pelleted diet, fed dry and wet. In P3 (22 to 34 days), chickens received WW and GW dry pellets of 4 and 6 mm diameter in 4 separate feed troughs. Ingredient compositions of all diets from day 8 to 34 were similar (350 g/kg wheat). Consumption from each feed trough was measured daily over a 6 h-period. In P1, a strong preference for wet feed was found of 90.7% of total air-dry feed intake ($P<0.001$). In P2, preferences over the 4 diets were WW wet (62.6%), GW wet (33.7%), WW dry (2.6%) and GW dry (1.1%). Separate statistical analyses were done for the 6 slowest growing pens (SGP) (64 g/d) and the 6 fastest growing pens (FGP) (70 g/d). A positive correlation between average daily gain (ADG) and WW wet consumption was found for SGP (r=0.88, $P=0.022$) and a negative correlation for FGP (r=-0.83, $P=0.042$). Moreover, the ADG of FGP correlated positively with dry WW intake (r=0.81, $P=0.050$). In P3, preferences over the 4 diets were WW 6 mm (33.4%), WW 4 mm (33.2%), GW 4 mm (24.3%) and GW 6 mm (9.1%). Only FGP showed a preference for 4 mm pellets (60.2% $P=0.005$). The ADG of FGP were negatively correlated with WW intake (r=-0.83, $P=0.039$); no significant correlations were found for SGP. It is clear that chickens strongly prefer wet over dry diets and a high intake of a wet diet in combination with WW seems to benefit the slowest growing chickens. WW is preferred over GW, however a somewhat higher GW vs WW intake seems to benefit the faster growing chickens. The preference for 4 mm over 6 mm pellets seems to depend on growth rate and diet structure.

Wheat varieties ranked differently based on growth and nutrient utilisation effects in broiler

O.A. Olukosi[1] and M.R. Bedford[2]
[1]SRUC, Monogastric Science Research Centre, Edinburgh, EH9 3JG, United Kingdom, [2]AB Vista, Marlborough, SN8 4AN, United Kingdom; oluyinka.olukosi@sruc.ac.uk

A 28-d experiment was done with 1,200 broilers to investigate, on a weekly basis, the effect of xylanase supplementation and wheat variety on growth performance, nutrient retention and development of digestive organs. On day 0, broilers were allocated to 8 treatments in a 4×2 factorial arrangement. The factors were 4 wheat varieties (WHV) supplemented with or without xylanase (XYL). Birds and feed were weighed on d 0 and every 7 d thereafter, excreta were collected over the last 2 d of each week. One representative bird per pen was euthanised each week and the weight of the gizzard and sections of the small intestine were taken. There was no significant WHV × XYL interaction for growth performance at any time except for feed intake in wk 3. Weight gain was significantly improved ($P<0.05$) by XYL on d 7 and 14 but FCR was improved by XYL on d 14 only. There was significant ($P<0.01$) WHV effect at all ages with broilers receiving wheat 2 having the greatest weight gain. There were no significant XYL effect or WHV × XYL interaction on nutrient retention during the study. There were significant ($P<0.01$) WHV effects on AME on d 7, 14 and 21 and on N retention on d 14, 21 and 28. Wheat 3 generally had lower ($P<0.01$) AME and greater N retention than the others. There were only marginal effects of the factors on digestive organs development especially beyond d 14. Birds receiving wheat 2 had a shorter ($P<0.05$) duodenum on d 7, lower gizzard:body weight ratio on d 14 and tended ($P<0.10$) to have shorter ileum and lighter empty gizzard on d 14. It was concluded that although the wheat varieties were similar in chemical profile they were different in ranking on the basis of their effect on broilers. The ranking based on body weight was similar to the ranking based on effect on digestive organ development and may be related to effect of the feedstuff on whole body energy expenditure.

Does the wheat cultivar or growing site affect broiler growth performance?

M.R. Azhar[1], S.P. Rose[1], M. Bedford[2], A.M. Mackenzie[1] and V. Pirgozliev[1]
[1]NIPH, Harper Adams University, Newport, Shropshire, TF10 8NB, United Kingdom, [2]AB Vista, Marlborough, SN8 4AN, United Kingdom; mazhar@harper-adams.ac.uk

The aim of this study was to examine the effect of wheat cultivar or growing site on nutritional value of wheat for broilers. Six current UK wheat cultivars (Leeds, Santiago, Lili, Trinity, Barrel, Basset) grown on 4 sites (Nottinghamshire, Yorkshire, Lincolnshire, Cambridge) were compared in an unbalanced block design. Wheat samples were analysed for dry matter, protein, ash, fat, gross energy, starch, soluble and insoluble non-starch polysaccharides (NSP), endosperm hardness, hagberg falling number, specific weight and kinematic viscosity. Eight hundred male Ross 308 broilers were allocated to 160 floor pens (5 birds in a pen). Seventeen diets were formulated including 670 g/kg of each wheat sample and 330 g/kg of balancer. The diets were made iso-nitrogenous by adding wheat protein isolate, replicated 8 times in a randomised block design. Feed intake (FI) and weight gain (WG) was recorded from 0-21 days. Data were statistically compared by ANOVA for unbalanced design using regression. The mean FI of broilers fed the 17 wheat samples ranged from 37.5 to 43.1 g/bird/day DM and WG 30.4 to 34.6 g/b/d. There were differences in FI and WG ($P<0.05$) between some of the 17 individual wheat samples, but further statistical analysis showed that these were not due to any consistent differences in FI between the 6 cultivars (39.8, 40.3, 41.3, 40.8, 40.0, 42.2 respectively, SED=2.08) or the 4 growing sites (40.5, 40.8, 39.6, 40.4 respectively, SED=1.68). Similarly, no difference ($P>0.05$) was found in WG due to cultivar (32.5, 32.9, 33.5, 33.1, 32.8, 34.3, SED=1.51) or site (33.2, 33.1, 32.4, 33.0, SED=1.22). Multiple linear regression analysis indicated no relationship ($P>0.05$) between broiler growth performance and determined AME, chemical composition and physical characteristics of wheat samples. The differences in FI and WG are economically important to poultry industry and require further investigation.

Effect of whole wheat inclusion and pellet diameter on pellet quality and performance in broilers

M.M.P. Raaijmakers, J. Van Loon, M.L. Elling-Staats, A.F.B. Van Der Poel and R.P. Kwakkel
Wageningen University, Animal Nutrition Group, P.O. Box 338, 6700 AH Wageningen, the Netherlands; miranda.elling-staats@wur.nl

Whole wheat (WW) is mainly fed to broilers along with a pelleted feed, which allows birds to select particles leading to inconsistent results. WW inclusion in a pellet may require larger pellets to maintain coarse structure and pellet quality. The effect of WW vs ground wheat (GW) inclusion and pellet diameter (4 vs 6 mm) on pellet quality and performance was investigated, using a 2×2 factorial design with 6 replicates per treatment (9 Ross 308 males per pen). An additional group was fed a reference diet (RD) containing GW in a 3 mm pellet diameter. From day 0-7, all birds received a crumble starter diet (250 g/kg GW). From day 8-14, one group received the RD and the other 4 groups a 4 mm pelleted diet, which included either GW or WW. From day 15-34, half of these 4 groups received the 6 mm pelleted diets (GW or WW). Formulations of all diets from day 8-34 were similar (350 g/kg wheat). Pellet durability, tested using a Ligno tester, was the highest in the RD (75.5%), tested with sieve size 2.1 mm. When tested with sieve size 3.6 mm, GW 6 mm had the highest durability (59.7%) and the lowest durability's were found using both sieves in WW 6 mm diets (60.8% and 36.6% respectively). Feed intake (FI) and feed conversion ratio (FCR) was higher for GW vs WW diets from day 8-14 (53.6 vs 51.1 g/d, $P=0.032$ and 1.314 vs 1.269, $P=0.038$, respectively). For RD the lowest FCR (1.240, $P=0.031$) and highest bodyweight gain (BWG) (44.0 g, $P=0.010$) was found. From day 15-34, FI and BWG was also higher for GW vs WW diets (154.9 vs 148.6 g, $P<0.001$ and 107.8 vs 103.9 g, $P<0.001$, respectively). Furthermore, FCR was better for the 4 mm pellet compared to the 6 mm pellet (1.408 vs 1.425, $P=0.014$); RD showed the lowest FI (145.5 g/d, $P<0.001$) and BWG (102.7 g, $P=0.011$), FCR remained unaffected. Overall, these results showed that 4 and 6 mm pellets vs 3 mm improve FI and BWG and that WW in pellets (vs GW) reduces FI and BWG, but improves FCR.

Whole grain feeding promotes energy utilisation more in sorghum- than wheat-based broiler diets

A.F. Moss, P.H. Selle and S.Y. Liu

The University of Sydney, Faculty of Veterinary Science, 425 Werombi Road, Camden, NSW 2570, Australia; amy.moss@sydney.edu.au

Wheat and, to a lesser extent, sorghum are the two commonly used feed grains in Australia for chicken-meat production. Whole grain feeding (WGF) is becoming increasingly accepted in countries where wheat is the dominant feed grain so, for this reason, the present study compares responses of broiler chicks to 12.5% whole versus ground barley in diets based on wheat, sorghum and a wheat-sorghum blend. Six nutritionally equivalent diets were offered to 7 replicate cages (6 birds per cage) of male Ross 308 chicks from 7 to 28 days post-hatch. These treatments were offered as an intact pellet containing 12.5% ground barley or as a mix of 12.5% whole barley and a balancing pelleted concentrate. The effects of dietary treatments on relative emptied gizzard weights, contents and pH, excreta dry matter and energy utilisation were determined. WGF increased relative gizzard weights by 22.5% (16.96 vs 20.77 g/kg; $P<0.001$) but reduced their contents by 13.7% (8.32 vs 9.64 g/kg; $P<0.025$). WGF reduced the incidence of dilated proventriculi from 4.76% to zero ($P<0.03$). WGF increased excreta dry matter by 13.3% (25.08 vs 22.13%; $P<0.005$) in birds offered wheat-based diets, which should aid in the prevention of wet litter. There were significant treatment interactions ($P<0.001$) for parameters of energy utilisation (AME, AME:GE ratios, AMEn). For example, WGF did not influence AME:GE ratios of wheat-based diets ($P>0.75$) but improved those of sorghum-based diets (0.729 vs 0.697; $P<0.001$) WGF notionally increased AMEn of wheat-based diets by 0.08 MJ (12.23 vs 12.15 MJ/kg) but by 0.70 MJ (11.52 vs 10.82 MJ/kg) in sorghum-based diets. The greater responses of sorghum to WGF reflect its inherently poorer energy utilisation relative to wheat, which is an underlying problem that needs to be addressed. However, the quality of ground sorghum for chicken-meat production may be enhanced by its incorporation into WGF regimes.

Inclusion of black soldier fly (*Hermetia illucens* L.) larvae fat in finisher broiler chickens diets

A. Schiavone[1], S. Dabbou[1], M. De Marco[1], M. Cullere[2], I. Biasato[1], E. Biasibetti[1], M.T. Capucchio[1], M. Meneguz[3], F. Gai[4], A. Dalle Zotte[2] and L. Gasco[3]

[1]University of Turin, Dept. of Veterinary Science, L.go Paolo Braccini, 2, 10095 Grugliasco (TO), Italy, [2]University of Padua, Dept. of Animal Medicine, Production and Health, Viale dell'Università 16, 35020 Legnaro (PD), Italy, [3]University of Turin, Dept. of Agricultural, Forest and Food Sciences, 10095 Grugliasco (TO), Italy, [4]National Research Council, Institute of Science of Food Production, 10095 Grugliasco (TO), Italy; achille.schiavone@unito.it

The objective of the present study was to evaluate the effects of a partial or a total replacement in finisher diets of soybean oil with *Hermethia illucens* (HI) larvae fat on growth performances, carcass traits and intestinal morphology of broilers chicken. At 21 days of age, a total of 120 male broiler chickens (Ross 308) were randomly allocated to three dietary treatments (5 replicates and 8 birds/pen). To a basal control diet (C), either the 50% or the 100% replacement of soybean oil with HI larvae fat was applied (HI50 and HI100 group, respectively). Diets were isonotrogenous and isoenergetic (CP: 21.1%; EMA; 13.3 MJ/kg)). Growth performances, were evaluated throughout the trial; then, the animals were slaughtered at the age of 48 days. Villus height (Vh) and cript depth (Cd) were measured (mm) on duodenum (D), jejunum (J) and ileum (I). Growth performances (FCR: 1.57, 1.58, 1.53; final body weight: 3,621, 3,577, 3,751), carcass traits (carcass yield: 70.8, 71.3, 70.6), and gut morphometric indexes (D-Vh: 2.53, 2.52, 2.38; D-Cd: 0.15, 0.17, 0.15; J-Vh: 2.02, 2.01, 2.04; J-Cd: 0.16, 0.17, 0.15; I-Vh: 1.74, 1.71, 1.56; I-Cd: 0.16, 0.15, 0.13) were not influenced by the dietary inclusion of HI larvae fat. Data obtained from this study suggest that the 50% and the 100% replacement of soybean oil with HI larvae fat in broiler chickens diets have no adverse effects on growth performance and do not impair gut morphology.

Application of selected insect fats in broiler chicken diets

D. Józefiak[1], B. Kierończyk[1], M. Rawski[1], A. Józefiak[2] and S. Świątkiewicz[3]
[1]Poznań University of Life Sciences, Department of Animal Nutrition and Feed Management, ul. Wołyńska 33, 60-637, Poland, [2]Poznań University of Life Sciences, Veterinary Institute, ul. Wołyńska 35, 60-637, Poland, [3]National Research Institute of Animal Production, Department of Animal Nutrition and Feed Science, ul. Krakowska 1, 32-083, Poland; damjo@up.poznan.pl

The aim of the present study was to examine how fats obtained by super-critical CO_2 extraction from *Tenebrio molitor* (TM) and *Zophobas morio* (ZM) affect growth performance and nutrient digestibility in broiler chickens. Two separate 28 d experiments were conducted, in both 1 day-old male broilers (Ross 308) were used. The birds were assigned to 2 groups in Exp 1 and 3 groups in Exp 2, (12 replicates/36 birds per treatment in both experiments). Birds were fed hot-pelleted (80 °C) isonutritive, soybean-maize based diets containing in control treatment 5% of soybean oil (SO). In experimental treatments SO was replaced by oils from TM in Exp 1 and TM or ZM in Exp 2. TiO_2 (0.3%) as an internal marker was used. The FI and BW of the chickens were measured. All raw materials samples were analyzed for crude protein, crude fat and crude fiber. In ZM oleic acid (37.5%), palmitic (32.5%) and linoleic (16,5%) were the dominant acids, in TM oleic (44%), linoleic (30%) and palmitic (16.5%). In both trials total replacement of SO by insect origin fats did not have negative effects on birds performance (1-28 d). In the Exp 1, TM fat from 7-21 d as well as in the entire experiment (1-28 d) lowered FI and improved FCR. The total tract crude fat digestibility was also higher in TM treatment at 7,14,21 and 28 d. In the Exp 2, birds fed SO oil were characterized by lower FCR value in 1-7 d. While ZM fat impaired FCR from 14-21 d. However, no differences among the treatments where observed in entire experimental period. Present data suggest that insect origin fats from *TM* and *ZM* can replace 100% of SO in young (1-28 d) broiler chickens.

Crude soybean lecithin as alternative energy source in broiler chickens

A. Vinado, L. Castillejos and A.C. Barroeta
Animal Nutrition and Welfare Service, UAB, 08193 Bellaterra, Spain; alberto.vinado@uab.cat

Crude soybean lecithin is a by-product of the soybean oil refining process, which represents an interesting and economical source of energy for broiler feeding. Lecithin contains a high proportion of phospholipids than can play an emulsifying role in poultry gut, improving dietary fat utilization. The aim of this experiment was to study the use of crude soybean lecithin, in replacement of soybean oil, as energy source in dietary broiler chickens. A total of 120 Ross-308 one-day-old female broiler chickens were randomly distributed in 30 digestibility cages and assigned to four experimental isonutritive treatments. A basal diet was supplemented with soybean oil at 3% (C: 12 replicates) and increasing amounts of crude soybean lecithin (L) were added in replacement of soybean oil: 1% (L1), 2% (L2) and 3% (L3) with 6 replicates per treatment. A 38-day growing period was conducted and two nutritional balances were performed between 9-10 d (starter period) and 36-37 d of age (growing-finishing period). Titanium dioxide (TiO_2) was used as an indigestible marker, at 0.5% in the diet. The inclusion of three levels of crude soybean lecithin in replacement of soybean oil did not modify the performance results. In the starter period, broiler fed C diet showed significantly higher feed AME than L diets ($P<0.01$) and higher total fatty acid (TFA) utilization coefficient than L2 and L3 ($P<0.01$). However, in the growing-finishing period, AME of C diet was not significantly different to L diets (C: 3,310 a; L1: 3,466 ab; L2: 3,249 b and L3: 3,254 b kcal/kg; $P<0.02$). Broilers fed lecithin supplemented diets (1, 2 and 3% level) showed similar percentage of digestibility of TFA, saturated FA and monounsaturated FA than those fed C diet. It is concluded that crude soybean lecithin can be used in growing-finishing broiler diets in replacement of soybean oil as alternative energy source.

Effect of the dietary supplementation of *Solanum glaucophyllum* on egg quality traits of aged hens

M. Zampiga[1], F. Calini[2], R. Losa[3], A. Meluzzi[1] and F. Sirri[1]
[1]University of Bologna, Department of Agricultural and Food Sciences, Via del Florio 2, 40064 Ozzano dell'Emilia, Italy, [2]Advisor to the Feed and Animal Industries, Largo E. Guevara de la Serna, 8, 48022 Lugo di Romagna, Italy, [3]Advisor to the Feed Additive Industries, Au Châtelard, 6, 1145 Bière, Switzerland; federico.sirri@unibo.it

The aim of this study was to test the effect of dietary supplementation of different standardized doses of *Solanum glaucophyllum* leaves (SG), a natural source of 1,25-dihydroxyvitamin D_3, on table eggs quality traits and productivity of aged laying hens. 300 Hy-Line brown hens of 70 wks of age, selected to be of similar body weight, were housed in 30 enriched cages, randomly divided into 3 experimental groups (10 replications each). All animals were fed the same diet for 4 wks, then submitted to the following treatments: A) commercial basal diet (BD) containing 3,000 UI vitamin D/kg feed, B) BD plus 100 mg of SG/kg feed, and C) BD plus 500 mg of SG/kg feed, delivering 1 and 5 mcg 1,25-dihydroxyvitamin D_3 kg/feed respectively, as levels already tested were much higher. At 0, 7, 14 and 21 wks into the trial, all the eggs laid in one day were collected and used for the quality traits evaluation. After 21 wks of treatments, corresponding to 95 wks of age, group C compared to group A and B showed higher eggshell percentage (9.84 vs 9.59 and 9.56%; $P<0.01$), thickness (0.366 vs 0.357 and 0.359 mm; $P<0.05$), density (84.1 vs 82.1 and 82.6 mg/cm2, $P<0.05$) and breaking strength (35.2 vs 33.7 and 35.0 N; $P=0.17$ with C and B different from A). Overall, egg deposition rate (65.4 vs 62.5 vs 61.94% for C, A and B respectively), feed conversion rate (2.558 vs 2.756 vs 2.625) and egg/mass (43.8 vs 40.9 vs 41.8 g/hen/d) resulted not statistically different among groups. Based on the results obtained in this preliminary study, the addition of SG to the hen feeding seems to be beneficial to counteract the detrimental effect of aging on some egg quality traits.

Performance and meat quality of two broiler strains fed diets containing potato waste meal (PWM)

A.A. Eldeek, M. Taher, M. Khalil and M. Elmalky
Faculty of Agric., Alex. University, Al Shatby, Alexandria, Egypt; eldeek@yahoo.com

Studies were done to investigate the effect of partially replacing potato waste meal (PWM) by yellow corn using levels of 0,25 and 50%, with and without commercial enzymes (Avizyme)on performance and meat quality of two broiler strains (Arbor Acers (AA) and Hubbard F15(HF)). Total number of 504 unsexed, 14 day- old broiler chicks, were distributed into six equal experimental groups for each strain in a factorial treatment design (2×3×2) during growth (15-28) and finishing (29-42) periods. Grower and finisher diets were formulated to be iso-energetic, and balance in amino acids. Results showed that (PWM) contained crude protein (7.4%), ether extract (4.1%), ash (1.6%), crude fiber (1.9%) and starch (58.6%), the ME content of PWM (3,221.6 kcal/kg), and was calculated as follows: ME(kcal/kg) = 53+38 (CP%+2.25×EE %+1.1×starch+sugar%) Significant difference was recorded of body weight gain between the two strains through the growth periods. At 42 days of age, the final body weight were 2,009.5 and 2,018.5 gm for AA and 1,854.67 and 1,870.17 gm for HF without or with enzyme mixture, respectively. Also a positive improvement was recorded as supplementing experimental diets with enzymes mixture. The presence of dietary PWM had highly significant effect on the amount of feed consumption and feed conversion ratio for both strains. Feeding 25% dietary tested material significantly increased the relative weight of carcass compared with that of the control group. No significant difference in color chroma and hue of breast meat was detected among the two types of strains breast meat. No difference in shearing force value of breast meat or cooking loss, were seen among AA and HF strains. Also, no significant differences were recorded on parameters of acceptability as judged by appearance, color, flavor, juiciness and tenderness. In conclusion, it is suggested based on economical profitability, using PWM up to 25% to partially replace yellow corn with enzymes mixture in broiler diets during grower and finisher periods.

Effect of using grape meal as natural antioxidant in Cobb 500 broiler feeding

M. Olteanu, D.R. Criste, D.T. Panaite, P.A. Vlaicu, M. Saracila and R.P. Socoliuc
National Research-Development Institute for Animal Biology and Nutrition, Nutritional Physiology Department, Calea Bucuresti nr. 1, 077015, Balotesti, Ilfov, Romania; alexandru.vlaicu@outlook.com

The work aimed to improve the oxidative status of the chicken meat using grape meal as natural antioxidant. The 5-week experiment involved 500, day-old Cobb 500 chicks assigned to two groups (C, E), housed in an experimental hall, with 23 hours light regimen. The chicks had free access to the feed and water. During the starter stage (0-14 days), both groups received the same conventional diet for this particular age. During the growth stage (14-35 days), diet C (based on corn, wheat, soybean meal and flax meal), differed from diet E, which was supplemented with 3% grape meal, as natural antioxidant. Five blood samples were collected from each group in the end of the experimental period and assayed for the cholesterol concentration. At the same time, 5 broilers/group were slaughtered and 5 breast meat and thigh meat samples were formed for each group. Part of each samples was chemically assayed immediately after slaughter (0 days), and another part after 7 days of refrigeration. The meat samples were assayed for pH, fat degradation indices and malondialdehyde (MDA) concentration. The pH values ranged between 6.41±0.06 in the thigh sample from group C and 6.25±003 thigh sample from group E, the differences being significant ($P \leq 0.05$). After 7 days of refrigeration, the malondialdehyde (MDA) concentration was lower in group E, both for the breast samples (0.111±0.045 mg/kg vs 0.139±0.088 in group C), and in the thigh samples (0.524±0.177 mg/kg, vs 0.688±0.244 mg/kg in group C). Serum cholesterol was also lower, 119.78±4.34 mg/dl, in group E, compared to 123.46±4.04 mg/dl, in group C.

Roles of sugarcane bagasse and corn particle sizes in nutrient digestibility and gizzard in broilers

S.K. Kheravii, R.A. Swick, M. Choct and S.-B. Wu
University of New England, School of Environmental and Rural Science, Elm Avenue, Armidale, NSW, Australia; swu3@une.edu.au

Measures to improve the digestibility of nutrients have been sought due to in-feed antibiotics being phased-out in poultry. Promotion of gizzard development by physical structure of feed ingredients or addition of dietary fibre is one such strategy with the hypothesis that larger ingredient particles and higher fibre improve the digestibility of nutrients. An experiment was conducted to evaluate the effect of sugarcane bagasse (SCB) and corn particle size on nutrient digestibility and gizzard development. A total of 336 Ross 308 male broilers were assigned in a 2×2 factorial arrangement of treatments with 2 particle sizes (coarse 3,576 µm or fine 1,113 µm geometric mean diameter) and 2 levels of SCB (0 g/kg or 20 g/kg). Each treatment had 6 replicate pens of 14 birds. The coefficient of digestibility for starch, protein and gross energy were measured from homogenised distal ileum digesta of three birds per pen at d 24. Titanium dioxide was used as an indigestible marker at a rate of 5 g/kg diet. The relative gizzard weight and gizzard pH were measured from two birds at d 35. Both ileal starch and protein digestibility increased in birds fed SCB compared to those with no SCB ($P<0.05$). Birds fed coarsely ground corn (CGC) had higher ileal protein digestibility ($P<0.05$) than those fed finely ground corn (FGC). A significant particle size × SCB interaction was observed for ileal gross energy digestibility ($P<0.05$). Addition of 20 g/kg dietary SCB showed improved ileal gross energy digestibility only in birds fed FGC. Broilers fed CGC had heavier relative gizzard weights ($P<0.05$) than those fed FGC. Relative gizzard weight was greater in birds fed SCB compared to those given no SCB ($P<0.01$). A particle size by SCB interaction ($P<0.01$) was observed for gizzard pH. SCB resulted in lower gizzard pH only in birds fed FGC ($P<0.01$). These findings demonstrated that SCB and CGC improved nutrients digestibility and gizzard function in broilers.

Tomato waste boiled as layer feed mixture

M. Mahata[1], T. Hidayat[1], G. Amalia Nurhuda[1], Y. Rizal[1] and A. Ardi[2]
[1]Andalas University, Nutrition and Feed Technology Industry, Faculty of Animal Science, Andalas University Campus, Limau Manis, 25163, Padang, Indonesia, [2]Andalas University, Agronomy, Faculty of Agriculture, Andalas University Campus, Limau Manis, 25163, Padang, Indonesia; mariamahata@gmail.com

An experiment was conducted to evaluate the effect of tomato waste boiled as raw material in layer diets on layer performances and egg quality. Tomato waste in this experiment was rejected mature tomato which was not sold by farmer. Tomato is wellknown as lycopen producer, and lycopen structure in tomato could be changed from trans-form to cis-form when it is boiled for 8 minutes at boiled water (100 °C). Cis-form lycopen is absorbed easily in poultry digestive tract, and could perform as antioxidant and for lowering cholesterol as well. Experiment was performed in completely randomized design with 5 different levels of tomato waste boiled powder (0, 3, 6, 9 and 12%) in the diet fed to 200 heads of Isa Brown layers with 80% hen day egg production (HDEP), and each tratment was replicated 4 times. The increasing of tomato waste boiled level in layer diet formulation replaced commercial concentrate, yellow-corn, and rice bran in diet formulation. Diets were arranged iso-protein (16%) and iso-caloric (2,600 kcal/kg). Measurements: feed consumption, HDEP, egg weight. egg mass, feed conversion, albumen height, eggshell thickness, eggshell strenght. Results in this experiment showed that the inclusion of all different levels of tomato waste boiled as feed mixture in layer diet did not affect feed consumption (116.5-121.6 g/d), HDEP (83.3-89.8%), egg weight (62-62.9 g), egg mass (51.6-55.9 g/d), feed conversion (2.18-2.19), albumen height (69.2-81,2 HU), eggshell thickness (0.409-0.430 mm), eggshell strenght (3.43-4.63), ($P>0.05$) significantly. In conclusion, tomato waste boiled as much as 12% could be included as raw material in layer diet for lowering layer commercial concentrate,yellow-corn,and rice bran utilization in diet formulation.

Evaluation of *Moringa oleifera* leaf in diets for laying hens and broilers

Y.M. C, S.G. Wu, W. Lu, J. Wang and G.H. Qi
Feed Research Institute, Chinese Academy of Agricultural Sciences, 12 Zhongguancun Nandajie, Haidian, Beijing 100081, China, P.R.; wushugeng@caas.cn

Two trials were performed to evaluate the inclusion of *Moringa oleifera* leaf (MOL) in diets for laying hens and broilers, respectively. In trial 1, 360 27-wk-old Hy-Line Grey layers were randomly allotted to four groups with 6 replicates of 15 birds supplemented with graded levels of MOL at 0, 5, 10, and 15%, respectively. The trial 1 lasted for 8 wks. All diets were formulated at a fixed level of crude protein of 16.5% and energy of 11.1 MJ of ME/kg. In trial 2, 360 1-day-old male Arbor Acres chicks were randomly divided into 6 dietary groups with 6 replicates of 10 birds supplemented with graded levels of MOL at 0, 1, 2, 5, 10 and 15%, respectively. The trial 2 lasted for 42 days. All diets were formulated at a fixed level of crude protein of 21% and energy of 12.35 MJ of ME/kg. In trial 1, no significant differences were observed in egg weight or feed intake among all groups ($P>0.05$). The birds in MOL15 group had higher feed conversion ratio (FCR) and lower egg production compared with birds of the control group ($P<0.05$). Layers in MOL5 had a deeper yolk color than those in control group ($P<0.05$). Supplementary MOL increased the activity of glutathione peroxidase ($P<0.05$). In trial 2, MOL quadratically decreased FCR, and supplementation 10% and 15% MOL groups decreased average daily gain. The MOL addition quadratically improved the value of a*, b*, pH, quadratically increased the content of C18:2, C18:3 and C20:4 fatty acids ($P<0.05$), but decreased the content of MDA in breast meat ($P<0.05$). In conclusion, dietary addition with 5% MOL could improve yolk color value without adverse effects on laying performance and egg quality. Dietary addition with 2% MOL could improve meat quality and meat antioxidant status without negative effects on broiler performance.

Apparent metabolic energy (AME) of a yeast culture protein product

N. Bernard[1], B. Renouf[2], C. Pedersen[3] and A. Bourdillon[1]
[1]Mixscience, 7 Avenue René Cassin ZI de Bellitourne, 53200 Aze, France, [2]Euronutrition, ZA du Bois de Teillay Quartier du Haut-Bois, 35150 Janzé, France, [3]Hamlet Protein, Saturnvej 51, 8700 Horsens, Denmark; cap@hamletprotein.dk

The energy level of the diet is the most significant factor determining feed intake in broilers. Therefore, when formulating diets, using least cost and maximum performance principles, the energy value of all feed ingredients is needed. Apparent metabolic energy (AME) is the standard measure of available energy in poultry used today. The aim of the experiment was to evaluate the AME value of HP 800 Booster (a yeast culture protein product, produced by Hamlet Protein, DK) to different SBM products by use of the European reference procedure for measuring the AME in feed for poultry. The model in short, eighteen adult roosters, were fed either a maize based control diet or a test diet. The test material replaced 30% of the maize. Maize had an incorporation rate of 97%, and the remaining 3% was minerals and vitamins. After a four day adaptation period, followed by a 12 h fasting period, total excreta was collected during the following 3.5 days and pooled. During the last 12 h of sampling, birds were fasted. Ingredients, diets and excreta were analyzed for DM, and gross energy (GE). Based on chemical analysis, and using the difference method, AME was calculated in test the ingredient. In our trial The AME values of control diet was 3,701 kcal/kg DM and for HP 800 Booster diet 3,458 kcal/kg DM. For control diet and test diet, the standard deviation of AME was found to be 14.2 and 7.7 kcal/kg DM, respectively. The AME value of HP 800 Booster was 3,033 kcal/kg DM and the ME/GE ratio was 62.1%. In comparison to table values published by INRA in 2002, the AME values of 46, 48 and 50% CP SBM are 2,568, 2,597 and 2,694 kcal/kg DM, respectively. The AME/GE ratios in the table are 55.2, 55.2 and 57.3% of same ingredients, respectively. It can be concluded that HP 800 Booster has a higher AME value than SBM.

Species-dependent response to adaptation length on AME of cereals determined by difference method

O.A. Olukosi[1], S.A. Adedokum[2] and J.O. Agboola[1,3]
[1]SRUC, Monogastric Science Research Centre, Edinburgh, EH9 3JG, United Kingdom, [2]University of Kentucky, Department of Animal and Food Sciences, Lexington, KY 40546, USA, [3]Wageningen University, Department of Animal Sciences, De Elst 1, 6708 WD Wageningen, the Netherlands; oluyinka.olukosi@sruc.ac.uk

Three experiments using 3-week old broilers, turkeys or 46-week old laying hens were used to study the influence of length of adaptation to experimental diets on metabolisable energy (AME) content of maize or barley. The birds were allocated to 6 treatments (2×3 factorial arrangement) with each treatment having 6 replicates and 3 birds per replicate. The factors were two test diets (wheat plus maize or wheat plus barley) and three lengths adaptation (10, 7 or 4 days). All the birds were kept on pre-experimental diets for at least 11 days prior to the start of feeding of experimental diets. The birds started receiving experimental diets 10, 7 or 4 days prior to the end of respective experiments; and excreta were collected on the last two days of each experiment. A wheat-soybean meal reference diet was used in each experiment and the AME of maize and barley were calculated in each experiment using the difference method. There were no diet type × adaptation length interaction on AME. Irrespective of the species, AME was greater ($P<0.01$) for maize compared with barley. The AME determined for maize was 13.5, 13.5 and 13.6 MJ/kg for broilers, turkey and layers, respectively. AME for barley was 12.2, 11.8 and 12.6 MJ/kg for broilers, turkey and layers, respectively. For broilers, AME tended ($P<0.10$) to decrease with shortening adaptation length whereas in turkey AME was lower ($P<0.05$) when birds had a 7-day adaptation compared with the other two adaptation lengths. There were no significant effects of adaptation length on cereals AME for the layers. It was concluded that the influence of adaptation length on AME of cereals that are different in fibre type and content is more significant in species with less physiologically matured digestive tract.

Growth performance and body composition of broilers affected by exchange of dietary fat by starch

T. Veldkamp[1], R. Dekker[1], J. Van Harn[1], A. Smit-Heinsbroek[2], A. Van Der Lee[2] and A.J.M. Jansman[1]
[1]Wageningen University & Research, Wageningen Livestock Research, De Elst 1, 6708 WD Wageningen, the Netherlands, [2]Agrifirm Innovation Center, Landgoedlaan 20, 7325 AW Apeldoorn, the Netherlands; teun.veldkamp@wur.nl

Dietary composition such as the concentrations of protein/amino acids, fat, and starch+sugar and their ratio, may affect the post-absorptive metabolism of energy, and protein and energy deposition in the body. In a 2×3 factorial block design, the effects of two dietary crude protein (CP) concentrations (high protein (HP) vs low protein (LP); 200/190 vs 170/160 g/kg in grower and finisher phase, respectively) and three dietary fat/starch concentrations (high fat (HF); fat and starch 120 and 350 g/kg, respectively, medium fat (MF); fat and starch 80 and 425 g/kg, respectively and low fat (LF); fat and starch 40 and 500 g/kg, respectively) in iso-energetic diets on the growth performance and body composition of Ross 308 broilers were studied (8 to 38 d). Overall, body weight gain (BWG) of broilers fed HP diets was significantly higher than BWG of broilers fed LP diets (59.6 vs 58.3 g/d; $P<0.001$) and feed conversion ratio (FCR) of broilers fed HP diets was significantly lower than FCR of broilers fed LP diets (1.65 vs 1.68; $P<0.05$). BWG of broilers increased as starch concentration in the diet was increased (HF: 55.3 g/d, MF: 59.5 g/d and LF: 62.1 g/d; $P<0.001$). FCR of broilers improved significantly as the dietary concentration of starch increased and of fat decreased (HF: 1.74, MF: 1.69, LF: 1.57; $P<0.001$). CP digestibility (77 vs 64%; $P<0.05$) and protein deposition as proportion of CP intake (63 vs 56%; $P<0.001$) in broilers fed LF diets was significantly higher than in broilers fed HF diets. It was concluded that dietary energy source and protein level in iso-energetic diets balanced for first limiting essential amino acids influence growth performance and body composition of broilers.

Effect of safety margin on ingredients with the use of NIR in laying hen performance

C. De La Cruz[1], E. Avila[2], B. Fuente[2], R. Santiago[1] and D. Leyva[2]
[1]Evonik Nutrition & Care GmbH, Rodenbacher Chaussee 4, 63457 Hanau-Wolfgang, Germany, [2]UNAM FMVZ, CEIEPAV, Tlahuac, 13300 Mexico City, Mexico; carlos.delacruz@evonik.com

NIR enables rapid and reliable assessments of the most important nutrients, which would otherwise require prolonged and costly wet chemical analysis. NIR combines speed in process and an accurate prediction, allowing an efficient evaluation of the variability and quality of the ingredients. A study was proposed in order to calculate the economic benefit to formulate based on nutrient values determined by the NIR and implications on production when safety margins (SM) are applied. SM is applied using standard deviation (SD) to a group of samples being a common practice to ensure nutrient availabilty in the feed and production outcome. A total of 480 layers (Bovans) were distributed in a completely randomized design into 4 treatments with 10 replicates of 12 birds each. Birds were housed in conventional cages. Before the diets were formulated, ingredients were assessed for crude protein (CP) and amino acids (AA) using NIR. Subsequently each ingredient was subtracted 0%, 25%, 50% and 100% of the standard deviation (SD) mentioned in the AMINODat 5.0 (feed Industry AA data base) for CP and AA. Diets based on sorghum-soybean meal-meat and bone meal were formulated to have 4 treatments: 0% SD; 25% SD; 50% SD; 100% SD. During 84 days production parametes were recorded. Egg quality was evaluated every 3 weeks at 40 eggs per treatment. Variables obtained were analysed using mean comparison (Tukey) Results obtained by productive performance and egg quality did not indicate differences ($P<0.05$). When experimental diet costs were calculated using a SM of 25% SD, diet cost increased by USD 4.71/ton. It can be concluded that the application of up to one SD in CP and AA values provide SM determined did not benefit productive parameters, meaning that the application of SM which increase feed cost is not necessary when a raw mateial evalution program is implemented.

Utilisation of near infrared reflectance spectroscopy to predict nutrient digestibility for broiler

M. Traineau, J.P. Metayer, B. Mahaut, F. Ammari and M. Vilarino
ARVALIS, SQV, Pouline, 41100 Villerable, France; m.traineau@arvalisinstitutduvegetal.fr

The Near-Infrared Reflectance Spectroscopy (NIRS) is an interesting tool for rapid prediction of the chemical composition of diet and feces in broilers. These predictions allow the estimation of digestibility faster and cheaper than with the reference methods. To go further, the development of NIRS calibrations was made to predict directly nutrients digestibility. A database was built with fecal nutrients values in one hand from the reference methods and from NIRS evaluation in the other hand. All these data allowed us to set up a digestibility prediction calibration. The NIRS predictions evaluate starch, nitrogen and energy digestibility for chickens. The aim of the trial was to validate the accuracy of the calibration equations in broilers fed with diversified diets: wheat (W), corn (C), triticale (T) or their mix (M) supplemented with soybean or rapeseed meal. Between 10 and 24 days old, 72 Ross PM3 broilers were fed with the experimental diets and at the end, feces were collected and analyzed. For all diets, the measured and predicted values were highly correlated for starch digestibility (95.8 vs 96.0%, R^2=0.94) and nitrogen digestibility (83.2 and 84.6%, R^2=0.82) but the correlation was slightly weaker forr energy digestibility (70.0 and 74.7%, R^2=0.62). The correlations between measured and predicted (R^2) were always significant ($P<0.01$) whatever the cereal in the feed (W, C, T and M, respectively), strongest for starch (0.99, 0.96, 0.95 and 0.87) and nitrogen (0.95, 0.95, 0.90 and 0.87) digestibilities and slightly lower, except for W, for energy digestibility (0.95, 0.70, 0.66 and 0.72). In conclusion, the prediction of nutrients digestibilities in broilers by NIRS, with various diet compositions, seem to be highly efficient for starch and nitrogen and promising for energy digestibility.

Near-infrared spectroscopy applied to characterize and to discriminate high protein sunflower meals

E. Bourgueil[1], J. Jachacz[2] and C. Gady[2]
[1]ADISSEO, Parc II, 92160 Antony, France, [2]ADISSEO, CERN, 03600 Malicorne, France; elisabeth.bourgueil@adisseo.com

180 High Protein SunFlower Meals (HPSFM) samples were collected in 2015. 116 samples out of them were identified as coming from France (FR), Spain (SP), Russia (RU) and Ukraine (UA).The production country of the remaining 64 was not determined (ND). All HPSFM were analysed for their concentrations in proximate, total and digestible amino acids (AA) using prediction models developed and validated beforehand using Near Infrared Spectroscopy (NIRS). A significant negative relation was found between crude protein (CP) and crude fiber (CF) contents (R^2=0.64). Lysine and methionine digestibility coefficients were not affected by CF. Concentrations in total and digestible lysine of the HPSFM database as a whole, ranged from 1.17 g/100 g to 1.52 g/100 g as fed with an average of 1.35 g/100 g and from 0.93 g/100 g to 1.28 g/100 g as fed with an average of 1.12 g/100 g respectively. A significant production country effect was shown on the total and digestible AA contents. A classification model based on discriminant analysis was performed with the 116 HPSFM properly identified. The ratio of digestibility coefficients of AA to CP allowed discriminating samples on the basis of their production countries (FR, SP, RU, UA).91% of the observations were correctly classified (83% when performing cross-validation procedure). The samples from UA were more difficult to separate into a group, since they tend to overlap the 3 other groups (FR, SP, RU). This model was used to determine the most probable production country of the 64 ND samples: 47, 15 and 2 HPSFM were respectively found from UA, FR and RU. The study showed that *in vivo*-based NIRS predictions combined with discriminant functions provided a promising model for evaluating the nutrient qualities and the production country of HPSFM.

Estimation amino acid apparent ileal digestibility of feedstuffs in broilers by meta-analysis

H. Salgado, F. Guay and M.P. Letourneau Montminy

Laval University, Animal Science, 2125 Rue de l'Agriculture, G1V0A6, Canada; marie-pierre.letourneau@fsaa.ulaval.ca

Optimizing nitrogen (N) use by birds needs a precise estimation of the amino acid (AA) value of ingredients. Given the within variation in AA digestibility in literature, the possibility of predicting AA digestibility based on proximal analysis of ingredient (e.g. crude protein, fiber, anti-nutritional factors) has been tested through meta-analysis tool. Two models has been performed: (1) Y = apparent digestible AA (AAdig, g/kg) and X = dietary CP (%); and (2) Y = AAdig and X = total analyzed AA (AAt, g/kg) in which the intercept is the endogenous losses (EL). A database of 49 publications has been used and was divided in 4 sub-databases: (1) cereals (barley, sorghum, wheat and corn) (n=47); (2) soybean meal (SBM) (n=38); (3) meat and bone meal (MBM) (n=34); and (4) database 1 plus faba beans and peas (n=65) to study EL. Study effect was random and ingredient effect in database 1 and 4 was fixed. In cereals database, prediction of AAdig by CP was accurate (R^2 from 0.62 for Met to 0.97 for Ala). The slope was similar between cereals except for Lys, Glu and Ser; in Lys corn and sorghum did not respond to CP compared to wheat ($P<0.001$) and barley ($P<0.001$) while for Glu and Ser only wheat responded to CP ($P<0.001$). In SBM and MBM, the study improved R^2 (respectively 0.09 to 0.91 and 0.31 to 0.68). More negative intercepts values and significantly different from 0 ($P<0.002$), except for Met ($P=0.237$) were systematically obtained for barley in cereal database indicating higher EL probably due to its high fiber content. When adding faba beans and peas, that contain more crude fiber, they systematically presented higher EL than cereals. Among the AA, intercepts were the highest for Glu, Leu, Ser and Thr, which are predominant in the ileal digesta. In conclusion, CP can be a predictor of digestible AA of most AA for cereals, but not for SBM and MBM. Also, total EL seems to be influenced by crude fiber given the high intercepts for barley, faba beans and peas.

Broiler performance is transgenerationally affected by dietary protein levels in breeder hens

S. Schallier[1], C. Li[1], J. Lesuisse[1], N. Everaert[2] and J. Buyse[1]

[1]KU Leuven, Department of Biosystems, Kasteelpark Arenberg 30, 3001, Belgium, [2]University of Liège, Gembloux Agro-Bio Tech, TERRA, Passage des Déportés 2, 5030 Gembloux, Belgium; seline.schallier@kuleuven.be

In mammalian species, the possibility of programming progeny through the maternal diet has been irrefutably demonstrated. However, far less is known about the possible programming effects in avian species. Therefore an experiment was conducted in which three generations of breeder hens (pure line A) were raised on different levels of dietary protein. The F0 generation was divided in a control group (C) on a standard diet and in a reduced protein group (RP) with a 25% balanced reduction in crude protein and amino acids during their entire lifespan. The female offspring of F0 was then subdivided in a C and RP group, resulting in 4 F1 breeder groups. Female progeny of these F1 breeders was then again raised as breeders of the F2 generation, which were all fed a C diet: C/C/C, C/RP/C, RP/C/C and RP/RP/C, defined as feed treatments for F0/F1/F2 generations. All breeder hens were raised according to standard management guidelines and fed to reach the target body weight (BW). The hens were artificially inseminated with semen from roosters on a standard diet. Male progeny of F2 breeders was raised as broilers for 5 weeks on C or 15% RP diets, resulting in 8 treatments. For the broilers on the C diet, RP in the F1 generation decreased broiler BW which was more pronounced when the F0 were reared on the C diet. On RP diets however, RP only in the F1 generation resulted in a higher BW than broilers descending from hens on RP in F0 and F1 or C in F0 and F1. For the broilers on the RP diet however it was shown that RP in the F0 generation increased the FCR in the F2 offspring. Furthermore, some organ weights were also affected by the different dietary breeder treatments in both C and RP fed broilers. Together, these results show that there are programming effects due to lower maternal protein in breeders and that they are carried over across multiple generations.

The influence of rapid protein or exogenous phytase on digestive dynamics in broiler chickens

H.H. Truong[1,2], A.F. Moss[1,2], P.V. Crystal[3], S.Y. Liu[2] and P.H. Selle[2]
[1]Poultry CRC, University of New England, P.O. Box U242, University of New England, 2351, Armidale, Australia, [2]Poultry Research Foundation, School of Life and Environmental Sciences, 425 Werombi Road, 2570, Camden, Australia, [3]Baiada, 642 Great Western Hwy, 2145, Pendle Hill, Australia; amy.moss@sydney.edu.au

Starch and protein digestive dynamics are pivotal to broiler performance as Liu *et al.* demonstrated that rates of starch and protein digestion rates govern feed conversion efficiency. Truong *et al.* found that phytase condensed or 'narrowed' starch:protein disappearance rate ratios in four small intestinal segments which were associated with superior weight gains. The objective of the present study was to compare the effects of rapidly digestible protein (casein) and exogenous phytase on broiler performance and the dynamics of starch and protein utilisation. The maize-based basal diet was supplemented with either 32 and 64 g/kg casein or 250 and 500 FTU/kg phytase activity. Each of the five dietary treatments were offered to 8 replicates (6 birds/cage) of male Ross 308 male chicks from 7 to 28 days post-hatch. The dietary inclusion of casein as a source of rapidly digestible protein in diets generated profound increases in both starch and protein digestibility coefficients. For example, casein linearly improved AME ($r=0.583$; $P=0.003$), ME:GE ratios ($r=0.859$; $P<0.001$) and N retention ($r=0.762$, $P<0.001$). Phytase supplementation linearly improved FCR ($r=-0.314$; $P<0.04$) and condensed starch:protein disappearance rate ratios in the proximal jejunum from 1.920 to 1.874 which is in agreement with Truong *et al.* Rapidly digestible protein generated more favourable digestive dynamics and linearly increased weight gain ($r=0.404$; $P=0.0501$. The digestibility of amino acids along the small intestine and concentrations of free amino acids in the portal and systemic circulations are being determined and should provide additional insights into starch and protein digestive dynamics of broiler chickens.

Energy and protein consumption and utilization of local female chicken reared under semi-scavenging

A. Adrizal, S. Syafwan, N. Noferdiman and T.M. Pasaribu
University of Jambi, Faculty of Animal Sciences, JL Raya Jambi, km. 15 (Mendalo Campus), 36361 Jambi, Indonesia; adrizala@yahoo.com

The present experiment aimed at estimating energy (ME) and protein (CP) needs of local female chicken raised under semi-scavenging system in the tropics. Five-day-old chicks were randomly distributed into 12 sheltered pens, 20 birds each. Two feeding methods (control and self-selection) were assigned to pens, so each treatment consisted of 6 replicates. The control received a control diet complying with the Hyline Brown Nutrient Requirements Standard, whereas the self-selection had access to the control and four other diets (high energy-high protein, high energy-low protein, low energy-high protein, and low energy-low protein diet). Feeds and drinking water were provided *ad libitum* to 20 wk of age. Feed consumption (FC), ME intake, CP intake, weight gain (WG), and concentration of dietary ME and CP were recorded weekly. Daily temperature and relative humidity in the morning (07:00), noon (12:00), and afternoon (17:00) were 22.0 to 25.2 °C and 68 to 98%; 24.0 to 36.0 °C and 40 to 82%; and 21.0 to 34.6 °C and 44 to 80%, respectively. Data were analyzed using Proc Mixed of SAS. The results showed that feeding method had apparent effect ($P \leq 0.001$) on ME and CP intake, dietary concentration of ME and CP, but not FC and WG. The effect of week and feeding by week interaction were also very significant. Weekly ME and CP intake of the self-selection group were greater ($P \leq 0.001$) than those in the control (1,161 vs 1,038 kcal/kg and 70.25 vs 65.12%, respectively). Dietary concentrations of ME and CP in the self-selection group was higher ($P \leq 0.001$) than those in the control (3,062 vs 2,844 kcal/kg and 186.52 vs 180.13 g/kg, respectively). The study suggested that self-selection feeding methods gave opportunity to hens adjusting their nutrient requirements. The ME and CP needs were likely greater than the current practices.

Metabolic and performance response of turkeys to different dietary levels and sources of methionine

J. Jankowski[1], K. Ognik[2], A. Czech[2], Z. Zduńczyk[3], K. Kozłowski[1] and J. Juskiewicz[3]
[1]University of Warmia and Mazury, 5 Oczapowskiego Street, 10-719 Olsztyn, Poland, [2]University of Life Sciences, 13 Akademicka Street, 20-950 Lublin, Poland, [3]Institute of Animal Reproduction and Food Research of the PAS, 10 Tuwima Street, 10-748 Olsztyn, Poland; kristof@uwm.edu.pl

The hypothesis was verified that dietary methionine (Met) improves the antioxidant and immune status of turkeys, and that the effect depends on Met inclusion level and its source. A 2×3 factorial completely randomized design was used to evaluate the effect of two dietary levels of Met originating from three sources. The experimental 816 female Hybrid Converter turkeys (6 groups with 8 replications per group, and with 17 birds per replication) were fed for four 4-week periods (1-16 wk of age) with wheat-soybean meal-based diets containing 3 Met sources: DL-isomer, L-isomer, and DL-hydroxy analog (DLM, LM and MHA, respectively), while the dietary Met content corresponded to the level recommended by NRC or it was increased by approximately 40% (recommended by some breeding companies). The increased dietary Met content resulted in higher final body weights of turkeys and improved FCR. Turkey feeding with diets with the higher Met content caused a decrease in plasma concentrations of glucose, triacylglycerols, and MDA as well as an increase in glutathione concentration and the ferric reducing ability of plasma (FRAP). The improvement in the blood redox status was accompanied by decreased IgA and increased IL-6 levels. As compared to DLM, MHA increased plasma SOD activity and glutathione concentration as well as decreased MDA and LOOH concentrations. In conclusion, dietary Met content, approximately 40% higher than that recommended by NRC, improved the growth performance and antioxidant status of turkeys, and simulated their immune system, regardless of Met source. Therefore, it seems advisable to gain a new deeper insight into this problem and be ready to revise the turkey Met requirements. This work was supported by the National Science Centre, Grant No. 2013/11/B/NZ9/02496.

Effect of methionine source at marginal and adequate methionine levels in turkeys

P.S. Agostini[1], P. Van Der Aar[1], V.D. Naranjo[2] and A. Lemme[2]
[1]Schothorst Feed Research, Lelystad, the Netherlands, [2]Evonik Nutrition & Care GmbH, Hanau, Germany; pagostini@schothorst.nl

Research suggested an average biological equivalency of the liquid hydroxy analogue of MET (HMTBA) of 65% compared to DL-MET (DLM; on product basis) for poultry meaning that 100 g of HMTBA could be substituted by 65 g DLM without affecting performance. This study was carried out to challenge this recommendation in turkeys at an optimal and sub-optimal dietary MET+CYS levels. 1,080 one-d-old BUT Big 6 male turkeys (69 g/bird) were housed in 24 floor pens (10 m2; 45 turkeys/pen). The trial consisted of 4 dietary treatments with 6 replicates (pens) each. A 6-phase feeding schedule was applied. A 2×2 factorial was used comprising two MET sources (DLM, HMTBA) and two dietary MET+CYS levels. Sub-optimal digestible MET+CYS levels were around 87% of the optimal levels which were set at 9.9, 9.0, 8.5, 7.4, 6.6, and 5.8 g/kg feed in phases 1 to 6. The amount of DLM added in both MET dose groups was 65% of that of added HMTBA (on product basis). Carcass composition was evaluated at 21 weeks of age in all animals. No interaction 'source × level' was observed from 0-21 weeks in any performance response ($P>0.05$). Growth rate of turkeys fed the sub-optimal MET diets was lower in this period (20.0 vs 19.4 kg, $P=0.03$) confirming that sub-optimal MET+CYS levels limited performance although no difference in feed efficiency was observed (2.661 vs 2.675, $P=0.63$). Neither a difference between DLM and HMTBA was observed for growth (19.64 vs 19.78 kg respectively, $P=0.55$) nor for feed efficiency (2.661 vs 2.674, $P=0.68$). No interaction was observed for slaughter and breast yields ($P>0.10$) but breast meat yield (% of carcass) was higher at higher MET level (37.2 vs 36.1%, $P<0.001$) and DLM led to higher yield than HMTBA (36.8 vs 36.4%, $P=0.02$). Results indicate that 65 g DLM can replace 100 g HMTBA both at adequate and marginal MET+CYS supply in male turkey diets without negative impact on performance. Even the opposite, breast meat yield was improved with DLM supplementation.

Different methionine sources and quality and shelf life of fresh broiler meat

A. Albrecht[1], D. Miskel[1], U. Herbert[1], C. Heinemann[1], M. Hebel[1], B. Saremi[2] and J. Kreyenschmidt[1]
[1]University of Bonn, Katzenburgweg 7-9, 53115 Bonn, Germany, [2]Evonik Nutrition & Care GmbH, Rodenbacher Chaussee 4, 63457 Hanau-Wolfgang, Germany; behnam.saremi@evonik.com

We studied effect of DL-methionine (DLM) and DL-methionine hydroxy analogue free acid (DL-HMTBA) in chicken feed on the meat quality and shelf life of breast filets. In total, 210 male broiler chickens (Ross 308) were tested in 7 groups with 30 animals each. The groups comprised three concentrations of each DLM and DL-HMTBA (0.04, 0.12, and 0.32% on equimolar basis) and a basal group. After slaughter and processing in a commercial slaughterhouse on day 35, the breast filets were transported under temperature controlled conditions and stored aerobically at 4 °C for 216 h. After specific time intervals, microbial load, pH-value, drip loss, cooking loss and color of the filets were measured. Sensory investigations were conducted and Purchase Decision and White Striping were assessed. In general, the initial microbial load of the different groups was below 2.5 log10 cfu/g at the beginning and below 8.5 log10 cfu/g after 192 h of storage. Mean pH values were between 6.1 and 6.4. Drip loss values were below 0.4% and the average cooking loss ranged between 22 and 28%. The filets showed a shelf life of 6 days. In comparison with the Basal group, methionine supplementation improved quality of broiler breast meat, led to higher pH-values and a higher water binding capacity and lowered the cooking loss. The color value L* showed a significant negative correlation to the methionine concentration supplemented. The influence of methionine supplementation on meat quality did not differ between methionine sources. White Striping was positively correlated to filet weight as well as color values, which significantly affected the Purchase Decision, the SI and thus the shelf life of the samples. In conclusion, higher methionine concentration in chickens diet, improves the meat quality parameters independent of methionine source.

DL-methionine and D-methionine are as efficient as L-methionine in broiler chickens

B. Saremi
Evonik Nutrition & Care GmbH, Animal Nutrition Research, Rodenbacher Chaussee 4, 63457 Hanau-Wolfgang, Germany; behnam.saremi@evonik.com

Although early studies reported no significant differences in utilization of DL-Methionine (DLM) and L-Methionine (LM), a few recent studies have concluded that the bio-efficacy (BE) of LM is higher than DLM. To overcome the experimental limitations shown in these studies (high variation and lack of a significant response) in 2 separate trials, we studied each time 1,210 broilers from 0-35 days. We compared DLM or D-Methionine (DM) to LM. Ross 308 male chickens were fed pelleted diet in a 3 phase feeding program. Eight different doses (0.03, 0.06, 0.09, 0.12, 0.15, 0.21, 0.27 and 0.36%) of either DLM, DM or LM were added to a basal diet deficient in Met+Cys (0.60, 053, and 0.47% in starter, grower, and finisher, respectively). All other amino acids and energy levels were formulated according to AminoChick®2.0 (Evonik, Germany). The non-supplemented group included 10 pens of 20 birds and the first two doses included 7 pens. A dose-response relationship was investigated in a non-linear regression model simultaneously for DLM or DM in comparison to LM. Body weight (BW) and feed intake (FI) were recorded, and body weight gain (BWG) and feed conversion ratio (FCR) were calculated. Carcass composition was measured in 4 slaughtered birds per pen at the end of trial. No difference was observed comparing LM to DLM with regard to BW, BWG, FCR, breast meat yield, and breast meat percentage of BW (a non-significant relative BE of 94, 89, 115, 102, and 96, respectively). Moreover, no difference was observed comparing DM to LM with regard to BW, BWG, FCR, carcass yield, carcass percentage of BW, breast meat yield, and breast meat percentage of BW (a non-significant relative BE of 100, 105, 110, 98, 116, 91 and 88, respectively). None of the BE values are significantly different from 100 and mean of BE values for comparison of LM to DLM and DM to LM were 99 and 101, respectively. Therefore, broiler fed with either DLM, DM or LM performed equally.

Comparision of DL- and L-methionine in corn-soybean-meal based broiler diets

A.A. Çenesiz, I. Çiftci and N. Ceylan

Ankara University, Animal Science, Ziraat Fak. Dışkapı/Altındağ, 06110, Turkey; anilcenesiz@gmail.com

In this study, it was aimed to compare effects of 2 supplemental Met sources (DL or L Methionine) on growth performance, carcass yield, feather and viscera ratio of broilers fed corn-soybean meal based diets. In this research, a total of 728 day old male Ross 308 chicks were weighed and randomly alloted to 7 treatments with 9 replicates in a randomized complete block design for 39 day. Experimental diets were prepared by addition of three levels (0.155, 0.310 and 0.455%) of synthetic DL and L-Met on basal diet containing 0.619, 0.555 and 0.523% digestible Met + Cys for starter, grower and finisher periods, respectively. Feed intake and body weight (BW) were recorded on day 11, 25 and 39. At the end of the trial, two chicks per replicate close to mean of each replicate were selected to determine carcass yield, feather and viscera weights. Met supplementation to the basal diet improved BW, feed conversion ratio (FCR), carcass and breast meat yield ($P<0.05$), regardless Met sources. Similarly, as percentage of BW, liver, pancreas and abdominal fat were reduced by addition of synthetic Met ($P<0.05$) regardless Met sources. Met supplementation and sources had no significant effects on feather percentage and mortality. The relative bioavailability of L-Met to DL-Met for BW, FCR and breast meat yield were found as 123.0, 91.5 and 88.0%, respectively and not statistically significant. It could be concluded that there is no significantdifferences on effectiveness between L-Met and DL-Met in broilers fed corn-soybean meal based diets.

Effect of sulfur amino acid sources on performance of chronic cyclic heat stressed finisher broilers

J. Michiels[1], E. Scarsella[2], M. Majdeddin[1,3,4], J. Degroote[1,3], N. Van Noten[1,3], L. Martens[5], A. Golian[4], A. Dalle Zotte[2] and S. De Smet[3]

[1]Ghent University, Department of Applied Biosciences, Valentin Vaerwyckweg 1, 9000 Ghent, Belgium, [2]University of Padova Agripolis, Department of Animal Medicine, Production and Health, Viale dell'Universita 16, 35020 Legnaro, Padova, Italy, [3]Ghent University, Laboratory for Animal Nutrition and Animal Product Quality, Department of Animal Production, Coupure Links 653, 9000 Ghent, Belgium, [4]Ferdowsi University of Mashhad, Centre of Excellence in the Animal Science Department, P.O. Box 91775-1163, Mashad, Iran, [5]Institute for Agricultural and Fisheries Research, Animal Sciences Unit, Scheldeweg 68, 9090 Melle, Belgium; joris.michiels@ugent.be

We showed that N-acetyl-L-cysteine (NAC), as a source of cysteine, improved performance of chronic cyclic heat stressed finisher broilers. Here, different sulfur amino acid sources were compared. The control diet had a ratio dig M+C to dig LYS of 0.73 (d25-41). Three experimental diets were prepared: control supplemented with NAC (2,000 mg/kg), L-cystine (1,479 mg/kg), or Ca-DL-HMTBa (2,168 mg/kg). Treatments were replicated in 9 pens with 20 Ross308 males each. A chronic cyclic heat stress model (34 °C, 50-60% rh for 7 h daily) was initiated at d28. One bird per pen was sampled on d29 (acute heat stress) and d41 (chronic heat stress) to determine malondialdehyde and activity of enzymes involved in glutathione metabolism in various tissues. ADG for Ca-DL-HMTBa supplemented birds was higher than control (90.8 vs 83.3 g/d; $P<0.05$), whereas for other treatments is was intermediate ($P>0.05$). F:G was not affected, but all supplemented diets showed app. 0.1 lower F:G. Enzyme activities were not affected by treatment. Liver malondialdehyde was lowered in Ca-DL-HMTBa supplemented birds compared to other groups on d29 ($P<0.05$), whereas it was higher in all supplemented birds as compared to control on d41 ($P<0.05$). In conclusion, additional Ca-DL-HMTBa showed beneficial effects on growth in chronic cyclic heat stressed finisher broilers.

Effect of beta-alanine and L-histidine on performance and muscle dipeptides in broilers

J. Michiels[1], M. Majdeddin[1,2,3], E. Claeys[2], A. Cools[4] and S. De Smet[2]
[1]Ghent University, Department of Applied Biosciences, Valentin Vaerwykweg 1, 9000 Ghent, Belgium, [2]Ghent University, Laboratory for Animal Nutrition and Animal Product Quality, Department of Animal Production, Coupure Links 653, 9000 Ghent, Belgium, [3]Ferdowsi University of Mashhad, Centre of Excellence in the Animal Science Department, P.O. Box 91775-1163, Mashad, Iran, [4]Eastman Chemical Company, Pantserschipstraat 207, 9000 Ghent, Belgium; joris.michiels@ugent.be

Beta-alanine is a non-protein beta amino acid and can be synthesized in the body. Beta-alanine and L-histidine are the precursors for the endogenous synthesis of the antioxidant dipeptides carnosine and anserine. We hypothesized that supplemental beta-alanine and L-histidine could increase breast muscle dipeptide content in broilers. A total of 720 one-day-old male Ross 308 broilers were allocated to 36 pens, with 20 birds each. Broilers were offered a starter (d0-10), grower (d10-25) and finisher (d25-38) diet. The basal diets contained 0.51, 0.48 and 0.47% L-histidine. The study had a 2×3 factorial design, i.e. L-histidine (0 and 0.25%) and beta-alanine (0, 0.015 and 0.030%) were added to the basal diets. At d38, one bird per pen was sampled to determine breast muscle dipeptides. L-histidine, beta-alanine and their interaction had no effect on performance in any rearing phase, or total period; except that L-histidine increased feed intake and F:G in the grower phase ($P<0.05$), and similar for total period ($P<0.1$). L-histidine increased carnosine (218 vs 152 mg/100 g; $P<0.05$) and anserine (711 vs 608 mg/100 g; $P<0.05$) in breast muscle, but beta-alanine did not. However, the interaction was significant for carnosine content suggesting that beta-alanine enhanced carnosine levels when L-histidine was not supplemented and, in contrast, lowered carnosine levels with L-histidine. In conclusion, L-histidine increased feed intake and F:G, and enhanced dipeptide levels in breast muscle; whereas beta-alanine showed no effects on performance but increased carnosine levels only when L-histidine was not supplemented.

Effects of dietary tryptophan levels on productive performance and blood parameters of laying hens

Z.S.H. Ismail, A.A.A. Abdel-Wareth and H.A. Hassan
South Valley University, Faculty of Agriculture, south Valley University, 83253 Qena, Egypt; hamde_202@yahoo.com

The objective of this research was to investigate whether dietary supplementation extra L-tryptophan to laying hen diet would influence productive performance, egg quality, blood biochemical parameters and antioxidant status in laying hens. One hundred and fifty, 25 weeks old, Hi-sex Brown laying hens were randomly assigned into five groups. Birds were fed a basal diet (213 mg/kg tryptophan) or basal diet supplemented with extra L-tryptophan at 150, 300, 600 and 1000 mg/kg diet.dietary. The results revealed that L-tryptophan levels (150, 300, or 1000 mg/kg) increased hen-day egg ($P<0.001$) by 6.63, 6.68, 8.99 and 9.89% than control, respectively. Although egg component did not show significant differences between different treatments, Feed conversion, egg quality in terms of egg shape index, haught unit, shell thickness were significantly improved by 1000 mg tryptophan/kg diet. Dietary supplementation with L-tryptophan exhibited a significantly positive effect on blood biochemical parameters and serum antioxidant indices. Serum superoxide dismutase activity, reduced glutathione concentration and total antioxidant capacity were significantly increased in groups fed diets supplemented with tryptophan. The malondialdehyde concentration was decreased ($P<0.01$) in treatment groups when compared to the control one. In conclusion, extra tryptophan supplementation up to 1000 mg/kg diet can be used as an effective feed additive to improve productive performance, egg quality, blood biochemical parameters and antioxidant status in laying hens.

Effect of glycine supplementation in low protein diets on water consumption in broilers

M. Hilliar[1], N. Morgan[1], G. Hargreave[2], R. Barekatain[3], S. Wu[1] and R. Swick[1]
[1]University of New England, Environmental and Rural Science, University of New England, Armidale, 2351, NSW, Australia, [2]Baiada, Sydney, 2000, NSW, Australia, [3]South Australian Research and Development Institute, University of Adelaide, 5371, SA, Australia; mhilliar@myune.edu.au

High dietary protein has been associated with high water consumption in poultry, having a negative impact on litter quality and bird health. Glycine is categorized as a non-essential amino acid, due to its involvement in protein and purine synthesis it may become limiting in low protein diets. The aim* of this study was to measure the effect of glycine supplementation in low protein diets on water consumption in broilers. Male day-old Ross 308 chicks (n=546) were raised in 42 floor pens, 13 birds per pen. Feed intake was recorded from d7 to d35 and water intake from d28 to d35. Diets were wheat-sorghum-soybean meal based and feed and water were provided *ad libitum*. The control treatment contained 21.7/19.8% CP for the grower and finisher phases respectively, and the remaining six treatments contained either 20/18% CP, 18.5/16.5% CP or 17/15% CP with or without glycine. Glycine was added to equal the total level of glycine in the control treatment and all essential amino acids were supplemented when limiting. The results supported that the reduction of dietary protein reduced water intake from d28 to d35, with 15% CP having a significantly lower water to feed intake ratio than the control ($P<0.05$). A significant increase in water to feed intake ratio was observed at 18% CP ($P<0.05$) with glycine supplementation but not at 16.5% CP and 15% CP. The results showed that the addition of glycine improved feed conversion ($P<0.05$) in birds fed 20/18% CP and in birds fed 18.5/16.5% CP, but had no significant effect in birds fed 17/15% CP. This study supports that the supplementation of glycine in low protein diets will improve performance while not affecting water intake.
*We would like to thank Evonik Industries and RIRDC for their financial support throughout this study.

Effect of leucine and valine level and glutamic acid-glycine in low-protein broiler grower diets

S. Golzar Adabi[1], N. Ceylan[1], I. Çiftçi[1], O. Kıyak[2] and V. Hess[2]
[1]Ankara University, Animal Science, Dışkapı, 06110, Turkey, [2]Evonik IndustriesAG, Rodenbacher, Postfach 1345, Germany; iciftci@agri.ankara.edu.tr

Effects of digestible Leucine(1.07, 1.50%) and Valine(0.64, 0.74, 0.84%) in a 2×3 factorial arrangement of total 6 treatments (T1-T6) were investigated in low protein grower diets. Additionally, T3 group contained 1.07% leucine and 0,84% valine(T3) was supplemented with 0.34% glycine and 1.32% glutamic acid(T7). So total 7 treatments with 12 replicates were tested with 840 broilers. The Leucine*Valine interaction was not significant ($P>0.05$) for body weight(BW), body weight gain(BWG) and feed intake(FI), but significant for FCR ($P<0.05$). At 1.07% leucine, only 0.84% dietary valine improved FCR, while both 0.74 and 0.84% valine improved at 1.50% leucine level. Glycine and glutamic acid supplementation had also resulted additional significant improvement in FCR compared to T3. Increasing dietary Valine level from 0.64 to 0,74 and 0.84% significantly increased BW and BWG($P<0.05$), although increasing leucine level did not. The Leucine*Valine interaction was found significant ($P<0.05$) for bone density and tibia breaking strength, both parameters were significantly increased by 0.74 and 0.84% valine level at 1.07% leucine, while just tibia breaking strength improved by 0.84% Val at 1.50% Leu. Glycine and glutamic acid supplementation significantly increased tibia weight and tibia breaking strength compared to T3. The present study obtained significant interaction between leucine and valine on FCR and bone development, so the balance between leucine and valine must be taken into consideration in low protein broiler diets. Glycine and glutamic acid supplemantation can also be beneficial in low-protein diets.

Influence of broiler breeder age on transfer of amino acids to the progeny

J.S. Santos, P.M. Martins, P.M. Rezende, A.F.B. Royer, A.C. Barbosa, M.B. Cafe, N.S.M. Leandro and J.H. Stringhini
Universidade Federal de Goias, Zootecnia, Avenida Esperanca, s.n., 74690900, Goiania, Goias, Brazil;
jhstring@uol.com.br

We aimed to quantify the transfer of amino acids to the progeny at different stages of broiler breeder life. A commercial nucleus composed by six sheds of Hubbard breeders was evaluated at 32, 42 and 52 weeks of age. At each age, four feed samples per shed were collected and 60 eggs were selected from a total of 460 eggs to obtain the yolk, as the rest were sent to incubation. After hatching, 30 newly hatched chicks were euthanized to obtain the yolk sac samples which, as the egg yolk samples, were lyophilized and sent to the laboratory, with feed samples, for amino acid quantification. Data were submitted to ANOVA by SAS statistical software, and Tukey test were applied at 5% of probability to compare the means. Breeder age influenced ($P<0.05$) the percentages of threonine, tryptophan, leucine, phenylalanine, glycine, serine, proline, alanine and glutamic acid. All these amino acids but tryptophan were highest found in the feed directed to the birds at 32 weeks of age, a result that was already expected since the birds feed undergoes nutritional adjustment according to the phase of the production cycle. Egg yolks at 42 weeks of age had lowest ($P<0.05$) percentages of methionine, cystine, methionine + cystine, threonine, tryptophan, arginine, isoleucine, valine, phenylalanine, glycine, serine, alanine, aspartic acid and glutamic acid, with lysine being the only amino acid to be found in the highest amount at this age when compared to 32 and 52 weeks old ($P<0.05$). Chicks from younger breeders showed in their yolk sac greater amounts of all of the evaluated amino acids ($P<0.05$). In general, regarding the analyzed nutrients, younger breeders showed more efficiency in transferring them to the egg. After hatching, the progeny from younger breeders were also more efficient in the absorption of these amino acids, considering the similarity of these nutrients quantities found in the different evaluated feed.

Guanidinoacetic acid supplementation improves feed conversion in broilers subjected to heat stress

M. Majdeddin[1,2,3], U. Braun[4], A. Lemme[5], A. Golian[3], H. Kermanshahi[3], S. De Smet[1] and J. Michiels[2]
[1]Ghent University, Department of Animal Production, Coupure Links 653, 9000, Gent, Belgium, [2]Ghent University, Department of Applied Biosciences, Valentin Vaerwyckweg 1, 9000, Gent, Belgium, [3]Ferdowsi University of Mashhad, P.O. Box 91775-1163, Mashhad, Iran, [4]AlzChem AG, Dr.-Albert-Frank-Str. 32, 83308 Trostberg, Germany, [5]Evonik Nutrition & Care GmbH, Rodenbacher Chaussee 4, 63457 Hanau-Wolfgang, Germany; maryam.majdeddin@ugent.be

Guanidinoacetic acid (GAA) is synthesized from Gly and Arg and the immediate precursor of creatine. The creatine (Cr) and creatine phosphate (PCr) system plays a critical role in energy metabolism. It is hypothesized that dietary supplementation with GAA will be beneficial to heat stressed finishing broilers. A total of 720 one-day-old male Ross 308 broilers were allocated to 3 dietary treatments with 12 replicates (20 birds each). Treatments were either 0, 0.6 or 1.2 g/kg GAA added to a corn/SBM diet and fed for 39 d. A chronic cyclic heat stress model (34 °C/50-60% rh for 7 h daily) was applied in the finisher phase (d25-39). One bird per pen was sampled on d26 (acute heat stress) and d39 (chronic heat stress) to determine levels of GAA, Cr and Arg in blood, and Cr, PCr and ATP in muscle. GAA at 1.2 g/kg decreased F:G compared to control in the grower phase (d10-25) (1.32 vs 1.35; $P<0.05$). In the heat stressed finisher period, 0.6 and 1.2 g/kg GAA reduced feed intake by 1.1 and 3.3% respectively and improved F:G (1.76, 1.66 and 1.67 for control, 0.6 and 1.2 g/kg GAA respectively, $P<0.05$). Examination of breast meat samples revealed at both sampling days significant ($P<0.05$) increases of PCr, Cr, and PCr:ATP ratio with increasing dietary GAA. GAA and Cr in blood were increased with increasing dietary GAA ($P<0.05$) at d26 and d39. Blood Arg increased with increasing dietary GAA (+18.2 and +19.9% for 0.6 g/kg GAA on d26 and d39, respectively, and +30.8 and +33.6% for 1.2 g/kg GAA, $P<0.05$) suggesting enhanced availability of Arg for other metabolic purposes than de-novo GAA formation.

Metabolomics revealed no major difference between L-methionine and DL-methionine

B. Saremi

Evonik Nutrition & Care GmbH, Rodenbacher Chaussee 4, 63457 Hanau, Germany; behnam.saremi@evonik.com

A 34-day feeding study was conducted to investigate the effect of DL-Met and L-Met at different concentrations. Basal diets were composed to meet nutrient recommendations (Cobb 500), except for methionine plus cysteine (SID Met+Cys%: 0.63, 0.56, and 0.50, for starter, grower and finisher diets, respectively). Each of the two methionine sources was supplemented at levels of 0.07, 0.14, 0.21, and 0.28% to the basal diet. Dietary treatments were fed to 10 replicate pens with 20 male broilers each. At day 34, one bird per pen was slaughtered and blood was collected in order to produce serum. Briefly, serum samples were extracted and split into equal parts for analyses on the GC/MS, LC/MS/MS, and Polar LC platforms. Profiling data from 726 identified metabolites served to characterize diet dependent biochemical events. In regards to sulfur amino acid metabolism, the related metabolites showed that both methionine sources behave similarly at both low and high doses. Subtle differences observed included significant up-regulation of homocysteine, N-acetylmethionine sulfoxide and methionine sulfone in L-Met vs DL-Met when supplemented at low doses. Homocysteine, N-acetylmethionine sulfoxide and methionine sulfone are derived from the oxidation of methionine. The glutathione metabolic pathway was not affected. TCA cycle was down regulated by L-Met in comparison to DL-Met at higher doses in what appears to be a disruption of normal metabolism. Both, DL-Met and L-Met, exerted similar effects in global metabolism. Thus, feeding L-Met provided no clear advantages over feeding DL-Met. The biochemical differences associated with various concentrations of methionine were very subtle, and these subtle differences pointed toward potential oxidative stress responses in DL-Met or L-Met supplemented birds due to higher growth rate of these birds. Overall, both the global and focused investigations of blood metabolome of birds fed diets supplemented with different methionine sources revealed only minor effects on their metabolism.

Validation of the recommendations in methionine + cysteine of broilers given by the eRNG

D.I. Batonon-Alavo and Y. Mercier

Adisseo France SAS, CERN, 6 Route Noire, 03600 Malicorne, France; dolores.batonon-alavo@adisseo.com

Updates of amino acids requirements are necessary with regards to the improvement of poultry genetics. Amino acids recommendations have been proposed by Adisseo by modeling daily weight gain as response to digestible amino acids intake (eRNg). This trial aims to validate the recommendations in methionine (Met) and methionine + cysteine (Met+Cys) in broilers using a quadratic plateau model. 720 Ross PM3 males chickens were reared from 0 to 28-d in floor pens, divided in two feeding phases: 0 to 14-d and 14 to 28-d. Birds were randomly allocated to 6 treatments with 8 replicates of 15 birds each. Treatments consisted in a basal diet deficient in Met and five treatments supplemented with graded levels of DL-Met in the 2 phases (0.07, 0.14, 0.21, 0.28 and 0.35%). For each phase, feed intake, body weight gain and feed conversion ratio were calculated. Dietary Met and Met+Cys contents were determined with the amino acid analyzer and allowed to calculate, for each treatment, the Met and Met+Cys intakes as a product between feed intake and the measured values in the diet. A quadratic plateau model was used to model daily weight gain and feed conversion ratio. During the whole experimental period, all performance criteria were significantly different from the control deficient diet and were improved with methionine addition until a plateau is reached. The Met requirement for maximizing weight gain on the 0-14 d period was 190 mg/d (i.e. 0.44% of feed), whereas the corresponding Met requirement for FCR was 243 mg/d (0.52% of feed). The Met+Cys requirement was determined at 405 mg/d and 436 mg/d (i.e. 0.82% and 0.90%), respectively for optimization of weight gain and FCR from 0 to 14 d. From 14 to 28 d, the Met requirement for weight gain was 595 mg/d (0.42%) while it was of 631 mg/d (0.53%) for FCR. Met+Cys requirement were 1,092 mg/d (0.83%) and 1,074 mg/d (0.91%) for weight gain and FCR for the 14-28 d period. These requirements are in lines with the recommendations of the eRNG.

Arginine requirement of broiler chickens

E. Esteve-Garcia[1], D. Khan[2] and C. Westermaier[2]
[1]IRTA, Animal Nutrition and Health, Centre Mas Bover, 43420 Constantí, Spain, [2]CJ Europe GmbH, Ober der Röth 4, 65824 Schwalbach, Germany; enric.esteve@irta.cat

L-arginine (L-Arg) is an essential amino acid in chickens. There is no much information available about its requirements in rapidly growing chickens. The objective of the present study was to determine the L-Arg requirement of broilers through supplementation technique. A total of 1,440 male chickens of the Ross 308 strain were distributed into 48 pens, 6.25 m2 each, at 30 per pen and assigned one of six diets with graded levels of Arg:Lys. The feeding program consisted of a starter diet between 0 and 14 days with 21.90% crude protein and 3,050 kcal ME/kg feed, grower phase between 14 and 28 days containing 20.00% crude protein and 3,100 kcal ME/kg feed. Arg:Lys ratios were 0.77, 0.85, 0.95, 1.05, 1.15 and 1.25 in all periods, and the Lys levels were 1.29 and 1.09% in starter and grower diets respectively. Chickens and feed were weighed at 14 and 28 days and performance was calculated for each period. Results were analyzed according to an exponential model and by two way ANOVA followed by a set of orthogonal contrasts to determine linear and quadratic responses to L-Arg supplementation. The birds showed a significant ($P<0.05$) response to Arg supplementation as compared to control. The exponential model did not fit to data, because there was a response beyond the asymptote. However, quadratic function showed a good fit to the data, but it was not clear from graphical presentation that the maximum response had been achieved. The maximum response for BW, F:G was achieved at Arg:Lys=1.15:1.00 for starter period, whereas, 1.25:1.00 for grower period respectively. It indicates a direct relationship of L-Arg demand with the age period in broilers. Results of the study suggest that the optimum ratio Arg:Lys is greater than the standard recommendation of 1.05.

Influence of dthr:dlys and CP on Ross 308 broiler performance, litter quality and foot pad lesions

A. Sacranie[1], L. Linares[1], W. Lambert[2] and E. Corrent[2]
[1]Aviagen Ltd, Nutrition, Newbridge, EH28 8SZ, Midlothian, United Kingdom, [2]Ajinomoto Eurolysine S.A.S., 153 Rue de Courcelles, 75817, Paris Cedex 17, France; asacranie@aviagen.com

The potential interaction between 7 different dThr:dLys ratios (ranging from 0.58-0.76) and 2 CP levels, low (starter=18-19%, grower=17-18%) or standard (starter=21-23%, grower = 19.5-21.5%) was investigated in 4,004 all male Ross 308 broilers from 0-24 d randomly distributed amongst 182 pens. Broilers exposed to the standard CP feeds, did not respond to elevated levels of dThr The lowest BWG and poorest FCR over the experimental period was recorded in birds fed the lowest dThr:dLys/Low CP dietary treatment however by 24 d the effect of low CP was compensated by dThr:dLys ratios greater than 0.61 ($P<0.001$). Optimal dThr intake was estimated for FCR and BWG following a quadratic model: for BWG and FCR at 10 d an optimal dThr:Dlys ratio of 0.62-0.66, equivalent to a maximum absolute dThr level of 0.74% was determined. For 10-24 d an optimal dThr level of 0.60-0.63 and 0.58-0.60 respectively, equivalent to a maximum absolute dThr level in the feed of 0.64% was determined. Birds exposed to the breeding company's recommended dThr:Lys ratio, 0.67, BWG from 0-24 d was not affected by dietary CP level. However, percentage of litter capping at 26 d was reduced from 55.8 to 47.8% in the lower CP treatments. As an indirect consequence, foot pad lesions incidence was dramatically decreased when reducing CP. At 26 d, percentage of birds presenting moderate foot pad lesion score was 21.8% and 4.5% in standard and low dietary CP level, respectively. For both the starter and grower feeding phases, a dThr:dLys ratio of 0.67 is sufficient to ensure good performance in diets formulated to standard or low CP levels. Performance of broilers fed low CP diets was similar to those fed standard CP diet while improving litter quality and reducing foot pad lesion incidence and severity.

Metabolizable energy and amino acids for female broilers

F.G. Perazzo Costa[1], M. Ramalho Lima[2], D.T. Cavalcante[1], L.E. Cavalcante[1], G. Ferreira[1], T. Moreira[1], M. Ladeira Ceccantini[3], J. Gonçalves[3] and R. Montanhini Neto[3]
[1]Federal University of Paraiba, Areia, Paraiba, 58397000, Brazil, [2]Federal University of the Southern of Bahia, Teixeira de Freitas, Bahia, 45987478, Brazil, [3]Adisseo Animal Nutrition, Sao Paulo, 01311000, Brazil; perazzo63@gmail.com

The aim was to evaluate levels of metabolizable energy (ME) and amino acids (AA) on performance of female broilers. A total of 1,050 female broilers Cobb were used in a 3×2 factorial design (3 levels of AA, and 2 levels of ME), or 6 treatments with 7 replicates with 25 broilers each, distributed in a completely randomized design. The ME based on the recommendations of the Brazilian Tables of Nutritional Requirements of Poultry (BrReq) with levels of 2,950 and 3,120 kcal/kg (1 to 10 d, 2,950 kcal/kg by BrReq), 3,000 and 3,290 kcal/kg (11 to 21 d, 3,000 kcal/kg by BrReq), 3,100 and 3,290 (22 to 35 d, 3,100 kcal/kg by BrReq), and 3,150 and 3,300 kcal/kg of ME (36 to 42 d, 3,150 kcal/kg by BrReq), with each ME level associated levels of AA, levels by BrReq, 5% increase, and 5% decrease. The amino acids used were Met, Lys, Thr and Val. From 1 to 21 d, in the interaction, the largest weight gain (WG) was observed in broilers with high level of ME and AA ($P=0.0029$). The same result was observed on feed conversion ratio (FCR) ($P=0.0403$). The WG was influenced by the level of AA and ME, with better results in the above BrReq levels. In the phase from 22 to 42 d, the effect was repeated in WG ($P=0.0191$), but the FCR had an influence only within the highest level of ME, where the AA is more efficient increase in this variable ($P=0.0427$). In the full phase, 1 to 42 days, the female broiler were heavier of treatment with higher ME and AA increase ($P=0.0019$), while the FCR had an influence only between those fed higher ME ($P=0.010$). In conclusion, it is recommended levels of 3,120, 3,290, 3,290, 3,300 kcal/kg, respectively for the 1-10, 11-21, 22-35, and 36-42 d, associated an increase of 5% of AA levels by BrReq to female broilers.

Metabolizable energy and amino acid for broilers chickens

F.G. Perazzo Costa[1], M. Ramalho Lima[2], D.T. Cavalcante[1], G. Castro[1], S. Hermenegildo[1], W. Correa[1], M.L. Ceccantini[3], J. Gonçalves[3] and R. Montanhini Neto[3]
[1]Federal University of Paraiba, Areia, Paraiba, 58397000, Brazil, [2]Federal University of the Southern of Bahia, Teixeira de Freitas, Bahia, 45987478, Brazil, [3]Adisseo Animal Nutrition, Sao Paulo, 01311000, Brazil; perazzo63@gmail.com

The aim was to evaluate the variation of metabolizable energy (ME) and amino acid levels in diets of male broilers. The birds, weighing 39.7±0.02 g, were divided into three treatments, arranged in a completely randomized design, with 10 replicates of 20 birds each. The treatments were control (ME=2,950/3,000/3,100/3,150 kcal/kg and Lys=1.31/1.17/1.07/1.01 g/100 g, respectively) in the evaluated phases: 1 to 10, 11 to 21, 22 to 35, and 36 to 43 d. Treatment (T) 2 had Lys at 1.15/1.16/0.98/0.83 g/100 g, with Met+Cys, Thr, and Val ratios with Lys 83/65/83, 78/62/75, 77/62/81, 76/70/92, respectively in the phases evaluated. T3 had an increase levels at 3,175/3,308/3,360/3,360 kcal/kg and Lys at 1.20/1.21/1.02/0.87 g/100 g, maintaining the amino acid: Lys ratio of T2. In the 1-10 d, T3 promoted higher WG (271.70a; $P=0.001$) and with better FCR (1.20c; $P=0.002$). In the 11-21 d, there was a significant effect, with similar results to the previous phase, occurring the same in the next phase, of 22-35 d. In the final phase, 36-43 d, treatment 3 remained more efficient than the others in the final weight (3,369.40a, $P=0.001$), WG (846.00a; $P=0.001$), and FCR (1.77c; $P=0.001$). In the complete phase, 1-43 d, treatment 3 proved to be more efficient, having a greater WG (3,329.79a; $P=0.001$), with a FI (5,147.65c, $P=0.001$) below the others and a better FCR (1.54c; $P=0.001$). Based on the results, it is recommended an increase in ME and amino acids, with levels at 3,175/3,308/3,360/3,360 kcal/kg and Lys levels at 1.20/1.21/1.02/0.87 g/100 g, with ratios Lys Met+Cys, Thr, and Val in 83/65/83, 78/62/75, 77/62/81, and 76/70/92, respectively in the evaluated phases 1-10/11-21/22-35/36-43 d old male broilers.

Metabolizable energy and amino acids for male broilers

F.G. Perazzo Costa[1], M. Ramalho Lima[2], D.T. Cavalcante[1], G.S. Lima[1], C. Lima[1], M. Neves[1], M.L. Ceccantini[3], J. Gonçalves[3] and R. Montanhini Neto[3]
[1]Federal University of Paraiba, Areia, Paraiba, 58397000, Brazil, [2]Federal University of the Southern of Bahia, Teixeira de Freitas, Bahia, 45987478, Brazil, [3]Adisseo Animal Nutrition, Sao Paulo, 01311000, Brazil; perazzo63@gmail.com

The aim was to evaluate increase or decrease levels of metabolizable energy (ME) and amino acids (AA) on performance of male broilers from 1 to 42 d. A total of 1050 male broilers Cobb were used in a 3×2 factorial design (3 levels of AA, and 2 levels of ME), or 6 treatments with 7 replicates with 25 broilers each, distributed in a completely randomized design. The ME based on the recommendations of the Brazilian Tables of Nutritional Requirements of Poultry (BrReq) with levels of 2,950 and 3,120 kcal/kg (1 to 10 d, 2,950 kcal/kg by BrReq), 3,000 and 3,290 kcal/kg (11 to 21 d, 3,000 kcal/kg by BrReq), 3,100 and 3,290 (22 to 35 d, 3,100 kcal/kg by BrReq), and 3,150 and 3,300 kcal/kg of ME (36 to 42 d, 3,150 kcal/kg by BrReq), with each ME level associated levels of AA, levels by BrReq, 5% increase, and 5% decrease. The amino acids used were Met, Lys, Thr and Val. From 1 to 21 d, body weight gain (P=0.0144) and feed conversion ratio (P=0.0257) were improved when the ME and AA were above the Control diet. The same effect was observed in the phase from 22 to 42 d in body weight gain (P=0.076). Considering the complete phase from 1 to 42 d, an increase of 5% of AA in the diet provided better body weight gain (P=0.0017) and feed conversion ratio (P=0.0196), especially with an increase of ME. In conclusion, it is recommended to feed at levels of 3,120, 3,290, 3,290 and 3,300 kcal/kg of ME, respectively from 1-10, 11-21, 22-35, and 36-42 d, associated with an increment of 5% levels by the BrReq.

Different forms of dietary energy and amino acids coordinately affect broiler feed intake and growth

S. Wu[1], R. Swick[1] and M. Choct[1,2]
[1]University of New England, School of Environmental and Rural Science, 1 Elm Ave, 2351, Australia, [2]Poultry Cooperative Research Centre, University of New England, 2351, Australia; shubiao.wu@une.edu.au

Two major components of feed required by animals are energy supplied through carbohydrates and fat, and amino acids supplied through protein. The balance of nutrients in feed are important for growth and feed efficiency (or FCR). In broiler chickens, efforts have been made to reduce protein and energy to minimum required levels while still achieving maximum growth and feed conversion thereby lowering the costs of production and residuals in environment. However, how the broilers respond to the nutrients in terms of the feed intake (FI), growth and feed conversion efficiency has not been fully investigated. In the present study the roles of metabolisable (ME) and net energy (NE) and amino acids on the performance of the birds were investigated with diets formulated to varying levels of energy and amino acid contents using 16 closed circuit calorimeter chambers. Male Ross 308 broilers were fed 16 diets for 3 d from 25 to 28 d of bird age to measure ME, heat production, growth and FI. The results showed that FI was negatively regulated by dietary energy, (ME or NE), and the amino acids methionine or threonine, while ADG was only related to the levels of amino acid contents in the diets especially lysine but not to energy levels. FCR, on the other hand, was affected by ME and NE, and a few amino acids in diet. Interestingly, lysine and threonine affected FCR while ME was included in the equation, while methionine, tyrosine and hydroxylysine appeared to affect FCR while NE was in the equation. It can be concluded that both dietary energy and amino acids need to be carefully controlled in the feed formulation to achieve optimal performance of broiler chickens particularly when consideration is given to the use of either ME or NE in the formulation. This may suggest different roles of different amino acids in their contributions to heat increments of feed.

Effect of carbohydrate type and amino acid density in early nutrition of broiler chicks

M. Vadiei[1] and I. Ciftci[2]
[1]Nutrex, Achteratenhoek 5, 2275 Lille, Belgium, [2]University of Ankara, Agriculture Faculty, Animal Nutrition Department, 06110, Dışkapı, Ankara, Turkey; mvadiei@gmail.com

This experiment was conducted to investigate the effects of dietary carbohydrate type and amino acid (AA) density in early nutrition on performance and GIT development of broiler chicks. In this study, 192 one-day-old male Ross 308 broiler chicks were randomly distributed to 4 groups with 6 replicates. With a complete 2×2 factorial arrangement, corn or dextrose based diets were formulated in higher or lower 12% of lysine levels recommended by chicken manual guide. Levels of other essential amino acids in diets were considered in ideal AA concepts according to lysine. Dextrose based diets increased body weight and feed intake from 0 to 13th days and improved feed conversion ratio for only first week ($P<0.05$). High AA density in diets also improved these performance traits ($P<0.05$) for first 13th days. Generally, interaction effects between carbohydrate type and AA density was not significant for performance criteria ($P>0.05$).Dextrose based diets depressed physical developments of proventriculus, gizzard, jejunum and ileum specifically for 7 and 13th days, but increased duodenum and liver weight ($P<0.05$). Low AA density in diets had detrimental effects on the physical development of digestive system ($P<0.05$) for 13th day. In the similar manner with physical development, morphological development of duodenum, jejunum and ileum were affected by carbohydrate type and AA density ($P<0.05$). With interaction effects, morphological developments of duodenum and ileum were depressed by lower AA density for dextrose based diets, but not affected for corn based diets. However, for jejunum, villus development was increased by higher AA density for corn based diets, but not for dextrose. There are no significant effects of carbohydrate type on the enzyme activities of pancreas. As a conclusion, dextrose based diets and high AA density in early nutrition of the broiler chicks had improved effects on performance, protein utilization and digestive system developments.

Effect of dietary energy, lysine and phosphorus levels on amino acid digestibility in broilers

N.K. Sharma[1], M. Choct[1], M. Toghyani[1], C.K. Girish[2] and R.A. Swick[1]
[1]University of New England, Armidale, 2351, Australia, [2]Evonik (SEA) Pte. Ltd., Singapore, 609927, Singapore; mchoct@une.edu.au

A 3-factor-3-level Box-Behnken design was used to study the effect of dietary digestible lysine (dLys, based on the ideal ratio) (9.5, 10.5, 11.5 g/kg), apparent metabolizable energy (AMEn) (12.77, 13.19, 13.61 MJ/kg) and available P (avP) (3.0, 4.0, 5.0 g/kg) levels on apparent ileal digestibility (AID) of amino acids (AA) in broilers. The design consisted of 15 treatments (3 center points) with 5 replications of 12 birds each. A total of 1,050 d-old Ross 308 male broiler chicks were allocated to treatment diets from d 14-34. On d 34, 3 birds were sampled from each pen to collect ileal digesta which were pooled per pen. The diet and digesta samples were analyzed for amino acid content by HPLC. The AID of AA was calculated using titanium dioxide as the indigestible marker. JMP statistical software v. 12.0.1 was used to generate the response surface plots. Response surface was fitted by first, second or third-degree polynomial regressions. The experimental units were pen means and a 5% level of probability was considered to be significant. dLys and avP levels in the diet had significant effects on AID of methionine ($P<0.001$) and threonine ($P<0.001$) but had no effect on AID of lysine, leucine, isoleucine, valine, arginine, histidine, phenylalanine, glycine and serine ($P>0.05$). AID of methionine was described by a first-order equation (adj. R^2=0.28, $P<0.001$) and was affected by dLys (linear) and avP (linear). Increase in the levels of dLys or avP increased the AID of methionine but the increase in AMEn had no effect ($P>0.05$) on AID of methionine. Similarly, AID of threonine was described by a first-order equation (adj. R^2=0.19, $P<0.001$) and was affected by dLys (linear) and avP (linear). The increase in the levels of dLys or avP increased the AID of threonine but the increase in AMEn had no such effect ($P>0.05$). These results indicate that increasing dLys or avP levels may improve AID of methionine and threonine.

Effect of probiotic, energy and digestible lysine-methionine ratio on egg production performances

K. Soisuwan and A. Airlang

Rajamagala University of Technology Srivijaya, Department of Animal Science, Faculty of Agriculture, Thungsong, Nakhon Si Thammarat, 80110, Thailand; ksoisuwan52@gmail.com

An experiment was conducted to determine the influence of dietary energy, digestible lysine-methionine ratio and supplementation probiotic on production performance and egg quality of laying hens in the late period of production. The experiment was designed as a 2×2×3 factorial arrangement with 2 dietary energy levels (2,800 and 2,900 kcal of ME/kg) and 2 levels of digestible lysine-methionine ratio (DLM; 0.81:0.44 and 0.97:0.53%) with supplementation of a commercial probiotic at the level of 0, 0.1 and 0.2%, respectively. Isa brown laying hens (n=540) in phase II (60 weeks of age) were randomly divided into 12 treatments (5 replicates of 45 hens per treatment). There was no interaction on production performance and egg quality between dietary treatments. Egg production, egg mass and feed conversion ratio were improved ($P<0.05$) when fed 2,900 kcal of ME/kg feed compared with those fed 2,800 kcal of ME/kg feed. Egg weight increased ($P<0.05$) and feed intake decreased ($P<0.05$) when DLM increased while feed conversion ratio improved ($P<0.05$) with increased DLM. It was also found that supplementation of probiotic had no effect on egg production performance and egg quality. Based on data of this experiment, it was concluded that brown laying hens in the late period of production (60-70 weeks of age) as reared in open-side houses in tropical climate required a concentration of 2,900 kcal of ME/kg and 0.97:0.53% of digestible lysine-methionine ratio while supplementation of probiotic had no effect on egg production performance and egg quality.

Environmental implications of decreasing dietary crude protein content in finishing broilers

B. Méda[1], P. Belloir[1,2], M. Lessire[1], H. Juin[3], E. Corrent[2], W. Lambert[2] and S. Tesseraud[1]

[1]INRA, URA, Centre Val de Loire, 37380 Nouzilly, France, [2]Ajinomoto Eurolysine S.A.S, 153 rue de Courcelles, 75017 Paris Cedex 17, France, [3]INRA, EASM, Le Magneraud, 17700 Saint-Pierre-d'Amilly, France; lambert_william@eli.ajinomoto.com

The aim of this study was to assess the environmental implications of decreasing CP content in broiler diets. Two trials with finishing broilers fed with diets with different CP content (19, 18, 17, 16 and 15%; 19, 17.5 and 16%, respectively) were carried out. From animal performance (growth, feed intake), a nitrogen balance was carried out to estimate retention and excretion. Manure was weighted and analysed for N content. From this data, nitrogen volatilization was estimated. Since animal performances were not impacted by the CP decrease, nitrogen retention efficiency was improved by 3.4% per CP point. As a consequence, nitrogen excretion was decreased by 12-14% per CP point. This phenomenon was also associated to a lower proportion of the excreted N lost through gas emission (volatilization). Furthermore, two scenarios of broiler production were built and assessed using Life Cycle Assessment (LCA). In the first scenario (S19, reference scenario), the CP content of the finishing diet was 19% whereas in the second one (S16), it was reduced to 16%. Animal performances were considered identical in both scenarios but nitrogen excretion of birds and volatilization from manure were lower in S16. Results showed that in S16, LCA impacts (per kg of live weight) were decreased by 8%, 7% and 1% respectively for climate change, eutrophication and energy use. For acidification, while the impact per ton of finishing diet increased by 4%, the reduced nitrogen excretion and volatilization compensated this negative effect, so that the impact per ton of live weight was reduced by 5%. These results confirm that decreasing dietary crude protein content is a relevant strategy to reduce the environmental impact of broiler production.

Reducing crude protein content in broiler feeds: impact on animal performance and meat quality

P. Belloir[1,2], B. Méda[2], M. Lessire[2], E. Corrent[1], W. Lambert[1] and S. Tesseraud[2]
[1]Ajinomoto Eurolysine S.A.S, 153 rue de Courcelles, 75017 Paris Cedex 17, France, [2]INRA, URA, Centre Val de Loire, 37380 Nouzilly, France; lambert_william@eli.ajinomoto.com

The aim of this study was to investigate the effect of decreasing CP content on animal performance and meat quality in growing-finishing broilers using an optimized dietary amino acid (AA) profile based on the ideal protein concept. 912 day-old PM3 Ross male broilers were reared together between 1 and 20 d of age. At d21, they were randomly distributed in floor pens (8 pens per treatment) and were fed with diets formulated with the ideal AA profile proposed by Mack et al., adjusted for Thr and Arg. More specifically, the dThr:dLys ratio was increased from 63 to 68 and the dArg:dLys ratio was decreased from 112 to 108. Three levels of crude protein (CP) were studied (19%, 17.5% and 16%). Birds were slaughtered at 35 days of age. There was no effect of dietary CP on animal performance (body weight gain, feed conversion ratio). Dietary CP content did not significantly affect breast meat yield but abdominal fat content was increased by the decrease in CP ($P<0.01$). Meat quality criteria responded to dietary CP content with higher ultimate pH and lower lightness and drip loss in the low CP diets. In conclusion, this study demonstrates that with an adapted AA profile, it is possible to decrease dietary CP content until 16% in growing-finishing male broilers, without altering animal performance and meat quality.

Effect of low dietary protein levels on performance, litter quality and footpad lesions in broilers

J. Van Harn[1], M.A. Dijkslag[2] and M.M. Van Krimpen[1]
[1]Wageningen Livestock Research, De Elst 1, 6708 WD Wageningen, the Netherlands, [2]ForFarmers N.V., Kwinkweerd 12, 7241 CW Lochem, the Netherlands; jan.vanharn@wur.nl

Providing diets with a reduced crude protein (CP) content that are supplemented with increased contents of free amino acids (AA) to cover the AA requirement might be helpful in reducing the soybean meal content of South American origin. Therefore, a broiler study was conducted, to investigate the effect of reduced dietary crude protein levels supplemented with free AA in grower (d11-28) and finisher diets (d29-35) on performance, litter quality and footpad lesions. A total of 884 day-old male Ross 308 broilers were distributed over 68 floor pens (0.75 m^2), and allotted to 1 of 4 dietary treatments. All birds received a standard starter diet (d0-10). CP level in the control grower and finisher diet was 20.8 and 19.8%, respectively, whereas CP level in the other treatments was reduced with 1, 2, or 3%, respectively, thereby decreasing SBM contents up to 10%. Low CP Diets were supplemented with free Lys, Met, Thr, Try, Val, Arg, Ile, and Gly to maintain contents of essential AA at the level of the control diets. Over the whole experimental period (d0-35), FCR was significantly ($P<0.05$) improved in birds fed the CP-2% (1.480) and CP-3% (1.481) diets compared to the control diets (1.518) and the CP-1% diets (1.512), whereas the other performance traits were not affected by dietary CP content. Reduction in dietary CP content linearly improved foot pad lesion score at d35 from 143 (control) to 39 (CP-3%). Breast meat-% of the birds fed control diet was significantly higher than those who receive the CP-3%, but similar to the CP-1% and CP-2% diets. Visual litter quality improved linearly with decreasing CP content. It can be concluded that CP-reduction in grower and finisher diets up to 3%, along with supplementation of essential AA, reduced breast meat-%, but improved FCR, litter condition and foot pad lesion score, thereby substantially reducing the use of imported SBM.

Effects of low protein diets on the production traits and carcass composition of meat type ducks

K. Dublecz, L. Pál, L. Wágner, F. Dublecz and O. Hegyi
University of Pannonia, Department of Animal Science, Deak F. u. 16, 8360 Keszthely, Hungary; dublecz@georgikon.hu

A total of 150 d-old Cherry Valley SM3 Heawy hybrid ducks were divided into 15 homogenous groups of 10 ducks each and placed in floor pens. Three dietary treatments were used with 5 replication. The experimental design consisted of a corn, wheat and soybean meal containing control diet (1) and two other treatments with 1 (2) and 2.5% (3) lower crude protein content. Beside lysine and methionine the protein reduced diets were supplemented with limiting amino acids of threonine and valine. Diets were formulated on standardised ileal digestible amino acids (SIDAA). Starter, grower and finisher diets were fed between days 1-7, 8-14 and 15-42 respectively. Live weight and feed intake of ducks were measured at the termination of each phase and feed conversion ratio (FCR) calculated on pen basis. At day 42, twelve birds per treatment were slaughtered and the carcass composition evaluated. The cumulative feed intake in treatment groups 1, 2 and 3 were 6,842, 6,574, and 6,846 g, the averages of cumulative weight gain 3,677, 3,639, and 3,740 g and those for FCR 1.87, 1.81, and 1.83 g/g respectively. None of the production traits were influenced by the dietary treatments. Regarding the carcass composition parameters significant differences were found only at the dressing percentage, where the average of treatment 2 (58.42%) was significantly lower than those of treatment 1 and 3 (60.0 and 59.98%). The protein reduction of diets failed to cause significant effects in the relative thigh and breast meat percentage. Lowering the protein content of the diets by 1 and 2.5% decreased the feed cost by 1.15 and 9.57 Euro cent and increased the profitability compared with the control treatment. It can be concluded, that using low protein, crystalline amino acid supplemented diets can be an option to decrease feed costs and reduce ammonia emission in duck production.

Effects of low crude protein diets balanced with supplemental amino acids on laying hens

C. De La Cruz[1] and V. Machander[2]
[1]Evonik Nutrition & Care GmbH, Rodenbacher Chaussee 4, 63457 Hanau, Germany, [2]International Poultry Testing Station Ustrasice, Ustrasice 63, 39002, Czech Republic; carlos.delacruz@evonik.com

A randomized design was used to conduct a study to evaluate the influence of dietary crude protein (CP) level on laying hen performance and egg quality. A total of 2,700 layers (Isa Brown) were distributed into 9 dietary treatments at the age of 21 weeks, layers were placed in 270 pens (300 per treatment) in 30 enriched cages. Birds were fed different CP diets with identical essential amino acid (AA) levels, crystalline Met, Lys, Thr, Trp and Val were supplemented according to an assumed ideal AA profile, at a fix dietary energy in a 3-phase feeding program (1; 21 to 32, 2; 33 to 40, 3; 41 to 52 wk of age). CP content of the diets of groups 1, 2 and 3 were 18%; 4, 5 and 6 were 17%; 7 and 8 were 16% and 9 was 15% for phase 1; treatments 1 was 18%; 2 and 4 were 17%; 3, 5 and 7 were 16%; 6 and 8 were 15%; 9 was 14% for phase 2; group 1 was 18%; 4 was 17%; 2 and 7 were 16%; 3, 5, 6, 8 were 15%; 9 was 14% for phase 3. Egg production and feed intake were recorded daily, egg weight (EW) weekly, egg quality was evaluated at week 32, 40, and 52; 30 eggs per pen were collected to determine eggshell strength (ES), eggshell thickness (ET), haugh units, yolk weight and yolk color (YC). Also, bodyweight (BW) was measured at week 20, 32, 40 and 52. Results were evaluated with ANOVA-Duncan test. Egg production nor feed intake hen–day did not show difference ($P<0.05$) when hens fed different CP diets among treatments, however EW showed lower weight when less CP were offered, BW did not change despite EW. Superior ES and ET were detected when lower CP were offered, led to 578 less broken eggs. YC increased in the low-protein groups. It was concluded that the application of the assumed AA profile diet can lead to reduced dietary CP without affecting the performance of laying hens from 21 to 52 wk of age. However EW, ES, ET and YC were influenced from hens fed the CP-reduced but without impairing BW.

Effects of dietary protein level and age at photo stimulation on reproduction in broiler breeders

R.A. Van Emous[1] and C.E. De La Cruz[2]
[1]Wageningen Livestock Research, Animal Nutrition, De Elst 1, 6708 WD Wageningen, the Netherlands, [2]Evonik Nutrition & Care GmbH, Animal Nutrition Services, Rodenbacher Chaussee 4, 63457 Hanau-Wolfgang, Germany; rick.vanemous@wur.nl

A 2×2 factorial design was used to conduct an experiment in which the effects of dietary crude protein level during lay and age at photo stimulation on reproduction traits in broiler breeder females was investigated. A total of 480 female and 64 male (Ross 308) breeders at the age of 20 weeks were placed in 16 pens. Birds were fed 2 different crude protein (high = HP and low = LP) diets with identical essential amino acids (AA) levels in a 3-phase feeding program (phase 1; 22 or 24 to 34, phase 2; 34 to 46, phase 3; 46 to 60 wk of age). Crude protein content of the diets of the LP group was approx. 1.4% lower in all phases. Birds were photo stimulated at 21 (early = ES) or 23 (late = LS) wk of age by increasing the day length from 8L:16D to 14L:10D in multiple steps. Hen-day egg production till 40 wk of age was not affected by different dietary proteins level, however, from 40 wk onwards birds fed the LP diets showed a lower hen-day egg production which resulted in 3 less total eggs. Incubation traits were not affected by CP level. From 45 wk of age onwards, BW was lower for the birds fed the LP diets, whereas egg weight was not affected. The ES birds showed an advanced age of sexual maturity (ASM, defined as age at 50% production) of 4.5 d compared to the LS birds, which resulted over the whole production cycle in 2.5 more total eggs. Despite a lower bodyweight between 25 and 30 wk of age of the ES birds, no effect on egg weight, fertility, hatchability and embryonic mortality was found between the ES and LS groups. It was concluded that feeding breeders a 1.4% lower crude protein diet decreased the number of eggs with no effects on incubation traits. Despite a lower BW during the start of the laying period, early photo stimulation increased the number of eggs with no effects on egg weight and incubation traits.

Replacing fullfat soy with different protein source on broiler performance and intestinal morphology

N. Puvača[1], O. Đuragić[2], D. Ljubojević[3], K. Pedrosa[4], C. Pedersen[4], S. Popović[2] and L.J. Kostadinović[2]
[1]Patent co. doo, Poultry Science, Vlade Ćetkovića 1a, 24211, Serbia, [2]University of Novi Sad, Institute of Food Science, Bulevar cara Lazara 1, 21000 Novi Sad, Serbia, [3]Scientific Veterinarian Institute 'Novi Sad', Rumenački put 20, 21000 Novi Sad, Serbia, [4]Hamlet protein A/S, P.O. Box 130, Saturnvej 51, 8700 Horsens, Denmark; nikola.puvaca@patent-co.com

Aim of research was to investigate the effect of full fat soy replacement with soybean meal, HP AviStart, corn gluten and fish meal on broiler productive results and intestinal histopathology. Experiment was conducted in completely controlled condition at trial facility of Educational and Research centre Patent IEC, Serbia. In total of 2,368 Ross 308 unsexed broilers have been assigned to 5 dietary treatments with 10 replicates, each. Control treatment (T1) of chickens were fed with full fat soy protein source, while in experimental treatments protein source were replaced: soybean meal (T2), HP AviStart (T3), corn gluten (T4) and fish meal (T5). Diets was isoenergetic and isonitrogenous. At the end of experiment (day 42 of life) chickens in experimental treatments T3 and T5 achieved significantly ($P<0.05$) higher final body masses (2,818.5 and 2,986.8 g) compared to the chickens in control and other treatments groups (T1: 2,642.3; T2: 2,164.6 and T4 2,732.1 g). Feed conversion ratio for the entire fattening period ranged from 1.52 (T3) to 2.15 kg/kg (T4) with recorded significant differences ($P<0.05$). The longest and widest intestinal villi of ileum at 14th day of sampling (575.9 and 218.8; 637.3 and 216.1 µm) with significant differences ($P<0.05$) was observed in chicken on treatments T3 and T5, compared to other treatments (T1: 532.6 and 201.3; T2: 512.8 and 199.4; T4: 496.3 and 202.6 µm). Based on the obtained results it can be concluded that partial substitution of full fat soy as protein source with HP AviStart and fish meal is a proper and effective way for increasing feed digestibility and intestinal absorption surface in order to improve broiler growth performance.

Improving protein efficiency of laying hens and broilers from nutritional and breeding perspectives

T. Veldkamp[1], Y. De Haas[1] and E.D. Ellen[2]
[1]Wageningen University & Research, Wageningen Livestock Research, De Elst 1, 6708 WD Wageningen, the Netherlands, [2]Wageningen University & Research, Department of Animal Sciences, Animal Breeding and Genetics, Droevendaalsesteeg 1, 6708 PB Wageningen, the Netherlands; teun.veldkamp@wur.nl

The demand for human edible protein sources will increase due to an increase in world population and wealth. This will result in an increased competition between humans and animals for high-quality protein sources. Therefore, increasing protein efficiency of livestock is an important challenge in livestock production. To improve protein efficiency of livestock, the animal prodution sector may: (1) efficiently use alternative protein sources or protein sources with a higher percentage human inedible proteins; and/or (2) use unimals that are more efficient with protein sources. The aim of the study was to investigate different future directions to improve protein efficiency of livestock. A desk study was performed but results were also obtained by workshops with stakeholders and pilot studies with available data of breeder organisations. Protein efficiency was defined as 'the amount of protein input to produce one kg of protein output'. To take into account competition for protein sources between humans and animals, protein output is corrected for whether protein is also directly suitable as human edible protein. To improve protein efficiency, it is important to consider both the animal output (for instance producing more eggs) and the protein input (use of human inedible protein sources). The most promising solutions are: (1) alternative protein sources; (2) precision feeding, taking into account genotype × nutrition interaction and nutrigenomics; and (3) selective breeding. A challenge will be the limited availability of datasets with phenotypic observations for protein efficiency. Close collaboration between nutritionists and geneticists is recommended in order to define a phenotype that can be used for protein efficiency and to optimise selection procedures for protein efficiency.

The effect of betaine and crude protein level on performance responses of broiler chickens

P.S. Agostini[1], B. Auer[2] and A. Gavrau[2]
[1]Schothorst Feed Research, Lelystad, the Netherlands, [2]Agrana Stärke GmbH, Vienna, Austria; pagostini@schothorst.nl

When betaine (BET) donates one methyl group to reduce homocysteine into L-methionine it becomes dimethylglycine. Glycine requirement has become an important issue in the last years due to the reduction of crude protein (CP) content of the diets. This study aimed to test the efficacy of three BET products in broiler diets with normal and low CP levels. 960 one-d-old Ross 308 male broilers were allocated in 8 treatments with 6 replicates of 20 birds (48 pens/2.2 m^2 each). Three BET products (ActiBeet L, ActiBeet VC and ActiBeet SD) were added in the starter (d 0-7) and grower diets (d 7-21) (700 and 600 ppm of active BET, respectively) containing either a normal or low CP level (22.3 and 21.0% in the starter and 21.5 and 20.2% in the grower phase). Diets contained a normal methionine (MET) level and no choline was added in the premix. A 4×2 factorial design (BET × CP) was used in this study, including a non-supplemented BET group (CD). A trend for a BET effect was observed for FCR from d 0-7 ($P=0.06$). ActiBeet VC and ActiBeet SD had better FCR than the CD (2.4 and 2.2% respectively). A significant and a trend interaction effect was observed for BWG ($P=0.004$) and FCR ($P=0.07$) from d 7-21, respectively. Birds fed low CP diets with either ActiBeet L, ActiBeet VC or ActiBeet SD had significantly higher BWG than the CD (4.6, 4,0 and 4,0% respectively). The increase in BWG has reflected in better FCR compared to the CD in low CP diets (2.0, 1.8 and 2.8% respectively). From d 0-21 a significant interaction effect was observed for BWG ($P=0.003$) and a trend interaction was observed for FCR ($P=0.10$) and for the European poultry efficiency factor (EPEF) ($P=0.07$). Supplementation of ActiBeet L, ActiBeet VC or ActiBeet SD to low CP diet led to 4.4, 3.7 and 3.7% higher BWG, 1.7, 1.8 and 2.7% better FCR and 6.2, 6.7 and 5.7% higher EPEF, respectively, than the CD. Supplementation of ActiBeet L, ActiBeet VC or ActiBeet SD in diets low in CP improved broiler performance up to d 21 of age.

Potential of betaine to replace methionine in broiler nutrition

M. Mueller[1], A. Lemme[1] and V. Machander[2]
[1]Evonik Nutrition & Care GmbH, Rodenbacher Chaussee 4, Hanau, Germany, [2]International Poultry Testing Station, Ustrasice, Tabor, Czech Republic; mario.mueller@evonik.com

Methionine (Met), choline, and betaine are metabolically linked in the homocysteine cycle. Within this pathway, betaine of dietary origin, or formed from its metabolic precursor choline, donors a methyl group and enables the methylation of homocysteine to form Met. This involvement of betaine in homocysteine methylation is claimed to have a Met replacement effect. The objective of this trial was to obtain updated results about the potential of betaine to replace supplemented DL-Met (DLM) in modern broiler nutrition. Thus, the effect of graded supplementations of betaine-HCl (BET), and DLM on live performance, and carcass traits of broilers were investigated at the International Poultry Testing Station Ustrasice, Tabor, Czech Republic. 7,020 male, day old Ross 308 chicks (feather sexed) were randomly allocated to nine dietary treatments with six replicates of 130 birds each. The treatments covered graded increase of supplemented BET and DLM in equimolar steps of 0.04%. DLM supplementation affected positively on final body weight (BW) ($P<0.001$). Opposite to this BET supplementation exerted no positive effect on final BW. Feed conversion ratio decreased from about 2.16 kg/kg in 0.00% DLM, via about 1.87 kg/kg in 0.04% DLM, to about 1.77 kg/kg in 0.08% DLM. Carcass, breast meat yield, and thigh yield have been affected by DLM supplementation rather than by adding BET to the feed. As DLM levels increased, parameters of slaughter performance improved significantly ($P<0.001$). A comparable effect could not be observed with increased levels of BET. This trial confirmed again that BET cannot replace DLM as an essential amino acid in broiler nutrition. In diets deficient in Met+Cys, but adequately supplemented with methyl group donors via choline, BET was not able to replace DLM or to take over any methionine function as an essential amino acid based on live performance, and carcass yield data.

No sufficient additivity of apparent prececal amino acid digestibility in broilers

W. Siegert[1], A. Helmbrecht[2] and M. Rodehutscord[1]
[1]University of Hohenheim, Institute of Animal Science, Emil-Wolff-Straße 10, 70599 Stuttgart, Germany, [2]Evonik Industries AG, Feed Additives Division, Rodenbacher Chaussee 4, 63457 Hanau, Germany; inst450@uni-hohenheim.de

We evaluated the additivity of apparent prececal amino acid digestibility (apAAD) of feed ingredients by investigating apAAD of corn (C), wheat (W), and soybean meal (S) separately and of diets containing CS and WS combinations of the same ingredient batches. Ross 308 birds were raised receiving a commercial starter diet until d 17. They were moved into 40 cages of 15 birds each for the duration of the experimental phase from d 17 to 22. Five diets containing the respective ingredient (C, W, S, CS, WS) as the only source of protein were offered *ad libitum* in 8 replicates each. Chyme was sampled from the terminal small intestine. The daily feed intake (DFI) was similar between the S, CS, and WS treatments. Compared to that, the DFI was considerably lower in the C treatment and higher in the W treatment. The ranking of apAAD between the diets was CS>WS>S>W>C for most amino acids. Predicted apAAD values of CS and WS were calculated based on the determined apAAD of C, W, and S. The difference between predicted and observed apAAD ranged from 4.1 to 10.0 percentage points for CS, and from 0.4 to 5.6 percentage points for WS. The predicted apAAD underestimated the observed for all amino acids in both CS and WS. The mean of all absolute differences was 6.8 and 2.6 percentage points for CS and WS, respectively. The differences in predicted and observed apAAD probably can be explained by basal endogenous losses which are considered twice in the predicted but only once in the observed values. The different fit of the predicted and observed apAAD of CS and WS probably was due to a different proportion of basal endogenous losses in the chyme samples because of the considerably deviating DFI of C and W. It is concluded that diet formulation based on apAAD of feed ingredients is not sufficiently precise to allow for an adequate prediction of the amino acid supply of broilers.

Modelling broiler performance with contrasting energy and lysine levels under different environments

J. De Los Mozos, A. Navarro-Villa, C. Alfonso and C. Torres
Nutreco, Trouw Nutrition R&D, Ctra. CM 4004 km 10.5, 45950 Casarrubios del Monte, Spain; j.delosmozos@ trouwnutrition.com

Energy and amino acids are the 2 nutrients with the largest contribution to the broiler feed price. Moreover, the response on broiler performance of these nutrients is influenced by environmental conditions such as temperature and humidity. Despite amino acid intake determines broiler performance, the effect varies according the dietary energy level and environmental conditions due to their effect on feed intake. 4,160 broiler chicks were randomly assigned to 80 pens, each pen consisted in 26 males and 26 females. The study design followed a 2 dietary energy densities [high vs low] × 3 Energy to dLys ratios [low, medium and high] × 2 environments [Control vs High Temperature and Humidity (HTH)] factorial. A 4 phase feeding program was applied using th following nutrient profile: starter (0-15 d) [3,050 vs 2,650 kcal] × [225, 248 and 275 kcal/g dLys]; Grower-1 (15-29 d) [3,100 vs 2,700 kcal] × [244, 268 and 299 kcal/g dLys]; Grower-2 (29-37 d) [3,150 vs 2,750 kcal] × [263, 289 and 321 kcal/g dLys]; Finisher (37-48 d) [3,200 vs 2,800 kcal] × [281, 309 and 344 kcal/g dLys]. Broilers under HTH had 6% lower feed intake, 5% lower daily growth ($P<0.05$) while FCR was unaffected. High dLys led to increased market weights (3.0%; $P<0.1$), reduced feed intake (4%; $P<0.1$), improving as consequence FCR by 7% ($P<0.05$). Broilers fed high levels of energy had lower feed intake (93.7 vs 109.5 g/d) and better FCR (1.505 vs 1.749). Carcass yield and breast meat yield increased with high dLys levels ($P<0.05$). No significant interaction between energy and dLys level was found. Broilers under Control or HTH environment responded similarly to energy and protein changes, however, the magnitude of the response was slightly different. In summary, broiler performance was optimised at similar energy and dLys levels disregarding the environment. Nevertheless, animal models combined with economic models may help nutritionists to select the most convenient nutrition to optimise business margins.

Adaptation of pancreatic and intestinal proteases to dietary protein level in chicken

V.G. Vertiprakhov, A.A. Grozina and E.N. Andrianova
Federal Scientific Center 'All-Russian Research and Technological Poultry Institute' of RAS, Ptitsegradskaya Str., 10, 141311 Sergiev Posad, Moscow Region, Russian Federation; andrianova@vnitip.ru

The adaptation of exocrine pancreatic function to dietary protein level was studied on six White Leghorn chickens at six months of age with chronic fistulation of the pancreas. The fistulation technique developed by Batoev and Batoeva involves fistulation of pancreatic duct, activation of pancreatic juice during the transition through an isolated section of the duodenum prior to juice sampling, and external anastomosis allowing the juice to enter the duodenum again when the sampling is performed. Intestinal enzymatic activities were determined in the samples of the digesta obtained using T-shaped fistulae of the ileum. The chickens were fed control diet 1 (16% CP) during first ten days of the trial; then the diet was diluted with wheat bran (13.6%) and soybean meal (9.4%) was replaced with sunflower cake (up to 11.3%) resulting in 14% CP in total diet 2 fed during another ten days. The activities of proteolytic enzymes were determined in the tests with casein hydrolysis under colorimetric control. The results showed that 2% reduction in dietary CP level significantly decreased the activities of proteases in pancreatic juice from 210 ± 18.4 to 152 ± 9.8 mg/ml/min (or by 27.6%), and in ileal digesta from 21.0 ± 1.39 to 13.8 ± 1.55 mg/ml/min (or by 34.3%). The proteolytic activities were found to decrease ten-fold or more during the passage through the intestine, due to the dilution of pancreatic juice with the intestinal digesta and the secrets of the digestive glands, and probably to partial absorption of the enzymes into the bloodstream. The reduction of proteolytic activity in pancreatic juice and intestinal chymus in response to the reduction in dietary CP level is one of the adaptive mechanisms to adjust intestinal digestion to any alterations in feed quality.

Dietary methionine alters global metabolites at the starter phase of meat-type chickens

S.E. Aggrey[1], M.C. Millfort[1], R. Rekaya[2] and B. Saremi[3]
[1]Univeristy of Georgia, CAES Campus, 114 Poultry Science Building, GA30602, Athens, USA, [2]Univeristy of Georgia, 348 Edgar L Rhodes Center, GA30602, Athens, USA, [3]Evonik Nutrition and Care GmbH, Rodenbacher Chaussee 4, 63457 Hanau, Germany; behnam.saremi@evonik.com

Methionine (MET) is the first limiting amino acid in a typical poultry diet and affects several molecular, cellular, immunological and metabolomic functions. The metabolomic mechanisms that underlie dietary methionine in poultry remain to be elucidated. We studied the metabolomic mechanisms that underlie a diet (DEF) deficiency in Met+Cys (0.77% in 0-10 day starter phases) and diets supplemented with 100% equimolar concentrations of LMET, DLMET or MHA-FA in Cobb 500 male chickens from hatch until 10 days of age. Protein crushed plasma samples collected at 10 days of age were used in a TSQ-Quantum Ultra triple quadrupole mass spectrophometer interphased with an 1100 HPLC system. The raw data was processed using apLCMS with xMSanalyzer (R packages). Pathway enrichment analysis (Mummichog) based on 429 differentially expressed features showed that vitamin B1, linoleate, glutamate, tryptophan, dorphin, vitamin D3, glutathione, bile acids, omega-3-fatty acid, histidine, valine, leucine and isoleucine degradation, and N-Glycan degradation metabolisms are the important significant pathways. The uric acid intensity for DEF, LMET, DLMET and MHA-FA was 16.62, 17.15, 6.49 and 15.92, respectively. The cyclic guanosine 3',5'-monophosphate intensity for DEF, LMET, DLMET and MHA-FA was 17.48, 17.52, 6.37 and 18.60, respectively. The gamma-glutamylcysteine intensity for DEF, LMET, DLMET and MHA-FA was 11.14, 15.14, 14.26 and 13.78, respectively. Methionine deficient birds had a different metabolic profile compared to methionine enrich diets. The intensity feature differences of metabolites among chickens fed on equivalent molar supplementation of LMET, DLMET and MHA-FA suggest that, these methionine sources on the metabolomics level, utilize some common and methionine-source-specific pathways for their metabolism.

Amino acid and sucrose appetites in feather eating and non-feather eating laying hens

S. Cho and E. Roura
The University of Queensland, St. Lucia, QLD 4072, Australia; s.cho2@uq.edu.au

Feather pecking outbreaks are still common in egg farms in spite of all the efforts to prevent it. Marginal deficiencies of nutrients in individual birds have been related to the induction of feather pecking events under commercial conditions. We hypothesize that the individual variation of the digestive and metabolic efficiencies in nutrient utilization result in specific nutrient appetites which ultimately initiate the feather eating behaviour. However, nutrient preferences have been rarely studied in poultry. This study aimed at testing nutrient appetite/preferences to assess the differences between feather eating (FE) and non-feather eating birds (NFE). 96 individually penned laying hens were used in a double choice test recently developed in our group. 12 amino acids (lysine, methionine, cysteine, tryptophan, glutamine, arginine, histidine, glycine, proline, serine, tyrosine, and alanine), and sucrose, which have been previously reported in feather pecking studies, were tested at three concentrations (0.2, 1, 5% for AAs and 1, 5, 17% for sucrose). At the end of the trial individual laying hens were divided into two groups after assessing feather consumption using gastro intestinal tract (GIT) contents. The consumption of the two containers was analyzed calculating a standard preference index and, it was compared to the random choice value of 50%. FE birds showed a significantly higher preference for Met and Lys than NFE birds ($P<0.05$). Our results are consistent with previous reports indicating that feather pecking was increased when hens were deficient in dietary Lys or sulphur amino acids (Met+Cys). In addition, NFE hens showed higher preference for sucrose ($P<0.05$). The sucrose preference in NFE birds could be related to a higher corticosterone levels compared to FE hens. Chronic stress has been previously associated with increased sucrose consumption. In addition, ongoing analysis of the stress hormones and chemosensory receptors mRNA abundance in the oral and GIT of FE and NFE laying hens will be reported.

Effects of feeding a bioactive extract from *Olea europaea* to broiler chickens on growth performance

J. Herrero[1], M. Blanch[2], J. Pastor[2], A. Mereu[2,3], I. Ipharraguerre[2,4] and D. Menoyo[1]
[1]Universidad Politécnica de Madrid, Producción Agraria, ETS Ingeniería Agronómica Alimentaria y de Biosistemas, 28040 Madrid, Spain, [2]Lucta S.A, Innovation Division, UAB Research Park, Edifici Eureka, 08193 Bellaterra, Spain, [3]Yara International ASA, Skøyen N-0213 Oslo Norway, P.O. Box 343, Norway, [4]Christian-Albrechts-University Kiel, Institute of Human Nutrition and Food Science, Hermann-Rodewald-Straße 6-8, D-24118 Kiel, Germany; j.herreroe@alumnos.upm.es

The present study aimed to investigate the effects of supplementing a bioactive extract from *Olea europaea* (EOE) in broiler diets on growth performance. For this purpose, 252 1 d broiler chickens (Ross 308) housed in floor pens were randomly assigned to 3 experimental groups (6 pens/treatment, with 14 birds/pen). Animals were fed with a common non-medicated starter diet (mashed) for 21 d, and from 22 to 42 d of age with their respective experimental diet (pelleted): a negative control with no additives (C), a positive control with 100 ppm of monensin (M) and the basal diet supplemented with 1000 ppm of an EOE. Feed intake and growth rate were monitored weekly throughout the trial. Data were analyzed as a completely randomized design, with the type of diet used as the main source of variation. The Tukey test was used to make pairwise comparisons between means. Improvement of average daily gain (ADG) was observed from 35 to 42 days of age in birds fed M and EOE diets compared to those fed the C ($P<0.01$; 102 and 97 vs 78 g/d respectively). Moreover, feed conversion ratio (FCR) during the last period was significantly lower in broilers fed the M and EOE diets compared to those fed the C ($P<0.05$; 1.98 and 2.10 vs 2.50). No significant differences on feed intake were observed among treatments. Performance of birds fed EOE or M was similar throughout the trial. In conclusion, the inclusion of the bioactive extract from *Olea europaea* has a positive effect on broiler chicken performance.

Effect of guanidinoacetic acid in corn and sorghum-based diets on broiler muscle pectoral myopathies

E.O. Oviedo-Rondón[1], H. Córdova[1], A. Sarsour[1], J. Barnes[1], P. Ferzola[1] and M. Rademacher-Heilshorn[2]
[1]North Carolina State University, Raleigh, NC, 27695, USA, [2]Evonik Nutrition & Care GmbH, Hanau, 63457, Germany; hcordov@ncsu.edu

Breast meat myopathies are a current problem in the poultry industry worldwide with potential nutritional implications. This experiment was conducted to evaluate the effects of supplementation of guanidinoacetic acid (GAA) in broilers fed corn or sorghum based diets on wooden breast (WB), white striping (WS), and spaghetti breast muscle (SM) myopathies. The treatments consisted of corn or sorghum diets with or without the addition of GAA (600 g/ton). A total of 800 male Ross 708 chicks were randomly placed in 40 floor pens with 10 replicates per treatment combination. At 51 and 55 d of age, 4 broilers/pen were processed and cut up. Breast meat was evaluated 9 h after slaughtering and 3 h after deboning for myopathies. Experts used scores from 1 to 4 for WB and WS severity, and presence or absence of SM. At 55 d, 2 broilers/pen were selected and samples were obtained for histology analysis. Data were analyzed as a randomized complete block design with grain type and GAA supplementation as main effects using GLIMMIX. No two-way interaction ($P>0.05$) effects were detected on any pectoral myopathy in both processing times. At 51 d of age, broilers supplemented with GAA had ($P<0.05$) double breast meat (0.30 vs 0.14) without WB (score 1) compared with broilers without supplementation. However, at 55 d, no effect of GAA supplementation ($P>0.05$) was observed and broilers fed corn diets had higher ($P<0.05$) WB (score 3) than broilers fed sorghum diets. Histological analysis showed no effects ($P>0.05$) of grain source or GAA supplementation on myopathy scores. WS and SM incidence were not affected ($P>0.05$) by treatments in both evaluations. In conclusion, supplementation of GAA acid in corn or sorghum based diets reduced WB at 51 d only, but no effects were observed in other myopathies or at 55 d. Broilers fed sorghum diets had lower incidence of wooden breast at 55 d of age.

Microalgal *Chlorella*-based additive in diets for broilers

E.N. Andrianova[1], I.A. Egorov[1], L.M. Prizyazhnaya[1], I.P. Uvarov[2] and O.A. Rozhkov[3]
[1]All-Russian Research and Technological Poultry Institute, Dept. of Nutrition, Ptitsegradskaya Str., 10, Sergiev Posad, Russian Federation, [2]'EcoFactor' Co. Ltd., Russia, Novosibirsk, 630112, Russian Federation, [3]State Veterinary Service, Novosibirsk, Novosibirsk, 630112, Russian Federation; andrianova@vnitip.ru

One of the most prospective directions in the search for efficient feed ingredients for poultry is microalgae and products of algal metabolism. There is a body of research regarding the efficiency of microalgal additives in animal diets; the technological difficulties related to the addition of liquid and suspended additives into diets, however, constrain their application. To our opinion, the most convenient technique for addition of microalgae into diets for animals and poultry is algalization of feeds when a feed is drenched with dense microalgal culture and then dried under soft and favorable conditions. Russian scientists designed a multi-purpose microalgal additive with lactic bacteria for animals and poultry. The production of microalgae in closed reactor can be realized even under the most severe conditions of Siberian winter. Controlled experiment on Cobb 500 broilers showed that supplementation of the diets with the algal additive (1% of total diet) since 1 or 7 days of age (treatments A1 and A7) improved live BW at 36 days of age from 1,938.74±27.68 g in control to 1,988.00±25.81 and 2,025.12±39.46 g, respectively, and FCR from 1.759 in control to 1.681 and 1.658. Digestibility of protein in A1 and A7 treatments was better by 0.30%, fiber by 1.76-1.86%, nitrogen assimilation by 0.67-0.70% compared to control; deposition of vitamins A, E, and B2 in liver was higher by 1.95-5.61; 1.43-1.95 and 1.60-1.05%, respectively. Trials in conditions of commercial poultry farms of Novosibirsk Region showed that the microalgal additive improved mortality rates by 3%, daily weight gains by 5-7%, EPEF by 8.4-11.7%. The supplementation of diets with this additive allows reduction in inclusion rates for animal-derived feedstuffs and vitamin-mineral premix.

Control of the poultry red mite by dietary essential oils: laying hens production and welfare

N. Puvača[1], A. Petrović[2], D. Horvatek Tomić[3], M. Marangi[4], T. Shtylla Kika[5], M. Vasiljević[1] and O. Sparagano[6]
[1]Patent co. doo, Poultry Science, Vlade Ćetkovića 1a, 24211 Mišićevo, Serbia, [2]University of Novi Sad, Faculty of Agriculture, Trg Dositeja Obradovića 8, 21000 Novi Sad, Serbia, [3]University of Zagreb, Faculty of Veterinary Medicine, Heinzelova 55, 10000 Zagreb, Croatia, [4]University of Foggia, Via Napoli 25, 71121 Foggia, Italy, [5]Agricultural University of Tirana, Koder Kamez, 1019 Tirana, Albania, [6]Coventry University, Vice-Chancellor Office, CV1 5FB Coventry, United Kingdom; nikola.puvaca@patent-co.com

The aim of research was to investigate the effect of thymol, carvacrol, eucalyptol and linalool mixture (EOm) in concentrations of 0.05% in laying hens nutrition on poultry red mite population reduction, productive results, eggs quality and hens welfare. Experiments were conducted on 25,000 laying hens. Experiments lasted 15 days, during which feed with addition of 0.05% of EOm was provided. At the end of the experiment significant ($P<0.05$) reduction of nymph (73.6%) and adult (67.5%) stages were recorded. Significant influence was not recorded ($P>0.05$) regarding the feed consumption (117.5 and 117.9 g/hen/day), feed conversion ratio (2.7 and 2.4 g feed/g egg) and egg mass (43.2 and 48.7 g/hen/day) while significant influence ($P<0.05$) in egg production was recorded (68.9 and 77.0%). Significant influence ($P>0.05$) in technological egg quality was not recorded, while ($P<0.05$) in organoleptic egg quality such as overall impression (5.91 and 6.26) was recorded. Supplement of 0.05% of EOm decreased severe feather pecking behaviour and improved the quality of the plumage of hens which led to improved hens welfare. Based on the positive results it can be concluded that the thymol, carvacrol, eucalyptol and linalool mixture can be used as dietary supplement with high efficacy in decreasing poultry mite population, increasing egg production and organoleptic quality of the eggs, and a good tool for improving hens welfare.

Effects of dietary nano-CuO administration on litter quality, tibia bone and excreta characteristics

A. Deldadeh, A. Tatar, M.R. Ghorbani and S. Salari
Ramin Agriculture and Natural Resources University of Khuzestan, Animal Science, Mollasani, Khuzestan, 6341773637, Iran; tatar@ramin.ac.ir

An experiment was conducted to determine the effects of dietary supplementation of nano-CuO. In total, 300 1-d. old Ross mixed sex broiler chicks were used in four treatments with 5 replicates and 15 chicks per each replicate. The experimental rations were prepared using 0, 50, 100 and 150 ppm Nano-CuO. All diets were formulated to meet NRC minimal nutritional requiremens. In this experiment, tibia bone and excreta characteristics were determined. The experimental design was completely randomized design. The results showed that Tibia bone weight, volume and proximal epiphysis diameter were increased significantly in control group compared to 150 ppm nano-CuO treatment ($P<0.05$). Also, supplementation of Nano-CuO did not have any significant effect on tibia dry matter and ash. Excreta characteristics such as pH, moisture and ash content have not been affected by experimental treatments but decreased litter moisture in all nano-CuO supplemented groups in comparison to control group ($P<0.05$). In addition, inclusion of different dietary levels of Nano-CuO did not have any significant effect on small intestine digesta pH and carcass characteristics of broiler chickens at 42 days of age.

The physiological response to nisin (E234) application in broiler chicken diets

B. Kierończyk[1], M. Rawski[1], S. Świątkiewicz[2] and D. Józefiak[1]
[1]Poznań University of Life Sciences, Department of Animal Nutrition and Feed Management, ul. Wołyńska, 60-637, Poland, [2]National Research Institute of Animal Production, Department of Animal Nutrition and Feed Science, ul. Krakowska 1, 32-083, Poland; bkieron@up.poznan.pl

The aim of the present study was to investigate the effects of nisin on broiler chicken growth performance, selected organs relative weights, activities of digestive enzymes, and selected blood parameters. Two separate 35-day-long experiments were carried out with use of 400 1-day-old male ROSS 308 chicks per experiment. The following treatments were applied; Exp 1: NA – no additives, SAL – salinomycin (60 mg/kg diet), NIS – nisin (2,700 IU/kg diet), SAL+NIS – salinomycin with nisin; Exp 2: NA – no additives, MON – monensin (100 mg/kg diet), NIS – nisin (2,700 IU/kg diet) and MON+NIS – monensin with nisin. The growth performance parameters were measured. Blood samples were collected to analyse levels of insulin, glucagon, leptin, ALT, AST, glucose, albumin, total protein, cholesterol, NEFA, and triglycerides. The activity of lipase, amylase, and trypsin was measured in the duodenum digesta. The weight and length measurements of the duodenum, ileum, jejunum, caeca, and pancreas were performed. Additionally, the weights of thymus, Bursa of Fabricius, and spleen were measured. The presented study highlight that nisin supplementation may increase the BWG more than 6%, as well as FI 3,211 vs 3,342 g, and improve FCR from 1.56 to 1.39 (1-14 d), in comparison to NA. Analyses of the serum indicated no influence of nisin addition on insulin, glucagon, and leptin concentrations. However, a decreased glucose level was observed in the nisin treated group. There were no significant differences in the case of thymus and spleen mass. However, nisin affects weight of Bursa of Fabricius, and decreased mass of ileum and length of jejunum. Nisin may be efficiently used in broiler nutrition as a novel growth promoting agent, with no negative influence on the bird's metabolism and immune status.

Heat stability of a probiotic containing viable spores of *Bacillus licheniformis*

V. Hautekiet[1] and M. Eeckhout[2]
[1]Huvepharma NV, Uitbreidingstraat 80, 2600 Antwerp, Belgium, [2]UGent, Valentin Vaerwyckweg 1, 9000 Gent, Belgium; veerle.hautekiet@huvepharma.com

B-Act® is a probiotic feed additive, consisting of viable spores of *Bacillus licheniformis*. A trial was conducted to examine the survival of a probiotic feed additive B-Act® under steam-pelleting conditions commonly used in the poultry feed industry. The diet was mainly wheat and soy based and formulated to meet the Ross 308 recommendations. One level of probiotic supplementation (1.6×10^9 cfu/kg feed) and three pelleting temperatures (75, 85 and 95 °C) were used. Conditioning time was 90 s. The feed was pelleted in a commercial press with pellet diameter of 2.0 mm. Ten samples of 200 g mash feed were taken at random. At each temperature during pelleting ten samples were collected at equal intervals over the complete run. Moisture content and temperature were closely measured. The enumeration of *B. licheniformis* in the feed was conducted using 10 g of each sample after homogenisation. The samples were serially diluted with peptone and the number of colonies were counted following 48 h incubation at 30 °C. Reached pelleting temperatures were similar to target values. The recovery count of *B. licheniformis* in the feed samples was 100% for all temperatures compared to the mash feed. The probiotic B-Act® can resist heat and high pressure, thus surviving the steam conditioning and pelleting process routinely used in the broiler industry.

Increasing broiler performance by supplementing viable spores of *Bacillus licheniformis*

V. Hautekiet[1] and S. Petkov[2]
[1]Huvepharma NV, Uitbreidingstraat 80, 2600 Antwerp, Belgium, [2]Biovet, Petar Rakov Street 39, 4550 Peshtera, Bulgaria; veerle.hautekiet@huvepharma.com

B-Act® is a probiotic feed additive, consisting of viable spores of *Bacillus licheniformis*. The aim of the study is to determine the performance of B-Act® supplemented broilers raised under commercial management. Three hundred sixty Ross 308 broilers (50% male-50% female) were housed in a standard environment. The broilers were fed a commercial diet and divided in two groups with 6 replicates each: one control group and one group supplemented with 0.5 kg B-Act® in starter, grower and finisher (1.6×10^9 cfu *B. licheniformis*/kg of feed). The trial lasted for 42 days. Daily weight gain during period 0-42 days was significantly increased with 6% for the B-Act® supplemented broilers (47.6 g/day/control broiler vs 50.5 g/day/B-Act® broiler). Feed conversion was significantly improved with 4% (1.97 for control broilers vs 1.89 for B-Act® broilers). Broilers raised under commercial conditions and supplemented with the probiotic B-Act® showed significant higher technical performance than control broilers.

Effect of *Bacillus licheniformis* on broilers raised under optimal management

V. Hautekiet[1] and S. Kaczmarek[2]
[1]Huvepharma NV, Uitbreidingstraat 80, 2600 Antwerp, Belgium, [2]Poznan University of Live Science, Ul. Wolynska 33, 60-637 Poznan, Poland; veerle.hautekiet@huvepharma.com

B-Act® is a probiotic feed additive, consisting of viable spores of *Bacillus licheniformis*. A trial was conducted to study the effect of dietary supplementation of B-Act® on the performance of broilers raised under optimal management. Nine hundred sixty male Ross 308 broilers were housed in a standard environment at Poznan University (Poland). The broilers were fed a commercial diet and divided in two groups with 12 replicates each: one control group and one group supplemented with 0.5 kg B-Act® in starter, grower and finisher (1.6×10^9 cfu *B. licheniformis*/kg of feed). Feed was mainly based on wheat and soybean meal. For the whole period (0-42 days) daily weight gain was significantly increased for birds receiving diets supplemented with B-Act® (63.8 g/day/control broiler vs 66.9 g/day/B-Act® broiler). Feed conversion was significantly improved with 4% (1.65 for control broilers vs 1.59 for B-Act® broilers). The results demonstrated that under optimal management and low environmental challenge B-Act® still significantly enhanced the performance of broilers.

Efficacy of a *Bacillus subtilis* and *Bacillus licheniformis*-based probiotic on performance of turkeys

A. Blanch[1], A. Berg Kehlet[1], F. Rudeaux[1], C. Aoun[2] and A. Schlagheck[2]
[1]Chr. Hansen A/S, Boege Allé 10-12, 2970 Hoersholm, Denmark, [2]Biochem Zusatzstoffe Handels- u. Produktionsges. mbH, Kustermeyerstrasse 16, 49393 Lohne, Germany; dkalbl@chr-hansen.com

It is hypothesized that the addition of a combination of *Bacillus subtilis* and *Bacillus licheniformis* together to turkey feed will help overcoming the performance loss associated with a diet reduced in energy, protein and amino acids. A total of 320 day-old male turkey poults were allocated to 2 treatments with 8 replicates per treatment. The diets were commercial-like diets (ranging from 2,750 kcal/kg and 26.5% protein in starter feed, and 3,230 kcal/kg, and 16.8% protein in finisher feed). Treatments were T1) Control, standard energy and protein levels and T2) *Bacillus* supplementation of 1.28×10^6 cfu/g feed (*B. subtilis:B. licheniformis*, 1:1) reduced in energy of 50 kcal/kg and 0.5% protein (AA/Protein ration constant). The duration of the study was from day-old to 126 days of age. At day 35 daily, weight gain was 5.3% higher for T2 than for T1 (49.7 and 47.2 g/d; $P<0.01$). There was also a tendency to an improved FCR in T2 by 3 points although not significantly (1.58 and 1.61). From day 36 to 84, FCR was significantly higher for T2 compared to T1 (2.02 and 1.97; $P<0.01$) and both weight gain (150.6 and 148.5 g/d) and feed intake (304 and 293 g/d) were numerically higher for T2. From day 85 to 126 gain was numerically higher for T2 (188.6 and 185.9 g/d), whereas feed intake was lower (535 and 537 g/d). Thus, FCR was numerically improved by 6 points (2.84 and 2.90). Overall performance from day-old to 126 days did not differ between the two treatment groups. In conclusion, these results indicate that feed formulation with the addition of double strain probiotic is optimized by reducing the energy 50 kcal and the crude protein 0.5 points, at constant amino acid: protein ratio.

Efficacy of a *Bacillus subtilis* probiotic on performance of chickens fed ME, CP and AA reduced diets

A. Blanch[1], M. Rouault[1], K. Männer[2] and J. Zentek[2]
[1]Chr. Hansen A/S, Boege Allé 10-12, 2970 Hoersholm, Denmark, [2]Freie Universität Berlin, Königin-Luise-Straße 49, 14195 Berlin, Germany; dkalbl@chr-hansen.com

Bacillus subtilis DSM 17299 (BS) strain is able to produce a wide array of digestive enzymes. In this sense, BS produces those enzymes potentially advantageous to digest feed in the gut. The objective of this study was to assess the effect of the addition of BS in ME-, crude protein (CP)- and amino acid (AA)-reduced diets on the performance of broiler chicks from 1 to 42 days. 735 birds were divided in six groups of 8 replicates each and 15 birds in each replicate. Each group got one of the following diets: T1 (basal diet without probiotics), T2 (T1-50 kcal/kg -0.3 points CP), T3 (T1-100 kcal/kg -0,6 points CP), T4 (T1+BS), T5 (T2+ BS), T6 (T3+BS). All diets had the same CP:AA ratio. The dietary inclusion level of BS was 1.6×10^6 cfu/g from 0 to 21 d and 8×10^5 cfu/g from 22 to 42 d. Body weight, body weight gain (BWG), feed intake (FI) and feed conversion ratio (FCR) were assessed on days 21 and 42. At 21 d, BWG was significantly lower ($P<0.01$) in T3 (820 g) than in T1 (889 g) and T4 (885), showing T2, T5 and T6 intermedium values (829, 844 and 833, respectively). The addition of BS in the diet numerically improved FCR at each nutritional level: T4 (1.368) vs T1 (1.384); T5 (1.388) vs T2 (1.826) and T6 (1.775) vs (1.837). On day 42, BWG was significantly ($P<0.01$) lower when ME, CP and AA were reduced, regardless of the addition of BS, although the probiotic supplementation numerically improved BWG at each nutritional level (T2: 2,786 g, T3: 2,742 g, T5: 2,835 g, T6: 2,800 g vs T1: 2,972 g and T4: 2,985 g). However, BS supplementation significantly improved overall FCR in nutritionally reduced diets (T5: 1.646 vs T2: 1.699, $P<0.01$ and T6: 1.663 vs T3: 1.715, $P<0.01$) and equalled it to the standard diet (T5: 1.646 and T6: 1.663 = T1: 1.645; none significant difference). In conclusion, the addition of BS efficiently compensated certain reductions of ME, CP and AA in broiler diets.

Far from senior – enabling old layers to perform to their full potential

K. Kozlowski[1], S. Kirwan[2], S. De Smet[2] and J. Jankowski[1]
[1]Department of Poultry Science, University of Warmia and Mazury in Olsztyn, 622 Olsztyn, Poland, [2]Kemin Europe N.V., Toekomstlaan 42, 2200 Herentals, Belgium; kristof@uwm.edu.pl

Layer hens are among the longest lived of the commercial poultry. One suggested reason for decreasing performance and shell quality in older hens is reduced gut health, making hens less efficient in terms of nutrient and mineral absorption. To test this hypothesis a total of 216 Lohman brown hens of 52 weeks of age were randomly allocated into three groups. Diets were identical apart from the addition of a defined cfu of *Bacillus subtilis* PB6 probiotic (T1 = control, T2 = 3.75×108 and T3 = 7.5×108 cfu/kg). The hens were fed until the age of 72 weeks. Hens were housed paired in cages, with nine replicates per treatment. Performance improved significantly ($P \leq 0.05$) (egg production in the final week was 76.69% for control and 83.04% for T3). No significant differences could be observed for egg weight across the trial duration. FCR decreased numerically though insignificantly as it decreased from 2.15 for T1 to 2.04 for T3. Despite the increase in productivity egg shell stability significantly ($P \leq 0.05$) improved after 4 weeks from 32.41 N (T1) breaking strength to 37.20 N for T3. Over the entire trial period T3 produced significantly ($P \leq 0.05$) more shell in grams: +7.4% compared to T1. Egg quality improved with yolk production increasing significantly by 7.9% ($P \leq 0.05$) from T1 to T3 and albumen height increasing significantly ($P \leq 0.05$) from 8.91 to 9.14 mm. At the end of this trial (72 weeks) the egg production of the old hens would have exceeded their body weight approximately 11 times (final average BW 2 kg; cumulative egg mass per hen housed 22.5-23.5 kg). If a hen can still lay at a rate of 86% with good shell quality in the final four weeks, it makes good economic sense to extend the laying period beyond 72 weeks.

A *Bacillus* spp. probiotic, avilamycin and their combination in broilers fed mixed grain diets

A.L. Wealleans[1], M. Sirukhi[2] and I.A. Egorov[3]
[1]Danisco Animal Nutrition, DuPont Industrial Biosciences, P.O. Box 777, Marlborough, SN8 1XN, United Kingdom, [2]Danisco Animal Nutrition, DuPont Industrial Biosciences, Stasovoy ul., 4, оф. А50, Moscow, Russian Federation, [3]All-Russian Research And Technological Poultry Institute, Ptitsegradskaya ul., 10, Sergiev Posad, Russian Federation; alexandra.wealleans@dupont.com

Alternative feed ingredients containing high levels of soluble NSPs are known to disrupt gut barrier function in broilers and increase gut pathogen burdens. This study investigated the effect of an antibiotic and multi-strain *Bacillus* spp. probiotic (DFM) on the performance and gut health of broilers fed a mixed grain diet. 800 Cobb 500 chicks were allocated to 4 dietary treatments (8 rep/trt; 25 birds/rep): control diet based on wheat/soy/maize/sunflower (NC), AGP (avilamycin, 180 ppm), DFM (150,000 cfu/g), or AGP+DFM. BWG, FI and FCR were measured at d0, 21 and 42. Duodenal, ileal and jejunal empithelial samples um were taken at d42 to determine villus height (VH), crypt depth (CD) and their ratio (VH:CD). Data were analysed using ANOVA in JMP 11; means separation was conducted using Tukey's HSD. Differences were considered significant at $P<0.05$. Ileal and caecal mucosal *Escherichia coli* and Lactobacilli counts were measured at d42. At d42, DFM and DFM+AGP birds were significantly heavier than NC and AGP treatments, with AGP birds heavier than NC (2,469 g NC, 2,512 g AGP, 2,583 g DFM, 2,581 g DFM+AGP). FCR followed a similar pattern (1.69 NC, 1.67 AGP, 1.63 DFM, 1.65 DFM+AGP). DFM and DFM+AGP significantly increased VH and CD in all gut sections compared to the NC; greatest VH:CD ratios were seen with AGP treatment. Reduced *E. coli* counts were seen with DFM and AGP+DFM treatments in both tissues compared to NC birds, with AGP reducing caecal counts only, while Lactobacilli counts were increased. Divergent histology and microbiology highlight the different, sometimes complementary, modes of action of AGP and probiotics for improving growth and efficiency in broilers.

Effect of synbiotic and prebiotic on productive performance of broilers under hot climate conditions

A.A.A. Abdel-Wareth[1], H.A. Hassan[1], W. Abdelrahman[2] and Z.H. Ismail[1]
[1]South Valley University, Animal and Poultry Production, Faculty of Agriculture, 83253 Qena, Egypt, [2]BIOMIN Holding GmbH, Erber Campus 1, 3131 Getzersdorf, Austria; a.wareth@agr.svu.edu.eg

This study was conducted to determine the effects of supplementation of a synbiotic or prebiotic in broiler chickens sorghum-soybean diets on growth performance, carcass yield, organs weights and serum metabolic profile under hot climate conditions. A total of three hundred 1-day old broiler chickens were randomly assigned to five dietary treatments, six replicates of 10 chicks each. A basal diet (control diet) was supplemented with 2 levels of a synbiotic (Biomin® IMBO, 1000 g or 1,500 g/ton of the starter diets and 500 g or 750 g/ton of the grower diets) or a prebiotic (Mannanoligosaccharide 500 g or 1000 g/ton of the starter diets and 250 g or 500 g/ton of the grower diets). The daily weight gain was significantly ($P<0.01$) increased while feed conversion ratio and abdominal fat measures were significantly decreased by the dietary inclusion of the synbiotic or prebiotic compared with the control corn-soybean-fed broilers. Interestingly, the addition of either prebiotic or synbiotic significantly decreased ($P<0.01$) the total cholesterol and Low-density lipoprotein (LDL) levels whereas the other biochemical parameters including total protein, GPT, GOT, T3 and T4 concentration were not statistically significantly changed. No differences were observed for carcass criteria in terms of dressing, heart, gizzard, spleen liver percentages among treatment groups. In conclusion, the synbiotic and prebiotic displayed a greater response as growth promoters compared to the standard diet in order to improve growth performance and decreased cholesterol concentration without adverse effect on carcass criteria of broilers under hot climate conditions.

Villus development and performance in broilers fed with a combination of prebiotics and probitics

A. Sozcu and A. Ipek

Uludag University, Faculty of Agriculture, Department of Animal Science, Bursa, Turkey, 16059 Bursa, Turkey; aipek@uludag.edu.tr

This study was performed to investigate the villus development in jejenum and growth rate in broilers fed with a combination of prebiotic and probiotics at 14 days of age. A total of 720 one-day old Cobb 500 broiler chicks were randomly assigned to four treatment including experimental groups as control, dose 1 (0,5 kg/tonnes), dose 2 (1 kg/tonnes) and dose 3 (2 kg/tonnes) diet with supplementation of a combination of probiotic and prebiotic. The combination of probiotic and prebiotic (Synerall™, Global Nutritech) included *Saccharomyces cerevisiae* (4×10^{12} cfu/kg), mannan-oligosaccharide (88.000 mg/kg) and glucan-oligosaccharide (96.000 mg/kg). Each experimental group consisted of six replicates containing 30 chicks (15 female, 15 male). Growth performance was determined for 1-14 days by following of live weight and feed conversion ratio. At 14 days of age, villus height, villus width, villus apparent surface area, crypt depth and the thickness of the *Tunica muscularis* of jejenum were measured. At 14 days of age, the live weight was higher in chicks fed with dose 2 and 3 diet with values of 530.6 g and 512.9 g, respectively, than other groups, whereas the live weight gain was higher in chicks fed with dose 2 and control chicks with values of 322.6 and 309.9 g, respectively ($P<0.01$). Feed conversion ratio was more efficient in chicks fed with probiotic and prebiotic in all dose groups, compared to control chicks (1.43, 1.38, 1.43 vs 1.54, respectively, $P<0.01$). Villus height and villus apparent surface area were found to be the highest in the chicks fed with dose 3 (1,685.5 μm and 245,558 mm^2). On the other hand, villus width was found to be similar among the experimental groups. In conclusion, probiotic and prebiotic supplementation to broiler diets had a significant effect on villus development, and therefore digestion and absorption activities, and eventually better growth performance.

Association of symbiotic and nutritive gel with broiler digestibility

F.V. Castejon[1], M.C. Alves[1], E.S. Fernandes[1], J.M.S. Martins[1], H.F. Oliveira[1], F.M. Ribeiro[2] and J.H. Stringhini[1,3]

[1]Universidade Federal de Goias, Departamento de Zootecnia, Escola de Veterinaria e Zootecnia, Avenida Esperanca, s.n. Campus Samambaia, 74690900, Goiania, Goias, Brazil, [2]Biomin Latin America, Tecnico-comercial, Estrada Prof. Messias José Baptista, 2007, 13432700, Piracicaba, Brazil, Brazil, [3]CNPq, Researcher, SHIS Bl.A-D, Lago Sul, 71605001, Brasilia, Brazil; jhstring@uol.com.br

A commercial symbiotic associated with a nutritive gel in the hatchery in broiler digestibility was studied. 400 male neonate Cobb 500 chicks were fed in the hatchery and allotted in battery cages, in 5 treatments (T1 –control; T2 – Gel in hatchery; T3 – Gel + symbiotic in hatchery; T4 – Gel + symbiotic in hatchery and symbiotic in water; T5 – symbiotic in water) and 8 replicates of 10 birds each. Symbiotic was offered for 3 d, during diet changes and once a week, so T4 received the symbiotic in the hatchery and d 2,3,7,10,11,12,14,22,23,24,28 (12 times) and T5 at d 2, 3, 4, 7, 10, 11, 12, 14, 22, 23, 24, 28 (12-times). All groups remained 24 h fasted of feed and water. the digestibility of dry matter, nitrogen, ash, ether extract and nitrogen retention in 3 different periods (7-11, 19-23 and 38-42 days of age) were evaluated. Statistical analysis was performed using ANOVA and Tukey test with Software R. The worst digestibility was obtained for T5 compared to T2 and T3 for all variables in starter period. In growing phase T2 and T3 showed the best ether extract digestibility (80.26 and 80.72 respectively), compared to T1 (75.39) and T5 (74.92). In finishing phase, EE digestibility was higher for T2 (90.44), T3 (89.99) and T4 (89.99) compared to control T1 (87.05) and symbiotic in water T5 (87.70). For ash digestibility, the best results were obtained for T3 (76.57) compared to other treatments. Thus, the symbiotic and gel in hatchery increased broiler digestibility.

Beta-glucans and mannoproteins impact immune response in an induced model of necrotic enteritis

C.N. Johnson[1], J.A. Byrd[2], M. Hashim[3], C. Bailey[3] and R.J. Arsenault[1]
[1]University of Delaware, Animal and Food Sciences, 531 S. College Ave, Newark, DE 19716, USA, [2]United States Department of Agriculture, 2881 F&B Road, College Station, TX 77845, USA, [3]Texas A&M University, Poultry Science, 101 Kleberg Center, 2472 TAMU, College Station, TX 77843, USA; johnsocn@udel.edu

Necrotic enteritis (NE) is one of the most common and costly diseases in modern broiler industry, having an economic impact exceeding 2 billion US dollars annually. Restrictions in antibiotic feed additives throughout the industry have resulted in an increased incidence. Finding effective antibiotic alternatives has become a priority. In this study, an experimental model of NE was induced and yeast cell wall components including, a yeast cell wall fraction (YCW), purified β-glucan (BG) and purified mannoproteins (MPT), were evaluated for their effects on challenged gut tissue. Chicken-specific immunometabolic kinome peptide arrays were used to measure differential phosphorylation signaling in three gut tissue segments. Data analysis revealed kinome profiles clustered predominantly by tissue, with duodenum showing the greatest relative signaling. BG, MPT and BG+MPT cluster together indicating similar gut tissue responses due to treatment. Considering peptides, significant differential phosphorylation was caused by BG, MPT and BG+MPT treatment. Gene ontology biological processes showed innate immune response and defense response. The regulation of immune response was altered in the duodenum (158 peptides) and jejunum (63 peptides). The immune response-regulating signaling pathway, the regulation of immune response, and innate immune response was altered in the ileum (139 peptides). In addition, there are unique phosphorylation events within the BG+MPT group which are synergistic. In conclusion, yeast component feed additive treatment separates the tissue response away from those of control and NE groups predominantly due to specific changes in immune signaling, pointing toward a prospective mechanism of disease amelioration in the NE model.

Effect of dietary yeast extract on performance and carcass characteristics of broiler chicks

M. Adam and A. Salih
University of Khartoum, Animal Nutrition, Khartoum, 13344, Sudan; merghani2007@yahoo.com

The objective of the present study was to evaluate the effect of different levels of Dietary Yeast Extract containing Beta-glucans and Mannan Oligosaccharides on broiler chickens performance. One hundred and twenty day-old unsexed Arbor Acres broiler chicks were brought from commercial hatchery. The chicks were randomly divided into three treatment groups. Forty birds per treatment, with ten birds per replicate. The birds were fed on three experimental diets, (A) basal diet (control), diet (B) supplemented with yeast extract 0.025% and diet (C) supplemented with yeast extract 0.05%. All diets were formulated to meet the nutrient requirements for Arbor Acres breed. Feed intake and body weight were recorded weekly and weight gain and feed conversion ratio were calculated. Eight birds from each treatment were randomly selected at 42 days of age, weighed and slaughtered for determining carcass dressing percentage. The result indicated that the addition of yeast extract at level 0.025 and 0.05% increased feed intake, live body weight and feed conversion ratio. There was no significant effect ($P \leq 0.05$) on dressing percentage.

Effectiveness of yeast cell wall in layers on intestinal and ovarian colonization of *Salmonella enteritidis*

C. Hofacre[1], M.A. Bonato[2], L.L. Borges[2] and G.F. Mathis[3]
[1]The University of Georgia, 953 College Station Road, Athens, GA 30602, USA, [2]ICC Brazil, Av. Brigadeiro Faria Lima, 1768 4C, 01451-909 Sao Paulo, Brazil, [3]Southern Poultry Research Group, Inc., 96 Roquemore Road, Athens, GA 30607-3153, USA; melina.bonato@iccbrazil.com.br

Salmonella enteritidis (SE) is currently the most common serotype associated with the human disease. In laying hens is especially important because colonize the ceca and internal organs, resulting in SE translocation to the ovary and as a consequence, can be found in eggs, by either ovarian or intestinal tract infection. The yeast cell wall (YCW) from *Saccharomyces cerevisiae*, highly concentrated in β-glucans and MOS has been demonstrated to be promisor. This study was designed to evaluate the use of YCW to mitigate intestinal and ovarian colonization by SE. For this, 200 Hyline W36 pullets at 10 wks of age were distributed in 2 treatments: control and YCW supplemented group (0.5 kg/ton). The birds were orally challenged by a Nalidixic acid-resistant strain of SE (3×10^9 cfu) at 16 wks, and 7 and 14 d post-challenge the ceca and ovary samples were obtained to observe the prevalence and enumerate SE (in ceca was by the MPN method at log 10). Statistical analysis was by Kruskal-Wallis test and Dunn's procedure ($P<0.05$). The ovarian prevalence was 41.7 and 4.2% from control group vs 33.3 and 2.1% in YCW supplemented group, at 7 and 14 d, respectively. The ceca prevalence was 93.8 and 47.9% by YCW vs 97.9 and 53.2% by control, at 7 and 14 d, respectively. The colonization (enumeration) was significantly reduced by YCW at 7 d (2.23 YCW vs 2.70 control, and 0.07 YCW vs 0.07 control, at 14 d). In summary, the SE challenge was effective in colonizing ceca and ovaries. The treatments with YCW did not reduce the prevalence of SE in the ceca, at this high challenge, but significantly reduce the level in those ceca. Also was effective in numerically lowering SE prevalence in ovaries at both 7 and 14 d post challenge.

Effect of yeast cell wall on the oral absorption of zearalenone in broiler chickens

M. Devreese[1], E. Gasthuys[1], S. Croubels[1], V. Marquis[2] and E. Auclair[2]
[1]Ghent University, Dept. of Pharmacology, Toxicology and Biochemistry, Salisburylaan 133, 9820 Merelbeke, Belgium, [2]Phileo Lesaffre Animal Care, 280 ave de la Marne, 59700 Marcq en Baroeul, France; v.marquis@phileo.lesaffre.com

Zearalenone (ZEA) is a mycotoxin implicated in reproductive disorders of farm animals. To reduce toxic effect of ZEA, an approach for decontamination in feedstuffs is the use of adsorbent materials. Safwall® is a specific yeast cell wall (YCW); his ability to bind mycotoxins has been demonstrated *in vitro* by a specific method for quantification of adsorption performance (>80%). Then, a model was developed *in vivo* for efficacy testing of ZEA-binder. The study was a cross-over study conducted on 8 broiler chickens. The 8 chickens received ZEA (5 mg ZEA/kg BW) at day 8 of the study, without binder. After a one day wash-out period, a dose of ZEA (5 mg/kg BW) was administered in combination with 0.05 g binder/kg BW. This was repeated for the two other doses tested (namely 0.2 and 0.5 g/kg BW) respecting each time a one-day wash-out period. Blood samples of the 8 chickens were taken. The time points of blood sampling were 0 min (just before administration) and 0.08, 0.16, 0.33, 0.5, 0.66, 0.83, 1, 1.5, 2, 3 and 4 h post administration. To evaluate the efficacy of the binder, ZEA and β-zearalenone (β-ZEL) plasma concentration–time profiles were set up and the toxicokinetic parameters were compared (area under the curve, maximal concentration, time at maximal concentration, elimination half-life, mean residence time and elimination rate constant). The relative oral bioavailability was evaluated as marker for efficacy of the mycotoxin binder. Results demonstrated a decreased oral bioavailability (more than 90%) of ZEA after concomitant administration with the YCW at the highest dose (0.5 g/kg BW). Indeed, all ZEA plasma concentrations were below the limit of quantification (LOQ), demonstrating a significant effect of the binder. We showed that Safwall® significantly reduces the absorption and oral availability of ZEA in the model.

Evaluation of a yeast culture protein concentrate in broiler diets with and without antibiotics

B.S. Lumpkins[1], G.F. Mathis[1], C. Pedersen[2] and D. Nelson[2]
[1]Southern Poultry Research, Inc, 96 Roquemore Road, 30607, Athens, GA, USA, [2]Hamlet Protein Inc., Saturnvej 51, 8700 Horsens, Denmark; cap@hamletprotein.com

The objective of the study was to evaluate the performance of broilers when fed a yeast culture protein concentrate (YCPC) in starter diets alone or in combination with an antibiotic feed program. A 42 d broiler floor-pen study was conducted using the following treatments: T1) control (commercial diet containing no YCPC or the anitibiotic bacitracin methylene disalicylate (BMD)) T2) control+ BMD 50 g/t (D0-42), T3) control + YCPC (D0-14) + BMD 50 g/t (D0-42) T4) control +YCPC (D0-14). A randomized block design with 12 replications of 50 birds per pen was used in pens with built-up litter from several flocks previously. Feed and water were available *ad libitum*. Bird weights and feed consumption (kg) by pen were recorded at study initiation, Days 14, 35, and 42. The data was analyzed statistically using SAS software with a P value of 0.05 to determine the level of significance. During the early period (0 to 14 d) there was no difference ($P>0.05$) between any of the treatments for BWG and FI, but the birds fed the diets containing YCPC had improved FCR regardless of the combination with BMD. At 35 d of age the birds fed the YCPC during the starter period had greater ($P<0.05$) BWG and improved FCR, which was most significantly seen in the YCPC only fed birds. Overall, (D0-42) there was no difference ($P>0.05$) between any of the treatments for FI, BWG and Mortality. However, the birds that were fed the YCPC with and without antibiotic maintained their improvement in FCR over the nonmed and BMD fed birds. Based on this study the presence of a yeast culture protein concentrate has the possibility of improving FCR in broilers fed diets with or without antibiotics.

Enhancing the efficiency of immunization via drinking water supplementation in broiler chickens

G. Csikó and O. Palócz
University of Veterinary Medicine, Department of Pharmacology and Toxicology, István utca 2, 1078 Budapest, Hungary; csiko.gyorgy@univet.hu

Administration of feed or drinking water additives in chicken flock may promote the health and improve the immune system of the animals. The fowl cholera caused by *Pasteurella multocida* is one of the major causes of economic losses in chicken industry. Vaccination against this microorganism is not entirely effective. The aim of our study was to enhance the protective effect of the *P. multocida* vaccine with drinking water supplements; fulvic acid (FA) and drinking water acidifier (DWA) in chicken stock. Fulvic acid and a drinking water acidifier (volatile fatty acids, amino acids, phosphoric acid and zinc and cooper salts) were used as water supplementers in our investigations. Fifty broiler chickens were divided into five equal groups. One group served as control. Four groups were administered via drinking water either with DWA or FA in dose of 0.1 ml or 25 mg/kg bw., two groups five and two groups for ten consecutive days, respectively. Each animal was immunized with *P. multocida* vaccine twice, first time at age of one month, and second time three weeks later. Blood samples were collected before and one week after the second immunization. *P. multocida* antibody titer in blood was determined by ELISA method. Compared to controls, the 10-day long DWA and both the 5- and 10-day long FA supplementation increased the level of P. multocida antibody titer significantly ($P<0.05$), after the first and the second immunization. The means±SE of titer values after first and second vaccination were 5,382±1,190 and 6,742±1,127 in controls; 12,146±2,881 and 15,234±2,564 in 10-day long DWA group; 14,728±2,738 and 19,659±1,780 in 5-day long FA group; 13,899±2,903 and 16,171±2,760 in 10-day long FA group, respectively. It is important to enhance the efficacy of vaccination to prevent the outbreak of infectious diseases. Supplementation of drinking water with either drinking water acidifier or fulvic acid appears to improve the effect of the *P. multocida* vaccination in chickens.

The energy effect of guanidino acetic acid in broiler nutrition

A. Yapontsev[1], A. Mitropolskaya[1], A. Klimenko[1] and I.A. Egorov[2]
[1]Evonik Chimia, Kozhevnicheskaya, 14, 115114 Moscow, Russian Federation, [2]VNITIP, Ptitsegradskaya 10, Sergiev Posad, Russian Federation; alexey.yapontsev@evonik.com

Guanidino acetic acid (GAA) is the metabolic precursor of creatine. The supplementation of broiler diets with GAA, commercially available as CreAMINO®, has repeatedly been proven to enable broilers to cope with a certain reduction of feed energy without any performance drop. Objective of this study was to test the effect of GAA supplemented to a common Russian diet containing sunflower meal and being energy-reduced. This trial was conducted at the FSUE Zagorskoye Experimental and Industrial Enterprise of ARSRTIPK of the Russian Agricultural Sciences Academy in Sergiev Posad, Russia. 450 day old Cobb 500 chicken were allocated to three dietary treatments, six replicates of 25 birds each. The treatments were: positive control – standard feed (T1), negative control – standard feed minus 50 kcal AME_N/kg (T2), and trial group – standard feed minus 50 kcal AMEN/kg, supplemented with GAA (T3). Feed intake in T3 was significantly lower compared to the other groups. Energy reduction in T2 resulted in less performance for all investigation parameters ($P<0.001$), whereas GAA supplementation in T3 enabled the birds to compensate lower energy levels, and to perform as good as T1. Selected results for T1, T2, and T3 are: body weight at day 35, g – 1,840, 1,790, and 1,871; feed conversion ratio, kg/kg: 1.63, 1.67, and 1.58, resp. Parameters of slaughter performance, such as carcass and breast meat weight, were affected a similar way. Within T2 all investigated parameters showed significantly lower values. Adding GAA to the feed (T3) enabled male broilers to perform on the level of T1, whereas the females achieved even higher carcass, and breast weights. These findings proved consistently that supplementation of GAA to an energy-reduced feed enables broilers to compensate the negative effects of this energy gap without any drop in performance. CreAMINO® addition to the feed offers a huge potential to save feed cost.

A protected benzoic acid on growth and intestinal *C. perfringens* and *E. coli* populations of broilers

R. Barea[1], G. Tosi[2], A. Caminiti[2], P. Massi[2], L. Fiorentini[2], M. Parigi[2] and S. Peris[1]
[1]Novus International, Neerveld 101-103, 1200 Brussels, Belgium, [2]Istituto Zooprofilattico Sperimentale della Lombardia e dell'Emilia-Romagna, Via Don E. Servadei 3E/3F, 47100 Forlì, Italy; silvia.peris@novusint.com

In recent years, there has been an explosion of interest in looking for alternative products (e.g. aromatic compounds) which can modulate the intestinal flora beyond the stomach barrier of the animal. In this study, the effect of a protected benzoic acid embedded in vegetable fat (PAC) has been evaluated on growth performance and intestinal *Clostridium perfringens* and *Escherichia coli* populations of broilers. The study was conducted in a commercial farm with 4 identical houses. A total of 47,000 ROSS 308 males and females were located in 2 groups (23,500 birds/group): (1) negative control (CTR); and (2) standard feed with 500 g PAC/MT of complete feed. Data was collected at market age. Regarding growth performance, statistical analysis was run considering 2 houses/treatment (assigning diverse weights according to the number of birds/house). The number of *C. perfringens* and *E. coli* colony forming units per gram (cfus/g) was counted from samples collected from the content of the ileum of 40 birds per treatment group. Regarding growth performance results, FCR showed a near-trend P-value ($P=0.11$; mean resulted 3.0% lower in PAC vs CTR). The number of *C. perfringens* cfus/g significantly decreased by 77.8% (= median of the posterior distribution) in the PAC group compared with CTR, whereas the number of *E. coli* cfus/g decreased by 97.3%. As a conclusion, this trial shows that supplementing broiler diets with a protected benzoic acid embedded in vegetable fat can positively impact the intestinal microflora by reducing clostridia and coliform counts in the gut. The use of this product, together with a proper farm management, can be a powerful, cost-effective solution to manage intestinal health challenges and animal welfare ensuring a profitable poultry production.

The impacts of encapsulated benzoic acid supplementation on gut bacterial status in broilers

M. Yousaf[1], F. Goodarzi Boroojeni[2], W. Vahjen[2], K. Männer[2], A. Hafeez[1], H. Rehman[1], S. Keller[3], S. Peris[4] and J. Zentek[2]
[1]University of Veterinary and Animal Sciences, 54000 Lahore, Pakistan, [2]Institute of Animal Nutrition, Free university Berlin, Königin-Luise 49, 14195 Berlin, Germany, [3]Novus Deutschland GmbH, Schwänheit 10, 34281 Gudensberg, Germany, [4]Novus Europe, Rue Neerveld 101, 1200 WS-Lambert, Belgium; farshad.goodarzi@fu-berlin.de

The impacts of encapsulated (vegetable oil matrix) benzoic acid (Avimatrix®, Novus) inclusion in broiler diets on performance and gut microbiota were studied. Eighty broilers were divided into 2 groups (8 pens per group). Birds in the treatment group received control diet plus 2 g/kg Avimatrix®. At d 35 pH, bacterial composition and activity were determined in the crop, jejunum, ileum and caecum of 1 bird per pen. Performance and pH were similar. While benzoic acid (BA) decreased rapidly in the proximal gut, the treatment increased BA in the entire gut. Total, L- and D-lactate in the crop (137, 65 and 72 vs 91, 52 and 39 µmol/g respectively) as well as D-lactate in the jejunum (72 and 39 µmol/g) were higher in the BA group, however BA inclusion reduced total volatile fatty acids (126 vs 146 µmol/g) in the caecum ($P<0.05$). Lactobacilli counts were changed by BA inclusion ($P<0.05$). Lactobacilli in the jejunum and ileum (8.8 and 9 vs 8.2 and 7.9 log10/g respectively) were increased by BA ($P<0.05$). The responses of *Lactobacillus* species to BA were different. While *L. amylovorus*, *L. acidophilus*, *L. johnsonii* and *L. reuteri* (particularly in the ileum) followed the course observed for total lactobacilli, *L. salivarius* was not modified. Inclusion of BA also increased clostridial clusters IV and XIVa in the jejunum (5.1 and 5.8 vs 4.6 and 5.2 log10/g respectively). Encapsulated BA modified the intestinal microbiota which can lead to the conclusion that the main beneficial mode of action of BA in the gut is the enhancement of lactic acid bacteria, which in turn may act as a vanguard against pathogens.

Encapsulating butyrates for targeted release – is it worth the effort?

S. Kirwan, N. Smeets and S. De Smet
Kemin N.V., Toekomstlaan 42, 2200 Herentals, Belgium; stef.desmet@kemin.com

The advantages of butyrates for a full health and absorptive performance of the GIT have been studied in detail, debate remains whether encapsulating butyrate is having an additional effect. In total, 432 Lohmann Brown laying hens aged 48 weeks were fed a diet with or without added butyrates until 72 weeks of age. Treatment T1 was the control group, and to T2 and T3 respectively, 500 g/t encapsulated calcium butyrate or 700 g/T free sodium butyrate was added. This equals a butyrate addition of 0/250/490 g per t of feed, respectively for T1, T2 and T3. Each group comprised eight cages with two hens each. Diets were based on wheat, triticale, soybean meal and corn. FCR, body weights, mortality, egg weights and laying rate were recorded during the trial. In addition, bone mineralisation and gut morphology were investigated at the end of the trial. The performance parameters differed across all groups, but no statistically significant ($P≤0.05$) differences could be established for egg production parameters (egg weights, laying rate), feed intake, or FCR. During the last period of the trial (68-72 weeks of age), the encapsulated butyrate led to a significantly better egg shell stability compared to T1. The average shell thickness was significantly higher for both butyrate groups compared to T1. The bone mineralisation study showed that T2 significantly ($P≤0.05$) showed the highest calcium concentration (23.49% Ca) compared to T1 (22.68%), whereas T3 (23.25%) was not significantly different from either of the other treatments. The concentration of phosphorus was numerically improved in T2 (10.86% P) and T3 (10.61%), compared to T1 (10.67%). Ca and P differed numerically between all groups (T1: 2.53 g Ca/1.19 g P; T2: 2.69 g Ca/1.24 g P; T3: 2.49 g Ca/1.13 g P). In conclusion, the addition of butyrates had an overall positive effect on egg production. However, in order to make significant improvements on mucosal structure and mineral absorption, butyrates have to be encapsulated.

Protected sodium butyrate and nutrient concentration on protein and energy utilization in broilers

M. Puyalto[1], C. Sol[1], J.J. Mallo[1] and M.J. Villamide[2]
[1]NOREL S.A., 28007 Madrid, Spain, [2]UPM, 28007 Madrid, Spain; mpuyalto@norel.net

The study was conducted to evaluate the effect of sodium butyrate protected with PFAD sodium salt (GUSTOR N'RGY) and three different nutrient concentration diets on protein and energy utilization in broilers. A 2×3 factorial design was used with a basal diet based on wheat, barley and soybean meal with three different nutrient concentrations and with or without additive. There were 6 treatments: CON (3,000 kcal AMEn/kg, 22.02% CP and 11.6 g/kg dig Lys) and CON-1 (CON with a first reduction of 60 kcal AMEn/kg and 2.3% of amino acids, AA), CON-2 (CON with a second reduction of 120 kcal AMEn/kg and 4.6% of AA), N'RGY (CON diet with GUSTOR N'RGY at 1 kg/t), N'RGY-1 (CON-1 diet with 1 kg of N'RGY/t) and N'RGY-2 (CON-2 diet with 1 kg of N'RGY/t). A total of 162 male Cobb broiler chickens, 14 days old, were used in the digestibility balance. Birds were placed in digestibility cages with 3 chickens per cage; there were 9 replicates/treatment. Digestibility balance lasted seven days, from 14 to 21 days of age. Total excreta was daily collected and weighed per cage on days 19, 20 and 21, frozen and oven dried. Crude protein retention and gross energy metabolizability (CPr and GEm) were calculated. Data were analyzed by two-way ANOVA using the GLM procedure of SAS. The inclusion of Gustor N'RGY did not affect CPr (58.87 vs 59.74%, $P=0.617$) but improved GEm (69.94 vs 72.55%, $P=0.022$). The reduction in nutrients concentration did not produce statistical differences in CPr ($P=0.164$) nor in GEm. There were no significant interaction between nutrient concentration and feed additive inclusion. It can be concluded that the use of GUSTOR N'RGY is able to improve energy utilization in chickens in diets with different nutrient concentration.

Coated butyrate to mitigate necrotic enteritis challenges: effects and mechanisms

T. Goossens[1], D.F. Ramírez[1] and M.H.H. Awaad[2]
[1]Nutriad, Hoogveld, 9200 Dendermonde, Belgium, [2]University of Cairo, Cairo University Rd, Oula, Giza, Cairo, Egypt; t.goossens@nutriad.com

Necrotic enteritis (NE) is a disease caused by intestinal tissue damage due to overgrowth of certain *Clostridium perfringens* strains. We evaluated the effects of butyrate in NE-challenged and non-challenged broilers, as it is known to trigger several responses associated with improved gut health. Moreover, the butyrate product tested was precision delivery coated (PDCB), meaning that butyrate can be delivered post-gastrically to the luminal side of the intestine, where it is hypothesized to be most effective in supporting gut integrity. 800 broilers were allocated to 4 treatment groups, each consisting of 10 replicates of 20 birds. The experiment was set up using a 2×2 factorial design: birds were supplemented with PDCB or not and were *Clostridium*-challenged or not. At day 21 birds were examined for pathological lesions of the small intestine. Zootechnical performance and hemagglutination inhibition (HI) titers against the antigens in the Newcastle Disease vaccin were monitored weekly, and organ weight at the end of the trial (d42). Hepatic IGF-1 levels were determined using qPCR at d35. Both in the challenged and unchallenged groups, PDCB-fed birds outperformed the corresponding control groups on live weight ($P<0.05$) and FCR. The necrotic lesion scores (X^2; $P<0.05$) suggest that the PDCB mitigates of intestinal tissue damage. In addition, PDCB-supplemented birds expressed more IGF-1 ($P<0.05$), which is reported to be related to body growth weight. The relative bursa weight was significantly increased in challenged birds that were administered PDCB ($P<0.05$). The HI titer against ND viral antigens were reduced in NE challenged birds ($P<0.05$); on the contrary, birds that received PDCB had significantly higher titers ($P<0.05$), which is suggestive for an PDCB-induced boost of immune responses. The results highlight some of the potential mechanisms underlying the performance improvement seen in birds that were given the PDCB.

Effect of guanidinoacetic acid in diets with poultry by-products on broiler live performance at 55d

E.O. Oviedo-Rondón[1], H. Córdova[1], A. Sarsour[1], D. Sacopta[1], D. López[1] and M. Rademacher-Heilshorn[2]
[1]North Carolina State University, Raleigh, NC, 27695, USA, [2]Evonik, Hanau, 63457, Germany; eooviedo@ncsu.edu

Guanidinoacetic acid (GAA) supplementation can improve broiler live performance in corn-soy based diets. This experiment was conducted to evaluate the effects of GAA supplementation in broilers fed corn-soy based diets with and without poultry by-products (PBP) on broiler performance. The treatments consisted of the inclusion of PBP in the diets at 0% and 5%, and either no GAA or supplementation of 600 g/ton added on top. A total of 1,280 male Ross 708 chicks were placed in 64 floor pens with 20 chickens/pen. At hatch, 14, 35, 48, and 55 d of age, BW and feed intake were recorded. BW gain and FCR were calculated at the end of each phase, and flock uniformity (CV%) at 55 d. Data were analyzed as a randomized complete block design in a 2×2 factorial arrangement with PBP inclusion and GAA supplementation as main effects. No interaction ($P>0.05$) effects were observed on BW or BW gain in any of the phases evaluated. GAA addition did not affect ($P>0.05$) feed intake of diets with or without PBP in any of the phases. Broilers that were fed corn diets without PBP were heavier and gained more weight compared to the broilers fed diets with PBP. Consequently, an effect ($P<0.05$) was observed at 0-14, 0-48 and 0-55 d on FCR. Chickens fed diets with PBP had worse FCR than broilers fed no PBP. FCR was affected ($P<0.05$) by GAA supplementation at 48 and 55 d. Broilers fed supplemented diets with GAA had better FCR ($P<0.05$) than the treatments fed diets with no GAA. Diets without PBP and GAA supplementation had slightly worse ($P<0.05$) flock uniformity with higher CV% of individual BW. However, GAA supplementation reduced ($P<0.05$) flock mortality from 48 to 55 d of age. PBP inclusion increased ($P<0.05$) mortality from 14 to 35 d of age. In conclusion, GAA supplementation improved broiler live performance in diets regardless of the inclusion of PBP.

Addition of guanidinoacetic acid can counterbalance energy reduction in broiler diets

H.N. Malins[1], N. Muley[1], V. Pirgozliev[2] and A. Lemme[1]
[1]Evonik Nutrition & Care, Rodenbacher Chaussee, 63457 Hanau-Wolfgang, Germany, [2]Harper Adams University, Harper Adams University, TF10 8NB Edgmond, United Kingdom; holly.malins@evonik.com

Guanidinoacetic acid (GAA) is produced via de novo synthesis from arginine and glycine as part of the creatine metabolic pathway. Supplemental GAA can be added to broiler diets to help meet daily requirements, so it was studied whether 600 g GAA/metric tonne feed can counterbalance dietary energy reduction in Ross 308 broilers, as GAA is deeply involved in energy metabolism. Trial length was 42 days, with *ad libitum* feed and water provided in starter (0-14), grower (14-25), finisher (25-42) phases. Diets represented UK commercial standards and included enzymes and coccidiostats. Four dietary treatments were used; 1. Negative Control (NC; -50 kcal AME_N/kg), 2. Positive control (PC; normal AME_N–levels of 3,000, 3,100 and 3,150 kcal/kg in starter, grower, finisher diets), 3. NC+ 0.06% GAA, 4. PC – 100 kcal AME_N kg + 0.06% GAA. Each treatment comprised 16 single sex replicates (8 female; 8 male) with 20 birds; 1,280 birds in total. Bird weight and feed consumption were measured by pen at 0, 14, 25 and 42 days. Dietary AME_N reduction by 50 kcal/kg (NC) resulted in numerically lower average final body weight (BW) and average daily gain (ADG) compared to PC which was more than compensated by GAA supplementation – even at energy reduction of 100 kcal/kg (Final BW: 2.782[b], 2.827[ab], 2.868[a], 2.868[a]; ADG: 65.3[b], 66.3[ab], 67.3[a], 67.3[a] for treatments 1, 2, 3, and 4, $P<0.05$). As significant differences occurred for BW development, feed conversion ratio was standardised to final BW with 3 points per 100 g difference to overall mean BW. Accordingly, NC+0.06% GAA outperformed the NC, while treatment 4 brought performance back to PC (1.687[B], 1.678[AB], 1.647[A], 1.671[AB], $P<0.10$ (tendency)). Results of this study indicate that an addition of 0.06% GAA to the diet can compensate a reduction of AME_N by 50 kcal/kg and even suggests a higher ME-reduction (100 kcal/kg) did not have detrimental effects on performance.

The energy effect of guanidino acetic acid in phytase-supplemented broiler diets

A. Ion[1], M. Mueller[2] and L. Stef[3]

[1]Evonik Nutrition & Care GmbH, Matasari 55, Bucuresti, Romania, [2]Evonik Nutrition & Care GmbH, Rodenbacher Chaussee 4, Hanau, Germany, [3]University of Timisoara, C. Aradului 119, Timisoara, Romania; ancuta-cosmina.ion@evonik.com

Creatine is a naturally occurring component in the animal's body and plays a central role in energy metabolism. Guanidino acetic acid (GAA), creatine's natural precursor, forms 96% of the feed additive CreAMINO®. Decreased feed conversion ratio (FCR), improved weight gain, and compensating an energy reduction of 50 kcal AME_N/kg compared to a standard feed have repeatedly been shown common effects of a GAA supplementation. Feed enzymes, in particular phytases, claim a similar energy saving potential. Therefore, it is of huge interest whether the GAA addition to an energy-reduced, phytase-containing diet is effective, too. The objective of this study was to check the energy sparing effect of CreAMINO® in a phytase-supplemented broiler diet on live performance. The trial was conducted in the research facility of University of Agricultural Studies and Veterinary Medicine of Banat, Timisoara, Romania. 240 male, day old Ross 308 chicks were allocated to three dietary treatments with ten replicates of eight birds each. The treatments, all phytase-supplemented, were: positive control – standard feed, negative control – standard feed minus 50 kcal AME_N/kg, and trial group as GAA-supplemented, negative control. The trial group performed better than the other treatments reaching a body weight (BW) of 2,301 g ($P<0.001$). Birds in negative control weighed 100 g less than positive control, and even 200 g less compared to the trial group. FCR was corrected to an average BW of 2,200 g. The trial group achieved an FCR of about 1.43 kg/kg being comparable to positive control, but ten points below that of energy-reduced group. The supplementation of CreAMINO® to an energy-reduced (- 50 kcal AME_N/kg), phytase-containing diet enabled broilers to perform at the level of positive control group for all parameters investigated. The animals in the negative control were not able to cope with the energy deficiency.

Sodium formate or sodium bicarbonate as dietary electrolyte balance regulator in broiler diets

J. Lorjé and S. Rengman

Perstorp AB, Neptuniag.1, 21120 Malmö, Sweden; sofia.rengman@perstorp.com

Sodium bicarbonate is commonly used to improve the dietary electrolyte balance (dEB) in poultry diets. Sodium formate is another source of highly available sodium with antibacterial properties. The objectives of the study was to evaluate the efficacy of diets containing sodium formate or sodium bicarbonate on performance and crop pH. Average daily feed intake, live weight, feed conversion ratio (FCR), feed conversion efficiency (FCE), crop pH and mortality was assessed. The bedding was 'spiked' with dirty litter from a commercial unit, to challenge the birds. 536 male Ross broilers were grouped, 11-12 birds/pen. 15 pens were allocated to each diet, commercial broiler diet containing sodium bicarbonate 1.5 kg/MT feed (T1), 1.3 kg sodium formate/MT feed (T2) or 5.9 kg sodium formate/MT feed (T3). The T3 diet had a higher sodium level and dEB (238) than T1 and T2 dEB (193) to assess the dEB effect. The diets were fed in two phases, phase 1 d 0 to 21 and phase 2 d 21 to 35. The birds were weighed on d 0, 21 and 35. Feed intake, body weight, FCR, FCE and mortality were calculated for each feeding phase and for the overall study period. On d 21 and 35 three broilers/pen were euthanized and crop pH was determined. During phase 1, T3 broilers had a higher feed intake when compared with T1 ($P<0.001$) and with T2 ($P<0.05$). T2 also had a higher feed intake than T1 ($P<0.05$). In the overall study period, feed intake for T3 was higher compared to T1 ($P<0.05$). On d 35, T3 broilers had a higher finishing weight 2.47 kg ($P<0.05$) compared with T1 2.38 kg and T2 2.37 kg. The crop pH readings were different on d 35 with T3 having lower pH readings compared with T1 ($P<0.001$). No differences in FCR, FCE and mortality (2.05%) was observed between the groups. Sodium formate in the diet can provide increased feed intake and live weight gain, especially during the first most critical weeks in broiler breeding. At d 35 crop pH levels was lower in the T3 groups compared with the T1 groups using sodium bicarbonate.

Salmonella remains an ongoing challenge, feed additives as a part of the solution

N. Tallarico, S. Kirwan and S. De Smet
Kemin N.V., Toekomstlaan 42, 2200 Herentals, Belgium; stef.desmet@kemin.com

Salmonella prevalence diminished but it is not gone. While the reduction of *Salmonella* in poultry across Europe has been very successful, recent outbreaks show smart *Salmonella* management remains very important for both poultry and humans. No control can rely only on biosecurity, biocides or management to produce safe and healthy poultry. The present study seeks to investigate whether feed additives can significantly reduce *Salmonella enteritidis*. In total, 80 ROSS 308 female broilers were housed in isolators in groups of 20. The birds were fed 0 kg (T1), 0.75 kg (T2), 1.0 kg (T3), 1.5 kg (T4) of a formiate and flavour based additive. After one week the birds were inoculated with a field strain of S. enteritidis at 106 cfu. The day after inoculation cloacal swabs were taken to establish colonisation. Subsequently pooled samples of faeces were collected on day 3, 5, 10 and 12 PI to investigate *Salmonella* shedding. At 7 days post infection (PI) 50% of the birds were euthanized for cecum and liver samples; the remaining birds were euthanized at the end of the second week. *Salmonella* shedding decreased directly in response to the inclusion of the additive. Correspondingly, *Salmonella* counts (log cfu/g) in the cecum decreased significantly ($P<0.05$) from T1 (5.12) to T2 (4.17) to T3 (3.43). The reduction to T4 (3.18) was not significant at this level. At the end of the experiment all groups showed significant decreases in *Salmonella* counts in the caeca corresponding to the dosage (6.25>4.12>3.02>2.01), respectively for T1 to T4. The liver samples below 1.00 log cfu/g were considered negative due to the analysis method used. At day 14 T1 was 4.72 log cfu/g, T2 was significantly ($P<0.05$) lower (1.34) than T3 (2.46), but higher than T4 (<1.00). This trial clearly showed the potential to control *Salmonella* all the way in the distal GI tract. A clear dose response of the *Salmonella* load could be observed, both in the faeces and the liver.

Essential oils and organic acids improved performance of necrotic enteritis challenged broilers

P.D. Steyl and C. Jansen Van Rensburg
University of Pretoria, Animal and Wildlife Sciences, Private Bag X20, Hatfield, 0028, South Africa; christinejvr@up.ac.za

With a reduction of antibiotic growth promoters (AGP) use in broiler feed, necrotic enteritis (NE) emerges as a common broiler disease worldwide. In this trial, feed additives and combinations thereof were tested against an AGP (zinc bacitracin) in broilers predisposed to conditions which stimulated NE. Apart from a negative control (no feed additives) and a positive control (AGP), three additional treatments were included. In the first treatment a direct fed microbial (*Bacillus subtilis*; DFM) was added to the feed, while a blend of essential oil compounds and organic acids (Biacid, Provimi) were added to the second treatment. The feed of a third treatment was supplemented with a mixture of essential oil compounds (Biacid Nucleus, Provimi) and DFM. Day-old, male Ross 308 broilers were randomly distributed in an environmentally controlled broiler house, with 12 replicate pens per treatment and 22 birds per pen at the start of the trial. Treatment diets were fed from day 0 for a 35 day grow-out cycle. At 10 days of age, birds received a coccidial vaccine (Immunocox, Ceva) at 10× the prescribed dosage and on day 14 they were orally inoculated with *Clostridium perfringens*. The broilers that received Biacid were significantly heavier at 28 days of age compared to birds from the negative control group (1,438 vs 1,385 g). Although body weight of the broilers at 35 days of age did not differ significantly between treatments, feed conversion ratio (FCR; g feed intake/g body weight gain) over the rearing period was significantly lower for broilers supplemented with Biacid (1.83) than broilers in the negative control (1.94). Broilers that received the AGP had a FCR of 1.87, while broilers from the DFM and DFM plus Biacid Nucleus treatments had FCRs of 1.87 and 1.84, respectively. It was concluded that Biacid, a combination of essential oils and organic acids, improved performance of broilers that were subjected to conditions favouring the development of NE in broilers.

Microencapsulated organic acids and essential oils as alternatives to antibiotic growth promoters

E. Vilarrasa and N. Torrent

Tecnología & Vitaminas, Pol. Ind. Les Sorts, p. 10, 43365 Alforja, Spain; register@tecnovit.net

Numerous alternatives to antibiotic growth promoters (AGP) have been considered over the years, such as organic acids (OA) and essential oils (EO), which have been shown to improve the intestinal-health status of animals. The objective of the present study was to test the efficiency of different microencapsulated products containing OA or a combination of OA and EO, as alternatives to AGP in broiler-chicken diets. For this purpose, 336 one-day-old male broiler chickens were randomly distributed in four treatments with seven replicates each. Treatment –C was a control diet without any AGP, treatment +C was the control diet plus monensin (125 ppm), treatment OA was the control diet with microencapsulated calcium butyrate (187.5 ppm) and fumaric acid (225 ppm), and treatment OA+EO was the control diet with the same organics acids described before plus microencapsulated cinnamaldehyde (30 ppm), carvacrol (3.75 ppm) and thymol (3.75 ppm). Chicks were fed in mash form for starter (1-21 d) and grower-finisher (22-38 d) phases. There were no statistical significant differences on growth performance among all treatments in starter phase ($P>0.05$). However, chicks fed microencapsulated OA+EO in grower-finisher phase had the same weight gain than those fed +C ($P>0.05$), and both treatments were statistically different than –C ($P<0.001$). No differences in feed consumption were found between treatments ($P>0.05$). This resulted in a better feed conversion ratio ($P=0.01$), a lower cost per kg of body weight ($P=0.03$), and a higher income over feed cost ($P<0.01$) of birds fed +C and OA+EO in comparison with those fed -C. Chicks fed only microencapsulated OA showed intermediate results between –C and +C. These data indicate that the combination of microencapsulated OA+EO achieves better performance results than only the supply of microencapsulated OA, probably due to the synergism exerted between both compounds. So, we conclude that the combination of microencapsulated OA+EO is an effective alternative to the use of AGP in broiler-chicken diets.

A mixture of medium chain fatty acids and phytogenic compounds will improve the broiler performance

M. De Laet, D. Hermans and S. Ayllón Ramos

Nuscience, Booiebos 5, 9031 Drongen, Belgium; mdl@nusciencegroup.com

It is known that medium chain fatty acids (MCFA's) have a positive effect on broiler performance when added to a broiler diet. Also phytogenic compounds have proven positive effects on broiler performance. The effects are wide: improved digestion, a better feed conversion ratio (FCR), decreased cases of dysbacteriosis, reduced mortality, better reproduction, with a higher profitability as end result. In this experiment, a mixture of medium chain fatty acids and phytogenic compounds was used to determine the effect on broiler performance. A total of 480 Ross 308 one-day-old male broilers were used. A completely randomized design was applied with two experimental treatments (8 repetitions per treatment, 30 birds per cage). A standard wheat/corn/soy based diet was used as a positive control (PC). The mixture of MCFA's and phytogenic compounds was dosed to the second treatment at 1.2 kg/MT in the starter-grower (day 0 until day 26), and at 0.6 kg/MT in the finisher (day 27 until day 39). For statistical analyses, SAS software is used. The supplementation of the mixture improves body weight significantly at 39 days (2,721 vs 2,553 g; $P<0.05$). The FCR was also significantly improved with addition of the specific mixture (1.55 vs 1.58; $P<0.05$). Average daily gain (66.2 g/day for the PC, 69.7 g/day for the mixture) and the feed intake (104.0 g/day for the PC, 107.9 g/day for the mixture) were also both significantly improved ($P<0.05$). It is concluded that a mixture of medium chain fatty acids and phytogenic compounds is an interesting functional feed ingredient for improving the performance of the broilers.

A precision delivery coated butyrate and a botanical product: application on broiler performance

D.F. Ramirez[1], T. Goossens[1], D. Jacob[1] and N. Mogyca[2]
[1]Nutriad, Digestive Performance, Hoogveld 93, 9200, Belgium, [2]Universidade Federal de Goiás, Depto de Zootecnia, Av. Esperança, Campus Universitário, 74690, Brazil; d.ramirez@nutriad.com

Good intestinal development, integrity and microbiota balance boost efficient digestion and absorption of nutrients. Many feed additives are developed aiming to enhance gut development, as for example the sodium butyrate. Besides, there is new evidence that components of botanical products, such as carvacrol and cinnamaldehyde, possess characteristics that optimize digestion and balance intestinal microbiota. We set up experiments using 784 day-old male Cobb 500® chicks, distributed across 4 treatment groups, each consisting of 7 replicates of 28 birds. First, we evaluate the effects of different concentrations (0.5, 0.75, and 1 kg/ton) of Precision Delivery Coated Butyrate (PDCB, ADIMIX®Precision) in starter feed on broiler performance. With increasing PDCB inclusion levels, we found an improvement in feed intake (FI) ($P=0.04$), body weight (BW) ($P=0.08$) and feed efficiency (FCR) ($P=0.05$) at day 21, and more weight gain ($P=0.05$) at the end of the trial (42 d). Secondly, we assess the potential of adding PDCB or a blend of botanical products (APEX®5) in the grower and finisher rations, on top of putting PDCB in the starter feed. The pre-starter (1-7 d) and starter rations (8-21 d) of all the treatments contained 1 kg/T of PDCB. In one treatment, no additives were added in the grower and finisher rations; a second treatment contained 0,5 kg/T PDCB in the grower feed and 0,25 kg/T PDCB in the finisher feed; a third treatment contained 150 g/T of the blend of botanicals during both the grower and finisher stages. We observed a higher weight gain ($P=0.01$) for the experimental groups containing PDBC in the growers and PBCD or botanical in the grower and finisher feeds at day 21., and an improvement of FI at day 42 ($P=0.01$). The best result on final BW was observed for the treatment combining PDCB in the starter feed with the botanical products in the grower and finisher rations.

Effect of a combination of eubiotics on broiler performance

A. Meuter[1], C. Paulus[1], A.L. Cotamo[2], C. Lozano[2] and D. Garcez[2]
[1]DSM Nutritional Products Ltd., Wurmisweg 576, 4303 Kaiseraugst, Switzerland, [2]DSM Produtos Nutricionais Brasil S.A, Av. Eng Billings 1729, 3003 São Paulo, Brazil; antoine.meuter@dsm.com

The rise of antimicrobial resistance against antibiotics is a worldwide concern affecting both human and animal health. In today's broiler production, ban of growth promoters (AGP) or development of antibiotic-free productions associated with a reduced usage of antimicrobial treatment on farm are common trends towards a more responsible use of antibiotics. Besides proper farm management, vaccination and biosecurity, optimum nutrition is an essential element to successfully reduce antibiotic usage. The choice of adapted feed additives such as eubiotics play a key role in ensuring gut health and good performances. This experiment aimed at evaluating the effect on the performance of broiler chickens of the combined use of CRINA® Poultry Plus, a formulation of essential oils and benzoic acid, and CYLACTIN®, an Enterococcus faecium strain (NCIMB 10415) probiotic. 1,440 Ross 308 male chicks were used and challenged with an anticoccidial vaccine against *Eimeria acervulina*, *Eimeria maxima*, *Eimeria tenella* given at 15 days of age by supplementing 3 times the recommended dose. The birds were fed with a corn soya based diet and received a shuttle anticoccicial program (narasin - nicarbazin from 1 to 21 days and salinomycin from 22 to 42 days of age). The birds were divided into the following 3 treatments with 8 replicates of 60 broilers each: a negative control (T1), a positive control containing the antibiotic growth promoter enramycin at 10 ppm (T2) and a eubiotic group including CRINA® Poultry Plus at 300 ppm and CYLACTIN® ME10 at 35 ppm (T3). T2 and T3 increased significantly breast meat yield compared to control (from 27.7% to 29.8% and 29.0% respectively, $P<0,05$).T2 also tended to improve weight gain similarly to T3. Those results showed that a synergistic combination of essential oil compounds, benzoic acid and the direct-fed microbial *Enterococcus faecium* NCIMB 10415 can effectively maintain optimum broiler performance in the absence of an AGP.

Effect of Ca pidolate and thymol-rich extract on egg quality and performance of laying hens

J. Estévez and S. Carné
Industrial Técnica Pecuaria, S.A. (ITPSA), Av. Roma 157, 08011 Barcelona, Spain; scarne@itpsa.com

A total of 96 Hy-line Brown laying hens of 60 weeks of age were used in a 42-d study to assess the effect of calcium pidolate and a thymol-rich extract on laying performance and egg quality. Animals were housed in battery cages of 0.336 m2 with individual feeders and nipple drinkers, and with automatic environment control and ventilation. Lighting regime was established at 16L:8D. A total of 32 cages with 3 animals per cage were used, with 8 replicates (cages) per treatment. A randomized complete block experimental design was used, with 4 treatments according to feed supplements added on top to a basal diet based on barley-wheat and soybean meal: Control (basal diet), Ca pidolate at 200 mg/kg (CaPid200), Ca pidolate at 400 mg/kg (CaPid400) and thymol-rich extract at 300 mg/kg (Thym). Egg production, BW and feed intake was obtained per replicate, and average individual daily feed intake and feed conversion was calculated. Shell breaking strength, shell percentage and thickness were assessed over 100 eggs per treatment, laid from 28 to 42 days of study. Calcium concentration (mg/g) and cholesterol content (%) was analysed in a sample of 7 eggs for control, CaPid400 and Thym treatments. Eggs were collected in the last week of study. No differences ($P>0.1$) between treatments were observed on laying performance. However, a dose-dependent tendency of improvement in the feed conversion was observed when Ca pidolate was added (Control, 1.950 ± 0.029; CaPi200, 1.920 ± 0.058; CaPi400, 1.858 ± 0.046; $P<0.1$). No effect ($P>0.1$) on parameters related to eggshell quality were observed. However, Ca pidolate tended to reduce cholesterol in yolk with respect to control (1.55 ± 0.01 vs $1.50\pm0.04\%$; $P<0.1$). Thymol-rich extract increased Ca concentration in the yolk a 8% with respect to control (1.44 ± 0.013 vs 1.57 ± 0.025; $P<0.05$). In conclusion, addition of calcium pidolate and thymol extract to the feed of laying hens might contribute to improve feed conversion while positively modifying the nutritional profile of eggs.

A phytogenic feed additive improves egg shell quality and tibia breaking strength in 90 weeks old laying hens

L. Jungbauer, A. Mueller and J.D. Van Der Klis
Delacon Biotechnik GmbH, Weissenwolffstrasse 14, Steyregg, Austria; leopold.jungbauer@delacon.com

The objective of this trial was to evaluate the effects of a phytogenic feed additive on egg shell quality and tibia breaking strength in laying hens. A total of 200 ISA Brown laying hens were assigned to 2 dietary treatments. Each treatment consisted of 20 replicates with 5 layers each. Birds were housed in enriched laying cages following EU requirements. Layers of the control group (CON) were fed a diet based on wheat, corn and soybean meal. Dietary Ca level was 35 g/kg during the first two weeks of the trial and reduced to 31 g/kg for the remaining trial period. Birds of treatment group (PFA) were fed the basal diet, supplemented with a blend of essential oils, herbs, spices and saponins at a dosage of 250 mg/kg diet. Production performance and egg breaking strength were recorded weekly (breaking strength was measured for 100 eggs per treatment). After the 9 week trial period (age of the laying hens: 90 weeks), from each repetition 1 bird having approx. mean replicate body weight was selected, killed and the left tibia removed for measuring breaking strength. For the whole trial period the addition of the PFA significantly ($P=0.030$) improved egg shell breaking strength from 4.3 to 4.6 (breaking force in kg). There was no significant difference in egg weight between the groups. Birds fed the phytogenic feed additive had a significantly ($P=0.035$) higher tibia breaking strength compared to the control group (256 N vs 205 N). The results of this study show that the tested phytogenic feed additive improves egg shell quality and tibia breaking strength in old laying hens and therefor improve profitability in egg production.

Effect of natural antimicrobial substances on Gram-negative bacteria and their efficacy in broilers

N. Roth, A. Kovács and A. Tacconi
Biomin Holding GmbH, Erber Campus 1, 3131 Getzersdorf, Austria; nataliya.roth@biomin.net

Outer membrane (OM) of Gram-negative bacteria protects cells from external agents. Permeabilizers (PS) destabilize the OM and increase the outer membrane's permeability to other antimicrobial agents. The purpose of the study was to test the potential of PS (Biomin® Permeabilizing Complex, BIOMIN, Austria) to weaken OM *in vitro* and its efficacy in broilers. An increase in the uptake of a hydrophobic probe, 1-N-phenylnaphthylamine was used to measure permeabilization of OM. The PS was shown to efficiently weaken and destabilize the OM of *Salmonella typhimurium* and *Escherichia coli*. The effect of the mixture (ACPS) of organic acids, cinnamaldehyde and the previously tested *in vitro* PS was evaluated in broilers. Growth performance as well as microbial population in the cecum were determined. A trial was carried out at the Nanjing Agricultural University (China), using 400 one-day-old healthy AA broiler chicks. Birds were assigned to two treatments with 4 replicate pens per treatment and fed commercial poultry diets. The control group diet contained no antimicrobial or natural growth promoters, whereas the trial group was fed ACPS (1 kg/t). The weight, feed intake and mortality of the birds were recorded. At 21 and 42 days of age, 9 birds were randomly selected for cecal sampling. Microbial analysis of samples at 21 and 42 days showed that numbers of *E. coli* and *Salmonella* spp. in the cecum of broilers were reduced ($P<0.05$) in the group fed ACPS in comparison to the negative control. Numbers of *Lactobacillus* spp. in cecum were higher ($P<0.05$) in the trial group than in the control group. Results showed that at 42nd day of trial body weight, average daily gain, feed intake were higher ($P<0.05$) and FCR and mortality were lower ($P<0.05$) in the trial group when compared to the control group. In conclusion ACPS positively influenced the intestinal microbiota and growth of broilers.

Utilization of plant extracts instead of chemical growth promotors and coccidiostats for broilers

F. Recoquillay, S. Kerros and B.D. Cadudal
Phytosynthèse, 57 Avenue Jean Jaures, 63 200 Mozac, France; sylvain.kerros@phytosynthese.com

After the ban of antibiotic growth promoters in 2006, alternatives such as plant extracts were developed to replace them. Consumers are now also asking for fewer chemical additives and this explains why a program using only plants extracts to replace growth promoters and chemical coccidiostats was designed. This trial was implemented to prove that it is possible to achieve the same performance with alternatives to chemical solutions. 13,200 one day old female chicks JA 657 were reared in an outdoor building intil slaughter(80-88 days old).All animals received the same feed and feeding program(Starter feed: ME 3,180 kcal/kg, CP 20.5%, available lysine 1.02%; Grower feed ME 3,120 kcal/kg, CP 15.5%, available lysine 0.75%; Finisher feed: ME 3,200 kcal/kg, CP14.5%, available lysine 0.66%). The trial consisted of 3 replicates of 2 control and 2 experimental groups of 1,100 chicks in the same building. Each group had automatic scales. Control groups were supplemented with Nicarbazin + Narasin (80 ppm) from D1 to D 23, then Monensin 100 ppm from D 24 to D 59. Experimental group received a feed containing a blend of 2 ppm phenols, 2 ppm diterpenoid lactone, 27 ppm phenylpropanoid, 4 ppm diallylsulfides from D1 to D11. Then, from D 11 to slaughter, these values were multiplied by 1,7. All the groups were slaughtered at around 85 days old in line with commercial policy. No clinical signs of coccidiosis were observed in any groups. The final body weight of broilers were similar in both groups (2.006 kg control vs 2.055 kg experimental). The average daily gain (ADG), mesured by automatic scales, was statistically similar (23.1+-0.8 g(C) vs 23.6+-0.7 g(E)). On the same time, the Feed Conversion Ratio was unchanged (2.94+-0.06(C) vs 3.02+-0.26 (E)). Finally, the performances of broilers were very similar. This type of program is very promising for the future, specifically in this context of reduction of antibiotic use.

Feeding influence of Mulberry leaves on performance of laying hens during the finishing period

S.H. Kim[1], H.G. Kang[1], S.B. Park[1], M.S.K. Sarker[2], C.I. Lim[3] and K.S. Ryu[3]
[1]Korean Poultry Research Institute, 321-11, Daeguanryung Maru-Gil, Pyungchang-Gun, 25342, Korea, South, [2]Bangladesh Livestock Research Institute, Poultry Production Research Division, Savar, 1341, Dhaka, Bangladesh, [3]Chonbuk National University, Animal Science, 567, Baekje-Daero, Duckjin-Gu, Jeonju, 54896, Korea, South; seon1288@gmail.com

A feeding trial was conducted to know the effect of dietary Mulberry Leaves (ML) on the performance of laying chicken during finishing period. Three hundred sixty Hyline Brown laying chicken of 74 weeks of age were distributed into four dietary treatments having five replications with 18 birds in each. The duration of the trial was 12 weeks between 74 to 85 weeks old during the summer season (June to August). The dietary groups were control (Basal diet), Basal diet+0.2% ML, Basal diet + 0.4% ML and Basal diet + 0.8% ML and nutrient content was CP 15% and ME 2,750 kcal/kg. The daily egg mass of 0.4% ML with Basal diet fed group bird was higher compared to control diet fed birds and the other two levels of dietary ML. Significantly highest ($P<0.01$) eggshell thickness (0.369 mm) was observed at 0.2% ML addition group compared to control (0.355 mm) and other two levels of ML. Egg shell and yolk color (7.19) were found significantly higher value with reduced serum total cholesterol value (187.82 mg/dl) compared to control (228.88 mg/dl). Serum HDL has increased (19.76 mg/dl) after addition of ML upto 0.4% in comparison to control diet (17.55 mg/dl). It may be concluded that Mulberry leaves can be added in the laying hen diet upto 0.4% at the finishing stage.

Use of a capsicum based additive improved performance of broilers fed diets containing animal fat

M. Blanch[1], G. Tedó[1], M. Francesch[2] and E. Roura[1,3]
[1]Lucta S.A., Innovation Division, UAB Research Park, Edifici Eureka, 08193 Bellaterra, Spain, [2]IRTA, Animal Nutrition and Welfare, Ctra. Reus-El Morell, km 3,8, 43120 Constantí, Spain, [3]Current address: The University of Queensland, Centre for Nutrition and Food Sciences, Queensland Alliance for Agriculture and Food innovation, St. Lucia, QLD 4072, Australia; marta.blanch@lucta.com

A previous study showed that the use of a capsicum oleoresin based additive (Luctarom Convert – LOM) was able to improve dietary fat digestibility in broilers. This study was conducted to evaluate the effect of LOM on growth performance of broiler chickens fed diets differing on its fat source. A total of 1,408 Ross 308 1 d-old sexed broilers (50% males + 50% females) were used until 38 d of age. The experiment was set up as a randomized complete block design with a $2\times2\times2$ factorial arrangement: 2 sexes (males and females), 2 fat sources (soybean oil–SB and animal fat–AF), and without or with LOM (500 ppm from 0-21 d and 250 ppm from 22-38 d). Each combination of factors had 4 replicates of 44 birds/pen. Basal diets differing in fat source were formulated to provide the same nutrient contents (3,100-3,150 kcal/kg, 1.1-1.0% dLys and 0.5-0.45% dMet from 0-21 d and 22-39 d, respectively). Body weight (BW), feed intake (FI), feed conversion ratio (FCR), and mortality wer e recorded at 21 and 38 d of age. Data were subjected to ANOVA. From 0-21 d, males showed significant higher ADG and FI, and lower FCR than females. The use of AF increased FI in comparison with SB (59.4 vs 58.3 g/d; $P<0.01$). Supplementation of LOM increased ADG (43.9 vs 42.8 g/d; $P<0.01$), and decreased FCR on average (1.34 vs 1.37). From 22-38 d and overall, the effect of sex and fat type confirmed the observations of the first period. In addition, a significant interaction between LOM and sex was observed ($P<0.05$) for FI. LOM increased FI ($P<0.05$) but only with males receiving AF diet. In conclusion, the use of LOM showed positive effects in terms of ADG and FCR during the first three weeks of the growing period.

Potential of a phytogenic feed additive against antibiotic growth promoter in broilers under stress

B.A.S. Syed[1] and S. Haldar[2]
[1]Biomin Holding GmbH, Austria, Sprottauer Str. 70, 53117 Bonn, Germany, [2]West Bengal University of Animal and Fishery Sciences, Kolkata, Bengal, India; basharat.syed@biomin.net

A study was conducted to assess performance, humoral immune response (HIR) against Newcastle disease (ND), cecal colonization pattern and fecal shedding of certain bacterial species in broilers under stress due to a real pathogen challenge, supplemented with a phytogenic feed additive (PFA) against an antibiotic growth promoter (AGP). One hundred twenty male day old Cobb 400 broiler chicks were randomly assigned to three treatments, each consisting of 8 replicate cages. Dietary treatments included a corn-soybean based control diet without growth promoter and the treatment diets containing either AGP or PFA. Despite oral challenge with *Salmonella enteritidis* and *Escherichia coli* supplementing broiler diet with PFA improved body weight gain and feed conversion ratio ($P<0.05$) compared to controls with an intermediate performance in the AGP group. HIR against ND was identical across the diets at 7 d, increased with age ($P<0.01$) irrespective of dietary treatments. At 21 d the ND hemagglutination inhibition titer was significantly higher in the AGP and PFA groups compared to control. Cecal population (10^5 cfu/g caecal digesta) at the end of the experiment on day 38 indicated significant reduction of *Salmonella* (18.48, 10.97), *E. coli* (20.17, 13.57) and *Clostridium* (9.97, 10.21) numbers respectively in the AGP and PFA supplemented groups ($P<0.01$) compared to the Control (30.18, 50.79, 15.87). The number of *Lactobacillus* increased in PFA group (25.50) in contrast to Control (10.44) and AGP (10.92) groups ($P<0.01$). *Salmonella* count in feces was lower at 0 h post challenge in the PFA group compared to the Control ($P<0.05$). At 24 h post challenge *Salmonella* and *E. coli* numbers were lower in the AGP and PFA supplemented groups than in the Control ($P<0.01$). Results depict that PFA could serve an effective alternative in broiler rearing even during stress situations.

Impact of drinking water additives on activity of drug metabolizing CYP450 enzymes in chickens

O. Palócz and G. Csikó
University of Veterinary Medicine, Department of Pharmacology and Toxicology, István utca 2., 1078 Budapest, Hungary; palocz.orsolya@univet.hu

During the treatment of poultry flocks there are some factors that need to be considered (e.g. nutrition, disease conditions, and concomitant use of chemicals). These factors may influence the activity of the xenobiotic metabolizing cytochrome P450 (CYP) enzymes. The interactions among the medicines and supplementers may lead to altered activity or safety of them. Sanguinarine (SN) and fulvic acid (FA) are commonly used as drinking water additives for several species of animals. The aim of the study was to determine the possible effect of two feed additives on the activity of major hepatic CYP enzymes in chicken. In the study 20 four-week-old broiler chickens were divided into five equal groups. Besides the control group two groups were administered with 5 and 25 mg/kg bw. SN and the other two groups were treated with 25 and 250 mg/kg bw. FA. The substances dissolved in water were applied to the chickens individually via crop probe for four consecutive days. A day after the last treatment day, the animals were humanely sacrificed. The activity of the hepatic CYP1A, CYP2C and CYP3A enzymes were determined by luminescence measurements. The high dose FA and both doses of SN treatment increased the activity of hepatic CYP1A enzyme. The RLU values were $1,211\pm111$ (mean±SE) in control group; $1,635\pm55$ and $2,623\pm99$ in 25 and 250 mg/kg FA groups; $2,173\pm325$ and $2,131\pm117$ in 5 and 25 mg/kg SN groups, respectively. Neither of the applied substances altered the enzyme activity of CYP2C and CYP3A compared to the values of control group; $7,143\pm895$ and 366 ± 32, respectively. The CYP2C subfamily has significance in drug metabolism of birds, smaller proportion of the active substances decomposed on the CYP3A and CYP1A subfamilies. Since the applied additives have not altered the enzyme activity of the CYP2C, these substances can be safely applied in chickens together with medicines, neither withdrawal time nor dose adjustment is necessary.

The efficiency of resin acid composition in turkeys nutrition

K. Lipiński[1], J. Vuorenmaa[2], Z. Antoszkiewicz[1], J. Kaliniewicz[1], M. Mazur[1] and Z. Makowski[1]
[1]University of Warmia and Mazury, Department of Animal Nutrition and Feed Science, Oczapowskiego 5, 10-718 Olsztyn, Poland, [2]Hankkija Ltd, Peltokuumolantie 4, 05800 Hyvinkää, Finland; krzysztof.lipinski@uwm.edu.pl

A 105-day experiment was conducted on 600 BIG 6 female turkeys to determine the effect of Progres (natural resin acids of Scots pine and Norway spruce, RAs) on the growth performance of birds, incidence of footpad dermatitis (FPD), litter quality, carcass quality, structure and function of the gastrointestinal tract – GIT (pH, viscosity, short-chain fatty acids; SCFAs). The experiment involved 600 turkeys divided into three groups in ten replications in each. The birds from each subgroup (20) were placed in a separate pen. The birds received a basal diet without RAs (T1), or diets supplemented with RAs at 0.5 kg/t (starter 1 and 2, finisher), and at 1.0 kg/t (grower 1 and 2) (T2), or diets supplemented with RAs at 0.75 kg/t (starter 1 and 2, finisher), and at 1.5 kg/t (grower 1 and 2) (T3). The inclusion of the Progres RA preparation in turkey diets tended to increase the body weights of birds by 1.6% (0.5-1.0 kg/t RAs) and 2.1% (0.75-1.5 kg/t RAs); $P \leq 0.10$. Dietary supplementation with RAs significantly improved the values of feed conversion ratio (FCR; $P<0.01$) and European Efficiency Index (EEI; $P<0.05$). The addition of RAs to the diets reduced the incidence of FPD at 9 and 15 weeks of age ($P<0.01$) and improved litter quality ($P<0.01$). The inclusion of the analyzed preparation in turkey diets had no effect on carcass quality, meat quality and most parameters of GIT structure and function. The experiment suggests that Progres (RAs) preparation improves the performance and wellbeing of turkeys.

Effects of functional sensory molecules on growth performance of broiler under two different density

J.F. Gabarrou[1], C.I. Lim[2] and K.S. Ryu[2]
[1]Laboratoires Phode, Z.I. Albipole, 81150 Tersac, France, [2]Chonbuk National University, Jeonju, 54896, Korea, South; jfgabarrou@phode.fr

Increasing stocking density could improve barn productivity, but would also increase stress and decrease growth performance. The important role of olfaction in the improvement of well-being could reduce the impact of stress reaction and therefore improve the birds' performances in stressful situations due to high stocking density. The effect of a sensory feed additive (SFA) based on plant extracts (VéOPremium, Laboratoires PHODE – 250 g/t), was investigated on 1,572 Ross broiler chick during 5 weeks in the CHONBUK experimental station. Birds were randomly distributed to 24 flocks of 51 or 80 birds each (14 birds/m^2 as the low density or 22 birds/m^2 as the high density. Live body weight (LBW), Feed Intake (FI) and Feed Conversion ratio (FCR) were measured weekly. For each treatment, from 10 birds blood was collected. Prestarter, starter and finisher diets in pellet form were fed from 1 to 7 d, 8 to 21 d and 22 to 35 d of age, respectively. Supplementation with SFA decreased more rapidly ($P<0.05$) vitellus sac weight (respectively 0.037 vs 0.072 g at 7 days) and significant ($P<0.05$) affected broilers' performance. Low stocking density resulted in better LBW (respectively 2,093 vs 1,973 g). FI was significantly reduced when reared under higher stocking density (3,049 vs 3,298 g, respectively). Under low density, SFA did nto result in a significant effect, but under high density, SFA allowed a significantly higher body weight (respectively 1,998 vs 1,947 g) and a lower FCR (respectively 1.57 vs 1.59). Serum Interleukin 2 content was significantly increased with addition of SFA (respectively 2.23 vs 1.08 µg/ml). It is concluded that under the conditions of this experiment, dietary addition of 250 g/t SFA significantly improves performance under high stocking density (22 birds/m^2), probably due to modulation of stress and immunity via interleukins.

Effect of qualified functional sensory molecules on laying hen's performances

J.F. Gabarrou[1] and M. Forat[2]
[1]Laboratoires Phode, Z.I. Albipole, 81150 Tersac, France, [2]Instituto Internacional de Investigacion Animal, Queretaro, Mexico; jfgabarrou@phode.fr

Modern egg production sometimes imposes stressful keeping conditions on animals. The role of olfaction is not given enough consideration in the improvement of well-being, although it may reduce stress and by this enhance hen's performance. The effect of a sensory feed additive (SFA) based on plant extracts (VéOPremium, Laboratoires PHODE – 250 g/t), was tested in 1,080 Bovans white hens grown from 14 days of age to 25 weeks at IIIA[(b)] experimental station. They were randomly distributed into 32 flocks of 36 animals each. Live Body Weight (LBW, n=200), Feed Intake (FI, and egg production (n=16) were recorded weekly. For each treatment, blood was collected from 45 birds for determining Gumboro antibody titer after vaccination at days 5 and 30 of age. During the rearing period (105 days), no significant difference was observed between the two groups for LBW (1,094±70 vs 1,095±85 g). The homogeneity of the pullets' LBW from the SFA group was significantly better ($P<0.02$) at transition time. The Gumboro immune response was significantly higher (+30 to +50%, $P<0.002$) over 75 days in the SFA pullets group, suggesting the establishment of a better immune system. During the start of the laying phase (W18 to W25), the laying rate (LR) increased more rapidly in the SFA flocks compared to the control ones. During the two last weeks of the trial (W24-25), the LR measured for the SFA group was significantly higher ($P<0.001$) than for the control group (92.5±4.6 vs 86.0±6.7%), eggs from SFA group were significantly heavier ($P<0.002$) than from the control, 53.6±0.9 vs 50.9±1.2 g. The Feed Conversion Ratio was significantly lower for SFA laying hens compared to control laying hens at the same period, 4.27±0.08 vs 4.64±0.09. It is concluded that SFA generates a positive sensory experience that makes the laying hens better prepared to be transferred to laying cages.

Growth promoting properties of a blend of protected aromatic compounds in growing broilers

S. Peris[1], P. Buttin[1], J. Jankowski[2] and K. Kozłowski[2]
[1]Novus International, Neerveld 101-103, 1200 Bruxelles, Belgium, [2]Faculty of Animal Bioengineering, Department of Poultry Science, University of Warmia and Mazury, 10719 Olstyn, Poland; silvia.peris@novusint.com

Benzoic acid is a flavouring organic acid reported to have antimicrobial activity that can lead to better production performance. The objective of the present study was to evaluate the benefit of a premixture based on a blend of aromatic compounds containing primarily benzoic acid embedded in a vegetable fat matrix (AVM = AVIMATRIX[®]) on production performance of broilers. 1,100 day-old Ross 308 chickens were allocated into 2 treatments of 25 pens with 22 broilers in each. A control diet, based on standard formulation for broilers, was compared with the same formulation with addition of 500 g AVM per ton of feed. The birds demonstrated a good health during the experiment with liveability ranging from 97.3% in control to 98.4% with the AVM treatment. During all growing phases the AVM treatment significantly outperformed the control treatment on both daily weight gain (DWG) and feed conversion ratio (FCR). Respectively from day 1 to 21: DWG 43.4 g vs 42.1 g, ($P<0.001$), FCR 1.256 vs 1.275 ($P=0.048$); from day 22 to 35: DWG 97.4 g vs 94.8 g ($P=0.011$), FCR 1.656 vs 1.689 ($P=0.045$); from day 1 to 35: DWG 63.4 g vs 61.7 g ($P=0.001$), FCR 1.481 vs 1.509, ($P=0.007$). There were no differences ($P>0.05$) in daily feed intake between treatment groups during all experimental phases. Under the conditions of this trial, the addition of this blend of protected aromatic compounds in the feed of broiler significantly improved the weight gain and feed efficiency compared to a control group.

Effect of replacing vitamin E by a blend of plant extracts on laying performace and egg quality

C. Alfonso-Carrillo, T. Van Kempen, F. Sánchez and A.I. García-Ruiz
Trouw Nutrition, Poultry Research Centre, Ctra. CM-4004, Casarrubios del Monte, 45950 Toledo, Spain;
c.alfonso@trouwnutrition.com

A total of 640 laying hens at 75 weeks of age housed in 40 cages were randomly divided into 5 groups to evaluate the effect of replacing vitamin E by Selko®AOmix on performance and egg quality traits. Diets were formulated to provide similar chemical composition only differing in the amount and source of antioxidants as follows: control diet with 10 ppm of vitamin E (Low) and 4 diets supplementing the basal diet with 30 (Medium-E) and 190 ppm (High-E) of vitamin E or with 30 (Medium-AO) and 150 ppm (High-AO) of antioxidant activity equivalent to vitamin E of Selko®AOmix. Layers received experimental diets during a total of 9 weeks, 7 weeks under standard conditions (25 °) and the last 2 weeks where animals were daily submitted to 5 hours heat stress (35 °C). Performance was studied weekly. At the end of both periods, the effect of diet on yolk fatty acids composition and egg quality, in fresh eggs and also in eggs after 28 days of storage was studied. Dietary antioxidant supplementation in standard conditions did not affect performance and egg quality traits. Performance and body weight of hens were impaired under heat stress conditions ($P<0.001$). Antioxidant level was more critical under heat stress conditions, since increasing levels, either with vitamin E or with Selko®AOmix, enhanced laying percentage and egg mass (72.3, 76.2 and 77.1%, and 48.4, 51.1 and 51.5 g/d/hen, as average, for Low, Medium and High, respectively; $P<0.05$). Regardless of antioxidant source, storage time did not affect yolk fatty acids composition, but decreased egg weight (7.6%) and Haugh Units (70.2%) and increased yolk and albumen pH (4.5 and 13.8%, respectively) ($P<0.001$). The results suggest that under standard conditions 10 ppm vitamin E is adequate. Under heat stress, higher levels of antioxidants are needed. And that vitamin E levels above 10 ppm can be safely replaced by Selko®AOmix, providing a higher profitability due to lower feed cost.

Effect of a polyherbal mixture on liver oxidative stress in broilers

J.F. Le Roux[1], C. Alleno[2], S. Olivera[1], R. Jones[1] and H. Borin[1]
[1]Nutritec International Sàrl, La Romanèche 13, 1163 Etoy VD, Switzerland, [2]Zootest, Parc Technologique du Zoopôle, 22440 Ploufragan, France; jf.leroux@nutritec.ch

During intensive growth the liver is potentially exposed to many insults such as xenobiotics and reactive oxygen species. This study was designed to measure the impact of an imposed oxidative stress from oxidized oil and excess ferrous ions during days 20 to 42 of growth in broilers. In total, 180 Ross 308 males were reared in a commercial farm until the age of 20 days, then transported in a specific unit and divided into 3 groups (A, B and C) of 6 cages of 10 broilers and kept until the age of 42 days. Individual weights were recorded at age of 20, 31 and 42 days, feed consumption per cage was measured at days 31 and 42, blood analysis and autopsies were performed at 42 days. From 20 to 31 days Group A received a control mash feed, Group B received the same feed supplemented with 10% oxidized soybean oil plus an excess of 0.5% of ferrous sulphate and Group C was fed same as B plus 250 g/T of polyherbal mixture (Livoliv). During the recovery period from 32 to 42 days, Group A and Group B received the control feed group and Group C the control feed with 250 g/T of Livoliv. Average starting body weight at 20 days was 856 g. Oxidative stress reduced final body weight non significantly at 31 days (1,796 g vs 1,854 g) and affected feed conversion non significantly (1.65 vs 1.62). Livoliv significantly increased body weight at 31 days, (1,886 g) and significantly improved feed conversion (1.49). After recovery, body weights at 42 days were 2,899, 2,871 and 2,923 g and feed conversion 1.83, 1.85 and 1.80 for Groups A, B and C, respectively, with insignificant differences. Liver weight as percent body weight showed no difference with 1.86, 1.84 and 1.80% for groups A, B and C, respectively. Alkaline phosphatase levels were 3,752, 5,306 and 4,934 UI/l for groups A, B and C, respectively. High levels of oxidized soybean oil and ferrous sulfate induced a reduction in zootechnical performance that 250 g/T of Livoliv compensates and even improves.

Tea polyphenols improved albumen quality and magnum morphology of aged laying hens

X.C. Wang, X.H. Wang, S.G. Wu, G.H. Qi, H.J. Zhang and J. Wang
Feed Research Institute, Chinese Academy of Agricultural Sciences, 12 Zhongguancun Nandajie, Haidian, Beijing 100081, China, P.R.; wushugeng@caas.cn

Albumen quality is one of great importance for customer performance. Laying hens in the late phase of production are under great oxidative stress, and albumen quality deteriorates as a result. Tea polyphenols, as a natural antioxidant, can scavenge reactive oxygen species. The objective of this study was to investigate the effects of dietary tea polyphenols supplementation on egg performance, albumen quality, albumen oxidation parameters, magnum morphology of aged laying hens. A total of 180 laying birds of 64 weeks of age were allotted into 2 treatments groups with 6 replicates of 15 birds each. The control group was fed corn-soybean meal diet (control), and the experiment diet with 200 mg/kg of tea polyphenols (TP200). The experimental diets were isocaloric (ME, 11.20 MJ/kg) and isonitrogenous (CP, 17.04%). The feeding trial lasted 10 wks. The results show that TP200 diet increases the egg production and feed conversion ratio from wk 5 to 10 ($P<0.05$). A higher albumen height and Haugh unit were observed in the TP200 group at wk 6, 8 and 10 ($P<0.05$). Compared to the control group, TP200 decreased protein carbonyl content and surface hydrophobicity in egg white, while sulphydryl content was increased at the end of feeding trial ($P<0.05$). The gland height, epithelial height and ciliated height of the magnum were higher in the TP200 group than that in the control group at the end of the trial ($P<0.05$), while the gland width of magnum had a trend to increase ($P=0.054$). Together, our results suggest that a supplement of 200 mg/kg TP may produce a positive effect on egg performance and albumen quality of aged laying hens. The positive effect might be associated with the inhibition of albumen oxidative damages and improved of secretion capacity in the magnum.

Oleuropein suppresses oxidative damage in avian skeletal muscle, possibly via SIRT1 upregulation

M. Kikusato, R. Shimao, H. Muroi and M. Toyomizu
Tohoku University, Graduate School of Agricultural Science, Aramaki-Azaaoba 468-1, Sendai, 980-0845, Japan; kikusato.m@tohoku.ac.jp

Oleuropein (Ole) is a major phenolic compound in *Olea europaea*. We previously showed that Ole induces the expression of avUCP and SOD2 genes, thereby reducing mitochondrial ROS generation in cultured avian muscle cells. The present study first evaluated the effects of Ole ingestion on muscle oxidative damage in broiler chickens. Thereafter, we investigated *PGC-1α*, *FoxO3*, and *SIRT1* involvement in *avUCP* and *SOD2* expression using a cell culture study to clarify the possible mechanism up-regulating the transcription of ROS-reducing genes. Animal study: Seven-day-old male chickens were fed either a diet supplemented with 0, 0.1, 0.5, or 2.5 ppm Ole for 2 weeks (n=8). Birds were killed at 22 days of age, *Pectoralis superficialis* muscle was collected, and the carbonyl protein content and mRNA levels of *avUCP*, *SOD2*, and *PGC-1α* were determined. Cell culture study: Isolated muscle cells were incubated in medium supplemented with 50 ppm Ole in combination with EX-527, a SIRT1 inhibitor. Thereafter, cells were collected to measure mRNA levels of the above genes as well as *FoxO3a* and *SIRT1*. Muscle protein carbonyl content was lower as a result of feeding Ole ($P<0.05$), and the Ole-0.5 group had higher muscle *avUCP*, *SOD2* and *PGC-1α* mRNA levels than did the control group ($P<0.05$). Cultured muscle cells supplemented with Ole exhibited higher mRNA levels of the above three genes as well as those of *FoxO3a* and *SIRT1* ($P<0.05$), each of which was restored to near-normal values by co-treatment with EX-527. The study demonstrated that Ole induces *avUCP* and *SOD2* genes expression, which may contribute to suppressing muscle oxidative damage. It further suggested that the ROS-reducing genes might be up-regulated by SIRT1-mediated FoxO3/PGC-1α transcriptional activation in avian muscle cells.

Effects of polyphenols in heat stressed broilers

K. Männer[1], H. Sauerwein[2], E. Von Heimendahl[3], J. Bartelt[3] and T. Diedenhofen[3]
[1]TU Berlin, Königin-Luise-Straße 49, 14195 Berlin, Germany, [2]University Bonn, Katzenburgweg 7-9, 53115 Bonn,
Germany, [3]Kaesler Nutrition, Zeppelinstrasse 3, 27472 Cuxhaven, Germany; elke.von.heimendahl@kaesler.de

Heat stress results in decreased performance and increased morbidity. Polyphenols may positively influence oxidative and immune status of animals, thus increasing performance. 180 1-day old broiler chicks were allocated to 3 treatments with 4 repetitions of 15 birds. Treatments were T1-control; T2-control plus heat stress (HS); T3-control plus HS and addition of a polyphenol product (PP) with 300 mg/kg feed. Birds were fed a starter from day 1-14 and a grower from day 15-42 according to GfE guidelines. Birds of T2 and T3 were exposed to acute HS (30 °C) from day 22-42. At day 42, birds were sacrificed and blood from *V. cutanea* and tissue samples from breast muscle were taken from 12 birds per treatment for further analyses. Already before HS period, PP tended to improve weight gain (+10%; 118 vs 107 g/d, T3 vs T1; $P=0.056$) from day 1-7. FCR was improved from day 8-14 (-7.1%; 1.25 vs 1.36 g/day, T3 vs T1; $P=0.018$) compared to T1. HS reduced body weight gain in T2 from day 22-28 compared to T1 (-35.3%; 388.4 vs 600.5 g; $P<0.001$). Feeding the PP numerically increased weight gain under heat stress (5.1%; 409 vs 388 g T3 vs T2). PP also resulted in improved oxidative status, indicated by reduced serum malonyl dialdehyde (1.48 nmol vs 1.58 nmol/l, $P=0.009$), increased superoxide dismutase (26.8 vs 23.2 U/ml, $P<0.001$) compared to HS without PP. It significantly reduced muscle advanced oxidized protein products and protein carbonyls indicating less tissue damage. Supplementation of PP tended to decrease Alanin-Aminotransferase (0.12 vs 0.14 µkat/l, $P=0.096$) and reduced Interleukin 6 (195 vs 266 pg/ml, $P=0.005$) compared to HS group. Results indicate that adding the PP product supports broilers during periods of HS.

Effect of a natural and a synthetic antioxidant on reproductive traits in Japanese quail breeders

M. Luna[1], F. Prado[2], R. Riveros[1] and C. Vilchez[1]
[1]Universidad Nacional Agraria La Molina, Nutrición Animal, Av. La Molina s/n La Molina, Lima 12, Peru, [2]Universidade
Estadual Paulista, Zootecnia, Via de Acesso Prof. Paulo Donato Castellane s/n, Jaboticabal, SP, 14870, Brazil;
faprado91@gmail.com

Different natural and synthetic antioxidants are available in the market to guarantee the quality of feed and to improve performance of birds by defending them against toxic products such as reactive oxygen species. Thus, the current experiment was conducted to compare turmeric powder (Curcuma longa) with a synthetic antioxidant on reproductive traits of Japanese quail breeders. A total of 972 female and 162 male Japanese quail breeders were placed in nine cages. Birds were fed three isocaloric (2.82 Mcal/kg) and isoprotein (21%) dietary treatments that consisted on: D1, a standard diet(SD) without supplementation of any antioxidant; D2, SD + 0.10% of synthetic antioxidant composed by ethoxyquin, butylated hydroxyanisole and other compounds, and D3, SD + 0.10% of turmeric powder as natural antioxidant during 70 days. A total of 11,880 eggs were incubated in five successive batches. Within each batch, metal plates with 60 eggs per treatment were divided into three replicates of 20 eggs each. Reproductive traits such as fertility (FERT), hatchability of fertile eggs (HFE) and hatchability of incubated eggs (HIE) were measured and expressed in percentage. Data collected were analyzed by PROC MIXED procedure of SAS under a randomized block design. Means were compared using Tukey-Kramer test. Supplementation of diets with turmeric powder or the synthetic antioxidant did not improve FERT ($P>0.05$). HFE and HIE from Japanese quail breeder fed diets supplemented with turmeric powder had better performance than those fed SD and D2 ($P<0.05$). In conclusion, turmeric powder showed beneficial properties that enhanced hatchability of both fertile and total incubated eggs.

Effect of rosehip addition to PUFA Ω3-high layer diets on hen performance and egg quality

P.A. Vlaicu[1,2], D. Dragotoiu[1], T.D. Panaite[2], A. Untea[2], M. Mitoiu[3] and M. Saracila[2]
[1]University of Agronomic Sciences and Veterinary Medicine of Bucharest, 59 Marasti Blvd, District 1, Bucharest, Romania, 011464, Romania, [2]National Institute for Biology and Animal Nutrition (IBNA), 1 Calea Bucuresti, Balotesti, Ilfov, Romania, 077015, Romania, [3]Institute of Biology Bucharest, Splaiul Independentei, No. 296, Bucharest, 060031, Romania; alexandru.vlaicu@outlook.com

A four weeks experiment was performed with 114 TETRA SL layers (29 weeks) assigned to 3 groups (C, E1 and E2) with 38 layers/group. The layers were housed in an experimental house with 16 h light regimen. The basal diet (C), with corn, rice bran, soybean meal and rapeseed meal as main ingredients, had 18.39% CP/kg and 15.904 MJ gross energy/kg. Flax meal (7%) was included in E1 and E2 diets to increase the polyunsaturated fatty acids (PUFA) level of the compound feeds. E2 diet was supplemented with 54 mg vitamin E/kg feed and 1.5% rosehip powder as antioxidants, while E1 diet was supplemented only with 3% rosehip powder. The average daily feed intake/layer was lower ($P \leq 0.05$) in groups E1 (115.41 g/day) and E2 (116.61 g/day), compared to group C (118.82 g/day). The average egg weight was significantly ($P \leq 0.05$) higher in groups E1 (60.18±0.203 g) and E2 (60.54±0.184 g), compared to group C (60.04±0.228 g). Egg quality was determined on eggs collected during the final experimental week. After the physical measurements on eggs (18 eggs/group), 6 yolk samples (3 eggs/sample)/group were formed for the antioxidant capacity determination. The Haugh unit values were higher in groups E1 (91.33±11.19) and E2 (90.53±13.71), compared to group C (84.49±12.68), but the difference was not statistically significant. No difference was observed in antioxidant capacity of the egg yolk between E1 and E2. However, both E1 and E2 had significantly higher ($P \leq 0.05$) antioxidant capacity than C (50.09±1.305 mM Trolox/g).

Effects of antioxidants in feed in heat-stressed broilers

T. Diedenhofen[1], H. Sauerwein[2], K. Männer[3], E. Von Heimendahl[1] and J. Bartelt[1]
[1]Kaesler Nutrition, Zeppelinstr. 3, 27472 Cuxhaven, Germany, [2]University Bonn, Katzenburgweg 7, 53115 Bonn, Germany, [3]FU Berlin, Königin-Luise-Str.49, 14195 Berlin, Germany; tanja.diedenhofen@kaesler.de

High performing animals are affected by various exogenous and endogenous stressors which may result in oxidative stress which is defined as an excess of free radicals and/or insufficient protection through antioxidants. Antioxidants as technological additives in feed have been demonstrated to prevent oxidative stress in animals, consequently leading to increased performance. 180 day-old male broiler chicks (Cobb 500) were randomly allocated to 3 treatments with 4 repetitions of 15 birds each. Treatments were NC = control; HS = control + heat stress; AO = HS + 300 mg antioxidant mixture per kg feed (Loxidan PG2+; Kaesler Nutrition GmbH, Germany). From day (d) 1-14 a starter diet, and from d 15-42 a grower diet was fed according to GfE recommendations. Chicken of HS and AO were exposed to heat stress (30 °C) from d 22-42. At d 42 the birds were slaughtered, blood samples from *V. cutanea* and breast muscle were taken from 12 birds per treatment for determining haematological and biochemical parameters. Heat stress resulted in significantly reduced feed intake and growth. Birds receiving AO showed improved performance before and under heat stress (daily weight gain without heat stress d 1-21 + 6.6% AO vs NC; with heat stress d 22-41 + 9.5% AO vs HS). AO significantly increased superoxide dismutase (SOD, AO 254 vs HS 232 U/dl, $P < 0.001$) and erythrocytes (AO 1.89 vs HS 1.76 T/l, $P = 0.005$) and numerally reduced leukocytes. The highest glutathion-peroxidase activity (AO > NC >> HS) was measured in breast muscle tissue in AO and thiobarbituric acid reactive substances (TBARS, AO 322 vs HS 426 mmol/g tissue, $P < 0.05$) were less than in HS. The results indicate that AO in feed prevented oxidative stress and its detrimental effects in broilers. Heat stress which can be quantified may be indicative for other stressors affecting animal health.

Impact of dietary omega-3 source and *Thymus vulgaris* extract on yolk fatty acid characteristics

M. Kamely and M. Karimi Torshizi
Tarbiat Modares University, Poultry Science, 14115-336 Tehran, Iran; m.kamely@gmail.com

Consumers' awareness of the health benefits of n-3 fatty acids is growing and is driving their demand for enriched food products. In order to develop a diet to obtain eggs enriched in omega-3 fatty acids with minimum disadvantages we conducted an experiment. In this study, effects of fish oil (FO) as omega-3 fatty acids source and thyme extract (TE) as a natural antioxidant supplementation in layer hen's diet, on egg yolk fatty acid profiles and yolk lipid oxidation were investigated. A 10 week trial was performed with one hundred, 58 weeks old White Leghorn laying hens distributed randomly to 5 experimental groups with 4 replications (5 birds each): (1) Control diet free from FO and TE; (2) 1.5% FO; (3) 1.5% FO + 300 ppm TE; (4) 3% FO; and (5) 3% FO + 300 ppm TE. Omega 3 fatty acids percentage increased with increasing FO level in the diet, while the n-6 percentage in yolk decreased by supplementing of FO to the diet ($P<0.01$). The n-6:n-3 ratios also declined from 25.96 in the control group to 9.18 and 2.13 by addition 1.5% and 3% FO, respectively. Thyme extract improved n-3 absorption and decreased n-6:n-3 ratio if compared to the same concentration of FO groups without thyme. Also EPA and DHA increased significantly ($P<0.01$) with increasing in dietary FO level ($P<0.05$), however, DHA enhancement was greater than EP. There were no significant differences on yolk color, but significant differences in sensory attributes for boiled egg samples were found between dietary treatments. The results showed that flavor quality decreased by addition of FO, however thyme extract improved that defects. Addition of 3% FO induced fishy aroma in boiled eggs ($P<0.05$) and declined overall acceptability. The main off-flavor and aroma of eggs produced by feeding 3% FO diets to hens has been described as 'fishy' by panelists.

The effects of dietary PUFA n-6/n-3 ratio, vitamin E and Se on egg production and quality in layers

A.S. Kavtarashvili, I.L. Stefanova, V.S. Svitkin and E.N. Andrianova
All-Russian Research and Technological Poultry Institute, Sergiev Posad, 141311, Russian Federation; andrianova@vnitip.ru

The influence of 6 different diets on productive performance and egg quality in White Leghorn layers was studied during 60 days with 6 replicates of 5 birds in separate cages per treatment; egg production and FCR were recorded for replicates, egg weight and composition were statistically analyzed using an egg as experimental unit. Control wheat-based diet 1 contained 3% of sunflower oil; concentrations of n-6 and n-3 PUFAs were 3.62 and 0.14% respectively, n-6/n-3 ratio 25.9:1, vitamin E 10 ppm, selenium (Se) 0.2 ppm. In diets 2 and 3 sunflower oil was substituted for flaxseed oil (3%) and 5% of flaxseed cake was added; concentrations of n-6 and n-3 PUFAs were 2.09 and 1.97%, n-6/n-3 ratio 1.06:1, vitamin E 100 ppm, Se 0.5 ppm. Diets 4, 5, and 6 were supplemented with 3% of flaxseed oil and 10% of flaxseed cake; concentrations of n-6 and n-3 PUFAs were 1.82 and 2.35%, n-6/n-3 ratio 0.77:1, vitamin E 150 ppm, Se 0.5 ppm. Se source in diets 2 and 4 was Sel-Plex, in diets 3 and 5 'DAFS-25' (Russia), in diets 1 and 6 sodium selenite. The egg production in treatments 1-6 was 47.5; 49.2; 49.0; 47.8; 49.3 and 49.4 eggs respectively; average egg weight 56.2; 58.7; 55.4; 55.8; 55.0 and 55.9 g; egg mass output per layer 2.68; 2.90; 2.72; 2.68; 2.72 and 2.76 kg; FCR 2.45; 2.22; 2.35; 2.31; 2.31 and 2.31 kg per 1 kg of egg mass; Se deposition in eggs per 100 g of egg contents 28.3; 61.6; 58.2; 61.8; 58.9 and 43.1 µg; vitamin E deposition 2.31; 4.47; 5.04; 6.40; 8.35 and 5.50 µg; total n-3 PUFA deposition 172, 767, 767, 807, 767 and 796 mg; n-6/n-3 PUFA ratio 12,7; 2,1; 2,0; 1,9; 1,7 and 1,8:1. Se content in eggs of treatments 2-4 was 1.5-2.2 times higher compared to control, vitamin E content 1.9-3.6 times higher, n-3 PUFAs 4.5-4.7 times (including α-linolenic acid 7.5-8.4 times, EPA 1.8-2.4 times, DPA 3.2-3.4 times, DHA 2.3-2.7 times higher compared to control). The study was financed by Russian Science Foundation, grant 16-16-04047.

Effect of Se and oils on laying hens productivity, egg quality parameters and content of Se

V. Buckiuniene, R. Gruzauskas and S. Bliznikas
Lithuanian University of Health Sciences, Veterinary Academy, Institute of Animal Rearing Technologies, Tilzes str. 18, 47181 Kaunas, Lithuania; vilija.buckiuniene@lsmuni.lt

In this study, effects of using different oil sources, organic and inorganic Se and vitamin E on laying hens productivity, egg quality parameters, Se and vitamin E accumulation. A total of 40 laying hens which were 22 weeks old were assigned to four treatment groups and fed with the experimental diets for 8 weeks. The trial was set up as a randomized complete block design with 2×2 factorial arrangement: 2 added oils (sunflower S and linseed L) and 2 sources of Se+ vit. E (organic O and inorganic I) and then four dietary treatments were used SO (sunflower oil+organic Se and vit.E), LO (linseed oil+organic Se+vit.E), SI (sunflower oil+inorganic Se+vit.E), LI (linseed oil+inorganic Se+vit.E). The laying hens were fed with compound feed 125 g per day. Daily were calculated and weighed all the eggs, every 14 days weighed and calculated feed conversion ratio, egg production intensity and egg numbers dynamics. The egg quality parameters were established by multifunctional automatic egg characteristics analyzer 'Egg Multi-Tester EMT-5200', hardness of eggshell – by 'Egg Shell Force Gauge MODEL–II' device, and thickness of eggshell – by electronic micrometer 'MITUTOYO'. Se and vitamin E concentration in egg yolk were determined by atomic absorption spectrometry, AOAC Official Method 975.43 method. The egg yolk colour intensity increased 33 and 21% ($P<0.05$) in SI and LI compared to the SO and LO treatments group. Se concentration in egg yolk decreased 25 and 12% ($P<0.05$) in SO and LO compared to the SI and LI groups. a-tocopherol increased 10 ($P<0.05$) in SI, and decreased 12% ($P<0.05$) LI compared to the SO and LO respectively. When analysed y-tocopherol concentration it's increased 25 in SI compared to the SO ($P<0.05$). The results of this study clearly demonstrate that inorganic Se and different oils had effect on vitamin E accumulation but had negative effect on Se concentration in egg yolk.

The effect of a combination of vitamin E and polyphenols on the meat quality of chicken

K. Bébin, D. Gardan-Salmon and M. Panheleux-Lebastard
Groupe CCPA, ZA du Bois de Teillay, 35150 Janzé, France; k.bebin@groupe-ccpa.com

The consumption of poultry meat is increasingly turning to cut-up and further processed products. Thus improving conservation and meat processing capacity are major concerns of investigation to better meet the market need. In this study, we investigated the effect of a combination of vitamin E (or alpha-tocopherol acetate) + polyphenols (G3) compared to a negative control with no added anti-ocydant (G1) and a positive control with vitamin E (G2) on oxidation aptitude and dripr losses of chicken meat. Each specific diet was distributed to 4 pens of 36 chickens Ross PM3 (144 chickens/diet) from 21 to 42 days. The start-up (0-10 d) and growing (10-21 d) diets were similar in the 3 groups. At 42 days, 20 chickens per group were slaughtered for meat quality measurements. Meat oxidation was lowered in the presence of vitamin E (G2) and the combination with polyphenols (G3) ($P<0.001$). TBARS analysis in the leg was significant improved in groups G2 and G3. Meat coloration (b note, Minolta) was higher for the group with polyphenol combination (G3) compared to G1 and G2 (respectively 18 vs 16.2 and 16.1; $P<0.05$). Meat technological qualities were assessed by drip loss measures, before and after cooking in the breasts. Both criteria were lower in G3 compared to the other groups. In conclusion, the combination of different antioxidant molecules (vitamin E and polyphenols) improved meat appearance and conservation (color and oxidation) as well as its technological quality (water loss). Taking prices of summer 2016, the supplementation cost was (base on the 100 scale): G1 = 0, G2 = 100 and G3 = 74. The return on investment based in the sale of filets (less water loss) was estimated to 10:1.

Effect of vitamin D supplementation on performance, vitamin D status and mineralization in broilers

P. Sakkas, T.R. Hill, S. Smith and I. Kyriazakis
Newcastle University, AFRD, King's rd, NE1 7RU, Newcastle on Tyne, United Kingdom; panagiotis.sakkas@ncl.ac.uk

The hypothesis was set up that performance and bone mineralization of two commercial lines will benefit from increasing vit D supplementation above commercial levels and by partial substitution of D3 by 25-OH-D3 (25D3). Male Ross 308 and 708 chicks, were reared separately in 48 pens, with 12 birds/pen. Birds were allocated to diets offering either low (LD; 1000), medium (MD, 4,000) or high levels of D3 (HD; 7,000) and medium levels with the majority of D3 substituted by 25D3 (25MD; 1000 D3+3,000 25D3 IU/kg). Performance was recorded at the end of the starter (d10) grower (d24) and finisher periods (d38). In addition, 3 birds per pen were dissected at the end of each period to assess femur ash percentage (PA) and serum levels of 25D3. Birds were gait scored (GS) at 37 days of age. Data were analysed with GLM using genotype, diet and their interaction for each of the three periods. Genotype and diet did not significantly interact for any of the variables. Breed 708 had significantly lower ADG across stages ($P<0.05$), higher FCR over the grower period ($P<0.05$), similar levels of 25D3 ($P>0.01$) but higher GS ($P<0.05$) than 308 broilers. Performance was not affected by diet. 25D3 levels were significantly affected by diet at the end of the starter and grower periods ($P<0.05$) being the lowest for LD and the highest for 25MD birds. Diet affected GS ($P<0.01$) being higher in LD than in 25MD ($P<0.05$). Femur PA was significantly affected by diet at the end of the starter and grower periods ($P<0.05$); it was significantly higher for 25MD than for LD birds ($P<0.05$) and for both HD and 25MD than for LD birds ($P<0.05$). In conclusion, differences in gait score between genotypes seem unrelated to the degree of bone mineralization and subtle differences in early performance. Offering 1000 IU/kg of feed impaired gait score, reduced mineralization, and vitamin D status with no effects on performance. Results do not suggest to supplement the level of vitamin D above the current maximum levels.

The effect of 1,25-dihydroxyvtamin D3 on chicken osteoblast differentiation *in vitro*

C. Chen and W.K. Kim
The University of Georgia, Poultry Science Department, 110 Cedar St, Athens, GA, 30602, USA; cxchen@uga.edu

In our previous study, the effects of 1, 25-dihydroxyvitamin D_3 (1,25D) on mRNA expression of key osteogenic genes *Col1a2, BMP2, Runx2, BSP* and *BGLAP*) of chicken osteoblast were tested at various time points until 7 d. However, the expression of *ALP* that codes alkaline phosphatase was missing from the study. Since synthesis of high level of *ALP* was a hallmark of bone formation and it would be important to understand the effect of 1,25D on osteoblast at a late cell stage. A study was conducted to explore the effect of 1,25D on *ALP* expression during osteogenesis and osteoblast differentiation at a late cell stage(14 d). Mesenchymal stem cells were derived from 2-day-old broiler bones and treated with control media (C), osteogenesis media (OM), OM with 1, 5, 10 and 50 nM 1,25D, respectively. The mRNA samples were obtained at 24 h, 48 h, 3 d, 7 d and 14 d. The expression of *Col1a2, BMP2, Runx2, BSP* and *BGLAP* at 14 d and *ALP* at all time points were tested. As for the early periods of differentiation (24 and 48 h), 1 and 5 nM 1,25D inhibited expression of *ALP* at 24 h and all treatments suppressed expression of *ALP* at 48 h compared with OM. As for the mid periods of differentiation (3 and 7 d), 10 and 50 nM 1,25D increased *ALP* expression at 3 d. At 7 d, 50 nM treatment increased ALP expression, but 5 and 10 nM treatments suppressed *ALP* expression. As for late stage (14 d), the expression of these markers became more interesting. 5 nM 1,25D increased *Col1a2* expression but decreased *BMP2* expression. 1 and 10 nM 1,25D decreased expressing of *Runx2* or *BGLAP*. Furthermore, all treatments decreased *ALP* expression. In summary, similar to the previous study, 1,25D mainly showed inhibitory effect on *ALP* expression during early and late differentiation stages but a stimulatory effect during mid stages. These results reconfirmed the effect of 1,25D on osteoblasts varied at different cell stages of osteoblasts and showed the complexity of 1,25D function on chicken osteoblasts and highlighted the idea of the time dependent effects.

Water solution to palliate heat stress effect on broiler performance and blood stress parameters

A. Saiz[1], A.I. García-Ruiz[1] and J. Mica[2]
[1]Trouw Nutrition R&D Poultry Research Centre, Carretera CM-4004 km 10.5, 45950, Casarrubios del Monte, Spain, [2]Trouw Nutrition R&D Registration and Patents, Veerstraat 38, 5830 AE, Boxmeer, the Netherlands; a.saiz.b@trouwnutrition.com

The objective of this study was to measure the effect of the supplement Farm-O-San AHS on performance (body weight, weight gain, feed intake, feed conversion rate and mortality) and blood stress markers (corticosterone and TBARS) of animals under heat stress conditions (35 °C during 4 hours per day). The study was executed at the Trouw Nutrition R&D Poultry research centre. The trial consisted of 4 treatments, which varied in the type and level of product added to drinking water: no product, 1 kg/1000 l Farm-O-San AHS, 2 kg/1000 l Farm-O-San AHS or 200 g/1000 l Farm-O-San Vitamin C. One thousand twenty four male Ross 308 broiler chickens were used. Performance (body weight, feed intake, mortality) was monitored from 0 to 36 days of life. Heat stress (35°) was applied during 4 hours/d throughout the entire experimental period. The remainder of the day animals were held under standard temperature. Animals which consumed the highest level of Farm-O-San AHS or Vitamin C showed higher weight gain, or feed intake ($P<0.01$) than the control animals. The lowest feed conversion rate was found in animals drinking the highest level of Farm-O-San AHS ($P=0.03$). Corticosterone levels were reduced ($P=0.02$) by the addition of Farm-O-San AHS. TBAR levels were not affected by the addition of the product. Farm-O-San AHS has a positive effect on broiler chickens performance under heat stress conditions, improving animal's growth and feed conversion.

Effects of a feed complex fed via drinking water on performance and health in heat-stressed broilers

K. Männer[1], H. Sauerwein[2], J. Yang[3], J. Bartelt[3] and E. Von Heimendahl[3]
[1]FU Berlin, Königin-Luise-Str.49, 14195 Berlin, Germany, [2]University Bonn, Katzenburgweg 7, 53115 Bonn, Germany, [3]Kaesler Nutrition, Zeppelinstr. 3, 27472 Cuxhaven, Germany; joohee.yang@kaesler.de

High temperatures have detrimental effects on broiler health and performance. Several feed additives may mitigate the adverse effects of heat stress (HS) via different modes of action in poultry. However, feed intake often decreases under HS whereas water intake does not. In the present study, the effect of providing a liquid supplementation (LS) containing L-carnitine, choline, and betaine via drinking water on performance and health in heat-stressed broilers was evaluated. In total 180 day (d) -old male broiler chicks were allocated to 3 experimental groups. Animals were fed ad libitum with a starter from d 1-14 and a grower diet from d 15-42. Drinking water was supplied ad libitum. From d 22-42, one group (T1) was kept at constant ambient temperature (23 °C) whereas T2 and T3 were exposed to HS (30 °C). T3 received in addition LS via drinking water during this period. At d 42, the animals were slaughtered and growth performance, as well as several blood and tissue (breast muscle) variables were assessed. The highest body weight was observed in T1 (d 42). Even though HS affected growth performance, feeding LS during the HS period (T3) improved total body weight when compared with T2 (2.12 vs 2.01 kg; $P<0.001$). T2 broilers had less erythrocytes (1.8 vs 1.9 T/l; $P=0.01$) and lymphocytes (31 vs 44%; $P<0003$) than the T1 animals, whereas T3 tended to have higher values of both (1.9 T/l and 33%) compared to T2. In T3 greater superoxide dismutase concentrations in blood (25.2 vs 23.2 U/ml; $P<0001$) and decreased levels of thiobarbituric acid reactive substances in breast muscle (366 vs 426 mmol/g tissue; $P<0.05$) were observed when compared with T2. These results suggest that a just-in-time supplementation which is easily possible via drinking water can counteract HS in broilers resulting in improved growth performance and oxidative defense.

Effects of a slow released betaine based product *in vitro* and in heat-stressed broiler chickens

S. Klein, J. Castier, M. Magnin and N. Brévault
Mixscience, Centre d'Affaires Odyssée, Cicé Blossac, BP CS 17228, 35172 Bruz, France; stephanie.klein@mixscience.eu

Supplementing broiler diets with betaine can improve performances, related to methyl donor and osmoprotective properties. Anhydrous betaine is especially interesting under heat stress. Nevertheless, its hygroscopicity makes it difficult to manipulate in feed plants. This study aims at validating the interest of a fat coated form of anhydrous betaine to improve its technological behavior, but also its bioefficiency. First, *in vitro* characteristics have been compared between free and coated betaine. In an artificial gastric juice, more than 95% of free betaine but only 44% of coated betaine are solubilised after 45 minutes; in intestinal juice, 84% of coated betaine is released after 6 hours. *In vivo* effects of coated betaine in heat stressed broilers have been investigated, as well. Day old male Ross PM3 chicks (n=160) allocated in 40 cages, with 4 birds per cage and 6 to 7 replicates per group, received one of the following diets: negative control (NC) without betaine, NC + 250, 500 or 1000 mg/kg anhydrous betaine (respectively AN250, AN500, AN1000), NC + 250 or 500 mg/kg coated anhydrous betaine (respectively BC250 and BC500). High temperature up to 30 °C was applied between 21 and 34 days. Heat stress induced variability of performance, but average daily gain (ADG) and feed conversion ratio (FCR) tended to be improved for all betaine groups compared to NC, especially during heat stress phase (+2.1 to +6.0 g/d ADG and up to -0.06 point FCR between 21 and 34 days). Finally, the best weight corrected FCR was observed for BC500 (-0.05 point compared to NC). Blood analysis in groups NC, AN500 and BC500 showed a reduction of creatine kinase and glutamic-pyruvic transaminase for both betaine forms, which may indicate a reduction of cellular injury related to hyperthermia, and an increase of glutathion peroxidase. This study confirmed the interest of anhydrous betaine for heat stressed broilers and showed that slow released coated betaine could be more efficient than free betaine.

Egg yolk pigmentation efficacy of apo-ester (10%) and a highly concentrated marigold extract (10%)

M. Umar Faruk[1], P. Jenn[1], J. Schierle[2] and F. Cisneros[2]
[1]DSM Nutritional Products, 1 Blvd d'Alsace, 68128, Village-Neuf, France, [2]DSM Nutritional Products, Wurmisweg, 4303 Kaiseraugst, Switzerland; murtala.umar-faruk@dsm.com

This 3-week egg yolk pigmentation trial compared the pigmenting efficiency of Apo-ester 10% (APE) and a highly concentrated marigold extracts 10% (LZ). 54-weeks old Lohmann Brown laying hens were fed with diets supplemented with 10, 20, 40 or 80 mg/kg diet of either APE or LZ in a completely randomized design. The control group was fed a common basal diet low (<5 mg/kg feed) in native xanthophylls. Laying performance and feed intake were measured. Carotenoid content of feed samples and egg yolks were determined by HPLC. Egg yolk colour was assessed using the YolkFan. No differences in daily feed intake, egg production and egg weight between the two carotenoids was observed. Measured egg carotenoids content and egg yolk color were four times higher with APE (y=0.0639x) and (y=-0.0012x+0.1566x) compared to LZ (y=0.0163x) and (y=-0.0012x^2+0.1676x) respectively. These results confirmed the higher efficacy of APE based products compared to LZ based products, irrespective of the LZ product concentration.

Influence of natural colorants on skin pigmentation and serum carotenoids of meat-type chicken

O.O. Adeleye and E.T. Oginni
University of Ibadan, Agric. Biochem. & Nut. Unit, Animal Science Dept., 200842, Nigeria; ooadeleye@rocketmail.com

The effect of natural colorants on live and post slaughter pigmentation, and serum carotenoids of meat-type chicken was investigated. One hundred and twenty-eight 1-day old meat-type chicks were allotted to four treatments, four replicates of eight chicks each. Chicks were fed a starter diet for 21 days and thereafter diets containing one of moringa (*Moringa oleifera*) and baobab (*Adamsonia digitata*) leaves, roselle (*Hibiscus sabdariffa*) calyces and orange (Citrus × sinensis) rind at 4% of the diet for 35 days. Pigmentation of the apterylum and shanks of live chicken were monitored weekly, while shank, breast, vent, back and abdominal fat pigmentation were measured post-slaughter. Skin pigmentation was scored using a DSM broiler fan. Serum carotenoids were measured as red and yellow pigments by a spectrophotometric method throughout the trial. Means were compared and separated within time using ANOVA and LSD ($P<0.05$). Higher live and post-slaughter skin pigmentation were observed in chicken on moringa diet and lower pigmentation in those on roselle diet. Shank pigmentation peaked at 21 d for chicken on baobab and roselle diets, and 35 d for chicken on the moringa and orange rind diets. Apterylum pigmentation peaked at 21 d for chicken on baobab and orange diets and 28 d for chicken on the moringa diet. Overall, pigmentation was higher in the shank > vent > breast. Dietary sources of carotenoids did not significantly influence yellow pigment deposition in the serum between 21-28 d of exposure while red pigment deposition in the serum was unaffected by the dietary treatments after 28 d of exposure. Yellow pigment composition of serum was significantly lowest ($P<0.05$) in chicken on moringa diet at 14 and 21 d (19.84 ± 0.15 and 18.54 ± 0.14 ppm). A similar trend was observed for red pigment composition of serum of chicken on moringa diet at 28 and 35 d (19.60 ± 0.13 and 18.36 ± 0.19 ppm). *M. oleifera* leaves improved live and post-slaughter pigmentation of meat-type chicken.

Dietary lysolecithin supplementation improves nutrient utilization in broiler chickens

A. Vinado[1], D. Sola-Oriol[1], M. Jansen[2], A. Karwacinska[2] and A.C. Barroeta[1]
[1]Animal Nutrition and Welfare Service, UAB, 08193 Bellaterra, Spain, [2]Kemin Europa NV, Toekomstlaan 42, 2200 Herentals, Belgium; alberto.vinado@uab.cat

Lysolecithin, a mixture of phospho- and lysophospholipids, shows emulsifier properties that can improve lipid digestion and increase dietary fat utilization. The aim of this experiment was to determine if dietary lysolecithin supplementation may improve nutrient utilization in broiler chickens. A total of 120 Ross-308 one-day-old female broiler chickens were randomly distributed in 24 digestibility cages (8 replicates/treatment). They were randomly assigned to three experimental treatments: T1, positive control; T2, negative control, 80 kcal AME/kg less than T1 (by reducing fat inclusion); T3, T2 diet supplemented with lysolecithin (250 g/T). Birds were reared till 42 d of age. The nutritional balance was analysed between 24 and 25 d of age. Titanium dioxide (TiO_2) was used as an indigestible marker, at 0.4% in the diets. No difference in final body weight (42 d) or feed conversion ratio was observed between the experimental treatments. Results from the digestibility balance showed that lower AMEn and nutrient utilization coefficients were observed for T2 than for T1 or T3 ($P<0.05$). Lysolecithin supplementation, on the other hand, improved the feed AMEn (T3) showing similar AMEn values to the positive control (T1: 3,062 and T3: 3,013 vs T2: 2,903 kcal/kg; $P=0.0001$). Moreover, similar utilization coefficients were observed for dry matter, organic matter and crude protein between chickens fed T1 and T3. Animals fed the lysolecithin (T3) showed higher apparent coefficients of absorption for saturated ($P=0.019$), mono-unsaturated ($P<0.001$) and poly-unsaturated fatty acids ($P<0.001$) than those fed T2 (2.61%, 1.67% and 3.04% improvement, respectively); but, not different than those fed T1. The present study confirms that dietary lysolecithin supplementation in fat/AMEn reduced diets can increase AMEn of the diet and improve nutrient utilization coefficients in growing chickens.

The impact of lysolecithin composition on nutrient digestibility in broilers

M. Jansen[1,2], F. Nuyens[1], J. Buyse[2], S. Leleu[3] and L. Van Campenhout[2]
[1]Kemin Europa NV, Toekomstlaan 42, 2200 Herentals, Belgium, [2]KU Leuven, Oude markt 13, 3001 Leuven, Belgium, [3]ILVO, Scheldeweg 68, 9090 Melle, Belgium; matias.jansen@kemin.com

Lysolecithin is used in feed to improve the digestion and absorption of nutrients, especially lipids. It contains a mixture of phospho- and lysophospholipids (PL & LPL) and varies in composition depending on its production and origin. In particular, lysophosphatidylcholine (LPC) is regarded to be the main active LPL. The impact of two different lysolecithins and purified LPC on nutrient digestibility and AMEn in broilers was investigated. Groups of 3 male Ross 308 chicks (24-28 days of age) were assigned to 18 cages (6 cages/diet, 1 cage = exp. unit) and fed basal diets with palm oil and supplemented with either 250 ppm lysolecithin A (T1), 250 ppm lysolecithin B (T2) or 9.8 ppm LPC (T3). The dosage of LPC in T3 was determined to provide identical amounts of LPC as in T1. The diets delivered 54.93, 44.58 and 0.0 g of PL and 29.70, 32.83 and 9.80 g of LPL (of which 9.8, 15.73 and 9.8 g were LPC) per tonne of feed, resp.. The LPL in T1 were much more saturated than those in T2. Nutrient digestibilities and AMEn were calculated based on 0.3% TiO_2 as tracer in the diet. Dry matter, crude fat and protein digestibility were higher ($P<0.05$) in T1 than in T2 or T3. Despite that 10% more LPL were added in T2 and identical amounts of LPC were added in T3, the crude fat digestibility was 10% higher in T1 than in T2 and T3 (72.6 vs 62.8 and 62.1%, resp.). Individual FA digestibility analysis revealed that these differences were observed for C14:0, C16:0, C18:0, C18:1, C18:2 and C18:3. This suggests that lysolecithins affect both the digestibility of palm oil (mainly saturated) and the lipids in other dietary ingredients (mainly unsaturated). The difference in nutrient digestibility resulted in a significantly higher AMEn for T1 than for T2 and T3 (3,361 vs 3,048 and 3,032 kcal/kg, resp.). In conclusion, the chemical composition (PL & LPL distribution and FA profile) of lysolecithin is determining its efficacy.

Lysophospholipids with monoglycerides and synthetic emulsifier enhance lipid digestion in broilers

M. Jansen, I. Mast, M. Di Benedetto and F. Nuyens
Kemin Europa NV, Toekomstlaan 42, 2200 Herentals, Belgium; matias.jansen@kemin.com

Lysolecithin, with lysophospholipids as its key component, is used in feed to improve the digestion and absorption of nutrients, especially lipids. The present work investigates if the addition of selected quantities of synthetic emulsifier and monoglycerides (MG) to lysolecithin as a mixture (LYSOFORTE® EXTEND) could enhance the mode of action of lysolecithin. Samples of rapeseed oil, oil with lysolecithin (0.5%) and oil with the mixture (0.5%) were assessed for emulsion stability (ES; 3 rep.). ES was significantly improved by the mixture. Samples of animal fat, fat with lysolecithin (1.2%) and fat with the mixture (1.2%) were subjected to *in vitro* hydrolysis (3 rep.). The free fatty acid (FFA) release rate (k; min[-1]) was the highest ($P<0.05$) in fat with the mixture ($k=15.06$) and the lowest in fat without ($k=10.00$). The digests were then applied to differentiated Caco-2 monolayers on permeable membranes to assess lipid absorption. The absorption of both MG and FFA increased ($P<0.05$) by more than 70% with the addition of lysolecithin or the mixture. Whereas the synthetic emulsifier and MG improved respectively emulsification and hydrolysis, lysophospholipids confirmed to be essential for improving lipid absorption. To confirm these findings, Ross 308 broilers (8 birds/pen, 9 pens/diet) were fed a positive control diet (PC) fulfilling all requirements, a negative control diet (NC) with lower AME (-60 to 80 kcal/kg) or a NC with 500 ppm of LYSOFORTE EXTEND dry (LED). At 14 days of age, birds fed LED diet had a higher ($P<0.05$) body weight (+18 g/bird), higher daily gain (+1.3 g/bird/day) and lower FCR (- 7 points) than birds fed the NC diet. Birds fed PC or LED had also a higher ($P<0.05$) carcass (+1%) and breast yield (+1.3%) than those fed NC. Birds fed LED diet also had significantly lower abdominal fat pad content (MIX: 0.91% vs PC: 1.19% and NC: 1.07%). The mixture thus confirmed to result in overall improved performance in broilers.

Effect of dietary Germanium supplementation on the performance of laying hens for finishing period

H.K. Kang[1], S.H. Kim[1], S.B. Park[1], C.I. Lim[2] and K.S. Ryu[2]
[1]Poultry Research Institute, 321-1 Daeguanryung-Myeong, Pyungchang-Gun, Gangwon-Do, 25340, Korea, South, [2]Chonbuk National University, Animal Science, 567, Baekje-Daero, Duckjin-Gu, Jeonju, 54896, Korea, South; seon1288@gmail.com

A trial was conducted to know the effect of dietary Germanium on the performance of laying chicken during finishing period. Three hundred sixty Hyline Brown laying chicken of 74 weeks of age were distributed into four dietary treatments having five replications with 18 birds in each. The duration of the trial was 12 weeks between 74 to 85 weeks old. The dietary groups were control (Basal diet + 0% Germanium), Basal diet+1% Germanium, Basal diet + 2% Germanium and Basal diet + 4% Germanium. Daily egg mass was found highest (48.52 g) after addition of 2% Germanium in the diet of laying chicken at finishing stage compared to control (47.22 g). Although egg shell color was affected with the addition of Germanium 4% (22.46) but 2% (24.15) level didn't reduce compared to control group. Haugh unit were found higher in all 3 phases (71, 81 and 85 weeks) after addition of 2% Germanium in the diet. Significantly highest ($P<0.05\%$) eggshell breaking strength (3.25 kg/cm^2) were observed after addition of Germanium at the age of 85 weeks of age. Addition of Germanium reduced the serum cholesterol level and increased HDL (21.29 mg/dl) at 2% dietary Germanium. Considering the findings it may be concluded that 2% Germanium product addition in the laying hen diet at the finishing stage.

A concentrate of humic substances for poultry

E.N. Andrianova[1], V.I. Fisinin[1], A.N. Shevyakov[1], S.A. Kirov[2] and Y.D. Shachnev[2]
[1]All-Russian Research and Technological Poultry Institute, Ptitsegradskaya Str., 10, 141311 Sergiev Posad, Moscow Region, Russian Federation, [2]Esson Co. Ltd., Nizhny Novgorod, 603005, Russian Federation; andrianova@vnitip.ru

Humic substances can be valuable additives for poultry diets due to the ability to form immobile and/or poorly dissociable compounds with toxic and radioactive elements, heavy metals, certain pesticides, hydrocarbons, and phenols, resulting in reduction of these pollutants in final poultry products. Humic substances can also chelate valuable trace elements and these chelated forms can be used in mineral premixes together with inorganic forms. Russian scientists from Nizhny Novgorod developed an advanced technology for production of dry and liquid humic concentrates as well as humate-chelated forms of Mn, Zn, Fe, Co, and Cu for supplementation of mineral premixes for animals and poultry. Trials on Cobb-500 broilers conducted at controlled conditions showed that the addition of liquid concentrate to drinking water (4 and 8 liters per 1000 liters) significantly ($P<0.05$) improved live BW at 36 days of age from 1,947.91 g in unsupplemented control to 2,064.47 and 2,037.38 g, respectively; FCR from 1.64 to 1.57 and 1.58; contents of heavy metals in meat were reduced. Supplementation of broiler diets with dried form of the concentrate (1000 and 2,000 ppm) from 1 to 36 days of age resulted in improvements in live BW at 36 days of age from 1,712.62 g in control to 1,856.77 and 1,891.59 g, respectively; FCR from 1.789 to 1.667 and 1.633. Partial substitution (30% of specified amount) of humate-chelated forms of trace elements for sulfate forms in premixes improved live BW from 1,942.13 g in control to 1,981.28 g; FCR from 1.759 to 1.640. Trials at a commercial turkey farm (on 13,585 BUT and BIG-6 birds) showed that supplementation of drinking water with liquid humic concentrate reduced mortality rate by 0.65%.

Effects of different levels of malt extract on performance and intestinal morphology of broilers

R. Vakili[1], N. Asadian[1] and M. Sedeghi[2]
[1]Islamic Azad University,Kashmar Branch, Department of Animal Science, Seyed Morteza Blvd.,IAU Campus, Kashmar, Iran, [2]Isfahan University of Technology, Department of Animal Science, Isfahan Province, Khomeyni Shahr, Daneshgah e Sanati HW, Isfahan, Iran; rezavakili2010@yahoo.com

The experiment was performed to evaluate the effects of different levels of malt extract on performance and intestinal morphology in broilers. Feed intake, body,weight gain,the creatine, glucose, total protein, influenza and Newcastle titer and villi height was measured. The results showed that at the age of 7 to 21 days, the addition malt extract had a significant effect on feed intake ($P<0.05$). In the treatment of of 0/2 percent malt extract in diet showed the highest weight gain. In this experiment, using level 0/2 percent malt extract in diet on AST and ALT levels were increased, but this increase was not significantally ($P>0.05$). In addition, our results showed that the use of 0/1 and 0/2 percent of malt extract in diet broiler any not significant effect on the creatine, glucose, total protein, influenza and Newcastle titer ($P>0.05$).The results of experiment showed the inclusion of malt extract improved performance and reduced villi height.

Session 04 Feed enzymes   Theatre 1

The action of xylanase in broilers depends on how the substrate is provided

A. Khadem[1], M. Lourenço[2], E. Delezie[2], L. Maertens[2], A. Goderis[3], R. Mombaerts[3], V. Eeckhaut[4], F. Van Immerseel[4] and G.P.J. Janssens[1]
[1]Ghent University, Heidestraat 19, 9820 Merelbeke, Belgium, [2]Institute for Agricultural and Fisheries Research, Scheldeweg 68, 9090 Melle, Belgium, [3]Nutrex NV, Achterstenhoek 5, 2275 Lille, Belgium, [4]Ghent University, Salisburylaan 133, 9820 Merelbeke, Belgium; evelyne.delezie@ilvo.vlaanderen.be

Beneficial effects of hydrolyzing water extractable arabinoxylans (WEAX) by xylanase (XYL) are attributed to digesta viscosity reduction and creation of prebiotic arabinoxylan oligosaccharides (AXOS). The importance of each effect was assessed assuming WEAX generate intestinal viscosity. Broilers were randomly allocated to 7 treatments with 8 replicates of 30 birds each. Broilers received WEAX inherent in the cell matrix (wheat-based diet; WHE) or purified in a corn-based diet (CON+WEAX), with and without XYL (*Bacillus subtilis*, 900U/kg feed). CON±XYL were the control. Viscosity reduction by XYL can be estimated by comparing CON+WEAX+XYL and CON+XYL+AXOS (Longlive) as the effect of purified AXOS might be comparable to that of AXOS created by XYL in CON+WEAX+XYL. Statistical analysis was done using a GLM followed by multiple pairwise comparison tests. XYL reduced digesta viscosity (Brookfield) in WHE, but not in the CON+WEAX. XYL improved performance for WHE versus no improvement with CON+WEAX and lower performance in CON+AXOS+XYL. CON+WEAX+XYL increased numbers of *Clostridium* cluster IV and XIV and gene copies for butyryl-CoA:acetate CoA-transferase (both by qPCR), consistent with the prebiotic activity of WEAX. Because viscosity and microbiome changes occurred together, it is not possible to conclude that viscosity reduction was XYL's only mode of action. As added WEAX did not induce viscosity, the relative impact of viscosity reduction by XYL remains inconclusive and warrants further study on the importance of WEAX in the cell matrix regarding viscosity. Use of dietary prebiotics or of prebiotics released by XYL in intestine seems promising, but effect on performance depends on the amount and type of WEAX and AXOS.

Effects of phytase, alone or combined with protease, on broiler performance and bone mineralization

E. Delezie[1], S. Leleu[1] and D. Feuerstein[2]
[1]ILVO, Scheldeweg 68, 9090 Melle, Belgium, [2]BASF, E-ENE/LW, 68623 Lampertheim, Germany;
evelyne.delezie@ilvo.vlaanderen.be

Phytases have been focus of research for longtime. However there are still uncertainties about the effect of phytase on nutrient requirement. Another important area is the effect on nutrient utilization by combining phytase with protease. Therefore a trial was performed to determine the bio-efficacy of a phytase whether or not combined with a protease. Both enzymes were introduced in 4 different negative control diets (NC) formulated by reducing the nutrient content of a positive control diet (PC) by four different matrix values (reduced Ca, phosphorus (P), energy and amino acid level). The experiment consisted of 14 dietary treatments (2,880 birds). Phytase was supplemented at 350, 500, 750, 1000, 1,500 and 2,000 FTU/kg and protease (serine protease by *Bacillus licheniformis*) at 400 mg/kg (30,000 PROT/kg). Reducing the nutrient level in a balanced way without phytase supplementation resulted in lower performance results compared to the PC ($P<0.001$). These diets also had a deleterious effect on bone mineralization as indicated by the lower tibia ash % on d25 and d39 ($P<0.001$) and increased number of animals with a reduced walking ability. Adding phytase to the diets could compensate for these shortcomings. By supplementing phytase and/or protease to the diets performance of the animals was comparable or even numerically higher compared to the PC fed birds. A higher dosage of phytase did not further improve the zootechnical results of the broilers at the 5% significance level so there was no additional effect. The effect of combining phytase and protease was not additive and no synergism was present. It can be concluded that by supplementing phytase, a reduction of the recommended P and Ca requirements by 25 to 30% did not negatively affect performance and bone development of broilers but even improved their Ca and total P retention. Additional supplementation of protease or increasing the phytase level did not further improve digestibility and performance.

Session 04 Feed enzymes   Theatre 3

No interactive effects of P, Ca, and phytase on precaecal amino acid digestibility in broilers

V. Sommerfeld[1], M. Schollenberger[1], I. Kühn[2] and M. Rodehutscord[1]
[1]University of Hohenheim, Institute of Animal Science, Emil-Wolff-Str. 10, 70599 Stuttgart, Germany, [2]AB Vista, Feldbergstr. 78, 64293 Darmstadt, Germany; v.sommerfeld@uni-hohenheim.de

The objective of this study was to distinguish between single and interactive effects of phosphorus (P), calcium (Ca) and phytase on precaecal crude protein (CP) and amino acid (AA) digestibility. Seven pens with 19 unsexed Ross 308 broiler chickens each were randomly allocated to each of the treatments on day 15. The treatments were arranged in a 2×2×2-factorial and included corn-soybean meal-based diets without (4.1 g P/kg DM) or with (6.9 g P/kg DM) monosodium phosphate supplementation, without (6.2 g Ca/kg DM) or with (10.4 g Ca/kg DM) limestone supplementation, and without or with supplementation of 1,500 FTU/kg of a modified, *Escherichia coli* derived 6-phytase (Quantum™ Blue). On day 27, digesta from a defined section of the lower ileum was collected, pooled on a pen basis and freeze dried. Samples were analysed for CP, AA and titanium dioxide, which was used as the indigestible marker. A three-way ANOVA was carried out using SAS 9.3. Significance was declared at $P<0.05$. Across all treatments, mean precaecal digestibility of AA ranged between 56% (Cys) and 84% (Met). The two- and three-way interactions were not significant for CP and AA digestibility values. However, precaecal digestibility of CP and all AA was significantly ($P<0.01$) increased between 2 (Arg, Met) and 6 (Cys) percentage units by phytase supplementation. Ca and P addition had no significant effect on precaecal CP or AA digestibility, except for Cys digestibility which was increased by 2 percentage units by P supplementation ($P=0.048$). The results give no clear indication for reduced dietary P and Ca to have an impact on precaecal AA digestibility. Phytase supplemented at the level used in this study increased CP and AA digestibility, possibly due to diminished phytate-protein-complexes, demonstrating an interesting strategy to improve protein utilization.

Efficacy of a endo-1,4-beta-xylanase/endo-1,4-beta-glucanase on nutrient digestibility in broilers

P. Ader[1], A. Tröscher[1] and M. Francesch[2]
[1]BASF SE, ENS/LD, 68623 Lampertheim, Germany, [2]IRTA, Animal Nutrition and Welfare, Crta. Reus-El Morell Km. 3.8, 43120 Constantí, Spain; peter.ader@basf.com

The effect of a NSP-hydrolysing enzyme preparation (NSPE), containing endo-1,4-β-xylanase and endo-1,4-ß-glucanase, on apparent total tract digestibility (ATTD) of dry matter, organic matter, fat, nitrogen, crude fibre and energy, AME content of the diet, and ileal digesta viscosity was evaluated in broiler chickens fed a maize-based diet. Male broiler chickens of the Ross 308 strain were allocated into 48 cages of 2 chickens for a total of 8 replicates per treatment (1 replicate=3 adjacent cages). A diet based on maize (57%), soybean meal (30%), animal fat (5%) and soybean hulls (3%) was tested without or with Natugrain TS (100 g/tonne feed providing 560 thermostable xylanase units (TXU) and 250 thermostable glucanase units (BGU)/ kg of feed). Performance from 14-24 d, ATTD for dry matter, organic matter, fat, crude protein, crude fibre and energy, and AME of the diet were measured from 21-23 d, using titanium dioxide as a marker. Additionally, viscosity of digesta from proximal ileum was measured at 24 d. Enzyme supplementation had no significant effect on performance of chickens from 14-24 d or on ileal digesta viscosity at 24 d of age. The NSPE increased ($P<0.05$) the ATTD for dry matter (66.2 vs 68.5%), organic matter (69.2 vs 71.4%), energy (70.0 vs 72.3%) and crude fibre (-2.7 vs 4.9%). The NSPE tended to increase fat ATTD from 70.5 to 74.3% ($P=0.13$) and had no effect on nitrogen ATTD. Enzyme supplementation increased ($P<0.05$) AME of the diet by 3.3% (from 2,869 to 2,963 kcal/kg). It was concluded that NSPE supplementation enhanced the nutritive value of a maize-based diet for broiler chickens in growing phase, by increasing the total tract nutrient digestibility and releasing additional energy from the diet.

Effect of a NSP degrading enzyme complex on layer performance: a meta-analysis

L. Nollet, S. Beckers, N. Soares and K. Bierman
Huvepharma NV, Uitbreidingstraat 80, 2600 Antwerp, Belgium; karel.bierman@huvepharma.com

Layers need efficient ways to extract more nutrients out of the diet, in order to reach their full egg laying capacity. Well-targeted feed enzymes offer nutritionists a tool to meet these crucial objectives. A multi-enzyme digestive complex produced by a non-GMO *T. citrinoviride* Bissett via a single Surface Fermentation process, has demonstrated to provide a consistent solution in dietary energy utilization. In the past years, plural efficacy trials in commercial laying hens have been performed by supplementing a xylanase-based complex of feed enzymes (Hostazym® X) on top of various types of layer diets (1,050 to 1,500 EPU of xylanase/kg). The applied enzymatic product is standardized on endo-1.4-beta-xylanase and due to its specific production on wheat bran it also contains different side-activities of hemicellulases and beta-glucanases. A meta-analysis was developed based on different studies (n=8) with first-cycle hens (from 20 weeks of age) and second-cycle hens (from 45 weeks of age) and with a trial period of 24 to 26 weeks. In all trials, laying hens were randomly distributed into cages and fed *ad libitum* wheat-, maize- or mixed diets (for 1 laying phase). Therefore the conducted trials disposed of practical diets with varying NSP fibre profiles. All animals had continuous access to water. In each trial daily egg mass production, egg weight, feed intake and feed conversion ratio (FCR) were continuously measured. On top supplementation of the xylanase-based enzyme complex yielded on average a +1.61 g (or 2.95%; $P<0.05$) higher daily egg mass per hen, and a -0.061 (or 2.97%; $P<0.05$) lower feed conversion rate, compared to a non-supplemented control group. Additionally, the enzyme treated laying hens showed on average a +1.13 g (or 1.90%;non-significant) higher egg weight compared to the non-supplemented animals. It could be concluded that Hostazym® X provides a consistent improvement in laying hen performance, independently of the cereal (NSP fibre) composition of the layer feed.

Effect of a NSP degrading enzyme complex on turkey performance: a meta-analysis

L. Nollet, S. Beckers, N. Soares and K. Bierman
Huvepharma NV, Uitbreidingstraat 80, 2600 Antwerp, Belgium; karel.bierman@huvepharma.com

Turkeys need efficient ways to extract more nutrients out of the diet, in order to reach their full growth potential and to maintain intestinal health. A multi-enzyme digestive complex produced by a non-GMO *T. citrinoviride* Bissett via a single Surface Fermentation process, demonstrates to provide a consistent solution in dietary energy utilization and allows nutritionist to reach these targets. In past years, plural efficacy trials in turkeys have been performed by supplementing a xylanase-based complex of feed enzymes (Hostazym® X) on top of various types of turkey diets (1,050 EPU of xylanase/kg). A meta-analysis was developed based on different studies (n=7) using one-day-old chicks until 12 or 18 weeks of age (resp. 8-10 kg and 12-20 kg of slaughter weight according to gender). In all trials, turkeys were randomly distributed inside the houses (floor pens) and fed *ad libitum* maize-, wheat- or mixed pelleted diets. Therefore the conducted trials disposed of practical diets with varying NSP fibre profiles. Each time body weight, feed intake and feed conversion ratio (FCR) were measured and determined at the end of each feeding phase. Also mortality figures were closely recorded throughout the entire grow-out period. On top supplementation of the xylanase-based enzyme complex yielded on average a +302 g (or 2.4%; $P<0.05$) higher end body weight and a -0.088 (or 3.5%; $P<0.05$) lower feed conversion rate (mortality adjusted for a 13 kg end body weight), compared to a control group. Furthermore, overall mortality rate was decreased from 5.8 to 4.9% for the enzyme treated animals (1,050 EPU of xylanase per kg of feed) compared to the non-supplemented animals. It could be concluded that standard supplementation (recommended dose) of the applied multi-enzyme complex (Hostazym® X, Huvepharma NV) provided a consistent improvement in bird performance, independently of the cereal (NSP fibre) composition of the turkey feed while reducing overall mortality rate by 15%.

Effect of a NSP degrading enzyme complex on broiler performance: a meta-analysis

L. Nollet, S. Beckers, N. Soares and K. Bierman
Huvepharma NV, Uitbreidingstraat 80, 2600 Antwerp, Belgium; karel.bierman@huvepharma.com

Well-targeted feed enzymes offer nutritionists a tool to increase performance in everyday broiler diet formulations. A multi-enzyme digestive complex produced by a non-GMO *T. citrinoviride* Bisset demonstrates to provide a consistent solution in dietary energy utilization. In past years, plural efficacy trials in broilers have been performed by supplementing a xylanase-based complex of feed enzymes (Hostazym® X) on top of various types of broiler diets (1,500 EPU of xylanase/kg). A meta-analysis was developed based on different studies (n=8) until 36 or 49 days of age. Above supplementation trials were combined with reformulation trials on the same enzyme product and -dosage, in order to calculate an overall energy equivalence value (kcal AME) based on technical performance and dietary energy levels fed. In all trials, birds were fed maize-, wheat- or maize-wheat based diets. Each time body weight, feed intake and feed conversion ratio (FCR) were measured and determined at the end of each feeding phase. Results show that the on top supplementation of the xylanase-based enzyme complex yielded on average a +66 g (or 2.9%; $P<0.05$) higher end body weight and a -0.053 (or 3.1%; $P<0.05$) lower feed conversion rate (mortality adjusted for a 2,250 g end body weight), compared to a non-supplemented control group. Considering the energy levels fed (range 2,875-3,100 kcal/feed) and the zootechnical performance in all conducted trials (on top or reformulated; n=14), the Metabolisable Energy (AME) equivalence of the multi-enzyme complex could be calculated to yield 104 kcal extra AME per kg of feed (+3.48% vs control), ranging from +85 kcal/kg for maize-soya based diets to 120 kcal/kg for wheat-based diets. It could be concluded that standard supplementation (recommended dose) of the applied multi-enzyme complex (Hostazym® X) provided a consistent improvement of bird performance, independently of the cereal (NSP fibre) composition of the broiler feed.

Efficacy of enzymes in wheat based diets depends on the non-starch polysaccharide content of wheat

N. Smeets[1], F. Nuyens[1], E. Delezie[2], L. Van Campenhout[3] and T. Niewold[3]
[1]Kemin Europa NV, Toekomstlaan 42, 2200 Herentals, Belgium, [2]ILVO, Scheldeweg 68, 9090 Melle, Belgium, [3]KULeuven, Kasteelpark Arenberg 30, 3001 Heverlee, Belgium; natasja.smeets@kemin.com

Feed enzymes are commonly added to wheat based diets to counter the possible anti-nutritional effects caused by non-starch polysaccharides (NSP). The aim of the following trials was to confirm earlier *in vitro* research, which showed that enzyme efficacy was related to the concentration and structure of NSP from wheat. Trial 1. Diets were formulated with three wheat cultivars (Centenaire – high NSP, Viscount and Rustic- medium NSP) and were tested with or without the addition of an enzyme (Kemzyme® Plus dry, 500 g/T) in a broiler balance trial. The diets (6 replicates) were fed to broilers housed in cages (5 birds/cage) from 13-26 d. Nutrient digestibilities and AME_n were calculated using a TiO_2 tracer in the diet. Trial 2. Diets were formulated with wheat (medium NSP) or with wheat mixed with of rye and barley (to mimic a high NSP diet) and were tested with or without the addition of an enzyme (Kemzyme® plus dry, 500 g/T). One-day-old broilers were assigned randomly to 32 pens (8 replicates, 30 birds/pen) and broiler performance during starter, grower and finisher periods was recorded. In the first trial, enzyme addition caused a significant increase in nutrient digestibilities and AME_n for the diet formulated with the high NSP wheat Centenaire. For the Rustic cultivar, only the digestibility of fat improved and for the Viscount wheat, no significant differences were observed. NSP digestibility tended to be higher in the high NSP wheat Centenaire, whereas the lowest digestibility was observed in the Viscount wheat cultivar. The latter could explain the lowest improvements with enzyme addition for the Viscount cultivar. Like in the first trial, the effect of enzyme addition in the second trial depended on the diet composition and significant improvements were only detected in the high NSP diet.

Performance, enzyme activity and nutrient digestibility of broilers fed cottonseed meal with enzymes

M.E. Abdallh, A.A. Omede, M.M. Bhuiyan and P.A. Iji
School of Environmental and Rural Science, University of New England, Armidale, NSW, 2350, Australia; mabdallh@myune.edu

A 3×3 factorial study examined the performance, endogenous enzyme activity and ileal digestibility of nutreints, including amino acids of birds fed diets containing graded levels of cottonseed meal (CSM) with or without microbial enzymes. Nine diets containing three levels of CSM (none, low (5, 10, 15%) or high (6, 12, 18%) in the starter, grower, and finisher respectively, supplemented with three levels (0, 100 or 150 mg/kg) of a new composite xylanase and beta-glucanase enzyme were used. Each treatment was randomly assigned to 6 replicates (10 birds per replicate). Feed intake (FI) and weight gain (WG) were recorded on d10, 24 and 35, and feed conversion ratio (FCR) obtained from the data, while enzyme activity and nutrient digestibility were measured on samples collected on d 10 and 24, and d 24 only, respectively. FI up to d 35 decreased ($P<0.05$) significantly with enzyme increasing. CSM supplemented with either 100 or 150 mg/kg enzyme improved ($P<0.05$) WG in grower phases, with the heaviest birds (1,514 g/b) observed in low CSM with 100 mg/kg enzyme group. Enzyme supplementation improved ($P<0.01$) FCR all through the growth phases. Enzyme improved ($P<0.05$) ileal crude protein and starch digestibility. High inclusion levels of CSM decreased ($P<0.05$) the digestibility of starch, but this was improved ($P<0.05$) by enzyme supplementation, showing an interaction between CSM and microbial enzyme on starch digestibility. The digestibility of arginine, glutamic acid and threonine improved ($P<0.05$) with increased CSM inclusion, and that of methionine with increased enzyme supplementation. At d 10, lipase activity was increased ($P<0.05$) by higher CSM inclusion, while on d 24, general proteolytic activity increased ($P<0.05$) when 100 mg enzyme was supplemented in the diet. These results indicate that relatively higher levels of CSM in diet supplemented with xylanase and beta-glucanase can improve growth and digestive physiology of broiler chickens.

Correlation among viscosity of feed, viscosity of small intestinal content and broiler performance

R. Mombaerts[1], A. Goderis[1], E. Delezie[2], A. Khadem[3], G. Janssens[3] and M. Lourenço[2]
[1]Nutrex nv, Achterstenhoek 5, 2275, Belgium, [2]ILVO, Scheldeweg 68, 9090, Belgium, [3]Ghent University, Heidestraat 19, 9820 Merelbeke, Belgium; ronny.mombaerts@kugegroup.com

For decades, the positive effects of xylanase in poultry feeds rich in wheat were explained by the viscosity reduction of the GIT. We investigated the relation between *in vitro* viscosity measurements of poultry feeds, corresponding viscosity of proximal and distal small intestinal contents (PSI and DSI) and bird performance. In total, 7 diets were formulated to be isocaloric, isonitrogenous and contained similar levels of water unextractable arabinoxylans. Dietary levels of water-extractable arabinoxylans (WE-AX) varied in quantity and type, going from low WE-AX levels in corn based diets (C) with and without bacterial endoxylanase produced by *Bacillus subtilis* (900 U/kg feed) (XYL,) to 3 types of feeds high in WE-AX: (1) C + purified WE-AX with and without XYL; (2) C + short chain arabinoxylan oligosaccharides with XYL; and (3) wheat based diets with and without XYL. A total of 1,680 male Ross 308 broilers were randomly distributed over 56 pens with 30 broilers per pen. Body weight and feed intake were recorded on days 13, 26 and 39 days of age. On day 34, total gut contents were collected from PSI and DSI from 1 chick per pen. Viscosity of gut contents was measured on the supernatant fraction, while feed viscosity was determined on full suspensions. The *in vitro* viscosity of the complete feed suspensions and those of the PSI and DSI supernatants were highly correlated ($P<0.0001$). On the other hand, there was a much lower correlation between the viscosity of the digesta and broiler performance. To conclude, it seems that WE-AX can increase viscosity of digesta but they're also fermentable. Their final effect on performance is a combination between those 2 factors and cannot be predicted by measuring only viscosity.

Response of turkeys fed wheat-barley-rye based diets to xylanase supplementation

G. González-Ortiz[1], K. Kozlowski[2], J. Jankowski[2] and M.R. Bedford[1]
[1]AB Vista, R&D, Woodstock Ct, SN8 4AN, Marlborough, United Kingdom, [2]Olsztyn University of Warmia and Mazury, Department of Poultry Science, ul. Oczapowskiego 5, 10-719, Olsztyn, Poland; gemma.gonzalez@abvista.com

The objective of this study was to evaluate the effect of xylanase on performance, nitrogen corrected apparent metabolizable energy (AMEn), caecal fermentation and microbiota profile of turkeys fed wheat-barley-rye based diets. Four hundred eighty female Hybrid Converter turkey poults (1-day-old) were placed in one of two experimental treatments: a control diet and the same diet supplemented with 16,000 BXU/kg of xylanase. Treatments had fifteen replicate pens per diet, with 16 birds each. Feed was supplied in four phases of three weeks per phase. At 0 and 84 d body weight was measured, and at 84 d feed intake (FI) recorded. Body weight gain (BWG) and feed conversion ratio (FCR) were calculated. Titanium dioxide was included in the diet as an indigestible marker, and on the last day of the study excreta samples were collected. The profile of volatile fatty acids (VFA) in addition to the microbial community structure by the percentage of guanidine and cytosine (G+C) method in the caecal digesta were analysed on d84. Control FCR (2.09) was improved on use of xylanase to 2.02 ($P<0.01$). Birds supplemented with xylanase tended to be heavier than the control birds (7,723 vs 7,560 g/bird; $P=0.07$). The AMEn in supplemented birds tended to be greater than that observed for the control birds ($P=0.120$). No differences were found between treatments in total VFA concentrations, but acetic and propionic acid were reduced as a percent of total VFA in xylanase supplemented birds ($P=0.047$ and $P=0.061$, respectively), and butyrate percentage was increased ($P=0.061$). Some regions of the G+C profile were statistically different, indicating a shift in the microbial community structure between treatments. Supplementation of turkey diets with xylanase improved performance, probably due to a better utilization of energy, with differences observed in fermentation and microbial profiles between treatments.

High dosing NSP enzymes for total protein and digestible amino acid reformulation in a broiler diet

K. De Keyser[1], L. Kuterna[1], S. Kaczmarek[2], A. Rutkowski[2] and E. Vanderbeke[1]

[1]AVEVE Biochem NV, New Business Development, Eugeen Meeusstraat 6, 2170 Merksem, Belgium, [2]University of Life Sciences, Animal Nutrition and Feed Management, ul. Wojska Polskiego 28, 60-637 Poznań, Poland; leni.kuterna@aveve.be

This research sheds a new light on the potency of non-starch polysaccharide (NSP) enzymes to save on total protein and digestible amino acids in broiler feed. Two NSP enzyme concepts were compared with concentrated protease in a broiler performance test. A standard (S) and negative control (NC, relative -3% total protein and -3% digestible amino acids) diet were formulated. NC was supplemented with 1 of 3 enzyme concepts: (1) 40,000 endo 1,4 β-xylanase XU/g (EC 3.2.1.8) and 9,000 endo 1,3(4) β-glucanase BGU/g (EC 3.2.1.6) (XG, double advised dosage); (2) 10,700 endo 1,4 β-xylanase XU/g (EC 3.2.1.8), 6,150 endo 1,3(4) β-glucanase BGU/g (EC 3.2.1.6) and 230 pectinase PGLU/g (EC 3.2.1.15) (XGP, double advised dosage); or (3) 650 neutral protease NHU/g (EC 3.4.21.14) (PRO, advised dosage). 400 day-old male Ross 308 broilers with (±40 g) were randomly allocated to 5 feeding groups of 80 birds (10 replicate pens with 8 birds) for a period of 42 days (2-phase feeding): S, NC, NC+XG, NC+XGP and NC+PRO. The 2 NSP enzyme concepts at double dosage enabled a relative -3% total protein and -3% digestible amino acids, without sacrificing on broiler performance compared to S or NC+PRO. The BWG and F:G for S and NC+PRO were similar ($P>0.05$) 2,686 g and 1.63, respectively 2,711 g and 1.62. There was no significant difference ($P>0.05$) between the latter and BWG and F:G of NC+XG and NC+XGP that were 2,698 g and 1.63, respectively 2,737 g and 1.61. NC resulted in 2,616 g BWG and 1.68 F:G and was lower ($P<0.05$) than NC+XGP and NC+PRO. This reformulation on protein and digestible amino acids allows for considerable cost savings of common broiler feed, in present case on average 4.32€ in starter and 2.72€ in grower feed (including enzyme cost, Belgian ingredient prices July 2015, EU antimicrobial growth promoter free feed).

Ability of carbohydrase to restore nutrient availability in a 3% sand-diluted broiler diet

P. Cozannet, R. Montanhini Neto, P.A. Geraert and A. Preynat

Adisseo France, 10 place du Général de Gaulle, 92160, France, Metropolitan; aurelie.preynat@adisseo.com

Current study investigated the effect of Rovabio® Advance, a multi-enzymes complex enriched in an endo-xylanase and arabinofuranosidase, on energy and ileal amino acid (AA) digestibility of a wheat/soybean-based diet in broilers with or without a 3% nutrient-diluted version using sand as an inert diluent. The effect of adding enzymes to these diets was studied in 120 broilers (13-26 d of age) using full factorial block design. The experimental period was divided into three phases: adaptation (7 d), excreta collection (3 d) and preparation prior to digesta collection (4 d). Digestibility of dry matter (DM), AA and gross energy (GE) were determined by analysis of feed, excreta and digesta. The data (n=120) were subjected to ANOVA with block (n=30), diet (n=2) and enzymes (n=2) as fixed effects. There was no difference in feed intake between the two diets. Faecal energy digestibility was similar at around 73% for diets without enzymes ($P=0.99$). Apparent metabolisable energy (AME) content was 3% lower in the diluted versus standard diet (14.0 and 14.4 MJ/kg DM respectively; $P<0.001$). Enzymes improved energy utilisation (+2.8%; 73.3 vs 75.4%; $P<0.001$), leading to a significant increase in AME content of 406 kJ/kg DM in the both diet. AME content of the diluted diet with enzymes was similar to that of the standard diet without enzymes ($P=0.98$) demonstrating the ability of enzymes to fully compensate for the 3% nutrient dilution. At the ileal level, AA digestibility was around 75% across all treatments. Addition of enzymes increased AA digestibility by 4.4% in average ($P<0.001$). In conclusion, multi-enzymes complex can restore nutrient availability of a diet diluted by 3%. The study also highlights the importance of considering the entire nutrient matrix when enzymes are supplemented to diets and offers further understanding of AA digestion characteristics in poultry feed.

Next-generation enzyme rich in xylanase and arabinofuranosidase enhance broiler feed digestibility

P. Cozannet[1], M. Kidd[2], R. Montanhini Neto[1] and P.A. Geraert[1]
[1]Adisseo France, 10 place du Général de Gaulle, 92160, France, [2]Center of Excellence for Poultry Science, University of Arkansas, Fayetteville, 72701, USA; pierre.cozannet@adisseo.com

Study was carried out to evaluate the effect of a multi-carbohydrase complex (MCC) rich in xylanase and arabinofuranosidase on nutrient digestibility in broilers. Energy utilization and digestibility of dry matter (DM), organic matter (OM), protein, starch, fat, and insoluble and soluble fibers were measured using the balance method. The experiment was carried out on 120 broilers (3-wk-old). Eight treatments were used to evaluate effect of arabinoxylan content and nutrient density with and without MCC (Rovabio® Advance). The graded content of arabinoxylan (AX; from 57 to 71 g/kg) obtained using various raw materials (i.e. wheat, barley, rye, wheat distiller). Diet-energy density modified with added fat (from 2,900 to 3,100 kcal/kg). Density and AX had a significant effect on most digestibility parameters. Addition of MCC resulted in improvement in the digestibility of DM, starch and insoluble and soluble fibers with average of 3.0, 3.0, 2.9 and 5.8% units, respectively ($P<0.001$) Interaction between MCC and AX content was significant for the digestibility of OM, fat, protein, and energy ($P<0.023$). Digestibility improvement with MCC for OM, protein, fat and energy were 2.3 and 6.0, 2.4 and 5.9 and 2.0 and 11.9% units for low and high AX diets respectively. Apparent metabolizable energy (AME) was increased by 101 to 212 kcal/kg for low and high AX content diets. Nutrient digestibility and diet AME were negatively correlated with AX content ($P<0.001$). Diets AME with and without the addition of MCC were successfully predicted by the diet digestible nutrient content with and without MCC ($R^2=0.87$; RSD=78 kcal/kg). Study confirms that the presence of AX reduces nutrient digestibility in broiler chickens. The dietary addition of MCC reduced deleterious effect of fiber and improved overall nutrient digestibility.

The interaction between dietary NSP levels and enzyme supplementation on broiler performance

L.P. Barnard and A. Belalcazar
DuPont Industrial Biosciences, Danisco Animal Nutrition, Innovation team, P.O. Box 777, SN8 1AA Marlborough, United Kingdom; luke.barnard@dupont.com

The aim of the study was to determine the effect of non-starch polysaccharide (NSP) content and enzyme treatment on bodyweight gain (BWG) and feed conversion ratio (FCR) in 42 d old broilers. A meta-analysis was carried out with replicate means data from 5 trials, totalling 2,880 birds. All experiments selected for the meta-analysis raised birds from 0-42 d and had a control and an enzyme treatment which was fed for the duration of the study. The enzyme treatment was a commercially available enzyme preparation which was applied to provide a guaranteed minimum of 300 U/kg xylanase, 400 U/kg amylase and 4,000 U/kg protease. Diets were based on corn/soybean meal, wheat/soybean meal or mixed corn/wheat and soybean meal. Two separate models were generated, for BWG and FCR, using the standard least squares model platform of JMP 12. In both models trial code was included in the model as a random effect. In the BWG model independent variables considered were: overall dietary NSP levels (weighted by feed intake for the different dietary phases), and enzyme treatment. The FCR model was similar but also included 42 d bird weight. In both models, linear and quadratic terms were explored as well as interactions between substrate levels and enzyme treatment. Only significant effects were included in the final models. There was a significant effect of enzyme treatment ($P<0.05$) and overall NSP level × enzyme treatment ($P<0.05$) on 42 d BWG. For 42 d FCR there was a significant effect of 42 d BW ($P<0.05$) and significant effects of enzyme treatment ($P<0.05$) and overall NSP level × enzyme treatment ($P<0.05$). In both cases in the absence of the enzyme increasing NSP level decreased 42 d BWG and increased FCR. However, in the presence of the enzyme increasing NSP level increased BWG and decreased 42 d FCR. The results demonstrated that xylanase, amylase and protease supplementation can alleviate the negative impact of increasing dietary NSP content and improve broiler performance.

Effects of an α-galactosidase and xylanase preparation on broiler performance

S. Carné, J. Estévez and J. Mascarell

Industrial Técnica Pecuaria, S.A. (ITPSA), Ave. Roma 157, 08011 Barcelona, Spain; scarne@itpsa.com

A total of 390 1-d male chicks (Cobb 500) were used in a 35-d study to evaluate the effect of an α-galactosidase and xylanase preparation (CAPSOZYME SB PLUS) on growth performance and feed efficiency of broilers. Animals were housed in 39 pens (1 m²/pen) with 10 animals per pen. Feeding was divided into 2 phases of 21 and 14 days, for starter and grower, respectively. Basal diets (AMEn: 12.1 to 12.6 MJ; CP: 21.6 to 20.5%; Lys: 1.20 to 1.14%) were based upon wheat, rye, corn and soybean meal. A randomized complete block experimental design was used, with 3 treatments according to the supplementation of the enzyme preparation on top of the basal diet: Control (T1), enzyme preparation at 350 mg/kg (T2; 17.5 U xylanase, 14 U α-galactosidase), enzyme preparation at 500 mg/kg (T3; 25 U xylanase, 20 U α-galactosidase). Each treatment was fed to 13 replicate pens. Body weight, body weight gain and feed consumption were recorded at the end of each feeding phase. Daily performance and mortality-corrected feed conversion were calculated. Mortality for the overall study was 8.5, 11.7 and 6.2%, for T1, T2 and T3, respectively, not differing between treatments ($P>0.1$). Final body weight was 2,391, 2,477 and 2,463 kg for T1, T2 and T3, respectively, and differed between T1 and T2 ($P=0.022$); a clear upward trend was observed between T1 and T3 ($P=0.054$). Daily weight gain was 67.2, 69.6 and 69.3 g/d for T1, T2 and T3, respectively; differences were assessed between T1 and T2 ($P=0.023$), and T1 and T3 ($P=0.053$). No difference was obtained in the daily feed intake. Likewise, feed conversion ratio did not differ between treatments. However, feed conversion differed between T1 and T3 when corrected for the registered mortality (1.58 vs 1.54; $P=0.029$); corrected feed conversion for T2 was 1.55. In summary, the enzyme preparation supplemented to diets based on wheat, rye, corn and soybean meal showed a beneficial effect on growth and feed conversion of fattening chickens.

Effect of mannanase and xylanase supplementation on broiler performance and carcass yield

Y. Ruangpanit[1], S. Attamnangkune[1], C. Traiprom[1], S. Cervantes-Pahm[2] and S. Charoensin[3]

[1]Kasetsart University, Department of Animal Science, Kamphaengsaen, Nakhon Pathom, 73140, Thailand, [2]Elanco Animal Health, Asia Pacific region, Pasig City, 1602, Philippines, [3]Elanco Animal Health, Bangkok, 10110, Thailand; ysungwa@hotmail.com

The experiment was conducted to determine the effects of combining xylanase and mannanase supplementation on broiler performance, carcass yield and litter quality. Two thousand, day old chicks (Ross 308), were divided into 4 dietary treatments. Each treatment consisted of 10 replications with 50 chicks each. The positive control (PC) diet (corn, wheat, and soybean meal) was formulated to contain 3,000, 3,100 and 3,200 kcal ME/kg in starter, grower and finisher diets, respectively. Three negative control diets (NC) were formulated to contain ME of 70 kcal/kg (NC1), 110 kcal/kg (NC2) and 150 kcal/kg (NC3) less than the PC diet. All diets contained phytase at 500 FTU per metric ton of feed, however, only the NC diets were supplemented with xylanase and mannanase. Birds were raised in an evaporative cooling house for a period of 35 days. A reduction in growth performance was observed when ME was reduced to 110 kcal/kg and 150 kcal/kg diet. However, feed intake, body weight gain and FCR of birds fed the NC1 diet did not differ ($P>0.05$) from the PC group. Chilled carcass weight yield ($P=0.0912$) and breast yield ($P=0.0942$) of birds fed xylanase and manannase (NC1, NC2 and NC3 diets) had a tendency to be higher than that of the PC group. Litter moisture at day 21 tended to be lower ($P<0.10$) in birds fed NC2 and NC3 but not at day 35. Under the conditions of this experiment, xylanase and manannase can be supplemented to broiler diet containing 70 kcal ME/kg less than the PC diet without compromising growth performance and it is likely to improve carcass weight yield and breast yield.

Processing of maize stillage with novel enzymes improves the nutritive value

D. Pettersson and M. Brøgger Pedersen

Novozymes A/S, Feed Ingredients and Nutrition, Krogshoejvej 36, 2880 Bagsvaerd, Denmark; danp@novozymes.com

Distiller's dried grains with solubles (DDGS) can be used as a feed ingredient, and to some extent replace the use of maize or soybean meal. Concerns related to the use of DDGS have mainly been related to an inferior quality of the protein and a high level of non-starch polysaccharides (NSP). The current trial was conducted to study the possibility to improve the nutritive value of DDGS with protease and fibre degrading enzymes. Freeze dried whole stillage was treated at 32° C for 72 hours without (two treatments) or with the supplementation of xylanase and arabinofuranosidase (xyl/araf). Following this the temperature was elevated to 80 °C for 6 hours and a thermo-tolerant protease was added to one of the enzyme treatments and to one of the controls, thereby providing four treatments in total. The material was freeze dried and fed to 40 adult roosters in a TME study (10 birds/treatment). The xyl/araf treatments increased the solubility of the xylans, measured as xylose release after acid hydrolysis, by about 23% and the soluble arabinoxylan fraction (arabinose+xylose), measured by polysaccharide analysis, increased from 1.5 up to 2.5% for treatments with the xyl/araf or xyl/araf and protease. The protease supplementation increased the crude protein solubility by 55%. The determined TME values (MJ/kg) were as follows: control 15.9, protease 16.2, xyl/araf 16.3 and xyl/araf and protease 16.4. It was evident that the pretreatment with the arabinoxylan degrading enzymes increased the solubility of this fraction in the DDGS and also that protease increased the crude protein solubility. Albeit the strong solubility effects the Rooster study only provided numerical improvements (about 3%) in TME content. Still, the data justifies further studies on enzymatic pretreatment of DDGS stillage.

Effect of feed structure, phytase and xylanase on ileal phytate degradation in broilers

I. Kühn[1], H. Whitfield[2] and K. Kozłowski[3]

[1]AB Vista, Feldbergstrasse 78, 64293 Darmstadt, Germany, [2]University of East Anglia, Norwich Research Park, Norwich, NR4 7TJ, United Kingdom, [3]University of Warmia and Mazury, Oczapowskiego 5, 10-719 Olsztyn, Poland; imke.kuehn@abenzymes.com

The objective of this study was to evaluate the effects of two phytase application rates, with or without added xylanase, on ileal inositol phosphate (InsP3-4) and inositol levels in broilers fed pelleted wheat soybean based diets in which the wheat component was milled or partially included as whole grains. 900 male Ross 308 broilers were distributed to 10 treatments (10 pens × 9 birds each). Diets were either adequate (PC) or reduced by 1.5 g/kg Pav and 1.6 g/kg Ca (CC). All CC diets contained 500 (CC500) or 1,500 (CC1500) FTU/kg phytase, with or without 16,000 BXU/kg xylanase (Xyl). At slaughter (d42), digesta, collected from the distal half of th ileum, was pooled from 4 birds per pen for InsP3-6 and inositol analyses. Blood samples from 2 broilers per pen were analysed for inositol. A two-way ANOVA was carried out and significance declared at $P<0.05$. Ileal InsP6 contents were greater in broilers fed whole grains compared with milled diets. Further InsP degradation was not influenced by feed structure. Ileal InsP6 levels in CC500 birds were 65% lower compared to PC broilers ($P<0.05$) and reduced by another 60% in those fed CC1500 diets ($P<0.05$). Xylanase did not influence ileal InsP6 levels when added to CC500 or CC1500 diets but increased ileal inositol levels by 19% when added to CC500 diets ($P<0.05$). Ileal inositol levels were highest in birds fed CC1500 and CC1500Xyl diets. Plasma inositol levels were lower in PC birds (77.8µmol/l; $P<0.05$) compared to those fed the enzyme supplemented CC diets (99.3 to 194.1µmol/l) and were highest in birds fed the diets CC1500 or CC500Xyl including whole grain. The results demonstrate that wheat structure and enzyme application influences gastro-intestinal InsP degradation to the point of the terminal ileum of broilers.

Effect of alpha-galactosidase and beta-glucanase on the carbon footprint from poultry feed mills

W.J. Ryan[1], D. Sizer[2], P. Griffiths[2] and S. Llamas Moya[1]
[1]Kerry, Global Technology and Innovation Centre, Millenium Park, Naas, Co. Kildare, Ireland, [2]Jacobs UK Ltd, 1180 Eskdale Rd, Winnersh, RG41 5TU Wokingham, United Kingdom; sara.llamasmoya@kerry.com

The environmental impact of agricultural activities is under the spot light since agriculture, forestry and other land use contribute 21% to global greenhouse gas (GHG) emissions, and consumers demand sustainable food at affordable prices. Feed contributes significantly to the overall profitability of poultry production, with the sustainable sourcing of ingredients being scrutinized along the entire supply chain. This evaluation assessed the effect of alpha-galactosidase and beta-glucanase (AGBL) on the GHG emissions from poultry feed mills. These enzymes increase the nutritional value of feed, particularly soybean meal (SBM) and cereal grains, reducing the need for expensive fat sources in poultry diets. This may be of relevance in geographies where soy is grown due to the direct impact on land use. Two variants of a typical broiler diet based on corn, wheat and SBM were considered. The main variable was the presence or absence of AGBL, which had an impact on the level of corn, SBM, soya oil and L-lysine used in the formulation (wheat content was fixed). Consequential life cycle assessment (LCA) was applied according to international standards (ISO 14040/44), including sensitivity analysis. This study defined appropriate GHG emission factors from publically available sources, considering the most applicable farming practices, crop geographical region and age of the data. Findings showed that adding AGBL allowed for reductions in GHG emissions from poultry feed mills in the order of 46.6 kg CO_2/MT feed. Considerable benefits of AGBL supplementation were apparent due to changes to direct land use as a result of the reduced demand for soya oil. This corresponded to a 10% reduction of GHG emissions from the modelled poultry feed inputs supplied to mills. The impact of land use changes from forest/grassland to cropland was the largest factor in terms of variability of results.

Performance of broilers fed starter diets (8-21 days of age) with animal by-products and protease

D.P. Carvalho[1], M.F. Pires[1], P.S. Assunção[1], R.A. Noleto[1], K.A. Teixeira[1], E.S. Fernandes[1], A.R. Ribeiro[1] and J.H. Stringhini[2]
[1]Universidade Federal de Goias, Departamento de Zootecnia, Escola de Veterinaria e Zootecnia, Avenida Esperanca, s.n., 74690-900, Goiania, Goias, Brazil, [2]CNPq, Researcher, UFG, 74690-900, Goiania, Goias, Brazil; jhstring@uol.com.br

Proteases in broiler diets tend to increase protein and amino acid digestibility. However, when animal by-products are included in the diets the efficacy of supplemental proteases needs to be further investigated. A total of 320 day-old Cobb chicks were allotted in heated batteries and distributed in 4 treatments in a factorial arrangement 2×2 (100% vegetal and/or 6.7% animal byproducts diets × protease) and 8 replicates of 10 birds each. Diets were based on corn soybean meal and bovine meat and bone meal, feathers and blood meal and viscera and bone meal, and supplemented with 0,05% of a protease (Cybenza®) according to the treatments and calculated according to the nutritional requirements and feed composition according to local Brazilian recommendations. A performance trial was conducted to measure weight gain, feed intake and feed conversion from 8 to 21 days of age. Data was analyzed as a completely randomized design with ANOVA procedure and Tukey test adopted to compare treatments. Broilers fed diets containing animal by-products showed best feed conversion from 14 to 21 days of age and from 1 to 21 days of age. Statistical interaction was observed from 14 to 21 days of age for final weight, weight gain and feed intake and the combination of animal by-products and protease showed the best results. Vegetal diets resulted in a lower level of performance and no effects of protease supplementation were noted. In conclusion, protease and animal by-products in broiler starter diets are effective to increase performance.

The effect of canola meal and protease supplementation on performance and egg quality of laying hens

S. Mirzaeigoudazri, S. Naderi Sahami Zamir, A.A. Saki and P. Zamani
Bu-Ali Sina University, Animal Science Department, Bu-Ali Sina University, Faculty of Agriculture, Hamedan-Iran, 65178-38695, Iran; saramirzaie0@gmail.com

This experiment was conducted to consider of replacing soybean meal (SBM) with canola meal (CM) on productive performance and egg quality of laying hens from 28 to 36 weeks of age. A total of 180 Hy-Line W36 laying hens were distributed into 6 treatments, 5 replicates and 6 hens in each. The experiment was based on a completely randomized 3×2 factorial design with 3 levels of canola meal: 0, 7.29, and14.59% (canola meal replaced 0, 20 and 40%, respectively, of protein from soybean meal; soybean meal was 27.41, 22.63 and 17.84% respectively) and 2 levels of protease (0 or 200 g/ton; 75,000 PROT/g; Ronozyme® ProAct). Six isoenergetic and isonitrogenous diets were formulated on digestible amino acid basis. Feed intake (FI), egg production (EP), egg weight (EW), egg mass (EM) and feed conversion ratio (FCR) were evaluated in this study. Shape index, Haugh units, egg shell thickness, shell weight and yolk color were measured at 36 weeks of age. Statistical analysis was conducted using an analysis of variance (ANOVA) followed by the Duncan multiple test ($P<0.05$). No interactions were found between canola meal and protease supplementation for any traits studied. For the entire experimental period, FI, EW, EP, EM and FCR were not affected by canola meal or protease supplementation ($P>0.05$). Haugh unit significantly decreased ($P<0.05$) at 20 and 40% substitution of CM in the diet (81.73 and 80.13; respectively) than control (87.72). Protease supplementation increased egg shape index (76.46 vs 75.27%) and egg shell thickness (0.562 vs 0.552 mm; $P=0.058$). In conclusion, canola meal can be used with a substituting ratio of 40% of the crude protein contributed by SBM in laying hen diets without any negative effect on bird performance. Egg quality was not affected by canola meal inclusion in the diet except for Haugh unit that was decreased. However, protease supplementation improved some egg traits.

Coated compound protease supplementation effects on broilers fed corn-soy or sorghum-soy based diets

X. Xu[1], H.L. Wang[1], L. Pan[1], X.K. Ma[1], Q.Y. Tian[1], Y.T. Xu[1], S.F. Long[1], Z.H. Zhang[2], X.S. Piao[1], M. Bieber[3] and D. Gonzalez-Sanchez[3]
[1]China Agricultural University, Ministry of Agriculture Feed Industry Centre, 100193, Beijing, China, P.R., [2]Kemin Industries CO, Sanzao Town, 519040, Zhuhai, China, P.R., [3]Kemin Europa NV, Toekomstlaan 42, 2200 Herentals, Belgium; david.gonzalezsanchez@kemin.com

The present study was conducted to evaluate the effects of coated compound proteases (CCP) from *Aspergillus niger*, *Bacillus subtilis* and *Bacillus licheniformis* on performance, nutrient utilization, gut morphology and carcass traits of broilers fed corn-soya or sorghum-soya based diets. A total of 256 one-day-old male Arbor Acres broilers were randomly allocated to 4 dietary treatments in a 2×2 factorial design which considered diet type (corn vs sorghum) and CCP supplementation (0 vs 1,200 U/kg). The feeding regime consisted on a starter diet fed from d 1 to 21 and a finisher diet fed from d 22 to 42. Broilers fed diets supplemented with CCP showed a significantly better average daily gain (ADG) ($P<0.05$) during the finisher phase and for the whole experimental period ($P<0.01$). Diet type had no effect on the performance of broilers. Apparent metabolisable energy (AME) and dry matter (DM), gross energy (GE) and nitrogen retention (NR) were determined at 2 periods of 3 d total collection, from d 19 to 21 and from d 40 to 42. At both periods AME, DM, GE and NR were all increased by supplementation with CCP ($P<0.05$). On d 21, the birds fed diets supplemented with CCP had significantly increased villus height and villus height:crypt depth ratio as well as decreased crypt depth in the duodenum, jejunum and ileum ($P<0.01$). CCP supplementation significantly increased breast muscle weight, pH24h value and decreased breast muscle drip loss ($P<0.05$). The present study confirms the CCP supplementation, independently of the diet type (corn vs sorghum), can improve ADG of broilers during finisher phase and overall period, and can improve AME and DM, GE and NR and gut morphology in the starter phase.

An *in vivo* assessment of coated compound and monocomponent proteases in broiler diets

S. Chandrasekar, D. Partha, Y. Bashir, M. Karthigan and S. Saravanan
Kemin Industries India, Chennai, 600058, India; chandrasekar.s@kemin.com

Use of protease with alternate protein sources is practiced in poultry for cost efficiency and reducing nitrogen excretion. Yet, diverse proteases may yield different responses. To enrich proteases, a compound protease was developed with acid, neutral and alkaline proteases produced by *Aspergillus niger*, *Bacillus subtilis* and *Bacillus licheniformis* respectively. Proteases were coated with pH sensitive polymers, to act in diverse microenvironment of GI tract. A study involving pelleted diets was conducted in broilers to assess the impact of coated compound and monocomponent proteases. The test diets were control, negative control and proteases. Each diet was tested with 10 replicates of 20 birds each. Control diet had a crude protein 22.6, 21.5 and 19.6% and dig. lysine 1.27, 1.20 and 1.01% for prestarter, starter and finisher respectively. Negative control had relatively 5% lesser dig. amino acids and 0.9% lesser CP than control. The diet was having corn, soybean meal (SBM), mustard seed meal and meat cum bone meal as base ingredients. Reformulation allowed to substitute relatively expensive SBM, synthetic amino acids with corn and other inexpensive protein meals. Birds fed control diet had a better growth than those fed negative control, regardless of enzyme supplementation. However, coated compound protease had shown FCR similar to control (1.57) in a reformulated diet, whereas negative control (1.61) and monocomponent protease (1.60) overlooked. In terms of EEF, control and coated compound protease elicited a closer response (336 and 331), whereas the other two groups showed a drop. In this study, coated compound protease permitted to substitute the expensive SBM, while maintaining feed efficiency and animal performance, might be due to improved protein digestion. It could be concluded that improving monocomponent protease to compound proteases is an efficient choice to sustain the feed material resources and efficiency in animal protein production.

Exogenous feed protease visibly degrades protein storage vacuoles from whole soybeans (*Glycine max*)

N.B. Pedersen
Novozymes A/S, Krogshoejvej 36, 2880 Bagsvaerd, Denmark; nbp@novozymes.com

The efficacy of a microbial S1 protease from *Nocardiopsis prasina* on protein digestibility has been demonstrated *in vitro* as well as *in vivo* in monogastrics. However, mechanisms behind this effect and how protein degradation may influence availability of other nutrients, such as phytic acid, is not well understood. Soybean proteins are located in protein storage vacuoles (PSVs), where they co-localize with phytic acid, making it interesting to analyze the availability of phytic acid after protease treatment. The aim was to investigate and visualize the activity of the microbial S1 protease on soybean protein and study the effect on solubilization of phytic acid. To visualize the effect of the protease, soybean cotyledon tissue was fixed and incubated with different concentrations of the protease in the range of 1.4-0.05 mg enzyme protein/ml. The surface of the PSVs within the cotyledon tissue was visualized using scanning electron microscopy. Full-fat soybean meal was incubated with similar dosages of the protease and the effect on phytic acid solubility was measured by high-performance ion chromatography. The microbial protease degraded soybean protein by attacking PSVs thereby increasing solubility of phytic acid. Protein degradation was visualized as holes in PSVs and at increasing concentrations the holes were larger and more abundant. These results confirm the efficacy of the protease to degrade soybean protein and demonstrate a direct effect on soybean PSVs. The results showed that the protease treatment lead to 9-10% more soluble phytic acid thus demonstrating an indirect effect of the protease on phytic acid solubility. These results indicate that the protease degrades PSVs by perforating them, thereby increasing surface area and internal access for other enzymes for further degradation. Hence, it is speculated that application of a protease in feed may increase not only protein digestibility but also increased availability of phytic acid which can then be hydrolyzed by feed phytases.

Effect of sorghum inclusion level and serine protease on nutrient digestibility of broilers

A.H. Sarsour[1], E.O. Oviedo-Rondón[1], H.A. Cordova[1], P. Ferzola[1] and N. Odetallah[2]
[1]North Carolina State University, Raleigh, NC, USA, [2]Novus Int., St. Charles, MO, USA; ahsarsou@ncsu.edu

Sorghum digestibility and energy value is variable due to kafirin proteins that encapsulate the starch. Two experiments were conducted to evaluate the effects of sorghum inclusion level and the addition of protease on digestibility at 14 and 34 d of age. Eight treatments were evaluated in 4×2 factorial arrangements of treatments with 4 inclusion levels of sorghum (0, 25, 50, and 100% replacement of corn) and presence or absence of protease (0 or 300 U/g feed), as main factors. Either 384 or 1,280 Ross 708 day-old male chicks were placed in 64 cages or floor pens with 6 or 20 chicks per cage or pen, respectively. Fecal collection was done at 13 and 14 d or 33 and 34 d of age and analyzed for DM, CP, energy, acid insoluble ash, and starch. Digestibility, AME and AMEn were calculated using this data. There was an interaction effect ($P<0.05$) on FCR. Addition of protease to corn diets improved ($P<0.05$) 0-35 d FCR, but no significant protease effect on diets with sorghum inclusion was observed. No interaction effects ($P>0.05$) or enzyme effect ($P>0.05$) were detected, and sorghum had a positive linear effect ($P<0.05$) on DM digestibility and energy utilization at 14 d. There was an interaction ($P<0.05$) on DM digestibility at 34 d, but no differences in diets with or without the enzyme. There was a quadratic effect of sorghum inclusion ($P<0.001$) on DM and protein digestibility, AME and AMEn. The worst values were observed in the 100% sorghum diets. Lastly, there was a negative linear effect of sorghum inclusion on starch digestibility ($P<0.001$). As sorghum inclusion increased by 1%, the starch digestibility decreased by 0.015% at 14 d and 0.022% at 34 d. Contrary to earlier research, this protease did not improve digestibility or energy utilization in these experiments, which might be attributed in part to the different variety of sorghum used. In conclusion, corn could be replaced up to 50% of the diet without affecting nutrient digestibility.

Influence of whole wheat inclusion and phytase on performance and nutrient utilisation in broilers

M.R. Abdollahi[1], F. Zaefarian[1], A.M. Amerah[2] and V. Ravindran[1]
[1]IVABS, Massey University, Private Bag 11 222, Palmerston North, 4442, New Zealand, [2]Danisco Animal Nutrition, DuPont Industrial Biosciences, Marlborough, Wiltshire, SN8 1XN, United Kingdom; f.zaefarian@massey.ac.nz

The effect of whole wheat (WW) inclusion and microbial phytase on performance and nutrient utilisation in broilers were examined. Three feeding methods of wheat and two inclusion levels of phytase were evaluated in a 3×2 factorial arrangement of treatments. A wheat-based diet was subjected to three different wheat inclusion methods: 622 g ground wheat (GW)/kg of diet and fed from d 1 to 21; 250 g/kg WW replaced GW (w/w) pre-pelleting and fed from d 1 to 21; and 250 g/kg WW replaced GW (w/w) pre-pelleting for the first 10 d and then post-pelleting from d 11 to 21. Two inclusion levels (0 or 500 phytase units (FTU) per kg of diet) of phytase were then used for these diets to develop six dietary treatments. A total of 288, one-d-old male broilers (Ross 308) were allocated to 36 cages and the cages were randomly assigned to the six dietary treatments. Weight gain of birds fed pre-pelleting and post-pelleting diets was highest and lowest ($P<0.05$), respectively, and those fed GW diets was intermediate. Feeding post-pelleting diets decreased ($P<0.05$) the feed intake compared with GW and pre-pelleting diets. Birds fed pre-pelleting and post-pelleting diets had similar feed per gain but lower ($P<0.05$) than that of GW diets. Supplementation of phytase increased ($P<0.05$) the weight gain and feed intake, and reduced ($P<0.05$) feed per gain. Nitrogen and calcium digestibility enhanced with post-pelleting WW inclusion ($P<0.05$). Supplementation of phytase increased ($P<0.05$) nitrogen and fat digestibility. Starch digestibility was increased ($P<0.05$) by phytase supplementation in pre-pelleting diets, but not in GW and post-pelleting diets. The present data suggest that there is opportunity for post-pelleting WW to enhance nutrient digestibility. WW inclusion and phytase had no additive effect on broiler performance. However, their effects on digestibility may be nutrient dependent.

Efficacy of a novel 6-phytase on long-term performance of laying hens

P. Ader[1] and M. Francesch[2]
[1]BASF SE, ENS/LD, 68623 Lampertheim, Germany, [2]IRTA, Animal Nutrition and Welfare, Ctra. Reus-El Morell, 43120 Constantí, Spain; peter.ader@basf.com

A study was conducted to evaluate the efficacy of a new 6-phytase (PHY, Natuphos E) on performance and egg quality in laying hens fed a maize-based diet, low in non-phytate P (NPP) from 20-55 wk of age. There were 4 dietary treatments, 24 replicates each, with 8 Hy-Line brown hens. Two diets were tested: one positive control (PC) diet (37 g/kg Ca and 3.2 g/kg NPP) and one negative control (NC) diet, without inorganic P supplementation (37 g/kg Ca and 1.0 g/kg NPP). The NC diet was supplemented with PHY at 100 and 200 FTU/kg feed. Performance and egg quality characteristics were recorded. The study was set up as a randomised complete block design and data were subjected to two-way ANOVA. Over the entire study, the low P diet without PHY reduced ($P<0.05$) egg production from 91.3% to 88.7%, average egg weight by -0.9 g, daily egg mass production by -2.4 g/hen/d, daily feed intake by -3 g/hen/d and final body weight (BW), and increased mortality ($P<0.05$), relative to the PC diet with adequate P supply. Phytase at 200 FTU/kg feed increased egg production by +2.2 percent points and egg mass by +1.5 g/hen/d relative to the NC ($P<0.05$). With PHY at 200 FTU/kg feed there were no significant differences on egg production, egg weight, egg mass, feed intake, FCR, BW change, and mortality of hens receiving the NC or the PC diet. Phytase at 100 FTU/kg feed also tended to increase egg production and egg mass, but egg weight and egg mass was not restored to that of the PC group. In the overall study, there were no significant differences between treatments on percentage of broken, faulty and dirty eggs, and Haugh units, egg yolk colour, eggshell weight and strength. In summary, results of this study suggest that the new PHY was efficacious in increasing performance of laying hens fed a P deficient diet, during the first 36 wk of the laying period. The results also suggest that 200 FTU/kg feed of PHY are required to restore performance to that of the PC group.

Validation of the nutrient matrix values for a new 6-phytase at different doses in broiler feeds

M. Francesch[1] and P. Ader[2]
[1]IRTA, Animal Nutrition and Welfare, Crt. Reus-El Morell, 43120 Constantí, Spain, [2]BASF SE, ENS/LD, 68623 Lampertheim, Germany; maria.francesch@irta.cat

A study was conducted to evaluate nutritional values of a new 6-phytase (PHY, Natuphos E) for broilers at different doses, using performance and tibia mineralisation. The study was a randomised complete block design of 8 treatments and 8 replicates of 22 male Ross 308 chickens. Treatments were: a positive control (PC) diet providing 4.1 (1-14 d), 3.9 (14-28 d) and 3.7 g (28-38 d) of non-phytate P/kg feed, with a constant tCa/tP ratio of 1.5; and 4 diets (M350, M500, M750 and M1000), with reduced energy, amino acids, P, Ca and Na, according to four corresponding matrix values given for PHY at 350, 500, 750 and 1000 FTU/kg feed (e.g for P 1.0, 1.3, 1.6 and 1.7 g NPP/kg feed reduction, respectively). Diets M500, M750 and M1000 were tested with/without PHY, and M350 diet only with PHY to verify the matrix. Performance (1-38 d) and tibia parameters at 38 d of age were measured. Data were subjected to ANOVA and regression analysis. From 1-38 d, reduced diets without PHY linearly decreased ADG and ADFI ($P<0.05$) and had no effect on FCR. Performance of chickens receiving reduced diets with PHY up to 750 FTU/kg was similar to that of the PC group. Chickens fed the M1000+PHY diet had greater ADG and ADFI ($P<0.05$) and similar FCR relative to PC ($P<0.05$). Reduced diets without PHY decreased tibia weight, ash % and weight, and breaking strength (BS) ($P<0.05$). Ash % and weight and BS reductions were linearly dependent on the matrix value of PHY applied ($P<0.01$). There were no significant differences on tibia DM, ash % and BS of chickens receiving reduced diets with PHY relative to the PC diet. Ash weight remained lower in chickens fed M350 and M500 diets with PHY compared to PC group ($P<0.05$), but not compared to M750 and M1000 diets with PHY. These results validated the nutrient matrix values for the PHY at 350, 500, 750 and 1000 FTU/kg, whereas in the latter case matrix values for energy and amino acids might even be underestimated.

Investigations on the effects of dietary phytase enzyme supplements in laying hens

L. Pál[1], F. Dublecz[1], K. Dublecz[1], F. Husvéth[1], L. Bustyaházai[2], J. Gyenis[2] and S. Janecskó[2]
[1]University of Pannonia, Georgikon Faculty, Department of Animal Science, Deák F. u. 16., 8360 Keszthely, Hungary, [2]UBM Feed Ltd., Fő u. 130., 2085 Pilisvörösvár, Hungary; pal-l@georgikon.hu

The present experiment was conducted to investigate the efficacy of two dietary phytase supplements (Ronozyme NP and Quantum Blue) on laying performance, retention and precaecal digestibility of phosphorus (P) in laying hens. Tetra SL layers of 20 weeks of age were placed in 48 individual cages and were randomly assigned to four experimental dietary treatments for 16 weeks (12 birds per treatment). The positive control (PC) diet was formulated to meet the nutrient requirements of hens containing 0.38% of available P and 3.80% Ca. The negative control (NC) diet contained 0.25% of aP and 3.63% Ca. The two phytases were added at the inclusion levels of 90 g/t (Ronozyme NP; 900 FYT/kg) and 60 g/t (Quantum Blue; 300 FTU/kg) to reach the aP and Ca concentration of the PC diet. According to our results, both phytase supplementation and the NC diet significantly ($P<0.05$) improved P retention of hens and reduced the P concentration of excreta after 16 weeks of experimental feeding compared to the PC diet. The diets containing the phytases resulted in similar precaecal digestibility of P, egg production (%) and egg weight values. Dietary treatments did not affect egg shell thickness and breaking strength. This study indicates that both phytase enzymes investigated can be used to improve precaecal digestibility and retention of P in layers.

Relative efficacy of two phytases based on bone ash of broilers with different body weight

S. Künzel[1], T. Zuber[1], J. Möhring[1], D. Feuerstein[2] and M. Rodehutscord[1]
[1]University of Hohenheim, Emil-Wolff-Str. 10, 70599 Stuttgart, Germany, [2]BASF SE, E-ENE/LW, 68623 Lampertheim, Germany; susanne.kuenzel@uni-hohenheim.de

The main objective of this study was to compare the effects of two phytase products and dicalcium phosphate (DCP) on bone ash weight of broiler chickens using regression analysis. A second objective was to evaluate the influence that number and body weight (BW) of broilers may have when evaluating bone ash data. Unsexed Ross 308 broiler chickens, 8 d of age, were randomly assigned (6 pens of 12 birds each) to 12 experimental diets: a low-P maize-soybean meal-based diet without DCP and phytase, 4 diets supplemented with DCP (0.6, 1.2, 1.8, 2.4 g P/kg from DCP), 4 diets supplemented with Natuphos E 5000 G (NE; 125, 250, 500, 750 FTU/kg) and 3 diets supplemented with Natuphos 5000 G (N; 250, 500, 1000 FTU/kg). After 2 weeks on treatments birds were euthanized. Four birds, representing different BW categories, were selected from each pen. The right tibia and foot were dissected, cleaned, and ashed. Using the MIXED procedure of SAS (9.3) linear regressions were calculated for the bone ash data (mg) plotted against the amount of P originating from DCP (g/kg) or the level of phytase supplementation (FTU/kg) simultaneously for all weight categories. Tibia ash linearly increased with added DCP and phytase. The estimated slopes of regressions did not significantly differ between the BW categories for NE ($P=0.154$), and N ($P=0.291$), and were calculated to be 0.40 (NE) and 0.19 (N). The slopes for DCP differed between BW categories ($P=0.027$) and varied between 120 and 162. Ratios between slopes of DCP and NE or N led to different results for the different BW categories. Evaluation of the foot ash data yielded very similar results to those obtained for tibia ash. Regression analysis indicated that NE was about twice effective as N in increasing bone ash content. Using only one single bird from a group does not yield representative results for bone ash responses.

Using P, Ca, ME, AA and Na contributions for phytase: a semi-commercial study

Y. Dersjant-Li[1], K. Van De Belt[2], C. Kwakernaak[2] and L. Romero[1]
[1]Danisco Animal Nutrition/DuPont IB, Box 777, Marlborough, United Kingdom, [2]Schothorst Feed Research, Box 533, 8200 AM Lelystad, the Netherlands; yueming.dersjant-li@dupont.com

A phytase that is highly active at stomach pH will effectively reduce the anti-nutritional effect of phytate and contribute to improved dietary energy and dig AA values. This study determined the economic benefit of using the contribution of AMEn, dig AA, P, Ca and Na for a *Buttiauxella* phytase. Three treatments were tested, with 5 replicated pens per treatment and 700 Ross 308 broilers (as hatched) per pen, in a complete randomised design. Birds were fed crumble or pelleted diets from 0-42 d in 4 phases. A positive control (PC) diet was based on wheat, corn and SBM, providing adequate nutrients for broilers (AMEn level of 2,800, 2,850, 2,925 and 3,000 kcal/kg, and dig lys of 11.5, 11.0, 10.0 and 9.5 g/kg in starter (0-10 d), grower 1 (11-21 d), grower 2 (22-35 d) and finisher phase (36-42 d), respectively). Two diets were made with reduced nutrient content based on the contribution of the *Buttiauxella* phytase at 500 or 1000 FTU/kg, and with the phytase added at corresponding dose. The nutrient reduction was done by exchange of $CaCO_3$, MCP, $NaCO_3$, poultry fat and synthetic Lys, Met, Thr and Val for diamol in phytase treatments, to keep consistent micro ingredients composition. Highest reduction of nutrients was for 1000 FTU in the starter diet (1.87 g P, 1.89 g Ca, 0.38 g Lys, 0.32 g Met, 0.31 g Thr, 0.38 g Val, 0.4 g Na and 74 kcal AMEn per kg diet). Water and feed were *ad libitum* available. From 0-42 d, BWG was 2,666, 2,706 and 2,687 g ($P=0.57$), FI was 4,461, 4,571 and 4,559 g ($P=0.10$), and FCR was 1.673, 1.689 and 1.697 ($P=0.18$) for the PC, 500 and 1000 FTU/kg phytase treatments respectively. The feed cost per kg BWG was €0.014 and €0.016 lower ($P=0.001$) with phytase treatments at 500 and 1000 FTU/kg, respectively, compared to PC. In conclusion, taking into account the ME and AA contribution together with P and Ca values for phytase is economically interesting in broiler production.

Performance of an *E. coli* 6-phytase on broiler feed rich in rapeseed meal

L. Nollet and K. Bierman
Huvepharma NV, Uitbreidingstraat 80, 2600 Antwerp, Belgium; karel.bierman@huvepharma.com

A broil trial was set up to evaluate the effect of OptiPhos on a feed rich in rapeseed meal. Broilers distributed over 3 treatments (6 pens per treatment with 26 birds/pen) were fed wheat/corn based feeds with rapeseed as major protein source, pelleted at 80 °C. A starter diet (d1-d5, 0.9% Ca and 0.42% available P (avP)) was fed to all treatments while grower (d5-d21, 0.60% Ca and 0.17% avP) and finisher diets (d21-35, 0.60% Ca and 0.17% avP) were supplemented with 0, 250 or 500 OTU OptiPhos® per kg. Technical performance was monitored per phase. Of each pen 3 birds close to average weight were selected at day 21. The right tibia bone was removed, pooled per pen and analysed on bone ash. At 35 days the ileal content from 6 birds per pen were sampled, pooled per pen and analysed for P and phytate isomers (IP3 to IP6). Adding OptiPhos® at 250 OTU/kg increased the end weight to expected level (2,418 g; $P<0.05$ vs control) while reducing feed conversion (corrected to 2,400 g of end weight) significantly by 0.10 to 1.49 ($P<0.05$). Increasing the level of OptiPhos® to 500 OTU/kg increased end weight additionally (+ 46 g; not significant.) without changing the feed conversion. OptiPhos® added at 250 OTU/kg decreased ileal P concentration (from 0.92 to 0.53 g/kg dry matter) and increased P digestibility from 41.7 to 52.1% ($P<0.05$). Increasing the dose to 500 OTU/kg increased ileal P digestibility further to 63.9% and decreased ileal P content to 0.39 g/kg DM ($P<0.05$). Adding OptiPhos® at 250 and 500 OTU/kg increased bone ash by 5.4 and 6.5% respectively ($P<0.05$). OptiPhos® added at 250 OTU/kg reduced the level of IP6 from 35.2 to 22.2 µmol/kg dry matter (37% reduction) while at 500 OTU/kg level was decreased to 13.8 µmol/kg dry matter (60% reduction vs control). It could be concluded from this trial that OptiPhos® at 250 and 500 OTU/kg performs very well on rapeseed meal diets showing increased performance, bone ash and P digestibility.

Different calcium values for phytase and its effect on bone ash and performance of young broilers

C. Kwakernaak[1], P. Plumstead[2] and Y. Dersjant-Li[2]
[1]Schothorst Feed Research, Box 533, 8200 AM Lelystad, the Netherlands, [2]DuPont Industrial BioSciences, SN8 1XN, Marlborough, United Kingdom; ckwakernaak@schothorst.nl

In efficacy studies the dietary supplementation of phytase always result in a clear increase of ileal P absorption and P retention, while such increase of ileal Ca absorption is not always clearly observed. Because phytate degradation by phytase also prevent complex forming with Ca in the birds small intestines, a Ca matrix value for phytase is commonly used in practise of around 1:1 (Ca:P) for poultry diets. The objective of this study was to test the effect of different Ca matrix values of a phytase in a practical broiler starter diet. Six experimental starter diets were fed as pellets to male broilers housed in 6 replicate floor pens, 29 birds per pen from 0-10 days of age. A maize based diet was used as positive control (PC/diet A: per kg feed: 2,800 kcal AMEn, 12.3 g d.Lys, 9.4 g Ca, 6.5 g P, 2.4 g phytate-P). PC diet reduced by 1.59 g P per kg feed and reduced Ca content of 0.85, 1.35, 1.77, 2.19, 2.69 g/kg feed in diets B to F, by exchange of MCP, limestone and diamol. 1000 FTU/kg of a Buttiauxella-phytase was added on top of diets B to F. Feed and water were freely available. At day 10, tibia of 4 birds per replicate were collected for ash measurements in fat free dry matter. Results were analyzed by ANOVA. Tibia-ash was not significantly affected by dietary treatment, but lowest value was found for the PC (449 g/kg DM), and numerically increased up to 463 g/kg DM for the diet with phytase and the highest Ca reduction. A positive linear relationship was found between Ca reduction levels and tibia-ash (R=0.97). Compared to PC (279 g), BWG was significantly increased ($P<0.05$) after phytase supplementation at the lowest reduction of Ca (Diet B). Diets D and F resulted in further significant increase ($P<0.05$) in BWG compared to diets B and C. Results showed the benefit of using a Ca matrix value for phytase, but implied that the optimum value is likely to depend on the Ca level used for the diet in practice.

Broiler age at trial start can influence phytase P equivalency

G.A. Gomes[1], A. Narcy[2], X. Rousseau[1] and R.A.H.M. Ten Doeschate[1]
[1]AB Vista, 3 Woodstock Court, Blenheim Road, Marlborough Business Park, Marlborough, SN8 4AN, United Kingdom, [2]INRA, Recherches Avicoles, Nouzilly, 37380, France; gilson.gomes@abvista.com

The aim of this study was to evaluate the effect of age at trial start on performance and bone parameters of broilers fed graded levels of available P (avP) and it's interactions with phytase supplementation. A total of 3,360 male day-old Cobb 500 broilers were assigned to 12 treatments with 8 replicate pens of 35 birds each in a factorial design with 4 dietary avP levels (0.45, 0.35, 0.25, 0.15%) and 2 age at start of trial. Birds either went on to treatment diets at d1 (D1) or d5 (D5) after being fed 0.45% avP from d1 to d5 (adaptation period). Two different dosages of phytase were employed (400 and 800 FTU/kg, Quantum Blue), and added over the top of the lowest avP diet. Crumbled diets were corn and soybean-meal based, and fed *ad libitum* up to 19 d. Body weight gain (BWG), feed intake (FI) and mortality were measured, and feed conversion ratio corrected for mortality was calculated. At 19 d of age, 3 birds per pen were euthanized, left tibias excised, and tibia parameters (breaking strength, diameter, length and weight, ash in grams per bone and ash percent) measured. Two-way ANOVA was performed (JMP Pro 13), and logarithmical regressions performed to assess P equivalency of phytase addition. Birds fed the lowest avP diets without phytase supplementation had reduced liveability, independently of the age at the start of the trial ($P<0.05$). Interactions were noticed on BWG, FI and tibia length, weight and ash% ($P<0.05$). Regression R^2 was improved on performance parameters, but reduced on tibia parameters, when D5 was employed. P equivalency calculations were 0.192 and 0.254% for 400 and 800 FTU/kg of phytase for D0, and 0.190 and 0.244% for 400 and 800 FTU/kg of phytase for D5. In conclusion, broiler age at trial start can influence phytase P equivalency, and different phytase doses showed to be efficacious regardless of bird age at trial start.

Morphometry, mineralization indexes and strength of tibias of broiler fed with phytase super-dosing

R. Riveros[1], E. Villegas[1], E. Mateo[1], K. Yupanqui[1], J. Inca[1], F. Prado[2] and C. Vilchez[1]

[1]Universidad Nacional Agraria La Molina, Nutrición Animal, Av. La Molina s/n La Molina, Lima 12, Peru, [2]Universidade Estadual Paulista, Zootecnia, Via de Acesso Prof. Paulo Donato Castellane s/n Jaboticabal, 14870 Jaboticabal, Brazil; faprado91@gmail.com

The objective of the experiment was to evaluate the effect of phytase superdoses on the morphometric, mineralization indexes and the resistance to compression of tibias of broiler of 21 days. Day-old Cobb 500 male chicks (n=100) were randomly placed in 20 cages (UE). Birds fed the following diets: T1, without phytase supplementation and T2, T3, T4 with 500, 1000 and 2,000 (superdose) in FTU/kg of phytase supplementation, respectively. The phytase was produced by Pichia pastoris with an *Escherichia coli* phytase gene. Two feeding phases were used: Pre-starter (d1 to 10) and starter (d11 to 21). The experimental diets were formulated to meet the broiler nutritional specifications; feed and water were available *ad libitum*. At 21 days, 10 birds per treatment were sacrificed and the right tibia was used to carry out the measurements of the following: Weight (TW), length (TL), diaphysis diameter (DD), volume (V), density (D) and compressive strength (CS). Shape (SI), Seedor (SeI), Quetelet (Q) and robusticity (R) indexes were calculated as mineralization indexes. The data were submitted to one-may ANOVA under a Complete Randomized Design using the IBM SPSS statistical program and the comparison of means was performed using Duncan´s MRT. The results showed that DD, V and D as well as the mineralization indexes (SI, SeI, Q and R) were not significantly ($P>0.05$) influenced by the dietary treatments. Tibias from birds under T4 showed the highest TW ($P<0.06$) and R ($P<0.05$) values than those from birds of the other treatments. In addition, the lowest ($P<0.06$) and the highest TL and SI values, respectively, corresponded to tibias from birds under T1. In conclusion, the use of high levels of microbial phytase in the diet has a positive effect on the hardness and strength of the tibia bone.

Effect of dietary NPP level on the performance and egg shell quality in the elongated laying period

A. Tischler, V. Halas and J. Tossenberger

Kaposvár University, Animal Nutrition, Guba Sandor str. 40, 7400, Hungary; tischler.annamaria@ke.hu

Egg shell quality problems occur towards the end of the laying period causing considerable economic loss. Our trial aimed to study the effect of two dietary non-phytin phosphorus (NPP) levels on laying performance and eggshell quality in the last 6 months of the long period (12-17 months) egg production. A total of 92 Tetra SL-LL layers were allocated into 4 dietary treatments: A=2.45 g/kg NPP (7.3 g/kg MCP + 0 FTU/kg); B=2.15 g/kg NPP (5.8 g/kg MCP + 0 FTU/kg); C=2.45 g/kg NPP (1.6 g/kg MCP + 300 FTU/kg); D=2.15 g/kg NPP (0 g/kg MCP+300 FTU/kg). Isocaloric feed AMEn/kg (11.6 MJ) with same CP (160 g/kg), Ca (38.2 g/kg) and amino acid content (8.2 g Lys, 7.3 g M+C per kg feed) were fed. Laying performance was characterized by intensity of egg production, feed intake and weight of eggs. Also, the eggshell quality parameters: strength and thickness as well as the rate of broken eggs were determined. Results show that dietary treatments had no effect on the laying intensity in the total duration of the trial, even though low NPP diet had a detrimental effect on the rate of egg production in the 12th, 14th and 17th months of the trial. Egg mass was higher in high NPP level treatment in the 13th, 14th and 15th months of laying period, and in the entire 6-month-trial. The rate of cracked shell or broken eggs was very high from month 15, particularly in hens fed high NPP diets formulated without enzyme. Phytase supplementation reduced the proportion of broken eggs from approximately 10% to 5% or below compared to MCP treatments in the last 4 months. In conclusion, dietary 2.15 g/kg NPP in layer feeds reduces somewhat the egg mass production, however, the loss of production due to the high rate of broken eggs is increasing if layer feed contains 2.45 g NPP/kg and dietary P is from MCP rather than phytate liberated P. It is recommended to use phytase supplementation at the last months of elongated laying period in order to improve the egg shell quality and reduce the rate of broken eggs.

Supplementation of exogenous phytase and inositol on performance of male broilers from 0 to 46 days

L. Linares, A. Sacranie and H. Willemsen

Aviagen Ltd., 11 Lochend Rd, EH28 8SZ Newbridge, United Kingdom; leoblinares@yahoo.com

The complete hydrolysis of phytate yields 6 Inositol Phosphate (IP) groups are IP1-IP6. The use of phytase at high levels, known as superdosing (>1,500 FTU/kg) leads to a more effective hydrolysis of IP groups. A trial was conducted to compare the possibility of superdosing phytase completely hydrolyse phytate with supplemental synthetic inositol (SI) on diets without phytase, evaluating performance of 1,600 male Ross 708 broilers. Birds were randomly distributed to 5 treatments (16 reps/trt): (1) Basal control (BC) Starter (0-10 d)-Grower (11-24 d)-Finisher (25-39 d)-Withdraw (40-46 d); (2) Basal + 500 FTU/kg (Phy500); (3) Basal + 2,000 FTU/kg (Phy2000); (4) Basal + 3,500 FTU/kg (Phy3500); (5) BC + SI at theoretical dose to release 150% of equivalent 3,000 FTU phytase/kg (Inos150%). All diets containing phytase were adjusted according to manufacturer's recommendations (-0.165% Ca, -0.15% available P and -0.03% Na). Live weight (LW) at 24 d was lower for Phy2000 compared to BC ($P<0.05$) and no difference was observed on feed conversion corrected for mortality (FCRmc) among treatments. At 32 d, LW increased with phytase dosages while FCRmc adjusted to 2 kg LW was reduced ($P<0.05$). Inos150% did not display the same performance benefits as the phytase treatments. Carcass yield was higher in Phy3500 compared to the control ($P<0.05$) and no significant differences were observed in breast yield among treatments. At 39 d, Phy2000 and Phy3500 had the highest LW and the lowest FCRmc adjusted to 3 kg LW ($P<0.05$). The BC treatment had the lowest carcass yield ($P<0.05$) and while there were no differences in breast yield among treatments at 39 d. All phytase treatments had the highest LW and the lowest FCRmc adjusted to 3.5 kg LW at 46 d ($P<0.05$), with the highest carcass yield ($P<0.05$) compared to BC. There were no differences in breast yield at 46 d. In summary, regular dose and superdoses of added phytase improved LW, FRCmc, and carcass yield in broilers at 32, 39 and 46 d. Supplementary SI did not have benefit on broiler performance or carcass traits.

Phytase relative activity in pH range of 1.5 to 6.5 and IP6 degradation rate: an *in vitro* study

T. Christensen[1], R. Mejldal[1], L. Romero[2] and Y. Dersjant-Li[2]

[1]Nutrition Biosciences ApS, DuPont IB, Edwin Rahrs Vej 38, DK-8220 Brabrand, Denmark, [2]Danisco Animal Nutrition, DuPont IB, Box 777, Marlborough, United Kingdom; yueming.dersjant-li@dupont.com

The standard phytase activity is measured at pH 5.5. In order to effectively reduce the anti-nutritional effect of phytate, a phytase should be highly active at stomach pH (e.g. 3.5) to degrade phytate quickly and completely. This *in vitro* study determined: (1) the activity of different phytases in a pH range of 1.5 to 6.5, relative to pH 5.5, and (2) the efficacy in degrading phytate in a corn/SBM diet or a phytate-lysozyme protein complex at pH 3.5. Six phytases were tested: a *Buttiauxella* sp. phytase (phytase B); three *Escherichia coli* phytases (phytase E1, E2 and E3); a *Citrobacter braakii* phytase (phytase C); and a Hafnia sp phytase (phytase H), with normalized activity at pH 5.5 for all phytases. In test 1, phytase was reacted with Na-phytate in a multibuffer solution (pH 1.5-6.5) at 37 °C for 60 minutes. In test 2, each of the phytases corresponding to 250, 500, 750 and 1000 FTU/kg, respectively, were incubated in a corn/SBM feed slurries in a pH 3.5 buffer preparation for 10 min at 37 °C. Phytate degradation was measured by anion exchange HPLC as relative to no phytase. In addition, the rate of lysozyme protein-phytate hydrolysis was evaluated as a decrease in turbidity of the complexed substrate at pH 3.5. Test 1 demonstrated that at pH 3.5, the activity relative to pH 5.5 was 188, 97,101,123,118 and 89% for phytase B, C, E1, E2, E3 and H respectively. With lysozyme protein-phytate complex in test 2, phytase B, C, E1, E2, E3 and H, respectively, degraded phytate by 95, 15, 33, 41, 9 and 19% after 4 minutes. In the corn/SBM feed slurry, phytase B, C, E1, E2, E3 and H, respectively, hydrolysed the phytate by 65, 48, 39, 39, 12 and 17% at 500 FTU/kg, and by 85, 75, 69, 71, 24 and 36% at 1000 FTU/kg after 10 minutes. In conclusion, the *in vitro* tests are useful tools in predicting efficacy of phytases *in vivo* and the efficacy strongly differs between the tested phytases.

Overdosing of phytase in broiler diets make sense

S.A. Kaczmarek[1], L. Nollet[2], K. Bierman[2] and A. Rutkowski[1]
[1]*Poznan University of Life Science, Department of Animal Feeding and Feed Management, Wolynska 33, 60-637 Poznan, Poland, [2]Huvepharma, Uitbreidingstraat 80, 2600 Antwerpen, Belgium; sebak1@up.poznan.pl*

The objective of this trial was to establish the effect of overdosing a phytase (OptiPhos® 2500 CT) in broiler chickens feed on performance. The trial was carried out with 960, 1 day old male Ross 308 chicken broilers in 4 treatments (control (CON) and 3 treatments with increasing dose of phytase; 500 (T1), 750 (T2) and 1 000 (T3) units of phytase activity (OTU) with 30 replicates per treatment and 8 birds per replicate. The total trial duration was 42 days with a 3 phase feeding system: starter (1-11 d), grower (12-24 d) and finisher (25-42 d). All diets were in mash form and were provided *ad libitum*. Available phosphorus level in all diets were: starter – 0.42%, grower – 0.31% and finisher – 0.28%. Ca levels in feed were: starter 0.9%, grower 0.7% and finisher 0.64%. The following activates of phytase were determined in experimental diets: starter 32, 565, 810 and 1,115 OTU/kg; grower: 50, 580, 905 and 1,250 OTU/kg; finisher: 27, 495, 680, 1,035 OTU for CON, T1, T2, T3 respectively. During the finisher period ($P<0.01$) and entire experiment ($P<0.05$) there was a linear BWG increase after phytase supplementation (up to + 90 g). During the starter period, but also over the entire experiment a decrease in FCR after phytase supplementation was observed ($P<0.05$), (starter; CON - 1.351 vs T3 - 1.232), (entire experiment; CON - 1.802 vs T3 - 1.753). There was a linear decrease (up to -0.066; $P<0.015$) in the FCR corrected for 2,850 g final weight (average end weight of all treatments) after phytase supplementation. The use of OptiPhos® at already double the normal dose already let to overdosing as it had beneficial effect on broiler chickens performance probably associated with phytate destruction.

Superdosing phytase in broiler feed containing xylanase makes sense

L. Nollet and K. Bierman
Huvepharma NV, Uitbreidingstraat 80, 2600 Antwerp, Belgium; karel.bierman@huvepharma.com

To answer the question if superdosing a phytase on a feed containing already a xylanase still gives beneficial response, a 35 d broiler trial was conducted using mixed-sex broilers distributed over 48 pens (66 broilers per pen). Feeds were corn/soy based (starter phase d0-d21, grower phase d21-d35). Treatments consisted of (1) a positive control feed, (2) a negative control feed by reducing the positive control feed by 100 kcal AME, 5% dig. protein, 0.16% Ca and 0.16% available P (avP), (3) the negative control feed + 1,500 EPU/kg from Hostazym® X + 250 OTU/kg from OptiPhos®, (4) as c but with 500 OTU/kg from OptiPhos®, (5) as c but with 1000 OTU/kg from OptiPhos® and (6) as (3) but with 1,500 OTU/kg from OptiPhos®. All data were subjected to ANOVA followed by a LSD-multiple range test to determine significance. Reducing energy, crude protein, Ca and avP without enzyme supplementation (= negative control) reduced performance(- 95 g end weight, +0.047 feed conversion, $P<0.05$). However, supplementing Hostazym® X at 1,500 EPU/kg and OptiPhos® at 250 OTU/kg restored performance to the level of the positive control, so compensating for the reduction in energy, protein, Ca and aP. Doubling OptiPhos® at 500 OTU/kg tended to increase ($P=0.08$) the end body weight with 43 g extra vs the OptiPhos® 250 OTU/kg treatment, while reducing the feed conversion by 0.02 (not significant). No extra technical benefits could be observed at higher phytase inclusion levels. An economic calculation showed a profit up to + 8.5 eurocent per broiler. It can be concluded from this trial that adding 250 OTU/kg from OptiPhos® and 1,500 EPU/kg from Hostazym® X compensated for a 100 kcal AME, 5% digestible crude protein, 0.16% Ca and avP reduction in feed formulation. in the present trial no extra statistically significant benefit on performance was observed at OptiPhos doses above 500 OTU/kg, demonstration that the double dosage already optimises the superdosing effect.

Effect of superdoses of a microbial phytase on the performance and abdominal fat in broiler chickens

R. Riveros[1], E. Villegas[1], E. Mateo[1], K. Yupanqui[1], J. Inca[1], F. Prado[2] and C. Vilchez[1]
[1]Universidad Nacional Agraria La Molina, Nutrición Animal, Av. La Molina s/n La Molina, Lima 12, Peru, [2]Universidade Estadual Paulista, Zootecnia, Via de Acesso Prof. Paulo Donato Castellane s/n Jaboticabal, 14870 Jaboticabal, Brazil; faprado91@gmail.com

The objective of the present study was to determine the effect of superdoses of microbial phytase on the performance and abdominal fat of broiler. Day-old Cobb 500 male chicks (n=100) were randomly placed in 20 cages (experimental units, EU) with five birds per cage. The treatments consisted of a control diet (T1) and other three diets (T2, T3 and T4) were formulated adding different levels of phytase enzyme 500, 1000, 2,000 FTU/kg, respectively, the latest was considered the superdose. The commercial phytase used is a derivative of Pichia pastoris with an *Escherichia coli* phytase producing gene. The feeding program consisted in two phases: Pre-starter (d 1 to 10) and starter (d 11 to 21). The experimental diets were formulated to meet the broiler nutritional specifications; feed (meal form) and water were available ad libitum. Body weight and feed intake were measured weekly and feed conversion ratio (FCR) and body weight gain (BWG) were calculated. At day 21, 10 birds per treatment were slaughtered to measure the abdominal fat content and expressed as a percentage of live body weight of the bird. The data were submitted to one-may ANOVA under a Complete Randomized Design using the IBM SPSS statistical program and the comparison of means was performed using Duncan´s MRT. BWG was not significantly influenced by treatments ($P>0.05$); however, T2 had a better FCR ($P<0.05$). As well as T2 and T4 had the lowest accumulation of abdominal fat ($P<0.05$). In conclusion, the inclusion of phytase favors feed efficiency, reduces the percentage of abdominal fat, although it does not show a difference between the recommended inclusion level and the superdoses.

Session 05 Nutrition and gut health   Theatre 1

Effect of arabinoxylo-oligosaccharides on intestinal short chain fatty acid production in broilers

N. Morgan, C. Keerqin and M. Choct
University of New England, Armidale, 2351, Australia; nmorga20@une.edu.au

Arabinoxylo-oligosaccharides (AXOS) are hydrolytic degradation products of arabinoxylans (AX) that can be fermented by the gut microbiota, thus potentially displaying prebiotic properties. This study examined the effects of AX and AXOS on ileal and caecal short chain fatty acid (SCFA) concentration of necrotic enteritis (NE) challenged broilers. Male day-old Ross 308 chicks (n=180) were fed a standard wheat-soybean meal based grower diet supplemented with either 2% AX, 2% AXOS or 2% AX with 16,000 BXU xylanase (Econase® XT 25, AB Vista Feed Ingredients) (AX+E) from d10-21. The AX was isolated from a starch milling by-product and AXOS produced by hydrolyzing this AX with 16,000 BXU xylanase. NE was induced in half of the birds by oral doses of 1 ml *Eimeria* species (*E. acervulina*, *Eimeria maxima* and *Eimeria brunetti*) at d9 and 2 ml field strain of *Clostridium perfringens* type A (approx. 107 cfu/ml) at d14. On d21, digesta was collected from the ileum and caeca from 10 birds per treatment and the SCFA content was determined by gas chromatography. Total ileal and caecal SCFA content was lower ($P=0.006$ and $P=0.003$, respectively) in birds fed AX compared to those fed AXOS or AX+E. Ileal formic acid was lower in the unchallenged birds fed AX or AX+E and higher in the challenged birds fed AX compared to any other treatment ($P=0.007$). Ileal acetic and caecal lactic acid levels were higher ($P=0.033$ and $P<0.001$ respectively) in birds fed AXOS compared to AX or AX+E. Caecal acetic and valeric acid levels were higher in birds fed AXOS compared to those fed AX, and isovaleric acid was lower in birds fed AX compared to those fed any other diet. Caecal lactic acid was also higher in birds fed AX+E compared to those fed AX. Ileal total SCFA and lactic, acetic and succinic acid contents were higher ($P<0.001$) in the challenged birds compared to the unchallenged birds. Caecal lactic, succinic and valeric acid contents were also higher in the challenged birds. These findings suggest that AXOS are effective prebiotics for poultry.

Small particle size wheat bran suppresses *Salmonella* by hosting a butyrogenic microbial network

K. Vermeulen, J. Verspreet, C.M. Courtin, S. Baeyen, A. Haegeman, R. Ducatelle and F. Van Immerseel
Ghent University, Salisburylaan 133, 9820 Merelbeke, Belgium; karen.vermeulen@ugent.be

A variety of compounds, including organic acids and probiotics, can reduce *Salmonella* colonization in the gut of broilers. But the feed industry is always looking for more economical ways to control *Salmonella*. One of these could be wheat bran, a highly concentrated source of insoluble fibre, which is partly fermented by the gut microbiota. We evaluated the effect of wheat bran on *Salmonella* in broilers. In 2 infection trials the supplementation of wheat bran with reduced particle size of 280 μm (WB280) resulted in a significant decrease in *Salmonella* cecal counts compared to birds receiving regular feed or feed supplemented with unmodified bran (1,690 μm (WB). We proposed an underlying mechanism based on a more efficient fermentation of WB280 into butyrate and propionate which led to lowered expression of invasion genes and ultimately a lowered invasion potential. Next, we analysed the microbiota adhered firmly to the bran. The relative abundance (RA) of *Bifidobacteriaceae* was significantly increased on WB and WB280. On WB280 the RA of *Lachnospiraceae* was significantly increased compared to WB. Both fractions, and specifically WB280, were less colonized by *Enterobacteriaceae* compared to the control. Significantly higher numbers of *Ruminococcus torques* and *Bifidobacterium pseudolongum* were associated with WB280. This may suggest that WB280 stimulates cross-feeding between the lactate producing *B. pseudolongum* and the lactate-consuming, butyrate-producing *R. torques*. The accompanied increased concentrations of butyrate may inhibit *Salmonella* and other *Enterobacteriaceae*. We conclude that particle size of prebiotic substrates (e.g. wheat bran) can influence the gut microbiota. Reduced particle size wheat bran can improve gut health and aid in pathogen control where untreated wheat bran may not have the same effect.

The effect of inulin and/or wheat bran in the diet during early life on intestinal health of chicks

B. Li, J. Leblois, B. Taminiau, L. Willems, Y. Beckers, J. Bindelle and N. Everaert
University of Liege, Passage des Déportés, 2, 5030 Gembloux, Belgium; bli@student.ulg.ac.be

Dietary fibers could improve the host's health by inducing favorable changes in intestinal microbiota. The complex intestinal composition of the microbiota develops early in life. In our study, the use of inulin and/or wheat bran during the starter phase was investigated in broiler chicks. Nine hundred and sixty 1-d old male broilers were fed four types of diet: 4% inulin (IN), 10% wheat bran (WB), 4% inulin+10% wheat bran (WB+IN) or a control diet without inulin and wheat bran (CON) for 11-d (6 pens per group). The body weight (BW, 10 per pen) was recorded on 7-d. Paraffin sections were used to measure villi (V) and crypt (C), and MUC2, Occludin, Claudin-1 were determined by RT-PCR. 16S rDNA high throughput sequencing was used to determine the microbiota composition in the ceca. On 7-d, BW of broilers fed the WB, WB+IN was significantly higher than that of broilers fed the IN, CON. The V/C in the IN+WB, WB groups were significantly higher compared to the IN, CON groups in the jejunum. In the ileum, the V/C was significantly higher in the IN+WB group compared to the CON group. Significantly higher expression of Claudin-1 in the jejunum was observed in the IN group compared to the WB, WB+IN groups, with the control group having intermediate values. In the ileum, the expression of Claudin-1 in the IN, CON groups were significantly higher than that in the WB, WB+IN groups. The relative abundances of *Anaerostipes* and *Faecalibacterium* were higher in the IN group compared to the other three groups. The relative abundance of *VadinBB60_unclassified* in the WB+IN group was higher compared to the other three groups and the relative abundance of *Defluviitaleaceae_unclassified* were lower in the WB+IN group compared to the CON group. In conclusion, WB and WB+IN as an ingredient of the starter diet could ameliorate growth performance and intestinal morphology during the starter period, while inulin had a positive effect on microbiota composition.

Effect of butyrate concentration in the GIT on innate and adaptive immune responses of broilers

P.C.A. Moquet[1], G. Konnert[1], A. Lammers[2], L. Onrust[3] and R.P. Kwakkel[1]
[1]Wageningen University and Research, Animal Nutrition Group, P.O. Box 338, 6700 AH Wageningen, the Netherlands, [2]Wageningen University and Research, Adaptation Physiology Group, P.O. Box 338, 6700 AH Wageningen, the Netherlands, [3]Faculty of Veterinary Medicine, Department of Pathology, Bacteriology and Avian Diseases, Salisburylaan 133, 9820 Merelbeke, Belgium; pierre.moquet@wur.nl

A complete randomized block design with 6 dietary treatment groups in 5 blocks was employed on 240 broilers to assess the effect of luminal butyrate concentration on innate and adaptive immune responses. Diets contained either no (control) or 1 g/kg butyrate protected by different means (unprotected, tributyrin, fat-coated, wax-coated and polymer-protected) to allow contrasts in luminal concentrations along the GIT. At d 21 and 22 post-hatch, 10 birds per group received intratracheally administered human serum albumin (HuSA; 0.5 mg/day). Natural and anti-HuSA IgM and IgG titers were measured at d 0, 3, 7 and 18 post challenge. HuSA-challenged birds were dissected afterwards. Contents of each GIT segment were analysed for butyrate concentration. At 42 d post hatch, remaining birds received an innate immunological challenge consisting of intraperitoneally administered lipopolysaccharide (LPS; 0.5 mg/kg BW). Body weight (BW), nitrogen and energy balance were measured in the 24 h post LPS challenge. Dietary intervention resulted in significant changes in butyrate concentration along the GIT $P<0.05$). Butyrate increased significantly natural IgG titers $P<0.05$) when present in the proximal GIT $P<0.1$) but not in the ileum and colon. Growth was positively correlated with natural IgG titers $P<0.0001$; $R^2=0.329$). Anti-HuSA Ig titers increased significantly with time $P<0.05$) and were numerically higher for butyrate-supplemented birds. Diets did not influence BW loss, feed intake and nutrient retention during the 24 h post LPS injection. This study indicates that the presence of butyrate in the proximal GIT has positive effect on humoral immunity of broilers.

Effects of butyrate on performance, cecal microbiome and intestinal immune-related genes of broilers

C. Bortoluzzi[1], A.A. Pedroso[1], J.J. Mallo[2], M. Puyalto[2], M.J. Villamide[2], W.K. Kim[1] and T.J. Applegate[1]
[1]University of Georgia, Athens, GA, 30602, USA, [2]Norel Animal Nutrition, Madrid, 28007, Spain; bortoluzzi.c@gmail.com

The objective of this study was to evaluate the effect of sodium butyrate (SB) on performance, cecal immune-related genes, and cecal microbiome of broilers when dietary energy and amino acids concentrations were reduced. One-day-old male Ross 708 broiler chicks (2,208) were fed dietary treatments in a 3×2 factorial design (8 pens/treatment; 46 birds/pen) with 3 dietary formulations (control diet, reduction of 2.3% of amino acids and 60 kcal/kg, and reduction of 4.6% of amino acids and 120 kcal/kg) with or without the inclusion of SB. Feed intake (FI), body weight gain (BW gain) and feed conversion ratio (FCR) were determined. At 28 d of age, cecal tonsils and cecal content were collected for gene expression and microbiota analysis using real time PCR and 16S RNA sequencing, respectively. SB improved the BW gain ($P<0.05$), without affecting FCR. The amino acid and energy reduction impaired BW gain by 6% ($P<0.01$), while the SB improved BW gain by 2% ($P<0.05$). An interaction effect ($P=0.004$) showed that SB improved the BW gain of birds fed the first level of nutrient and energy reduced diets. SB and the nutritional density of the diets modified the structure, composition and predicted function of the intestinal microbiota. The nutritionally reduced diet altered the imputed function performed by the microbiota, and the SB supplementation was able to reduce these variations, keeping microbial function similar to that observed in chickens fed a control diet. Dietary supplementation of SB to broiler chickens modulated the immune measures in the cecal tonsils; wherein SB upregulated the expression of ubiquitin-editing enzyme A20 in broilers fed control diets ($P<0.05$) and increased IL-6 expression ($P<0.05$). SB partially recovered BW gain of birds fed nutrient and energy reduced diets, modulated the cecal microbiota and demonstrated immune modulatory effects.

Broiler performance and gut health as affected by dietary lactylates and coated calcium butyrate

M.H. Van Den Brink[1], W. Merckx[2], S. Massart[2], G.J. Bouwhuis[3] and E. Jochems[4]
[1]Twilmij, Houtbeekweg 4, 3776 LZ Stroe, the Netherlands, [2]ZTC Leuven, Bijzondere Weg 12, 3360 Lovenjoel, Belgium, [3]GvP Emmen, Atlantis 39, 7821 AX Emmen, the Netherlands, [4]Corbion, Arkelsedijk 46, 4206 AC Gorinchem, the Netherlands; mvdbrink@twilmij.nl

The effect of lactylates either or not combined with calcium butyrate was evaluated in a broiler trial. 576 male chicks (Ross 308) were randomly divided over 36 pens with 6 treatments. The trial setup, a 3×2 factorial design, contained the diets with and without lactylates (0.1% Aloapur CP or 0.2% Aloapur, Corbion) either or not combined with coated calcium butyrate (0.05% Genial BUTYGAN, Twilmij). The broilers were fed a 3 phase program, with wheat-soybean meal-corn based diets and nutrient levels according to CVB recommendations. Enzymes and coccidiostats were excluded from the diets. Birds were housed at commercial density and had *ad libitum* access to feed and water. Average pen weight and feed intake were recorded at 1, 14, 28 and 38 days of age. Feed conversion ratio (FCR), average daily growth (ADG) and average daily feed intake (ADFI) were calculated for each period and the total period. Litter quality and foot pad lesions were scored visually at day 21, 28 and 38. At day 28, one broiler per pen was sacrificed for determination of gut morphology parameters. All data were analyzed by GLM procedure (SPSS). Inclusion of lactylates numerically improved FCR (day 28-38 and day 0-38) by decreasing ADFI ($P<0.10$). The positive effect of lactylates on FCR seems to be most pronounced in the diets supplemented with coated calcium butyrate; there was an interaction between lactylates and calcium butyrate in the period 14 to 28 days ($P<0.05$). Broilers which received the calcium butyrate supplemented diets showed numerically increased ileal villus lengths and crypt depths ($P<0.10$). Both lactylate and calcium butyrate supplementation improved the overall histological gut scores ($P<0.05$).

β-hydroxy-β-methyl butyrate and conjugated linoleic acids stimulating immune system in broiler

A. Saki, V. Khoramabadi, P. Zamani and A. Ashoori
Bu Ali Sina University, Animal Science, Bu Ali Sina University, Hamedan, 6517833131, Iran; dralisaki@yahoo.com

Growth performance, carcass quality, immune system and hematological parameters were examined by the effect of dietary β-hydroxy-β-methyl butyrate (HMB) and conjugated linoleic acids (CLA) in broiler chickens. Total of 300 day old chickens (Ross 308) were randomized across 20 floor pens and reared up to 42 day. Treatment includes: (1) a normal-protein diet (NPD) was formulated according to the Ross 308 catalog by corn and soybean meal; (2) A reduced-protein diet (RPD) was prepared with dietary protein reduced by 12 g/kg (1.2%) relative to the NPD; (3) HMB (3 gr/kg diet) + RPD; (4) CLA (5 gr/kg diet) + RPD; and (5) CLA + HMB + RPD. The results have shown significantly higher Body Weight Gain (BWG) and improved Feed Conversion Ratio (FCR) by HMB during the grower or the whole period in broiler chicken ($P<0.05$). In addition, more breast muscle yield was found by CLA+HMB in broiler at 21 days of age ($P<0.05$). Fat storage was significantly reduced by HMB supplemented treatment at 42 days. Serum antibodies titer of Newcastle disease was maximum by treatment supplemented with CLA+HMB compared to the control (RPD) in day 27. Lymphocyte increased, Heterophil (H) and H/L ratio decreased by feeding CLA in comparison with the control, at day 21 ($P<0.05$). White Blood Cells (WBC) was elevated by HMB and CLA+HMB treatments, as well as Red Blood Cells (RBC) and Hemoglobin (Hb) were significantly increased by HMB supplemented diet ($P<0.05$). This study have suggested that dietary supplementation of HMB and CLA improved growth performance, stimulated the breast muscle development, and decreased the abdominal fat deposition in broiler chickens.

Protective effect for broiler chickens of monovalerin and the antibiotic BMD in NE challenge model

C.L. Hofacre[1], G.F. Mathis[2] and R. Sygall[3]
[1]University of Georgia/PDRC, Population Health, 953 College Station Road, Athens, GA 30602, USA, [2]Southern Poultry Research, 96 Roquemore Road, Athens, GA 30607, USA, [3]Perstorp BU Feed & Food, Waspik, B.V., Netherlands Antilles; chofacre@uga.edu

The broiler chicken disease Necrotic Enteritis (N.E.) is most often a multifactorial disease of intestinal insult and *Clostridium perfringens* (C.P.). Short chain fatty acids have been effective for both clinical and subclinical N.E. Valerins, being glycerolesters of valeric acid, may provide benefits of triggering the physiological responses associated with improving intestinal health by the short chain fatty acid minus any negative effects associated with non-esterified short chain fatty acids. A 32 cage study evaluated 4 in-feed treatments with 8 replicates of day old Cobb male broiler chickens: No additive/No C.P. challenge; No additive C.P. challenge; BMD 50 g/ton/C.P. challenge; Valerins 1.5 kg/mt/C.P. challenge. The N.E. challenge model used was *Eimeria maxima* gavage at 14 days, C.P. (~10^8 cfu/chick) gavage on days 19, 20, and 21. Intestine lesion scores (LS) were performed per Hofacre et al. on Day 21. Feed consumption and body weight (BW) were evaluated on days 14 and 28. Clinical N.E. results were the no additive/no C.P. had 0%[c] N.E. mortality and L.S. were 0.0[c]; no additive/C.P. 43.8%[a] N.E. mortality and L.S. 1.3[a]; BMD N.E. mortality 7.8%[c] and L.S. 0.8[b]; Valerins N.E. mortality 17.2%[b] and L.S. 0.9[b]. Subclinical N.E. results Day 0-28 were no additives/no C.P. BW 1.015 kg[a] and mortality adjusted (adj.) FCR 1.526[c]; no additive/C.P. BW 0.845 kg[b] and adj FCR 1.966[a]; BMD BW1.044 kg[a] and adj FCR 1.548[bc]; Monovalerin BW 1.064 kg[a] and adj FCR 1.675[b]. In conclusion, this study demonstrated, as expected, antibiotic BMD had the greatest impact in reducing effects of *C. perfringens* for both clinical and subclinical N.E. In a program without antibiotics, Valerins alone in feed significantly prevented both clinical and subclinical N.E.

Improved feed efficiency by MCFA is related to decreased number of bacteria and Lactobacilli in ileum

L.L.M. De Lange[1], N. Van Stralen[2], A. Diekenhorst[2] and P. Agostini[1]
[1]Schothorst Feed Research, P.O. Box, 8200 AM Lelystad, the Netherlands, [2]Chaincraft/NOBA Vital Lipids, Hornweg 61, 1044 AN Amsterdam, the Netherlands; ldlange@schothorst.nl

Medium chain fatty acids (MCFA) especially lauric acid (C12:0) have a strong antimicrobial effect against Gram-positive bacteria. Microbes in the content of the small intestine of broilers are dominated by Gram-positive bacteria like lactobacilli. The objective of this experiment was to study the effect of different MCFA on feed conversion ratio (FCR) and microbial profile (MP) in the small intestine. Eight grower diets, differing in feed viscosity (low vs high) and fat source (1% MCFA as C6:0 (C6), 50/50 mixture of C8:0 and C10:0 (C8+10) or C12:0 (C12), replacing a mix of 1% soybean oil and animal fat (control diet, CD) were fed to broilers from day 7 to 28, resulting in a 2×4 factorial experiment. The microbiota is studied at day 24 by qPCR to quantify the number of all bacteria and by V4 16S rDNA amplicon pyrosequencing to study the MP. In this abstract, only the results of the main effects of MCFA on FCR and MP are reported and discussed. A significant improvement of the FCR compared to the CD was observed in diets containing C6, C8+10 and C12 (1.361 vs 1.334, 1.326 and 1.333 respectively). Only C12 decreased the total numbers of bacteria in the gut significantly compared to the CD (4.77 vs 5.32 10log fg/μl). The MP showed that 93% of the studied sequences were from the genus *Lactobacillus*, and that only C8+10 and C12 compared to CD reduced the numbers of Lactobacilli. C8+10 and C12 caused a shift from L. gasseri to L. aviarius. Although lactobacilli are non-pathogenic, they consume valuable nutrients as amino acids, sugars and starch. This parasitism makes that feed is utilised less well by the broilers. This explains most probably why the long MCFA improve FCR. The positive effect of a short MCFA as C6:0 or caproic acid on FCR remains by this experiment unexplained, but might be related to the effect on Gram-negative bacteria or the immune system.

Investigation of possible synergistic effect of using formic acid and plant essential oils in broilers

R. Taherkhani
Payame Noor University, Department of Animal Science, Faculty of Agriculture, 19395-3697, Iran; r_taherkhani@pnu.ac.ir

The possible synergistic effect of using formic acid (FA) and plant essential oils (EO) in broiler chickens drinking water was investigated. Performance and gut microflora were assessed from day old to 42 d of age. The experiment was carried out using a completely randomized design with factorial arrangement (2×3). Factors were included formic acid (0, 1000 and 2,000 ppm) and EO (0 and 250 ppm) level which were administered through drinking water. Both FA and EO improved performance criteria but their combination failed to create a synergistic effects. Chicks received FA supplemented water had significantly lower numbers of *Clostridium perfringens* and forms. Administration of EO also significantly lowered numbers of pathogenic bacteria (*C. perfringens* and coliforms) while did not affect lactobacilli population. Results obtained in our study suggest a synergistic effect of simultaneous using of FA and EO in reducing gut pathogenic bacteria.

The addition of *Scutellaria* and curcuma plant extracts improve the feed efficiency of chickens

M. Panheleux-Lebastard and M. Mireaux
Groupe CCPA, ZA du Bois de Teillay, 35150 Janzé, France; m.panheleux@groupe-ccpa.com

In the context of antibiotic reduction, anti-oxidant and anti-inflammatory ingredients have been widely investigated and proven to bring health benefits. The objective of this study was to evaluate the effects of a combination of plant extracts (*Scutellaria baïcalensis* and Curcuma) with recognized anti-oxidant and anti-inflammatory properties on growth and feed efficiency of broilers. Three groups of 8 replicates with 4 ROSS 308 male chickens were fed during 35 days with the control diet, the control diet supplemented with the combination of vitamin E (100 ppm) and selenium (0.2 ppm) or the control diet supplemented with the combination of plant extracts *Scutellaria* and curcuma). The zootechnical performances: body weight, average daily gain, feed consumption and feed conversion ratio were measured at each feed change. There was no difference between the groups in terms of growth performance. Feed conversion ratio was statistically improved in the 'Scutellaria and Curcuma' group during the start-up and growth period compared to the control group ($P=0.006$). The vitamin E and selenium association treatment intermediate between the two other treatments. Although not statistically significant at 35 days, the feed efficiency difference was being maintained until slaughter. The addition of plant extracts significantly improved the feed conversion ratio of chickens between 0 and 21 days under experimental conditions. The antioxidant and anti-inflammatory properties of the selected plant extracts influenced the metabolism by improving the feed conversion ratio.

Effects of dietary aromatic plants given to broilers reared under heat stress

R.D. Criste, T.D. Panaite, C. Soica, C. Tabuc, M. Saracila and M. Olteanu
National Research Development Institute for Animal Biology and Nutrition (IBNA), 1 Calea Bucuresti, Balotesti, 077015, Ilfov, Romania; mihaela.saracila@yahoo.com

Researches conducted over the past two decades showed that the vegetable feed additives (plants, plant extracts, spices) may have beneficial effects on the health state of the digestive tract. The 35-d trial used 120 day-old Cobb 500 chicks with an average body weight of 38.15 g/chick, assigned to four groups (C, E1, E2, E3). The broilers were reared under 31 °C average air temperature and 36% air humidity. The starter (1-21 days) and grower (22-35 days) diets had the same basic formulation (corn, rice, wheat, rapeseeds meal, corn gluten and soybean meal). The experimental diets differed, in both stages, from the control diet by the supplements of: 2% artemisia (E1), 2% oregano (E2) and 2% rosehip powder (E3). During the growing period, unlike the diet formulation for the control group, the diet formulations for the experimental groups didn't have coccidiostats. Six chicks per group were slaughtered at 21 and 35 days and bacteriological assays were performed on their intestinal content. The average body weight of the broiler chicks at 21 days was significantly different ($P \leq 0.05$) between group C (850 g/chick) and groups E1 and E3 (782.38 g/chick and 775.91 g/chick, respectively). There were not different between the average body weight of groups C and E2 (792.73±76,48 g). The weight gain during the starter stage (1-21 days) displayed significant differences only between group C (26.95 g/day/chick) and group E1 (24.5 g/day/chick). The gain during the growth period (22-35 days) was not significantly ($P > 0.05$) different between the 4 groups. The determinations for Enterobacteriaceae, *Escherichia coli*, *Salmonella* and Lactobacilli in the intestinal content of the slaughtered chicks were not different between the four groups. Throughout the entire experimental period, the droppings were adequate, dry.

Effects of olive leaves extract on nutrient utilisation and intestinal parameters in broilers

J. Leskovec, A. Levart, J. Salobir and V. Rezar
University of Ljubljana, Biotechnical Faculty, Department of Animal Science, Groblje 3, 1230 Domžale, Slovenia; jakob.leskovec@bf.uni-lj.si

The aim of this study was to evaluate the effect of diets supplemented with the olive leaves extract on the apparent total tract digestibility of some nutrients and of energy and various intestinal and excreta parameters in broilers fed diets, enriched with linseed or walnut oils. Olive fruits and leaves *Olea europaea* L.) contain various secondary plant metabolites (hydroxytyrosol, oleuropein, tyrosol, etc.) that exert anti-microbial, antioxidant, anti-parasitic, anti-inflammatory effects, and may have other favourable physiological functions. Dietary supplementation with the olive leaves extract can therefore affect many physiological systems and improve the health and productivity of broilers. Twenty-four 20-day-old commercial broilers Ross 308 were housed individually in balance cages and divided in 4 groups of 6 animals. Animals were fed a linseed or walnut oil-enriched diets supplemented (OLIVE) or not (CONTROL) with the olive leaves extract equivalent to 1% weight ratio of the dry leaves in the feed. The effects of the olive leaves extract supplementation on the apparent total tract digestibility of dry matter, crude protein, fat, ash, and energy, as well as on the intestinal viscosity, volatile fatty acids (VFA) concentration and excreta malondialdehyde (MDA) were examined. The results showed that animals fed linseed oil enriched diets had increased concentration of MDA in the excreta for 196% ($P < 0.001$) compared to walnut oil. In the animals fed with the linseed oil-enriched diets, MDA concentration in the excreta, supplemented with olive leaves extract, was for 35% lower ($P < 0.001$). There were no effects on the nutrient utilization and on the intestinal viscosity and VFA concentration. We have shown that supplementing a broilers diet with olive leaves extract can have a positive effect on the antioxidative status in the intestine, but does not affect nutrient utilisation and intestinal fermentation.

Improvement of performance and gut parameters of broilers using a new *Bacillus subtilis* probiotic

V. Jacquier[1], A. Nelson[2], L. Rhayat[1], P.A. Geraert[1], K.S. Brinch[3] and E. Devillard[1]
[1]Adisseo France SAS, CERN, Commentry, France, [2]Novozymes Biological Inc., Salem, VA, USA, [3]Novozymes, Animal health & nutrition, Bagsvaerd, Denmark; lamya.rhayat@adisseo.com

The gut microbiota of chicken is a complex community of microorganisms that interplay with the host in a symbiotic relationship. This equilibrium can be weakened, generating dysbiosis, especially when broilers are exposed to stressful conditions or diseases. Probiotics are live micro-organisms which, when administered in adequate amounts, confer a health benefit on the host. This study aims to determine the effects of a new *Bacillus subtilis*, strain 29784, on performance, commensal microbiota and intestinal morphology of growing broilers. A 42-d study was conducted on 2,000 d-old Cobb 500 male broilers, fed with corn-soybean meal based diets as following: (1) Control or (2) Control + *B. subtilis* (1×10^8 cfu/kg of feed). Body weight gain, feed intake, feed conversion ratio, and mortality were measured. At 42 d, digestive contents and tissues were collected for metagenomics and morphology studies. At 42 d, the performance of the broilers fed *B. subtilis* were higher than those of the control group, with +5.7% for BWG ($P=0.001$) and -5.4% for FCR ($P<0.0001$). Comparison of gut communities at cecal level showed 7 significant genus-level differences between the 2 groups as measured by relative abundance. Amongst the most important differences, *Ruminococcus* ($P=0.037$) and *Lachnoclostridium* ($P=0.029$) were increased in treated birds vs control. Both genus are known as butyrate producer, an important microbial metabolite to promote gut health. Microvilli were significantly longer in ileum (+18%, $P<0.001$) and cecum (+17%, $P<0.001$) of *B. subtilis*-fed birds compared to controls, suggesting an increase in surface of absorption. The inclusion of *B. subtilis* 29784 in diet brought benefits to broilers in terms of performance. This effect might partially be explained by a positive change in intestinal microbial ecosystem and morphology.

Reducing necrotic enteritis lesions by administration of *Bacillus licheniformis*

V. Hautekiet[1] and J. Verbeke[2]
[1]Huvepharma NV, Uitbreidingstraat 80, 2600 Antwerp, Belgium, [2]Poulpharm, Prins Albertlaan 112, 8870 Izegem, Belgium; veerle.hautekiet@huvepharma.com

Necrotic enteritis, caused by enterotoxigenic *Clostridium perfringens*, affects the gut health of industrial poultry worldwide. Probiotics can be vital to prevent the disease. B-Act® is a probiotic feed additive, consisting of viable spores of *Bacillus licheniformis*. The study describes the effect of the probiotic B-Act® in preventing lesions during induced necrotic enteritis. In this floor pen trial, broilers were supplemented with B-Act® (6.4×10^9 cfu *B. licheniformis*/kg feed) from starter to finisher. Necrotic enteritis was experimentally induced by feeding a high protein diet, a coccidiosis challenge and *C. perfringens* inoculation. Next, zootechnical performance and intestinal lesion scores of the supplemented birds were compared with an infected untreated control (IUC) group. The mean body weight was numerically higher but not significantly different in B-Act® treated birds compared to the IUC group. However, B-Act® supplemented birds were less likely to show higher necrotic enteritis lesion scores compared to the IUC group. Under the present study conditions, B-Act® reduced the pathology caused by experimentally induced necrotic enteritis.

Bacillus licheniformis improves broiler performance during necrotic enteritis infection

V. Hautekiet[1], G. Mathis[2] and C. Hofacre[3]
[1]Huvepharma NV, Uitbreidingstraat 80, 2600 Antwerp, Belgium, [2]Southern Poultry Research, 96 Roquemore Road, Athens, GA 706-354-1980, USA, [3]University of Georgia, Poultry Diagnostic and Research Center, 953 College Station Road, Athens, GA 30602, USA; veerle.hautekiet@huvepharma.com

B-Act® is a probiotic feed additive, consisting of viable spores of *Bacillus licheniformis*. The objective of the 42 d floorpen study was to determine the benefit of feeding the probiotic B-Act® to broiler chickens in order to reduce *Clostridium perfringens* (CP) induced necrotic enteritis. A randomized block design with 8 replications of 50 birds per pen was used. The treatment groups were: 1. No additive, no CP (NMNI); 2. no additive, CP (NMI); or 3. B-Act® (6.4×10^9 cfu *B. licheniformis*/kg feed), fed continuously in feed. All chicks were vaccinated at hatch with a commercial coccidia vaccine. On d19, 20 and 21 all birds, except NMNI were challenged with CP (1×10^8 cfu/bird). On d21, five birds per pen were scored for NE lesions (scoring 0-3). Bird weights and feed consumption were measured on d21, 35, and 42. All weights were in kg and significance was set at $P<0.05$. This study reproduced clinical Necrotic Enteritis (NE) (9.3% NE mortality for NMI). The adjusted feed conversions at all weigh periods were significantly improved for B-Act® compared to NMI (d0-21 (1.53 vs 1.55), d0-35 (1.66 vs 1.73), and d0-42 (1.75 vs 1.83). Average weight gains were significantly improved for B-Act® compared to NMI at d35 (1.59 vs 1.40) and d42 (2.19 vs 1.97). B-Act® fed bird's d42 FCR and weight gain were statistically equivalent to the NMNI birds. NE was reduced by feeding B-Act® with significantly lower % NE mortality (1.9% vs NMI 9.7%) and NE lesion scores (0.58 vs NMI 0.83). The feeding of B-Act® demonstrated significant improvements in performance as well as reducing necrotic enteritis in coccidia vaccinated broilers.

A probiotic strain of *Bacillus subtilis* reinforces intestinal epithelial barrier function

R. Martín-Venegas[1,2], A. Pradilla[2], M.T. Brufau[1,2], L. Rhayat[3], E. Devillard[3] and R. Ferrer[1,2]
[1]Institut de Recerca en Nutrició i Seguretat Alimentària (INSA), Universitat de Barcelona, Recinte Torribera Av. Prat de la Riba, 171, 08921 Santa Coloma de Gramenet, Spain, [2]Departament de Bioquímica i Fisiologia, Facultat de Farmàcia i Ciències de l'Alimentació, Universitat de Barcelona, Av. Joan XXIII 27-30, 08028 Barcelona, Spain, [3]Adisseo Inc., 2 Rue Marcel Lingot, 03600 Commentry, France; raquelmartin@ub.edu

The intestine possesses fundamental functions in metabolism and protection from exogenous agents. In poultry, one of the tools to help to maintain the correct homeostasis of this organ is the use of probiotics, especially under intestinal inflammation stress when the epithelial barrier function is compromised. The objective of this study was to investigate the effect of *Bacillus subtilis* 29784 on epithelial barrier function using intestinal Caco-2 cells. Cells were cultured onto polycarbonate filters (Transwells: 12 mm diameter and 0.4 μm pore size) and maintained during 21 days to allow cell differentiation. Then, the bacteria (MOI 1 and 10) were added to Caco-2 cell cultures, both in non-disrupted and in disrupted barrier condition induced by the incubation with TNFα (200 ng/ml). Paracellular permeability, as an indicator of epithelial barrier function, was assessed by measurement of transepithelial electrical resistance (TER) and paracellular fluxes of FD4, a specific substrate for this route. In non-disrupted Caco-2 cell monolayers, the incubation with *B. subtilis* 29784 significantly enhances epithelial barrier function by increasing TER. Moreover, the incubation with *B. subtilis* 29784 prevents the epithelial barrier disruption induced by TNFα, by increasing TER and decreasing the paracellular fluxes of FD4. In conclusion, *B. subtilis* 29784 reinforces epithelial barrier function in intestinal Caco-2 cells by improving paracellular permeability, in non-disruptive as well as in disruptive conditions.

Presence of *Enterococcus faecium* M74 in gut of young chicks after *in ovo* application

L. Skjoet-Rasmussen, T. Styrishave, A. Blanch and D. Sandvang
Chr. Hansen A/S, Boege Allé 10-12, 2970 Hoersholm, Denmark; dklisk@chr-hansen.com

Providing probiotics to embryos before hatching may improve bird performance after hatching. *Enterococcus faecium* M74 was injected into fertile broiler eggs on day 18, being the objective of this study to investigate the presence and level of M74 in newly hatched and 7-day-old chickens using Pulsed Field Gel Electrophoresis (PFGE) typing. From six 1-day-old (app. 12 hours post-hatch) and six 7-day-old chickens, the following samples were retrieved: Yolk sac (YS), caecal tonsils (CT), and the rest of the intestinal tract (IT). The samples were diluted 10-fold and plated out on TSA blood agar and *Enterococcus* selective agar. Counting of cfus was done using the selective agar results. In general, a high number of uniform bacterial cultures was revealed. When visually inspecting the colonies, many looked homologous and had the colony morphology of *Enterococcus*. One randomly collected colony from each sample was cultivated to ensure purity in preparation for PFGE typing. The YS, CT and IT isolates from the 1- and 7-dayold chickens and the M74 reference strain were typed by PFGE with the rare cutting restriction enzyme SmaI. *E. faecium* M74 was found in high concentration in both YS and intestinal samples of the 1-day-old and 7-day-old chickens tested. The M74 strain was found in high concentration in YS (2×10^8–6×10^9 cfu/g), CT (2×10^6–7×10^8 cfu/g) and IT (5×10^4–1×10^8 cfu/g) samples compared to the initial inoculation dose (2×10^5 cfu/g), indicating that the strain has been multiplying in the animals. Isolates with M74 PFGE pattern were found in samples from both 1-day-old and 7-day-old chickens. The prevalence of M74 was highest in 1-day-old chickens (88%) as compared to 7-day-old chickens (67%). In conclusion, *in ovo* injection of *E. faecium* M74, the probiotic strain M74 was found in high concentration in YS, ST and IT of both 1-day-old and 7-day-old, demonstrating that the embryos ingested orally this probiotic before hatching.

The effect of encapsulated indigenous probiotics bacteria supplementation on broiler gut health

S. Harimurti, M.S.I. Pradipta, A. Priyono and W. Hadisaputro
Faculty of Animal Science Universitas Gadjah Mada, Animal Production, Jalan Fauna No. 3 Bulaksumur, Yogyakarta 55281, Indonesia; sriharimurtibasyar@yahoo.com

Supplementation of encapsulated indigenous probiotics lactic acid bacteria on intestinal histomorphology and goblet cell density of broiler chickens was investigated. A total of 48 day old chick Lohmann strain broilers reared during 21 days were divided into four treatment groups, namely T0, T1, T2, and T3. The T0 group was raised with unsupplemented probiotics, while T1, T2, and T3 were supplemented with multistrain probiotics at different level 0.5 g; 1.0 g; 1.5 g/kg feed, respectively. The result showed that feed consumption was not affected by probiotic supplementation whereas weight gain and feed conversion ratio were significantly affected ($P<0.05$) by the treatments. Furthermore, probiotic supplementation treatments had the highest and widest villus in intestinal epithelium than control and increased cell goblet density in the jejunum. In conclusion, supplementation of encapsulated indigenous probiotics lactic acid bacteria stimulate proliferation of intestinal epithelium in broiler chickens at starter phase in order to improve the performance and gut health.

Modulation of intestinal microbiota and lipid metabolism in mule ducks

M. Even[1], S. Davail[1], K. Gontier[1], A. Tavernier[1], M. Houssier[1], M.D. Bernadet[2] and K. Brugirard-Ricaud[1]
[1]INRA, NuMéA, 371 Rue du ruisseau, 40004 Mont de Marsan, France, [2]INRA, UEPFG, Domaine d'Artiguère route de Haut Mauco, 40280 Benquet, France; maxime.even@univ-pau.fr

The intestinal microbiota is a complex environment and plays an essential role in host metabolism and physiology. Previous studies from the laboratory allow us to characterize the implantation of ducklings' microbiota and its diversity before and during overfeeding. It was demonstrated an initial plasticity and a potential role of the *Lactobacillus* genus during overfeeding. In order to understand the functional role of the microbiota and in particular the *Lactobacillus* genus on the hepatic steatosis in ducks, we supplemented ducks with probiotics. Because of initial plasticity and potential role during overfeeding, we decided to perform the supplementationfrom hatching to the end of overfeeding with a *Lactobacillus salivarius* strain isolated from ileal content of an overfeed duck (groupA) and a mix of strain isolated from chicken (groupB). A control without supplementation was used to determinate probiotics effects (groupT). Measurement of livestock performances, immune and metabolism gene expression and gut microbial population were performed at 15 d of age and after overfeeding (96 d of age). The liver weight represented almost 10% of BW at the end of overfeeding period. No effect of probiotics strains could be highlighting on livestock performances. The lipogenesis genes expression, as FasN, increase up to 200 fold during overfeeding, independently of the experimental group. Pro-inflammatory gene expression, as LITAF, decrease up to 3 fold in over-fed duck. The main order in ileal content before overfeeding is *Clostridiales* and represent more than 96% of total sequences. After overfeeding, the main orders are *Lactobacillales* and *Clostridiales* which respectively represented 67% and 30% of total sequences. This experiment did not improve the performance of ducks during overfeeding. These results highlight the difficulty of carry out an effective supplementation.

Isolation of lactic acid bacteria from poultry and their anti-pathogenic activities

N. Vieco Saiz[1], F. Gancel[1], R. Raspoet[2], E. Auclair[2] and D. Drider[1]
[1]University of Lille, Institut Charles Viollette EA7394-ICV, Avenue Paul Langevin, 59655, France, [2]Phileo, Lesaffre Animal Care, 59700, Marcq en Baroeul, France; nuria.viecosaiz@etudiant.univ-lille1.fr

The reduction of antibiotic growth promoters has lead to the prevalence of infectious diseases, such as necrotic enteritis. The emergence of this disease seems to be related to high mortality rate and economic losses. For these reasons, health organizations are currently promoting the use of alternative strategies to antibiotics in order to maintain animal production and welfare. Several beneficial bacteria have the ability to prevent the pathogenic bacteria colonization and reduce disease incidence. Because of their GRAS status and capacity to survive in the gastrointestinal tract, lactic acid bacteria (LAB) have demonstrated antimicrobial activities based on the secretion of organic acids, H_2O_2 or bacteriocins. Therefore, the aim of this study was to isolate LAB, mainly lactobacilli from the chicken's ceaca and to study their antimicrobial potential. Ceaca from 3 broiler chickens were aseptically removed, homogenized in PBS where after dilutions were plated on LAB selective LAMVAB (MRS + vancomycin 20 mg/l) or Elliker media, resulting in 1.12 and $3.23 \, 10^8$ cfu/g respectively. Colonies were purified and further selected based on Gram + staining and absence of catalase activity. Further identification was carried out by MALDI-TOF and RAPD-PCR, resulting in 28 *Lactobacillus salivarius*, 14 *Lactobacillus reuteri*, 2 *Streptococcus lutetiensis*, 2 *Streptococcus alactolyticus*, and other Lactobacilli such as *Lb. gallinarium, Lb. gasseri, Lb. jonhsonii* and *Lb. antri*. The antimicrobial activity assessment performed on the 50 strains with the slab method revealed an anti-*Clostridium perfringens* activity for most of these strains. Future research will now focus on the mode of action and *in vivo* evaluation of the efficacy of these strains.

The effect of the addition of different probiotics in diets for chickens with necrotic enteritis

F. Jin[1], A.B. Kehlet[1], A. Blanch[1], H. Kling[1] and M. Sims[2]
[1]Chr. Hansen A/S, Boege Allé 10-12, 2970 Hoersholm, Denmark, [2]Virginia Diversified Research Corp, 1866 E. Market St., Suite 327, Harrisonburg, VA 22801, USA; usfrji@chr-hansen.com

A stable intestinal flora is fundamental in managing *Clostridium perfringens* (CP) overgrowth. Hence a study was run to assess the effect of the addition of five probiotics (with different strains) on broilers in a *C. perfringens* (CP) challenge model. 2,160 day-old chickens were allocated to 6 treatments with 12 replicates/treatment. The treatments were: Control (C), PBA (*B. licheniformis* 1.6×10^6 cfu/g feed); PBB (*B. subtilis* + *B. licheniformis* 1.28×10^6 cfu/g feed, 1:1); PBC (*B. subtilis* 1.0×10^4 cfu/g feed); PBD (*B. subtilis* 1.6×10^6 cfu/g feed); PBE (*B. licheniformis* 1.6×10^6 cfu/g feed). Birds and feed were weighed at day 21 and 42. On day 17, 1 ml of a CP culture (1×10^9 cfu/g) was given to each bird by oral gavage. On day 21, 3 birds from each pen were euthanized and intestines were examined for lesion scores on a 0-3 scoring system. At day 42 PBB (2,174 g), PBA (2,116 g), and PBD (2,041 g) were significantly heavier than C (1,910 g). PBB birds were significantly heavier than PBC (1,997 g) and PBE (2,014 g) birds. PBA chickens were significantly heavier than PBC ones. There were no significant difference between PBC and PBE compared to C. All probiotics significantly improved FCR at d 42 compared to C (2.251) in the following order: PBB (1.847) < PBD (1.863) < PBA (1.867) < PBE (1.887) < PBC (1.926). PBA (2.98%), PBB (6.55%) and PBD (6.55%) resulted in significantly lower mortality ratios than C (22.92%), PBD (13.39%) and PBC (21.43%). All probiotic-supplemented diets significantly reduced NE lesion score (1.139-1.444) compared to the control (1.972). In conclusion, these results demonstrated that, in general, the addition of *Bacillus*-based probiotics has the ability to improve performance and diminish the severity of lesions in CP challenged broilers. The efficacy of each probiotic does not depend on its species, but clearly on the specific strain selected for that probiotic.

A β-galactomannan-rich product combined with *L. plantarum* restores intestinal barrier function

M.T. Brufau[1,2], J. Campo-Sabariz[2], S. Carné[3], R. Ferrer[1,2] and R. Martín-Venegas[1,2]
[1]Institut de Recerca en Nutrició i Seguretat Alimentària, UB, Av. Prat de la Riba, 171, 08921 Santa Coloma de Gramenet, Spain, [2]Facultat de Farmàcia i Ciències de l'Alimentació, UB, Av. Joan XXIII, 27-31, 08028 Barcelona, Spain, [3]ITPSA, Av. de Roma, 157, 08011 Barcelona, Spain; teresabrufaubonet@ub.edu

Salmosan (S-βGM), a mannan oligosaccharide product extremely rich in β-galactomannans with several intestinal health-promoting properties in animal nutrition, is able to protect Caco-2 cell cultures from *Salmonella* Enteritidis invasion and to improve epithelial barrier function. To further investigate the potential prebiotic role of S-βGM, we studied the effect of this product in the presence of *Lactobacillus plantarum* (LP) in an *in vitro* model of intestinal inflammation. The model used consists in the co-culture of differentiated Caco-2 cells with macrophages differentiated from THP-1 cells. The macrophages were stimulated with LPS from *S.* Enteritidis (250 ng/ml), and the differentiated Caco-2 cells were incubated in the absence or presence of S-βGM (500 µg/ml), LP (MOI 10) or a combination of both. Cytokine production (TNFα, IL10 and IL6) and paracellular permeability, assessed by transepithelial electrical resistance (TER), were measured. After 24 h of LPS stimulation, TNFα production was significantly increased by 10-fold, whereas IL10 and IL6 levels were not modified. The combination of S-βGM and LP reduced TNFα production to non-stimulated cell values and significantly increased IL10 and IL6 levels (5- and 7.5-fold, respectively). After 48 h of LPS stimulation, TER was significantly reduced up to 25% by LPS and the addition of S-βGM or LP alone did not modify this variable, whereas the combination of both restored TER to values of non-stimulated cells. Moreover, S-βGM was able to induce an increase of 5-fold in LP growth. In conclusion, we have demonstrated that S-βGM combined with LP protects epithelial barrier function by modulation of cytokine secretion thus giving an additional value to this MOS as a potential symbiotic.

Water-soluble short-chain arabinoxylans improve growth performance and intestinal health of broilers

N. Yacoubi[1], L. Saulnier[2], E. Bonnin[2], E. Devillard[1], L. Rhayat[1], R. Ducatelle[3] and F. Van Immerseel[3]
[1]Adisseo, 6 Route Noire, 03600 Commentry, France, [2]INRA, B.P 71627, 44316 Nantes, France, [3]Faculty of Veterinary Medicine UGent, Salisburylaan 133, 9820 Merelbeke, Belgium; estelle.devillard@adisseo.com

Carbohydrate-degrading enzymes are used to improve broiler performance. Their mode of action is complex and not fully understood. We studied their effect on wheat grain by characterizing the different water-soluble fractions produced with and without enzymatic treatment. These fractions were then produced at pilot scale and their effects on animal performance and intestinal health were investigated. These fractions were incorporated in a wheat-based diet (0.1% w/w) to feed broilers during the first 2 weeks post-hatch and their effects on intestinal production of short chain fatty acids (SCFA), diversity of microbiota and inflammation parameters were studied. The enzymatic treatment increased (+19%; $P<0.05$) the amount of water-soluble arabinoxylans (AX) and reduced their molecular weight (176 vs 49 kDa). Degradation products were short chain AX polymers (SC-AX) with an average degree of polymerization of 54 and not oligosaccharides. The SC-AX significantly increased body weight gain by 14.7% ($P<0.05$) during the first week and by 5.4% ($P<0.05$) during the second week. The abundance of butyrate-producing bacteria in the ceca was increased ($P<0.05$)(i.e. Lachnospiraceae by 5% and Ruminococcaceae by 2.5%). Accordingly, the concentration of SCFA, mainly butyrate and acetate, increased in the ceca (×1.6; $P<0.05$). In addition, the T-lymphocyte infiltration decreased (-50%; $P<0.05$) in the cecal and ileal mucosa while the L-cell density increased (+68%; $P<0.05$) in the ileal epithelium. These results indicate that SC-AX stimulated the growth of butyrate producers and decreased inflammation in the intestinal tract. The MEP appears to generate pre-prebiotic compounds (SC-AX) that stimulate the growth of beneficial bacteria, and further improve intestinal status.

Influence of a yeast cell wall product on microbiological traits of broiler chickens

H. Vodde[1] and J. Zentek[2]
[1]MIAVIT GmbH, Robert-Bosch-Str. 3, 49632 Essen-Oldb., Germany, [2]FU Berlin, Königin-Luise-Str. 49, 14195 Berlin, Germany; heike.vodde@miavit.de

Mannan oligosaccharides & β-D-glucans are main components of a yeast cell wall and can influence intestinal microbiota of broiler chickens. However, results of different studies are inconsistent. Studies regarding effects of yeast cell wall components on microbiota in the crop are rare. Therefore, a study was conducted to investigate whether a yeast cell wall product can exert a dose-dependent impact on the composition and activity of microbiota in the crop, ileum and caeca of broiler chickens. A total of 669 male broiler chickens were used in two consecutive trial runs. Four experimental groups were fed a yeast cell wall product (ImmunoWall®, MIAVIT GmbH) in rising concentrations (0.05, 0.10, 0.20 and 0.30%) over the whole fattening period. The fifth experimental group served as control. All animals received a starter (d 1-14) & grower pelletized compound feed (d 15-35) based on soybean meal, maize and wheat. On day 35, one broiler chicken from each experimental unit was slaughtered and digesta samples from crop, ileum & caeca were taken. Via qPCR lactobacilli, bifidobacteria, enterobacteria & *Escherichia/Hafnia/Shigella* spp. were quantified. These samples were also used for determination of pH-value & bacterial metabolites. The data were analysed by kruskal-wallis-test & polynomial contrast analysis. A positive dose-dependent effect was observed for the lactobacilli in crop- and ileum digesta ($P=0.010$ and $P=0.002$). Similar trends were found regarding lactate & acetic acid concentration in crop and ileum, while inverse dose-dependent effects were noticed in pH-value of crop digesta ($P=0.001$). The yeast cell wall product had no impact on the composition of microbiota in the caeca. It is possible that carbohydrates of the yeast cell wall were fermented by lactobacilli in the upper intestinal tract and thus could indicate a better intestinal health. Health benefits would needed to be further assessed.

Effect of yeast cell wall on corticosterone and intestinal villi in crossbred Thai native chickens

Y. Theapparat[1], P. Rodjan[1], P. Boonyoung[2], N. Roekngam[3], S. Kongtong[1] and J. Lamai[1]
[1]Faculty of Veterinary Science, Rajamangala University of Technology Srivijaya, Department of Veterinary Biomedical Science, Thung Yai, Nakhon Si Thammarat, 80240, Thailand, [2]Faculty of Science, Prince of Songkla University, Department of Anatomy, Hat Yai, Songkhla, 90110, Thailand, [3]Faculty of Science, Prince of Songkla University, Department of Biochemistry, Hat Yai, Songkhla, 90110, Thailand; yongyuth.theap@gmail.com

The experiment was carried out to investigate the effects of yeast cell wall (*Saccharomyces cerevisiae*) supplementation on corticosterone level in plasma and intestinal villi in crossbred native Thai chickens. 1,500 day-old chickens (3-strains) mixed sex chicks, with 5 replicates of 50 birds (25 males and 25 females), were assigned to a 2×3 factorial in CRD with three levels of yeast cell wall (0, 0.625 and 0.125%) supplemented in starter ration (1-28 days of age) and in finisher ration (29-56 of age and 57-84 days of age). According to the results at 84 of age, sex influenced the duodenal villi height in chickens which females showed significantly higher villi than males ($P<0.05$). Supplementation of yeast cell wall into diet with 0.125% had significantly lower crypt depth in jejunum and significantly increased villus height in ileum as compared to the control group ($P<0.05$). In addition supplementation of *S. cerevisiae* cell wall had significant sex × diet interactions on crypt depth and villus height:crypt depth ratio in duodenal mucosae compared with those of the control diet without yeast cell wall ($P<0.05$). However, supplementation of yeast cell wall on circulating corticosterone levels in plasma was not significantly different ($P>0.05$) compared with the control diet. Overall, the results suggest that *S. cerevisiae* cell wall supplementation has the certain potential to enhance nutrient intestinal histomorphology of crossbred native Thai chickens as rearing in a tropical climates.

Changes in immune organs in response to dietary prebiotic oligosaccharides or enzyme supplementation

A.D. Craig[1], M.R. Bedford[2] and O.A. Olukosi[1]
[1]SRUC, Monogastric Science Research Centre, SRUC, Edinburgh, EH9 3JG, Scotland, United Kingdom, [2]A B Vista, Marlborough Business Park, Marlborough, Wiltshire, SN8 4AN, England, United Kingdom; allison.craig@sruc.ac.uk

The aim of this 21-day study with 384 male broilers was to investigate how the addition of carbohydrases or prebiotic oligosccharides affected organs associated with immune function and whether this was dependant on cereal type. The treatments were organised into a 2×4 factorial arrangement with six replicates per treatment. The factors were diet type (wheat or barley based) and four additives (no additive, carbohydrases at 16,000 U or 32,000 U, or prebiotic oligosaccharide). Wheat diets were supplemented with xylanase or xylo-oligosaccharide (XOS). Barley diets were supplemented with β-glucanase or galacto-oligosaccharide (GOS). On day 0 the chicks and feed were weighed and two birds per pen were selcted and euthaised. The weight of spleen, empty gizzard and bursa were recorded and length and weight of sections of small intestine were measured. There was no treatment effect on weight gain or FCR. There was a significant diet type × additive interaction ($P<0.05$) for relative bursa weight with lower bursa weight in diet plus XOS. There was significant ($P<0.01$) diet type × additive interaction for relative ileum length (i.e relative to total small intestine length). Ileum length was shorter ($P<0.001$) in wheat diets supplemented with 16,000 U of xylanase whereas xylanse at 32,000 U or XOS had no effect. In barley diets, 16,000 U of β-glucanase decreased ($P<0.001$) ileum length but ileum length increased ($P<0.001$) when 32,000 U of β-glucanse was added but GOS had no effect. It was concluded that the addition of XOS in wheat diets had a positive effect on organs of immune function. However, the enzymes and oligosaccharides had marginal effect on growth performance in diets adequate in nutrients and energy.

Broilers fed β-mannanase display reduced gut feed-induced immune response signaling

R.J. Arsenault[1], M.H. Kogut[2], B. Carter[3] and J.T. Lee[4]
[1]University of Delaware, Department of Animal and Food Sciences, 531 South College Ave, 19716, Newark, Delaware, USA, [2]United States Department of Agriculture, 2881 F&B Road, 77845, College Station, Texas, USA, [3]Elanco Animal Health, 2500 Innovation Way, 46140, Greenfield, Indiana, USA, [4]Texas A&M University, Poultry Science, 101 Kleberg Center, 77843, College Station, Texas, USA; rja@udel.edu

β-mannans found in broiler feed are known to cause physiological effects that are hypothesized to be related to gut inflammation. Previous studies have shown that the incorporation of β-mannanase in the diet results in improvements to performance parameters related to gut health and feed conversion. Here we report a potential signaling mechanism of β-mannan activity on the gut of commercial broilers and how this signaling changes with β-mannanase supplementation. Two doses of β-mannanase (200 and 400 g/ton of feed) with and without additional β-mannan (3,000 ppm) were tested at 3 time points (d 14, d 28 and d 42 post-hatch) in commercial broilers. Jejuna were collected from 5 birds from each treatment group and time point. Each sample underwent chicken-specific peptide array kinome analysis to determine signal transduction profiles. Cluster analysis of the kinome data showed that birds clustered by age and predominantly by whether β-mannanase had been included in the diet. Pathway analysis showed that the inclusion of additional β-mannan into the diet resulted in increased signaling related to immune response (7 immune processes of $P<0.05$). β-mannanase in feed eliminated these immune response (0 processes of $P<0.05$), indicating that the feed-induced immune response within the jejuna had been eliminated by the breakdown of β-mannan. We also saw statistically significant changes in specific metabolic (glucose, mTOR) and gut function (tight junction) pathways in birds fed β-mannanase. These observed signaling changes in the gut of β-mannanase fed birds are likely the mechanism for the enhanced performance and feed conversion observed in birds given β-mannanase in their diets.

Effects of β-mannanase on intestinal health analyzed in 17 European experiences

K. Poulsen, K.T. Baker, T. Kwiatkowski and K.L. Watkins
Elanco Animal Health, Plantin en Moretuslei 1, 2950 Antwerp, Belgium; poulsen_karl@elanco.com

Antinutritive β-mannan fibers from common feed ingredients provoke a wasteful immune response, which has been demonstrated to cause intestinal inflammation and reduce growth performance of broilers. The use of a β-1,4-mannanase enzyme to degrade these fibers has been demonstrated to reduce intestinal inflammation[1,4] and reduce the detrimental impact of intestinal infections in broiler challenge models. 17 trials conducted under field (14) and field-like (3) conditions were selected to evaluate the effects of a β-1,4-mannanase (Hemicell®) on intestinal health. All the trials selected in the evaluation contained control diet is included (similar diets with and without the enzyme had been used during similar time periods) and if the lesion scoring data already were available in Elanco's HTS (Health Tracking System) database. All lesions were scored according to a robust scoring system. 2,613 average, healthy birds were euthanized and necropsied during the trials for collection of lesion scores related to bird health and welfare, and 23 parameters related to intestinal health were combined in an intestinal integrity index (I2). Statistically significant improvements were demonstrated on the I2 index (Control=94.15 and β-mannanase (T)=95.01; $P<0.001$), Excessive intestinal fluid (C=0.174 and T=0.131; $P<0.01$), Gross coccidiosis lesions due to *Eimeria acervulina* (C=0.442 and T=0.341; $P<0.001$) and *Eimeria maxima* (C=0.121 and T=0.108; $P<0.05$), Litter Eater (C=0.193 and T=0.153; $P<0.01$), and pododermatitis (C=0.928 and T=0.818; $P<0.001$). The evaluation demonstrated that the use of a β-1,4-mannanase enzyme to degrade β-mannans my improve intestinal health.

Performance and microbiota activity in broilers fed organic feed mixtures supplemented with enzymes

P. Konieczka, I. Bachanek and S. Smulikowska
The Kielanowski Institute of Animal Physiology and Nutrition Polish Academy of Scieces, Instytucka 3, 05-110 Jablonna, Poland; p.konieczka@ifzz.pl

An experiment was run to evaluate the influence of diets formulated to fulfill the regulations concerning organic livestock production, and two non GMO feed enzymes (A: 1,4-β-xylanase, 1,4-β-glucanase,1,3(4) β-glucanase and B: fungal β glucanase, pentosanase, hemicellulases) on performance and bacterial enzymes activity in broilers. Control (C) diet was based on wheat and soybean meal (SBM). In experimental diets (E1 and E2) a mixture of cold-pressed rapeseed cake, yellow lupin, pea and potato protein concentrate (PPC) was substituted by SBM, and the AME value was equalized with rapeseed oil. In diets C and E1 amino acids were adjusted to chicken requirements with pure amino acids. In E2 diet sulphur amino acids only were equalized by increasing PPC content. Each diet was prepared without or with either A or A and B enzymes. Total 162 Ross 308 females were fed 1 of 9 pelleted diet from 9 to 35 d of life. At the end the ileal and caecal digesta were collected and analysed for bacterial enzymes activity (α- and β- galactosidase, α- and β- glucosidase and β-glucuronidase). In birds fed C, E1 and E2 diets BWG was 2.10, 1.98 and 1.92 kg, respectively ($P<0.001$), FCR was 1.46, 1.55 and 1.62 kg/kg, respectively ($P<0.001$). The supplementation with one or both enzymes had no effect on performance. The glycolytic enzyme activity in ileum and caecum increased in birds fed E1 and E2 diets ($P<0.01$) compared to C, addition of enzyme A increase activity α- and β- galactosidase in ileum ($P<0.05$), α- and β- glucosidase and β-glucuronidase in caecum ($P<0.01$). It may be concluded that SBM could be substituted in broiler feed mixtures by home-grown feedstuffs, but effect on performance is negative, especially without amino acids supplementation. Addition of feed enzymes probably make some feed components more available to gut bacteria.

GH11 xylanase increases prebiotic oligossach. in wheat bran favouring butyrate-producing bacteria

J.L. Ravn[1], D. Pettersson[1], F. Van Immerseel[2], R. Ducatelle[2] and N.R. Pedersen[1]
[1]Novozymes A/S, Animal Health and Nutrition, Kroegshoejvej 36, 2880 Bagsvaerd, Denmark, [2]Gent University, Bacteriology and Avian Diseases, Salisburylaan 133, 9820, Belgium; jzr@novozymes.com

Alternative solutions to optimize intestinal health in monogastric animals have become essential since the ban of antimicrobials in animal feed. In this study, the prebiotic potential of a commercial feed GH11 xylanase was investigated *in vitro*. Enzymatic degradation of insoluble arabinoxylan (AX), present in wheat bran cell walls, was visualized using microscopy. Enzymatic generation of arabinoxylooligosaccharides (AXOS) were analysed by non-starch polysaccharide (NSP) analysis, mass spectrometry (MS) and carbohydrate chromatography to investigate how AXOS glycan complexity and enzyme dosage affect fermentation patterns in a wheat-based diet. Using a 10 mg EP/kg dosage of xylanase, AXOS with an average degree of polymerization (avDP) of 10 were generated, while using a higher enzyme dosage (50 mg EP/kg) avDP shifted to 4-8. Wheat bran incubated with or without xylanase was simultaneously fermented by broiler cecal bacteria *in vitro* and short chain fatty acid production was monitored. A small but significant increase in butyrate production by addition of xylanase was shown to be dose-dependent. Butyrate-producing bacterial genera Faecalibacterium and Intestinimonas were significantly increased in fermentation reactions while Bacteroidetes levels were significantly lowered. Supernatants from fermentation reactions of wheat bran incubated with and without xylanase and cecal microbiota were tested in an intestinal epithelial layer permeability assay using Caco-2 cells stimulated with LPS. The xylanase addition to the bran incubated with cecal content of broilers reversed LPS-induced epithelial layer resistance losses. The GH11 xylanase was able to degrade wheat bran AX to yield low avDP AXOS, that can be fermented by cecal microbiota resulting in microbiota shifts and beneficial effects on transepithelial resistance *in vitro*.

The impacts of enzymatic pre-treatment and fermentation of pea on gut bacterial activity in broilers

F. Goodarzi Boroojeni, W. Vahjen, K. Männer and J. Zentek

Institute of Animal Nutrition, Freie Universität Berlin, Königin-Luise-Str. 49, 14195 Berlin, Germany; farshad.goodarzi@fu-berlin.de

This study examined the impacts of different inclusion levels of native, enzymatically pre-treated or fermented peas on gut bacterial activity in broilers. For pre-treatment, pea was mixed with water (1:1) containing 3 enzymes, AlphaGal™ (α-galactosidase), RONOZYME® ProAct (protease) and VP (pectinases), and incubated for 24 h at 30 °C. For fermentation, the water contained 2.57×10^8 Bacillus subtilis (GalliPro®) spores/kg pea and incubation time was 48 h. Nine diets were produced by supplying 10, 20 and 30% of the protein with native and processed peas. Diets were allocated to 72 pens (8 pens per diet). At d 35, pH and bacterial metabolites in the crop, gizzard, ileum and caecum of 1 bird per pen (8 birds per treatment) were determined. Inclusion of pre-treated pea in diets reduced pH (4.6 vs 5.0) and increased acetate, ammonia, L-, total lactate and metabolites in the crop compared with native groups (4.5, 17, 27, 51 and 91 vs 2.8, 11, 16, 32 and 56 µmol/g respectively). Broilers fed fermented pea had higher L-, D-, total lactate and metabolites in the crop compared with those fed native pea (25, 26, 51 and 85 vs 16, 16, 32 and 56 µmol/g respectively). Broilers received fermented pea had higher ammonia (6.3 µmol/g) and i-valerate (0.12 vs 0.07 and 0.08 µmol/g) in the crop compared with other groups. Both processes had no effect on bacterial activity in the gizzard, ileum and caecum, except for an increase in the gizzard L-lactate with inclusion of pre-treated pea (27 vs 16 µmol/g). Increasing level of pea products in the diets reduced acetate (14.3, 13.9 and 13.9 µmol/g) and propionate (0.12, 0.09 and 0.03 µmol/g) in the crop as well as D-lactate in the caecum (0.81, 0.22 and 0.40 µmol/g). In conclusion, while inclusion of both types of processed peas in broiler diets increased metabolic activity of bacteria in the crop, the impacts of enzymatic pre-treatment seemed to be relatively more favorable.

Health-related response of broilers fed diets containing raw, full-fat soybean meal and protease

M.M. Erdaw[1], R.A. Perez-Maldonado[2] and P.A. Iji[1]

[1]University of New England, Animal Science, Armidale, NSW, 2351, Australia, [2]DSM, Singapore, Animal Health and Production, 30 Pasir Panjang Road #13-31Mapletree Business, City Singapore 117440, Singapore; piji@une.edu.au

A 2×3 factorial study (protease: 0 or 15,000 PROT/kg and raw full-fat soybean meal (RSBM), replacing commercial SBM at 0, 15 or 25%, equivalent to 0, 45 and 75 g/kg of diet, respectively) was conducted to examine the performance and welfare of broiler chickens. Microbial phytase (2,000 FYT/kg) was uniformly added to each diet, and replicated six times, with eight birds per replicate. Birds were raised in climate-controlled rooms using sawdust as the bedding material and offered starter, grower and finisher diets between hatch and 35 d of age. There were statistical ($P<0.05$) differences in feed intake (FI) and body weight gain (BWG) with increase in RSBM, but FCR (1-35 d) was unaffected. Over 1-24 d, neither RSBM nor protease supplementation affected ($P>0.05$) mortality, footpad dermatitis or intestinal lesions in birds. On 24 d, the weight, length, width and strength of tibia bone were reduced in chickens on high RSBM levels, but this was not statistically significant at 35 d of age. Neither RSBM nor protease levels significantly ($P>0.05$) affected the DM, Ca and P contents of tibia at 35 d of age. On days 24 ($P<0.05$) and 35 ($P<0.01$), Ca concentration in the litter was reduced by high levels of RSBM, but P content was not affected. On days 24 ($P<0.05$) and 35 ($P<0.01$) N content in litter was also increased with increase in RSBM content. Protease supplementation significantly ($P<0.05$) increased the uric acid concentration in the litter at 35 d of age, but the reverse was the case for ammonia concentration. The concentration of inositol in blood plasma at 24 d of age was reduced (not significant) by up to 11.2 and 3.3% with increased levels of RSBM and protease supplementation, respectively. Overall, the results of this study indicate that there are no major health-related risks associated with the replacement of commercial SBM with RSBM at up to 25% in broiler diets.

Effect of fermentable protein in broiler diets on performance and microbiota

D.M. Lamot[1], E.A. Soumeh[1], A.M. Amerah[1], J.E. De Oliveira[1], V.L. Mcintosh Jr[2] and H. Enting[1]
[1]Cargill Animal Nutrition Innovation Center, Veilingweg 23, 5534 LD, the Netherlands, [2]Cargill Animal Nutrition Innovation Center, 10383 165th Ave NW, Elk River, MN 55330, USA; ahmed_amerah@cargill.com

Fermentable (FRM) protein is defined as the amount of undigested protein, corrected for part that may escape hindgut fermentation. This study examined the effect of dietary FRM protein (1.4 vs 2.3%) on growth performance and gut microbial profile (MP) of broiler chickens fed two levels of crude protein (CP; 22.4 and 20.8% from 1 to 14 and 20.4 and 18.8% from 14 to 35 d of age, respectively; low CP diets were supplemented with AA to maintain similar levels as high CP diets), with or without antibiotics as growth promoter (AGP). This resulted in a 2×2×2 factorial design. Feed intake and BW were determined and gain to feed ratio (G:F) was calculated. Litter quality was monitored weekly from day 21. At 35 d, the ceca MP was characterized from 2 birds/pen. Over the entire period (1-35 d) feeding a lower CP diet reduced G:F compared to a higher CP diet. A higher FRM CP diets resulted in a lower ADG and G:F compared to the lower level. Addition of AGP increased G:F compared to the unsupplemented diets. An interaction between CP and AGP inclusion was found for ADFI. AGP inclusion lowered ADFI in the low CP diets and increased it in the birds fed high CP diets. A lower FRM CP decreased probability for a worse litter score from 21 d onwards, whereas a lowered CP level only resulted in a reduced probability at 35 d. Although the MP differed between birds fed low or high CP levels, the differences due FRM CP level were more pronounced. Differences in the MP were reduced in the presence of AGP. In conclusion, it can be hypothesized that the changed MP has contributed to the impaired performance at high FRM CP levels. Based on these results, we propose that FRM CP can be used to formulate broiler diets and formulation based on this concept change nutrient and substrate reaching the lower gut, resulting in performance differences that are linked to changes in intestinal MP.

Effect of diet density on growth performance and intestinal development of broiler chickens

D.M. Lamot, A.M. Amerah and H. Enting
Cargill Animal Nutrition Innovation Center, Veilingweg 23, 5534 LD, Velddriel, the Netherlands; ahmed_amerah@cargill.com

This study examined the effect of diet density, fed until different ages, on broiler growth performance, intestinal development and nutrient absorption using an 8×2 factorial design. Eight levels of dietary fat (ranging from 3.5 to 28%) were fed until either d7 or 21 of age. Treatments fed different diet densities until d7 received a 3.5% dietary fat for the remainder of the period. Increase in dietary fat level was parallelled by an increase in AA, minerals and premix ingredients to maintain a constant ratio to energy. Diets were pelleted using extrusion and fat was added using vacuum coating. At d7, chickens were sampled to measure intestinal length and weight. Jejunum samples were processed for histological analysis. On d 8 and 15, the nutrient absorptive capacity of the intestinal tract was determined using D-xylose. Feeding of increased density levels from d 0 to 7 resulted in an upward quadratic response for BW and ADG, and a downward quadratic response for ADFI and FCR. Duodenum and ileum weights linearly increased, whereas a downward quadratic response was found for duodenum, jejunum and ileum length as diet densities increased. As villi length and crypt depth increased with increased diet densities, the ratio was not affected at d7. Absorption of D-xylose at d8 linearly lowered as diet density increased. Continued feeding of increased diet densities from d7 to 21 resulted in an upward quadratic response for BW and ADG and a downward quadratic response for ADFI. FCR from d7 to 21 linearly lowered. Observed effects of feeding increased diet densities on ADG and FCR from 0 to 7 d of age disappeared from 7 to 21 d of age when birds switched to a low density diet (3.5% dietary fat). No effect was found for absorption of D-xylose at d15. In conclusion, feeding increased diet densities to broiler chickens results in a higher ADG and lowered FCR, but mainly during the period that these diets were fed. Broiler chickens seem to adapt intestinal development based on the diet density level.

Haematology and serum biochemical, indices of broiler chicks raised with different source of water

E. Ehebha, T. Eikaehor and S.E. Okosun

Ambrose Alli University Ekpoma, Edo State, Animal Science, Ambrose Alli University Ekpoma, Edo State, 234, Edo state, Nigeria; theoehebha@yahoo.com

seven weeks water trial was conducted to assess the effect of various sources of water on the haematology and serum bio-chemical indices of broiler chickens.105 broiler were used for the experiment with 21 birds each randomly selected & assigned to 5 water treatments (Rain water, Borehole water, River water,Mud water & Rain water treated with chlorine) in a Completely Randomized Design (CRD).Each group was replicated three times with 7 birds/replicate. Result on haematological indices revealed that MCV was significantly ($P<0.05$) influenced with highest 126.53 for birds on mud water followed by similar values 120.03 & 120.00 recorded among birds on borehole water & river water followed by 118.97 in rain water similar to 118.33 in rain water treated with chlorine. MCH was significantly highest ($P<0.05$) a mong birds on MW with the value 42.47,statistically similar to 42.13,in BHW followed by 41.27 in RNW, followed by similar values 40.87 & 40.80 recorded in birds on RNW & RWC respectively. Serological studies showed that total protein was significantly ($P<0.05$) highest 4.03 in RNW, followed by similar values 3.97 in RVW, 3.67 in BHW & 3.63 in RNW, 3.40 is least value in RWCr. It was therefore concluded that administering RVW & RWC improved the blood & health status of broiler chickens. The physico-chemical analysis of sampled water was done & results suggests drinking water can't be totally free of these elements. The highest turbidity in MW might be due to frequent stirring by fetchers, livestock and contaminants. BHW had the higest level with total hardness in MW in this research. This might be due to differences in geographical locations;it could also be as a result of high level of soil mineral content, especially Ca,Mg and Cl in the water.

Session 05 Nutrition and gut health   Poster 37

Are broiler intestinal health problems connected to the age related succession of their microbiota?

S. Ranjitkar, B. Lawley, G. Tannock and R.M. Engberg

Aarhus University, Animal Science, 8830 Tjele, Blichers Allé 20, Denmark; ricarda.engberg@anis.au.dk

A feeding trial with maize based feed (MBF) and maize based feed supplemented with 30% crimped kernel maize silage (CKMS) was conducted to study the age-related microbial community composition in the broiler GI-tract. Contents from crop, gizzard, ileum and ceca were collected on d. 8, 15, 22, 25, 29 and 36. Microbial diversity was analysed by 454 pyrosequencing of the 16S rRNA-gene. No difference was observed between MBF and CKMS birds and the data from both groups were pooled. Each of the GI-segments was characterized by a specific microbiota. Lactobacillaceae (mainly *Lactobacillus*) were dominant at all ages and in all segments except for the ceca, where Ruminococcaceae and Lachnospiraceae were most abundant. A significant variation in the bacterial diversity was seen between d 15 to d 22, where a remarkable increase of *Lactobacillus salivarius* at the expense of other *Lactobacillus* species was observed in the crop (5% to 37%), gizzard (3% to14%) and ileum (1% to 29%). Further, *Clostridium* increased in ileal content (1%-18%) as birds grew older. Both *Clostridium* and *L. salivarius* are known to de-conjugate bile acids resulting in broiler growth depression via impaired fat digestion. Accordingly, an increased bile acid de-conjugation was observed at that time point. At d 29, a sharp decline in bacterial diversity was observed which may indicate changes in the GI environment affecting the bacterial populations during this period. After the withdrawal of antibiotic growth promoters (AGPs) with effect on Gram-positive bacteria, the frequency of gut related problems in broiler production e.g. dysbacteriosis and necrotic enteritis has increased with a peak in the period from 20-28 days, reflecting the time point of profound changes of the GI microbiota. In the absence of AGPs, the substantial increase of *L. salivarius* and *Clostridium* in the ileum may be a contributing factor to the increased frequency of GI problems.

Growth and nutrient utilization of broiler on mash and pelleted whole-sorghum

M. Mabelebele[1], R.M. Gous[2] and P.A. Iji[3]
[1]University of South Africa, Agriculture and Environmental Sciences, 75 Chistiaan de Wet, Florida Campus, Johannesburg, 1724, South Africa, [2]University of Kwa-Zulu Natal, School of Agricultural, Earth and Environmental Sciences, Pietermaritzburg, Scottsville, South Africa, [3]University of New England, School of Environmental and Rural Science, Armidale, Australia; mabelebelem@gmail.com

A 4 (0, 25, 50 or 75% inclusion level) × 2 (pellet or mash) factorial array in a completely randomized design having six replicates per treatment, with 9 birds per replicate was used in this study. Whole sorghum inclusion did not affect ($P>0.05$) the feed intake, body weight or FCR of broiler chickens at 1-35 days. Pelleting increased ($P<0.05$) feed intake and body weight of broiler chickens during 1-24 days. The interaction between factors was significant for body weight at 24 ($P<0.0021$) and 35 ($P<0.0019$) days of age. Relative gizzard weight and pH were lowest ($P<0.05$) in broiler chickens offered mash diets between hatch and 24 d of age. Feed conversion ratio between hatch and 35 d increased ($P<0.035$, quadratic effect) with an increase in whole sorghum and levelled off with higher inclusion rates. Broiler chickens offered mash diets had significantly higher ($P<0.05$) gross energy and crude protein digestibilities than those on pelleted diets. Although higher levels of whole sorghum inclusions enhanced the gizzard development, the performance of birds offered these levels were not affected. Overall, the results showed that pelleted diets were superior to mash diets.

Effects of rye inclusion on immune competence related parameters in broilers

M.M. Van Krimpen[1], M. Torki[2] and D. Schokker[1]
[1]Wageningen University & Research, Wageningen Livestock Research, P.O. Box 338, 6700 AH, the Netherlands, [2]Razi University, Animal Science Department, P.O. Box 6715685418, Kermanshah, Iran; marinus.vankrimpen@wur.nl

An experiment was conducted to investigate the effects of dietary inclusion of rye, a model ingredient to increase gut viscosity, between 14 and 28 days of age on immune competence related parameters and performance of broiler. A total number of 960 one-day-old male Ross 308 chicks were weighed and randomly allocated to 24 pens (40 birds per pen), and the birds in every 8 replicate pens were assigned to one of three experimental diets including graded levels, 0, 5, and 10% of rye. Tested immune competence related parameters were composition of the intestinal microbiota, genes expression in gut tissue, and gut morphology. The inclusion of 5% or 10% rye in the diet (d14-28) resulted in decreased performance and litter quality, but in increased villus height and crypt depth in the small intestine (jejunum) of the broilers. Relative bursa and spleen weights were not affected by dietary inclusion of rye. In the jejunum, no effects on number and size of goblet cells, and only trends on microbiota composition in the digesta were observed. Dietary inclusion of rye affected expression of genes involved in cell cycle processes of the jejunal enterocyte cells, thereby influencing cell growth, cell differentiation and cell survival, which in turn were consistent with the observed differences in the morphology of the gut wall. In addition, providing rye-rich diets to broilers affected the complement and coagulation pathways, which are parts of the innate immune system. These pathways are involved in eradicating invasive pathogens. Overall, it can be concluded that inclusion of 5 or 10% rye to the grower diet of broilers had limited effects on performance. Ileal gut morphology, microbiota composition of jejunal digesta, and gene expression profiles of jejunal tissue, however, were affected by dietary rye inclusion level, indicating that rye supplementation to broiler diets might affect immune competence of the birds.

Stimulation of antioxidant defences in broilers supplemented with pelleted SOD-rich melon

F.B. Barbe[1], A.S. Sacy[1], J.C. Carillon[2], E.C. Chevaux[1] and M.C. Castex[1]
[1]*Lallemand SAS, Animal Nutrition, 19, rue des Briquetiers, 31702 Blagnac cedex, France,* [2]*Bionov, Research, 939, rue de la Croix Verte, 34090 Montpellier, France; fbarbe@lallemand.com*

The stimulation of antioxidant defences can be investigated by Western Blotting of endogenous antioxidant enzyme production in different organs. In the current study, this robust analysis was used to evaluate this endogenous antioxidant stimulation in broilers supplemented with pelleted SOD-rich melon pulp concentrate (MPC), along with parameters of the immune system (weight, diameter and lesions of Bursa of Fabricius, as a key organ involved in the immune system of poultry). 27 Ross PM3 broilers were individually housed in standard battery cages and randomly allotted in 3 groups (n=9 animals/group) fed the following pelleted diets (pelleting process was performed at 65 °C during 10 seconds) during 18 days: control (coated with 1% soya oil: R1), pelleted (pelleted at 50 g MPC/ton and coated with 1% soya oil: R2), coated (coated with 50 g MPC/ton: R3). At day 18, the antioxidant status was investigated by SOD expression in the Bursa of Fabricius and in the intestine (Western Blotting) and by the analysis of NADPH oxidase activity in the gastrocnemius muscle (chemiluminescence). SOD proteomic expression was increased by 22% in R2 and R3 groups for the Bursa of Fabricius and by 37% and 125% ($P=0.001$) for the intestine in R2 and R3 groups, respectively, compared to R1 group. Moreover NADPH oxidase activity was decreased by 43% in R2 and R3 groups in the gastrocnemius muscle, compared to R1 group ($P=0.001$). This stimulation of antioxidant defences was observed along with increased weight and diameter of Bursa of Fabricius and decreased percentage of damaged follicles ($P<0.05$) in both supplemented groups. This study provides additional information regarding the stimulation of antioxidant defences and the protection of the immune system in poultry supplemented with SOD-rich melon pulp concentrate, without altering birds production performances.

A polyssacharidic extract of Ulva sp stimulates *in vitro* innate immunity in broilers

M. García[1], P. Nyvall-Collen[1] and N. Guriec[2]
[1]*Olmix SA, ZA du Haut du Bois, 56580 Brehan, France,* [2]*Universite de Bretagne Occidentale, 3 rue des archives, 29238 Brest, France; animalcare.ts@olmix.com*

Since microbial resistance to antibiotics is still increasing worldwide, more and more measures are taken to reduce antibiotic use. One example is the French plan Ecoantibio 2017, which aims at reducing the use of antibiotics in veterinary medicine by 25% within the next 5 years. This requires new strategies to be developed to support broilers' immunity. Here we evaluate the effect of an extract of Ulva sp. harvested in Brittany (France), and rich in sulphated polysaccharides (MSP), on avian heterophils and monocytes *in vitro*. Heterophils and monocytes were purified from peripheral blood of 28 day-old, Ross 308 broilers raised in standard conditions. MSP induces the activation of heterophils via improvement of oxidative burst and glucuronidase activity in a dose and time-dependent manner ($P<0.05$). In parallel, the ability of MSP to induce NO production in monocytes was assessed and it was also occurring in a time and dose-dependent manner. Both cell types responded in a range of MSP dose similar to the one of a usual activator. MSP allows *in vitro* stimulation of both heterophils and monocytes; being both cell types major players in broilers' innate immunity. The mode of action of MSP is currently under study. The potential practical applications of this algal extract could be: enhancement of immunity and improvement of animal performance and the ultimate goal: the reduction of the use of antibiotics in animal production systems.

Intestinal mucin dynamics in different chicken strains fed diets with yellow mealworm inclusion

I. Biasato[1], E. Biasibetti[1], A. Sereno[1], F. Gai[2], L. Gasco[3], A. Schiavone[1,4] and M.T. Capucchio[1]
[1]University of Turin, Department of Veterinary Sciences, Largo Paolo Braccini 2, 10095, Grugliasco, Turin, Italy, [2]Institute of Science of Food Production, National Research Council, Largo Paolo Braccini 2, 10095, Grugliasco, Turin, Italy, [3]University of Turin, Department of Agricultural, Forest and Food Sciences, Largo Paolo Braccini 2, 10095, Grugliasco, Turin, Italy, [4]University of Turin, Institute of Multidisciplinary Research on Sustainability, Via Accademia Albertina 13, 10100, Turin, Italy; ilaria.biasato@unito.it

The study evaluated gut mucin composition in chickens fed diets with *Tenebrio molitor* (TM) meal. A total of 140 female medium-growing hybrids and 160 female and male broilers were divided into 2 (basic feed and 7.5% TM inclusion) and 4 (basic feed and 5,10 and 15% TM inclusion) dietary treatments, respectively. Ten birds/diet were slaughtered at 97 (hybrids), 40 (female broilers) and 53 (male broilers) days of age. Neutral, acidic sialylated and acidic sulfated mucins were evaluated by histochemistry on gut samples and mucin staining intensity was semiquantitatively scored. Mucin staining intensity in crypts of hybrids significantly depended on mucin type, gut segment and crypt fragment, while gut segment and villus fragment affected it in villi. Histochemical findings of female broilers were similar, even if TM inclusion influenced them in villi. Histochemical findings of male broilers were also similar, but TM inclusion affected them in crypts. TM5 female and male broilers showed higher mucin secretion among TM animals in villi and crypts, respectively. Crypts overall showed higher production of neutral and acidic sialylated mucins, lower mucin secretion in caecum and greater mucin storage in base fragment. Greater mucin production in ileum and in base fragment was also overall observed in villi. Physiological intestinal mucin dynamics were preserved in all the animals and TM5 seemed to improve mucin secretion in broilers.

Isoquinoline alkaloids improve performance and egg quality in the late stage production in layers

A. Pastor[1] and H.-H. Hsieh[2]
[1]Phytobiotics Futterzusatzstoffe GmbH, Wallufer Straße 10a, 65343 Eltville, Germany, [2]National Pingtung University of Science and Technology, Department of Animal Science, 1 Shuefu Road, Neipu, Pingtung 912, Taiwan; a.pastor@phytobiotics.com

Feed additives strive to support animal performance and well-being. Especially in laying hens during the late stage of production specific phytogenic compounds can be a valuable tool to optimize performance. A standardized blend of plant-derived isoquinoline alkaloids (IQs, Sangrovit® Extra) is known for its anti-inflammatory properties and its positive effects on gut health. The aim of the study was to evaluate the effect of IQ supplementation in laying hens during the late stage of production. 96 laying hens (Hy-Line) were subjected to two treatments at the age of 70 weeks for 6 weeks: (1) Control based on corn and soybean meal, no additive; (2) IQs (120 g/t Sangrovit® Extra). 12 hens/replicate were used in 4 replicates/treatment. Animals were kept in cages. Performance, feed intake and egg quality parameters were measured weekly. At the end of the trial 4 birds were randomly selected from each treatment and sacrificed for evaluation of villus height and villus height : crypt depth ratio in jejunum and ileum. Supplementation of IQs improved egg production significantly ($P<0.05$) compared to animals of the control group (70.9 and 74.7%, respectively). Furthermore, egg mass was positively influenced, if IQs were applied (320 and 338 g, respectively, $P>0.05$). Birds fed IQs had a significantly lower FCR in comparison to hens of the control group (2.23 and 2.13, respectively, $P<0.05$). Haugh Unit, egg shell strength and yolk color were numerically higher in hens fed IQs ($P>0.05$) contributing to a higher product quality. In compliance with earlier studies, villus height and villus height : crypth depth ratio was significantly influenced ($P<0.05$) in birds fed IQs compared to control birds. In conclusion, isoquinoline alkaloids supported laying hens in the late stage of production and offer an economical solution to improve performance and egg quality in laying hens.

Efficacy of an algo-clay complex on decreasing mycotoxins liver toxicity on broiler

M. García, M. Rodríguez and J. Laurain
Olmix SA, ZA du Haut du Bois, 56580 Brehan, France; animalcare.ts@olmix.com

The aim of this study was to measure the efficacy of an algo-clay complex on T2-HT2 toxins, fumonisins and aflatoxins individual liver toxicity. Three trials were conducted by the Samitec Institute of Analytical (Brazil). 1,080 broilers chickens (Cobb 500) were used in total for the 3 trials. In each trial, 360 animals were allocated to 5 treatments with 6 replicates for the 5 test treatments or 12 replicates for the control treatment. Each group contained 10 animals. The study was run from day 1 to day 21. 3 trials were set up allowing to test each mycotoxin individually at a contamination level of 2.8 ppm for aflatoxins, 100 ppm for fumonisins and 2 ppm for T-2/HT-2 toxins. Treatments differed by the presence of each individual mycotoxin, alone or with an inclusion of the algo-clay complex at 2.5 kg/ton or 5 kg/ton. Performance and liver parameters were measured: feed intake (FI), body weight (BW), individual relative liver weight (RWL), Sphinganine-to-Sphingosine ratio (Sa/So) for fumonisins, Lamic/Samitec Index (LSI) for aflatoxins and total plasma proteins (TPP). The inclusion of 0.50% of algo-clay complex in the diets containing mycotoxins significantly improved FI and BW compared to diets containing mycotoxins only ($P \leq 0.05$). In each study, the RWL of the birds receiving 0.50% of the algo-clay complex, was significantly improved when compared to those receiving mycotoxins only. The inclusion of 0.50% of algo-clay complex in the diets containing 2.8 ppm of aflatoxins improved significantly the LSI compared with those diets containing aflatoxins only ($P \leq 0.05$). The inclusion of 0.25% and 0.50% of algo-clay complex in the diets containing 100 ppm of fumonisins diminished significantly the Sa/So compared with those from the birds fed with fumonisins only ($P \leq 0.05$). The inclusion of 0.25% and 0.50% of algo-clay complex improved TPP ($P \leq 0.05$). The algo-clay complex decreased significantly ($P \leq 0.05$) the deleterious hepatic effects and performance losses caused by very high level of 3 types of mycotoxins on broiler.

The effect of mycotoxicosis on broiler breeders productivity

A. Koppenol[1], J. Buyse[2], J. Lesuisse[2], C. Li[2] and S. Schallier[2]
[1]Impextraco NV, Heist op den Berg, 2220, Belgium, [2]KU Leuven, Dept. of Biosystems, 3000 Leuven, Belgium; astrid_koppenol@hotmail.com

Myctoxins are unavoidable contaminants in feeds exerting harmful effects on animals. Information is scarce regarding mycotoxicosis experimentally induced in broiler breeders. In this research broiler breeders were artificially contaminated with ochratoxin (OTA) and zearalenone (ZEA) to investigate the effect on their reproductive perfomance. ZEA affects the reproductive tract and sex hormone-sensitive receptors, whereas OTA exerts its toxicological effect in production losses. In total 160, 40-wk-old, pure line grandparent stock Ross broiler breeders were allocated to 16 pens and contaminated during 6 wks with 0, 200 ppb OTA, 500 ppb OTA and 500 ppb ZEA, respectively, resulting in 4 replicates per treatment. Feed intake was restricted to maintain optimal body weights. Eggs were collected daily to evaluate egg quality and incubation parameters. After 6 wks, 4 animals per treatment were euthanized to collect organs. Increasing OTA concentrations led to decreasing laying ratios (-29% at 200 ppb and -46% at 500 pp OTA), increasing liver weights (+2.2% at 200 ppb and 3.3% at 500 ppb OTA), inreased occurrence of blood pots (+18%), decreased fertility (-24.3%) and hatchability (-22.2%) and increased early embryonic mortality (+50.8%). ZEA contamination led to decreased laying ratios (-18%), lower liver (-6.8%) and spleen (-7%) weight, higher kidney weight (+1.7%) and decreased fertility (-4.6%). In conclusion, even if poultry better tolerate ZEA compared to pigs, adverse effects on reproductive performance were demonstrated. OTA not being an estrogenic mycotoxin, also affected broiler breeder reproductivity.

The use of biomarkers in the battle against mycotoxicosis in poultry

A. Koppenol[1] and L.F. Caron[2]
[1]Impextraco NV, Heist op den Berg, 2220, Belgium, [2]Universidade Federal do Parana, Department of Basic Pathology, Parana, Curitiba, Brazil; astrid@impextraco.be

Diagnosis of mycotoxicosis in poultry is difficult. When clinical signs are detected, it is already too late as mycotoxins are immunosuppressive creating opportunity for infectious disease to manifest. Biomarkers are great tools to find significant effects as early as possible in the animals reaction against mycotoxicosis. To evaluate the ffect of naturally contaminated feed, 96 male Cobb 500 day-old-broilers were assigned to a control or a contaminated diet until 28 days of age (48 animals over 6 pens per treatment). The contaminated diet was formulated by replacing control corn by a naturally *Fusarium* contaminated corn, resulting in a contamination of 17 ppm FB1+FB2. Blood was collected from 8 animals per treatment and used to quantify circulating lymphocytes through flow cytometry, activity of aspartate transaminase, gamma glutamyl transferase, alkaline phosphatase and levels of uric acid, total protein, albumin, globulin, total leucocytes count and hematocrit, as well as free sphinganine to sphinogosine ratio. Results showed that fumonisins had detrimental effects in broilers, resulting in decreased Ht (-9.5%), decreased TLC (-8.7%), increased Alb/Glb (+13.4%) and SA:SO (+79.4%) values in the blood. The immune response to fumonisins was also clearly demonstrated by a significant effect on the amount of circulating helper T-lymphocytes (+7.2%), regulatory T-lymphocytes (-13.0%) and terminally activated cytotoxic T-lymphocytes (-15.0%). The number of circulating monocytes and macrophages was also significantly decreased (-4.1%), showing the immunosuppressive properties of fumonisins. In conclusion, the use of biomarkers clearly demonstrated significant changes in an early stage and hence are a promising tool to be used to evaluate *in vivo* effects generated by mycotoxins on health status and production losses.

Multi-mycotoxin occurrence in European poultry feed and imported raw materials in 2016

V. Starkl[1], S. Schaumberger[1], P. Kovalsky[1], T. Jenkins[1], M. Sulyok[2] and U. Hofstetter[1]
[1]BIOMIN Holding GmbH, Austria, Erber Campus 1, 3131 Getzersdorf, Austria, [2]IFA Tulln, Department for Agrobiotechnology, Konrad Lorenzstrasse, 3430 Tulln, Austria; verena.starkl@biomin.net

The Biomin Mycotoxin Survey has since 2004 provided a global analysis of over 40,000 samples of finished feed and feed commodities. This study presents multi mycotoxin analysis results for 46 samples of 2,016 European poultry finished feed. The analysis screened for over 380 fungal secondary metabolites analyzed by LC-MS/MS (Spectrum 380®). There were significant imports of South American soybean and corn for European poultry feed. Results are presented for the major established mycotoxin risks from those raw material sources. In the finished feed samples, a mean average of 43 different mycotoxins and fungal metabolites were found per sample. Brevianamid F was detected in all samples and the majority of samples also contained tryptophol (98%), beauvericin (98%), and enniatin b (96%) with mean average of positives 56 ppb, 712 ppb, 31 ppb and 16 ppb, respectively. The more commonly known mycotoxins like zearalenone (ZEN) were found in 89%, fumonisin B1 (FB1) in 78%, and deoxynivalenol (DON) in 59% of poultry feed with average values of 136 ppb, 317 ppb and 505 ppb, respectively. The masked mycotoxin Zearalenone-sulfate and DON-glucoside were also found in 78% and 33% of samples with average of 400 ppb and 84 ppb, respectively. Out of 214 Brazilian soybean samples (normally considered lower mycotoxin risk) 91% were contaminated with up to 5,680 ppb of DON, 74% were contaminated with ZEN and 62% with T-2 toxin among other mycotoxins. Brazilian corn was mainly contaminated with FB1: 95% contamination, maximum of 18,860 ppb FB1 and an average value of 2,647 ppb. Multi-mycotoxin analysis elucidates the occurrence of a wide variety of fungal metabolites some of which are highly prevalent. More research is warranted on the emerging mycotoxins and their impact on poultry production.

Mycotoxin contaminated diets affect immunity parameters of broiler chickens

S. Schaumberger[1], S. Masching[1] and B. Doupovec[2]
[1]BIOMIN Holding GmbH, Erber Campus 1, 3131 Getzersdorf, Austria, [2]BIOMIN Research Center, Technopark 1, 3430 Tulln, Austria; simone.schaumberger@biomin.net

Mycotoxins in feed can have negative effects on immunity of animals. The *Fusarium* mycotoxins fumonisins (FUM) and deoxynivalenol (DON) are of major interest given their common co-occurrence on feeds worldwide. The aim of the feeding trial was to elucidate the effects of a mycotoxin-contaminated diet on immune parameters in broiler chickens and to investigate the efficacy of a counteracting strategy. Broiler chickens were randomly assigned to 4 groups with 8 replicate pens (40 birds per pen). Naturally mycotoxin-contaminated corn was used to prepare the feed with the following mycotoxin levels. The negative control group (A) had <200 µg/kg FUM and <100 µg/kg DON, the FUM contamination group (B) had 1000 µg/kg FUM, the combination groups had 1000 µg/kg FUM + 200 µg/kg DON (C) and 1000 µg/kg FUM + 200 µg/kg DON + 0.1% mycotoxin counteracting additive (D). Parameters evaluated were performance parameters, antibody titers (IBD, NDV, IB) and liver parameters. The trial lasted for 35 days. The contaminated diets had no negative effect on performance parameters within the trial period. No acute intoxication was observed at the mycotoxin contamination levels provided in feed. IBD antibody titers were significantly decreased in group B (246 IU/L) compared to the other groups (group A 285 and group C 346, respectively). This effect was counteracted by the additive (374 IU/l). Whereas, IB titers were significantly lower in group A and group C (411 and 551 IU/l) compared to the other groups (B 1,027 and D 852 IU/l). Liver parameters investigated showed negative trends in mycotoxin groups B and C compared to groups A and D. To conclude, a combination of mycotoxins can impair immune response and show a negative impact on liver health in broiler chickens. The feed additive was able to counteract these negative effects. The effect of the low levels of mycotoxins in broilers suggests that guidance levels for the mycotoxins should be reevaluated.

Effect of excessive calcium feeding on health of gastrointestinal track in broiler breeder hens

S. Honarbakhsh[1], M. Zaghari[2] and B. Sang[2]
[1]University of Tehran, Department of Animal and Poultry Science, College of Aburaihan, Pakdasht, Tehran, 33916-53755, Iran, [2]University of Tehran, Department of Animal Science, Karaj, 31587-11167, Iran; honarbhk@ut.ac.ir

A completely randomized design was conducted to evaluate whether it is useful to feed excessive calcium (Ca) to broiler breeder hens which are feeding semi-mash feed instead of pellet, or not. A total of 56 broiler breeder hens (Ross 308) were caged individually into 4 treatments, with 14 replicates. The first treatment, was the basal diet without excessive Ca. The second treatment, had the basal diet with 5 g of excessive Ca. The third treatment, had the basal diet with 10 g of excessive Ca and the fourth treatment had the basal diet with 15 g of excessive Ca. All of the treatments received their excessive Ca at 4 pm. Feeding excessive Ca, had no significant effect on egg production (%), hen-house egg production (eggs/bird/cum.), hen-house hatching egg production (eggs/bird/cum.), egg weight, clutch size, clutch interval, number of intervals between clutches, body weight, egg shell quality (thickness and strength), crop microflora, gizzard weight, kidney weight and blood factors (Ca, Fe and Zn). The amount of blood factors was different at 11 am and 11 pm ($P<0.05$). The pH of crop and ileum was affected by interaction between level of excessive Ca and time of sampling (11 am, 3 pm and 7 pm). The level of pH at 11 am was higher than two next times of sampling. Excessive Ca had no significant effect on pH of gizzard, proventriculus, duodenum and jejunum. In conclusion, feeding excessive Ca intake had no effect on performance and health status of gastrointestinal track, in broiler breeder hens feeding semi-mash feed.

Effect of vit D source and Ca and P adequacy in coccidia infected broilers

I. Oikeh[1], P. Sakkas[1], D.P. Blake[2], T.R. Hill[1] and I. Kyriazakis[1]
[1]Newcastle University, Newcastle upon Tyne, NE1 7RU, United Kingdom, [2]Royal Veterinary College, Hertfordshire, AL9 7TA, United Kingdom; i.oikeh1@newcastle.ac.uk

Coccidian infections reduce fat soluble vitamin status and bone mineralization. We hypothesized that vit D supplementation in the form of 25-OH-D3 (25D3), which is utilized more efficiently than D3, would promote improved bone mineralization and performance on Ca and P deficient diets, in the presence of coccidiosis. Day old male Ross 308 broilers were randomly allocated to diets with either 25D3 or D3 (4,000 IU/kg of feed). During the grower period (d10-24) birds were allocated to diets with adequate total Ca and available P (H; 8.7:4.4 g/kg) or marginally deficient levels (L; 6.1:3.1 g/kg). On d12 birds were infected (I) or not (C) with 7,000 *Eimeria maxima* oocysts. Each treatment group consisted of 6 replicate pens with 6 birds/pen. Average daily gain (ADG) and feed conversion ratio (FCR) were measured during period 1 (d13-18) and period 2 (d19-24). At the end of each period, one bird/pen was sampled and dissected to assess tibia breaking strength (BS) and ash as a proportion of BW at dissection, ash percentage (PA) and plasma Ca and P. Infection significantly reduced ADG and increased FCR ($P<0.0001$) in both periods 1 and 2, reduced PA, plasma Ca and P ($P<0.005$) on d18, and BS, ash and PA ($P<0.001$) on d24. H birds had higher BS ($P<0.05$), ash ($P<0.0001$) and PA ($P<0.001$) on d18, and higher ash ($P<0.0001$) on d24. Level interacted with source for tibia ash ($P<0.05$) and PA ($P<0.05$) on d24; birds on H diets offered 25D3 had the highest values. Vit D source interacted with infection status for Ca level ($P<0.05$) on d18; C birds on 25D3 had the highest values. In conclusion, infection and offering L diets reduced bone mineralization. However, offering 25D3 increased bone mineralization irrespective of infection status, whilst there was no effect on performance.

Effect of source and level of vitamin D supply in coccidia infected broilers

P. Sakkas[1], I. Oikeh[1], D.P. Blake[2], T.R. Hill[1] and I. Kyriazakis[1]
[1]Newcastle University, AFRD, King's rd, NE1 7RU, Newcastle upon Tyne, United Kingdom, [2]Royal Veterinary College, Pathology and Pathogen Biology, Hatfield, AL9 7TA, United Kingdom; panagiotis.sakkas@ncl.ac.uk

Coccidian infections reduce fat soluble vitamin status and bone mineralization. We hypothesized that benefits of increased dietary vit D supplementation or supplementing with 25-OH-D3 (25D3), which is absorbed more efficiently, would be more pronounced in the presence of a coccidiosis infection. Male Ross 308 chicks were randomly assigned to diets with low (L) or high (H) vit D levels (1000 vs 4,000 IU/kg) supplemented as D3 or 25D3. At d11 of age birds were orally inoculated with water (C) or 7,000 (I) *Eimeria maxima* oocysts. Each treatment group consisted of 6 replicate pens with 7 birds/pen. Pen ADG and FCR were calculated over period 1 (d1-6), 2 (d7-10) and 3 (d11-14) post infection (pi). At the end of each period, one bird/pen was blood sampled and dissected to assess tibia breaking strength (BS) and ash as a proportion of BW at dissection, and ash percentage (PA). Infection significantly reduced ADG and FCR over period 1 and 2 ($P<0.05$), ash ($P<0.0001$) on d14, BS on d10 ($P<0.05$) and d14 ($P<0.0001$) and PA at all timepoints ($P<0.001$). Birds on H or 25D3 diets had reduced FCR ($P<0.05$) and on 25D3 also increased ADG ($P<0.05$). Both H and 25D3 diets increased PA on d10 and d14 ($P<0.05$), BS on d6 ($P<0.05$) but only 25D3 increased ash ($P<0.05$) on d10 and d14pi. Level and infection interacted for FCR ($P<0.05$) with infected birds on L diets showing the highest FCR. Level, source and infection interacted for ash on d14pi being the lowest for I, LD3 birds. In conclusion, supplementing with high level of vit D and 25D3 improved performance, and bone mineralization, irrespective of infection status. Low levels of D3 adversely affected performance and mineralization to a higher degree in I birds.

Zinc amino acid complex is associated with improved intestinal health parameters in broilers

A. De Grande[1,2], S. Leleu[1], R. Ducatelle[2] and F. Van Immerseel[2]
[1]ILVO (Institute for Agricultural and Fisheries Research), Scheldeweg 68, 9090 Melle, Belgium, [2]Ghent University, Department of Pathology, Bacteriology and Avian Diseases, Salisburylaan 133, 9820 Merelbeke, Belgium; annatachja.degrande@ugent.be

As zinc plays an essential role in many biological processes and acts as an important catalytic component of proteins, it can be a beneficial nutritional element in the absence of in-feed antibiotics in broilers. The mechanisms by which zinc improves gut health in broilers are not yet clear, although effects on the gut wall structure and the microbiota composition are likely involved. As organic zinc sources are considered to be more bio-available, a trial was conducted to evaluate and compare the effect of organic and inorganic zinc sources. The trial consisted of two treatments: either a zinc amino acid complex (Zn-AA) or zinc sulphate ($ZnSO_4$) were added to a challenge feed (60 ppm), with 10 replicates (30 broilers per pen) per treatment. The treatment group receiving the organic Zn-AA showed a significantly lower feed conversion ratio (d0-10: 1.149 vs 1.172, $P=0.029$) compared to birds receiving $ZnSO_4$. Intestinal health was evaluated by measuring villus length and crypt depth on hematoxylin-eosin stained sections. A significant higher villus length (d10: 8.5%, $P=0.048$ and d28; 12%, $P=0.012$) was observed for the Zn-AA treatment group. Moreover, the cecal and ileal microbial composition was analysed using 16S rDNA sequencing and QPCR. The relative abundancy of *Enterobacteriaceae* was decreased in ileal samples for the Zn-AA treatment group, which might indicate a reduction in non-beneficial members of this family, such as *Escherichia coli*. Supplementation with Zn-AA designates a positive impact on intestinal health with improved villus morphology and an effect on specific bacterial communities, such as the *Enterobacteriaceae* family, typically associated with poor intestinal health.

The effect of coccidiosis and decreased feed intake on expression of nutrient transporter genes

K.B. Miska and R.H. Fetterer
ARS, USDA, ABBL, 10300 Baltimore Ave, BARC-EAST, Beltsville, MD 20705, USA; kate.miska@ars.usda.gov

Coccidiosis caused by *Eimeria* is endemic to poultry operations and results in decreased feed intake, diarrhea, and decreased weight gain. Our goal was to determine the effect that *Eimeria* maxima causes on the expression of genes that encode peptide and amino acid transporters (AATs). We also wished to determine whether decreased feed intake contributes to the change in gene expression by including a pair fed group of broilers, which were not infected but were fed the same amount as infected chickens. Male Ross broilers were used for the study that comprised of three groups: 1. not infected, 2. infected, and 3. not infected pair fed group. Chicks were infected with 1000 oocysts of *Eimeria maxima* at 21 days of age. Feed consumption was obtained daily, and at days 0, 3, 5, 7, 10 and 14 post-infection (PI) six birds were euthanized, and a portion of the ileum was removed for qRT-PCR. Infected birds had significantly decreased feed consumption between days 6-9 PI. At day 7 PI infected birds had a 45% reduction in weight gain, and pair fed birds had a 32% reduction in weight gain. The feed conversion ratio at day 7 PI of infected birds was 2.2 while that of pair-fed birds was 1.7, compared to 1.5 in uninfected birds. We can conclude that growth parameters were more affected in infected birds than in pair fed birds. By measuring expression levels of nutrient uptake and processing genes via qRT-PCR we determined that genes encoding proteins located at the brush border of the ileal gut epithelium were most affected by infection as well as change in feed intake. The expression of AATs: BOAT, bO+AT, EAAT3, PepT1 in infected birds decreased sharply at the height of infection, however in birds that were pair fed an increase in expression of bO+AT and PepT1 was observed, and little change was seen in expression of BOAT and EAAT3. We conclude that changes in expression of nutrient uptake are distinct between coccidia infected birds that experience decreased feed intake compared to birds that experience limited feed intake but no infection.

Impact of direct feed microbial on gut health, and eggs quality from free-range hens in late cycle

R.D. Malheiros, R. Crivellari and K.E. Anderson
NCSU, Prestage Poultry Sci. Depart., 2711 Founders Dr. Scott Hall, 27695-7608, Raleigh, NC, USA; rdmalhei@ncsu.edu

Probiotics is used as a feed additive to improve the gastrointestinal tract of poultry. In free range chickens is common the intestinal parasites infestation (PInf). Better intestinal morphology is connected with better animal performance. This study was to evaluate direct feed microbial (DFM; Star-Labs Inc., Clarksdale, MO, USA) on hens in later stage of production, to improve intestinal health, egg quality, and production criteria. The study was conducted at the NCDA Station, Salisbury, NC. We used 400 commercial egg hens, during a late cycle, housed in a Free-range system, divided in two treatment groups of 200 hens for each group, 50 hens per replicate. Four replicates of control feed and four replicates provided (DFM) at 3 lbs/ton, fed *ad libitum*. From 89 to 109 wk of age egg production and mortality data were collected. In each period eggs were evaluated for external and internal egg quality. At 109 wks, 10 hens/replicate were euthanized for tissue sample and PInf evaluation. The trial was a completely randomized design, Student's T test were used to compare DFM vs Control (Ct), and ChiSquare to compare the PInf. In the overall DFM vs Ct, has greater HH% (56.54 vs 50.30), Daily Egg Mass (40.33 vs 37.17 g egg/d), and better feed conversion (.359 vs310 g egg/g feed). The mortality trended lower in the DFM group ($P>0.05$). In external or internal egg quality, just the vitelline membrane elasticity was better in DFM hens (5.23 vs 4.85 mm). The intestinal villi high (vh), tip wide (tw), bottom wide (bw), crypt depth (cd), muscular thickness (mt), globet cells count (gc), vh/cd, and villi surface (vs) were not affected by the inclusion in the DFM. The round worm population tender lower in the DFM group. Cecal worms were shown to have a significant reduction ($P=0.0396$) in the DFM group. In conclusion, the use of the DFM in the free-range hens in late cycle, have some beneficial effect on eggs per HH%, in the internal egg quality, and strong reduction in the ceca parasites population in laying hens.

Session 06 Mineral nutrition   Theatre 1

Early life adaptation of broilers to low dietary phosphorus

M.M. Van Krimpen[1], E. Willems[2] and J. Van Harn[1]
[1]Wageningen Livestock Research, Animal Nutrition, P.O. Box 338, 6700 AH Wageningen, the Netherlands, [2]Agrifirm Innovation Center B.V., P.O. Box 20018, 7302 HA Apeldoorn, the Netherlands; marinus.vankrimpen@wur.nl

Several studies investigated broiler's adaptation in later life to early phosphorus (P) deficiency. To validate earlier findings, a study was conducted to determine the effect of deficient P supply to broilers in early life on a wide range of parameters in later life. The experiment was carried out in a 2×2 factorial completely randomized block design with in total 1,920 male Ross 308 broilers. Broilers were fed a 4-phase feeding program: pre-starter (d0-4), starter (d5-10), grower (d11-21), and finisher diet (d22-38). Pre-starter diet was also provided from hatch to arrival on the farm. From hatch to d4, birds were fed a low P diet (retainable P (rP) content 2.0 g/kg), or a rP-adequate diet (4.0 g/kg). During starter and grower phases, all birds received adequate rP diets (4.0 and 3.1 g/kg, respectively). Thereafter, they were fed a rP-adequate finisher (2.9 g/kg) or low-rP diet (2.0 g/kg) till the age of 38 d. In all diets, Ca:rP ratio was kept constant at 2.25. Diets did not contain microbial or intrinsic phytase. At d4, blood P level was reduced in birds fed low-P prestarter (1.0 vs 2.3 mmol/l), indicating a stage of P deficiency. Contrary to our expectations, precaecal P digestibility at d37 was reduced in birds fed the low-P prestarter (63.7 vs 65.8%). The prestarter did not affect contents of ash, Ca, and P in the carcass at d4, d21, and d37, whereas tibia ash content remained unaffected as well at d21 and d37. Tibia breaking strength at d21 was increased in birds fed the low-P prestarter (173 vs 158 N), but breaking strength was not affected by prestarter at d37. Feed intake and body weight gain were reduced in birds fed the low-P prestarter, but FCR remained unaffected. It can be concluded that the expected adaptation of broilers in later life to low dietary P supplementation in early life could not be confirmed in the current experiment.

Dietary manganese supplementation modulated eggshell ultrastructure in laying hens

Y.N. Zhang, G.H. Qi, J. Wang, S.G. Wu and H.J. Zhang
Feed Research Institute, Chinese Academy of Agricultural Sciences, 12 Zhongguancun Nandajie, Haidian, Beijing 100081, China, P.R.; qiguanghai@caas.cn

The effect of dietary manganese (Mn) supplementation on eggshell ultrastructure of laying hens was explored by 2 trials. In trial 1, we examined the effect of different supplemental levels and sources of Mn on shell quality and ultrastructure. A total of 1,080 46-wk-old hens were randomly allocated into 9 groups that were fed a basal diet or supplemented with inorganic ($MnSO_4 \cdot H_2O$) or organic (amino-acid-Mn, 8.78%) Mn at 40, 80, 120, or 160 mg per kg of feed. Each group had 8 replicates of 15 hens. In trial 2, we evaluated the mechanical and ultrastructural changes during eggshell formation by using the optimal levels of organic and inorganic Mn. A total of 270 62-wk-old hens were randomly allocated into 3 groups and fed a basal diet or supplemented with 120 mg Mn per kg feed from $MnSO_4 \cdot H_2O$, or 80 mg Mn per kg feed from amino-acid-Mn. Each group had 6 replicates of 15 hens. In trial 1, dietary inorganic and organic Mn supplementation increased eggshell breaking strength and thickness, and affected mammillary and effective thickness. The supplementation of organic Mn increased fracture toughness by decreasing the width of mammillary knobs. In trial 2, in neither was the elasticity of their shell membranes, measured during the nucleation and mammillary knob formation stages, affected by dietary Mn ($P>0.05$), whereas the breaking strength of the eggshells was greater at the linear and terminate deposition stages ($P<0.05$). Ultrastructural changes during the eggshell formation indicated that dietary Mn supplementation increased the nucleation-site and mammillary-knob densities, decreased the mammillary thickness, and increased the proportion of effective thickness and total thickness of the eggshells ($P<0.05$). Overall, dietary Mn supplementation can improve eggshell breaking strength thickness, and can modulate mechanical and ultrastructural changes during eggshell formation. The mammillary and palisade layers are the crucial structure affected by Mn.

Comparison of Selenium bioavailability in laying hens fed different organic Selenium sources

F.B. Barbe[1], A.S. Sacy[1], E.C. Chevaux[1], S.P. Poulain[2] and M.C. Castex[1]
[1]Lallemand SAS, Animal Nutrition, 19, rue des Briquetiers, 31702 Blagnac cedex, France, [2]Aveyron Labo, 195, rue des Artisans, 12031 Rodez, France; fbarbe@lallemand.com

Selenium (Se) is an essential micronutrient in livestock feed, which can be provided by different forms: mineral (sodium selenite) or organic forms (Se-yeasts or synthetic forms). This study was conducted in laying hens, which represent a well-known model of Se absorption. Ninety-six 30-weeks old ISA Brown laying hens were allotted in 4 groups receiving iso-Se diets (0.2 ppm) during 55 days. Se was provided by different Se sources: sodium selenite (SS), Se-yeast enriched in organic Se (A: 63% selenomethionine) or 2 forms of synthetic organic Se (SM1, SM2: 100% selenomethionine). Selenium concentration and quantity were measured in eggs (D0, D34) and in pectoralis muscle (D55). Haugh units (HU) and the egg water loss were measured after 10 storage days at 24 °C on the eggs laid on the last trial day (D55). Compared to SS, the different organic Se sources increased the Se transfer in the whole egg (+95%, +52%, +32% - A, SM1, SM2) and in the albumen (+133%, +82%, +56% - A, SM1, SM2) ($P<0.05$). In the yolk, only the Se-yeast improved Se transfer (+27%), while it was reduced for SM1 (-2%) and SM2 (-11%), compared to SS ($P<0.05$). Se concentration in pectoralis muscle was increased mainly by A and SM1 (482, 910, 964, 788 µg/kg DM - SS, A, SM1, SM2 - $P<0.05$), which also positively affected egg freshness (26.6, 36.9, 36.6, 29.7 UH - SS, A SM1, SM2). These results demonstrate that Se-yeast was the most efficient Se source to increase Se deposition in muscle and its transfer to the eggs, whose freshness was also improved, when compared to synthetic sources of organic Se. The combination of different seleno-amino acids (SeMet, SeCys), provided only by Se-yeast, appeared therefore more favourable than a single supply of SeMet to increase Se bioavailability in laying hens.

Model development for selenium enrichment in broilers following Se-yeast supplementation

F.B. Barbe, A.S. Sacy, E.C. Chevaux and M.C. Castex
Lallemand SAS, Animal Nutrition, 19, rue des Briquetiers, 31702 Blagnac cedex, France; fbarbe@lallemand.com

The aim of this trial was to develop a practical model to determine selenium (Se) bioavailability in different organs of broilers supplemented with Se-yeast. 30 Ross PM3 chicks were randomly assigned to one of 2 pens (15 broilers/pen) receiving one of the following treatments: sodium selenite (SS: 0.5 ppm total Se) or Se-yeast supplemented at 0.2 ppm Se (SY: 0.5 ppm total Se) during 32 days. Se analysis was then performed in the serum, the pectoralis muscle, the feathers and the Bursa of Fabricius (BF) at d14 (7 broilers/pen) and d32 (8 broilers/pen). At d14, SY significantly increased Se level in the muscle (SS: 149, SY: 349 µg/kg FM, $P=0.001$), the feathers (SS: 955, SY: 1,372 µg/kg FM, $P=0.004$) and BF (SS: 244, SY: 288 µg/kg FM, $P=0.004$). At d32, SY also significantly induced Se enrichment in the serum (SS: 190, SY: 221 µg/kg FM, $P=0.029$), the muscle (SS: 149, SY: 380 µg/kg FM, $P<0.001$) and BF (SS: 206, SY: 271 µg/kg FM, $P=0.006$). Interestingly, the kinetics of Se enrichment was different between the organs studied: Se in the serum still increased between d14 and d32 in SY group ($P=0.027$), while there was no significant difference between d14 and d32 for Se content in the muscle, BF and the feathers. On the contrary, Se level decreased in the feathers between d14 and d32 in SS group ($P=0.043$), while there was no significant difference between d14 and d32 for Se content in the serum, the muscle and BF. The Se level therefore reached a plateau in the muscle, BF and the feathers from d14 of SY supplementation, while serum constitutes a body pool of Se still enriched by SY supplementation after 32 days. This model provides valuable information on Se bioavailability and kinetics in broilers, which can differ according to the analyzed organs.

Meta-analysis of various selenium sources on tissue enrichment and performances in broiler chickens

M. Briens, J. Jachacz, A. Neves-Mayer and Y. Mercier
Adisseo France S.A.S., 10 place du Général de Gaulle, 92160 Antony, France; mickael.briens@adisseo.com

A meta-analysis integrating the results of 13 controlled randomized broiler chicken studies (92 data lines) containing 2-hydroxy-4-methylselenobutanoic acid (HMSeBA), and other Se sources was conducted. It aimed to identify parameters influencing tissue Se concentration, growth performances or mortality. Examined factors were Se source-products (none, selenite, glycinate, HMSeBA, Se-yeasts, selenomethionine (SeMet)) or Se source-types (none, mineral, organic); feed Se supplementation levels (0; [0; 0.2]; [0.2; 1.25] ppm), challenge (standard conditions or specific nutritional/environmental challenge). Growth performances and mortality were considered for whole study period: 0-7; 0-21; 0-42 or 0-49 d; and tissue Se concentrations from end of period collections. For each study a sodium selenite treatment was used as reference (SS_ref) for improvement calculation compared to other treatments (Trt_X) for tissue Se concentrations and growth performances ([Trt_X – SS_ref]/SS_ref x 100). Results showed that organic Se source-types increased breast muscle Se concentrations compared to mineral ones ($P<0.050$). HMSeBA and SeMet were significantly more efficient to increase muscle Se concentration compared to other Se source-products ($P<0.001$). Increasing feed Se supplementation levels resulted in a significant improvement of breast muscle Se content ($P<0.001$). Similar Se sources-types effects were observed for liver Se concentrations and liver SeMet concentrations. Interestingly, Se supplementation on the range [0.2; 1.25], gave an improvement of feed conversion ratio and body weight gain compared to the range [0; 0.2], on average by 1% and [-1.7; 4.0%], respectively ($P<0.010$). No consistent observations were obtained on mortality. Those results confirmed a higher Se transfer in tissues for organic Se forms compared to mineral ones. They also provide indications that increased Se supplementation can positively influence growth performances.

Regulation of Se metabolism, selenogenome, & selenoproteins by a new organic Se compound in broilers

L. Zhao[1], L.H. Sun[1], J.Q. Huang[2], D.S. Qi[1], M. Briens[3] and X.G. Lei[4]
[1]Huazhong Agric Univ, Wuhan, Hubei 430070, China, P.R., [2]China Agric Univ, Beijing, 100083, China, P.R., [3]Adisseo France S.A.S., 10, Place du Général de Gaulle, 92160 Antony, France, [4]Cornell Univ, Ithaca, NY 14853, USA; mickael.briens@adisseo.com

A new organic Se compound, 2-hydroxy-4-methylselenobutanoic acid (SeO), was more bioavailable than Na_2SeO_3 (SeNa) or seleno-yeast (SeY) to chicks. This study was to compare regulations of Se deposition, expressions of selenogenome and selenoproteome, and antioxidant status in tissues of chicks by the three sources of Se. Day-old male broilers (n=6 cages/diet, 6 chicks/cage) were fed a Se deficient, corn/soy-based diet (BD, 0.05 mg Se/kg) or the BD added with SeNa, SeY or SeO at 0.2 mg Se/kg for 6 wk. While SeO led to the highest ($P<0.05$) concentrations of total Se in pectoral and thigh muscles among the 4 diet groups, both SeO and SeY produced ($P<0.05$) higher selenomethionine and lower selenocysteine concentrations in the liver and pectoral muscle than SeNa. Compared with SeY or SeNa, SeO exhibited similar effects on growth performance and all redox status measures in plasma and tissues except for a moderate improvement of muscle thioredoxin reductase activities at day 21. Only SeO upregulated ($P<0.05$) the expression of *Selenos* and *Msrb1* in the liver and thigh muscle over the BD group. Furthermore, SeO improved ($P<0.05$) the expression of *Gpx3*, GPX4, SELENOU, and SELENOP over the SeNa group and the expression of *Selenop*, *SepSecS*, GPX4, and SELENOP over the SeY group in various tissues. In conclusion, SeO illustrated a unique ability to enrich selenomethionine and total Se, to induce the expression of *Selenos* and *Msrb1* mRNA, and to enhance the production of GPX4 and SELENOP proteins in the tissues of chicks.

Effects of manganese sources on physical and chemical parameters of bone in broiler chicken

P. Shokri, S. Ghazanfari and S. Honarbakhsh
University of Tehran, Department of Animal and Poultry Science, College of Aburaihan, Pakdasht, Tehran, 33916-53755, Iran; honarbhk@ut.ac.ir

Manganese (Mn) is an essential trace element in animals, and is necessary to normal bone formation, enzyme function and carbohydrate metabolism in poultry. It is well established that manganese is essential for normal bone development in the young chick. In the present study, the effects of different manganese sources and levels on tibia bone characteristics [bone manganese content, ash bone mineral content, bone morphology (length and diameter), and tibia bone strength parameters] were compared. A total of 480 birds (1 day old) were assigned to 10 treatments, 4 replicates of 12 birds each. The treatments included control (basal diet without manganese) and basal diet supplemented with manganese-sulfate, manganese proteinate and manganese-nano-Max (that was synthetized based on nanochelating technology) at level of 60, 90 and 120 mg/kg of diet. The data were analyzed using two-way ANOVA with Mn concentration and Mn source as the main effects. In addition, 1-way ANOVA was performed and all means were compared using the Tukey test. Samples of birds (one per pen) were killed at the same ages at 42 days of age; Ash bone mineral content, bone morphology, and tibia bone strength parameters were not affected by dietary treatments. Relative to the control group, the bone manganese content were significantly higher in groups supplemented with manganese. Obtained results indicate that use of manganese was not affected physical parameters, but it improved chemical parameters in the tibia bones of broiler chickens.

Effect of manufacturing method on uptake of a novel silicon supplement and bone strength in broilers

S. Prentice[1], E. Burton[1], C. Perry[1], D. Scholey[1] and M. Bedford[2]
[1]Nottingham Trent University, Brackenhurst, NG25 0QF, United Kingdom, [2]AB Vista, Marlborough, SN8 4AN, United Kingdom; sophie.prentice@hotmail.com

Supplementing poultry diets with silicon (Si) could improve skeletal integrity but most forms of Si are not bioavailable. This trial examined efficacy over time of a novel bioavailable Si supplement by assessing absorption and bone strength. 288 male Ross 308 broiler chicks were fed 2 phase (starter = D0-21, finisher = D21-35) wheat and soybean meal mash diets formulated to meet the requirements of the age and strain of the birds. Dietary treatments all had 500 FTU Phytase added to replicate a commercial diet and were supplemented with Si as follows: No Si (Control); control plus 1000 ppm Si made weekly (Si-W); control plus 1000 ppm Si made in one batch prior to the trial (Si-1B); control plus 1000 ppm Biosil (Biosil). Each diet was fed to 9 replicate pens of 7 birds. Performance parameters were collected weekly, and on D14, D21 and D35 serum and tibia bones were collected post-mortem from 2 birds per pen. Serum was analysed via ICP-OES for Si content to assess absorption, and bones tested for strength using a 3-point bend rig to assess efficacy. One-way ANOVA was used to determine the effect of dietary treatment on all parameters. No significant performance differences were recorded between the diets, but Si levels in serum were significantly increased in birds fed both Si-W and Si-1B compared with Biosil and the control. There were no significant differences in tibia strength at D14, but at D21 birds fed Si-1B had significantly stronger tibia than those fed Si-W, Biosil and the control, and at D35 the birds fed Si-1B had significantly stronger tibia than those fed Biosil and the control. This suggests that the novel Si is being absorbed at higher levels than other supplements (Biosil), it has a positive impact on bone strength compared to Biosil, and it retains its efficacy when produced in one batch so does not require weekly manufacture.

Effect of zinc oxide sources and doses on broilers under heat stress

M. Zaghari[1], M. Riahi[1], S. Durosoy[2] and A. Romeo[2]
[1]University of Tehran, University College of Agriculture and Natural Resources, Daneshkadeh Street, 3158777871 Karaj, Iran, [2]Animine, 335 Chemin du Noyer, 74330 Sillingy, France; sdurosoy@animine.eu

Heat stress is a main concern for poultry producers: in warm environments, feed intake is reduced and mortality increased. Zinc is an essential trace element commonly supplied as zinc oxide or as zinc sulfate in poultry diet. It would be able to improve the immune response and the nutrient digestibility in heat-stressed broilers. The aim of the study was to evaluate the effect of a potentiated ZnO source on heat-stressed broilers, compared with a standard ZnO source. 1,200 male broilers (Ross 308) were used for the experiment. They were divided into 40 pens (30 birds/pen). After 21 days of age, from 1 until 5 pm, experimental house's temperature was kept between 28 and 34 °C. Basal diets were formulated and supplemented with 100 ppm of Zn from ZnO (standard practice, NC) and 3 dosages of Zn from potentiated ZnO (HiZox®): 75 ppm (HZ75), 100 ppm (HZ100) and 125 ppm (HZ125). Growth performance was measured at 7 days, 14 days, 21 days, 28 days and at the end of the experiment (42 days). The mortality was recorded throughout the trial. Skin resistance was evaluated at 42 days: 80 birds were slaughtered and an incision about 2 cm of length was made in the region between the thigh and the back; after defeathering, the incision was again measured and the difference before and after defeathering was recorded as skin tearing. There were no significant differences in growth performance. Concerning the mortality, the potentiated ZnO tended to reduce the number of dead birds: around -3% for HZ75 and HZ125 ($P<0.06$) and around -2.5% for HZ100, compared to the group NC (4.3%). The skin resistance was also improved in groups fed the potentiated ZnO: 3.0 cm for HZ125, compared to 6.7 cm for NC ($P<0.07$), 3.6 for HZ75 and 3.9 for HZ100. In conclusion, whatever the dosage, the potentiated ZnO was more effective than standard ZnO to decrease heat stress mortality and to improve skin resistance.

Effect of two zinc oxides on the growth performance of broilers stocked at a high density

A. Romeo[1], A.S. Farjam[2], M.A. Hossain[2], S. Durosoy[1] and I. Zulkifli[2]
[1]Animine, 335 Chemin du Noyer, 74330 Sillingy, France, [2]Universiti Putra Malaysia, Universiti Putra Malaysia, 43400 UPM Serdang, Selangor, Malaysia; sdurosoy@animine.eu

Zinc sulfate ($ZnSO_4$) and zinc oxide (ZnO) are commonly supplemented in poultry diet in order to satisfy the requirements of the birds. In swine, ZnO is generally used at high dosage for its growth promoting effect and its positive impact on gut health. The aim of the study was to evaluate the effect of a potentiated ZnO source at different doses on the growth performance of broiler chickens stocked at a high density. A total of 1,024 day-old broiler chicks (Cobb 500) were assigned to 32 floor pens, with 32 birds/pen at a density of 15 birds/m^2. The basal diet was based on corn, soybean meal and palm oil. Experimental diets were supplemented with 60 ppm of zinc from $ZnSO_4$ (negative control NC) and three doses of zinc from a potentiated zinc source (HiZox®: 60 ppm (HZ60), 90 ppm (HZ90) and 120 ppm (HZ120). Growth performance (bodyweight, feed intake) was recorded at 14 days, 21 days and at the end of the trial (42 days). Broilers fed $ZnSO_4$ did not achieve the standard bodyweight of Cobb 500 at 21 days. In comparison, the potentiated ZnO source improved growth performance. Weight gains from 1-21 days were significantly ($P<0.05$) greater in HZ60 (+6.2%), HZ90 (+8.2%) and HZ120 (+10.9%) broilers. Feed conversion ratios were significantly ($P<0.05$) better in HZ60 (-7.5%), HZ90 (-8.6%) and HZ120 (-10.0%) chickens. Performance, however, was not significantly ($P>0.05$) different among the dietary groups on day 42. In conclusion, the potentiated ZnO at 60 ppm significantly improved feed conversion ratios and weight gains of broilers raised in high density during the starter period, compared to $ZnSO_4$ at the same dosage. In addition, increasing the supplementation level of the potentiated ZnO numerically increased growth performance.

The effect of organic and inorganic zinc sources on performance and egg quality in laying hens

S. Mirzaeigoudazri, A. Mehrabani Mamduh, A.A. Saki and H. Ali Arabi
Bu-Ali Sina University, Animal Science Department, Bu-Ali Sina University, Faculty of Agriculture, Hamedan-Iran, 65178-38695, Iran; saramirzaie0@gmail.com

This study was conducted to evaluate the effect of dietary zinc sources in inorganic or organic form on productive performance and egg quality of laying hens from 28 to 36 weeks of age. A total of 150 Hy-Line W36 laying hens were distributed in a completely randomized design as a factorial management 2×2+1 with 5 treatments, 5 replicates and 6 hens in each. The dietary treatments consisted of a control diet (corn-soybean meal) without zinc supplementation, or the control diet supplemented with 80 or 120 mg/kg Zn as Zn-Methionine or Zn sulfate. Egg production (EP), egg weight (EW) and feed intake (FI) were recorded daily and weekly, respectively. This information was used to calculate egg mass (EM) and feed conversion ratio (FCR). Egg shape index, Haugh units, egg shell thickness, shell weight and yolk color were evaluated at 36 weeks of age. Statistical analysis was conduced using an analysis of variance (ANOVA) followed by the Duncan multiple test ($P<0.05$). No interactions were found between level and Zn source in any of traits studied. In the entire experimental period, FI, EW, EP, EM and FCR were not affected by levels or Zn sources ($P>0.05$). Egg shape index (76.74 vs 75.27%; $P<0.05$) and egg shell thickness (0.561 vs 0.526 mm; $P<0.0001$) were significantly increased by Zn-Methionine than Zn sulfate. However, egg characteristics were not affected by Zn levels in the diet. Egg shape index was higher in control than other treatments ($P<0.05$). Results of this study have demonstrated that laying hens performance was not affected by source or level of Zn supplementation in the diet. Therefore, it is not necessary to increasing zinc in 120 mg/kg of diet. The dietary inclusion of Zn-Methionine could be utilized more effectively on egg shape index and egg shell thickness compared to Zn sulfate.

Effects of different trace mineral sources on performance and trace mineral excretion of broilers

M. De Marco[1], M.V. Zoon[2], C. Margetyal[1], C. Picart[1] and C. Ionescu[2]
[1]Neovia, Vannes, 56000, France, [2]Pancosma S.A., Geneva, 1218, Switzerland; mdemarco@neovia-group.com

The study aim was to compare the effect of Cu, Fe, Zn and Mn supplementation with different trace mineral sources on growth and mineral excretion of broilers. Metal chelates of glycine, hydrate (GC) were compared to inorganic sources (sulphates and oxides, SO) and metal chelates of amino acids, hydrate (AC). 4,800 one day-old broilers (Ross PM3) were randomly assigned to 6 treatments, each consisting of 8 pens (1:1 sex ratio). A diet without trace mineral supplementation served as a negative control (NC). NC diet served as a base for the other experimental diets. Positive control (PC) included a full dose of trace minerals (FD) as SO: 10 ppm Cu, 30 ppm Fe, 80 ppm Zn and 100 ppm Mn, GC½ = NC+½FD as GC, GCFD = NC+FD as GC, AC½ = NC+½FD as AC, ACFD = NC+FD as AC. Feed intake (FI), body weight (BW), average daily gain (ADG) and feed conversion ratio (FCR) were determined per pen. At day 31, samples of feces were collected in 3 pens per treatment. At day 35, 50 birds were slaughtered and carcass yields were recorded. Differences were tested by one-way ANOVA, followed by Tukey's post hoc test ($P<0.05$). At 35 days, chickens fed NC showed a lower FI, ADG and BW than PC ($P<0.001$, $P=0.001$, $P=0.001$). Between GC½ and GCFD only numerical differences of FI, ADG and BW were found and PC performances were achieved by using GC½. By comparing AC½ and ACFD, it was also possible to note a numerical improvement of FI, ADG and BW increasing mineral supplementation, but only ACFD enabled to achieve PC performance. FCR was not affected. Breast weight and breast yield were affected by treatments ($P=0.026$, $P=0.021$). GCFD showed the highest breast weight and breast yield, however GC½ was enough to reach and even exceed the breast characteristics of PC. ACFD was just enough to reach the breast characteristics of GC½. GC½ and AC½ lead logically to a significant intermediate mineral excretion between NC and PC ($P<0.001$). Mineral excretion with AC was always higher than with GC.

Effects of replacing inorganic by organic chelated minerals on performance parameters in broilers

S. Peris[1], A. Bourdonnais[1], N. Senkoylu[1], B. Renouf[2] and N. Bernard[2]
[1]Novus Europe, Neerveldstraat 101-103, B-1200 Brussels, Belgium, [2]Euronutrition, Experimental Station, 72240 Saint Symphorien, France; silvia.peris@novusint.com

Recent studies indicate that trace minerals provided as organic chelated sources are more bioavailable than their inorganic salts. The aim of this trial was to test the effect of replacing inorganic Zn, Cu and Mn by reduced levels of organic minerals chelated with methionine hydroxy analog (Mintrex). 3 treatments consisted of: Positive Control (PC): added with inorganic Zn, Cu and Mn at 80,16, 80 ppm respective levels; Negative Control (NC): with half level of the PC, same mineral sources; and Treatment group (TG): with organic Zn, Cu and Mn as Mintrex at half level of PC, into starter diet (0-14 d). In grower diets (15-35 d) inorganic and organic minerals were reduced to 64/16/64 ppm in the PC and 32/8/32 ppm in the NC and TG. 30 cages with 4 1 d old male broiler chicks were randomly allocated to each treatment from 0 to 35 d of age (10 cages/treatment). At 21 and 35 d birds were inspected for Food Pad Dermatitis (FPD) by randomly selecting 2 right legs from each cage. Live weight (LW), average daily gain (ADG), feed (FI) and water intake, FCR, mortality, and quality and dry matter content of feces were determined at 21 and 35 d of age. Supplementation of organic Zn-Cu-Mn into basal diet at half level of PC group significantly ($P<0.05$) increased LW (2,634 vs 2,475 g) and ADG (74.1 vs 69.5 g/d) of TG compared to NC at 35 d. Differences were not significant between PC and TG for the same parameters (2,569 g and 72.2 g/d for PC). No significant ($P>0.05$) differences were detected among the groups in relation to FCR, water/feed, mortality and FPD. However, numerical improvement was evident in some of the parameters despite half inclusion dose of the organic minerals. The results of this experiment demonstrate that organic Zn, Cu and Mn can replace broiler starter and grower diets at half level of their inorganic salts without adverse effects on performance, feces quality and FPD.

Effect of dietary organic minerals and antioxidant on broiler carcase quality at day 35

M.S. Bekker, P. Paspisanu, X. Arbe Ugalde, V. Kuttappan and S. Asad
Novus International, Technical, 20 Research Park Drive, Saint Charles MO 63141, USA; matthew.bekker@novusint.com

This study was conducted to determine the effect of different sources and levels of trace minerals and an antioxidant on carcase quality in broiler birds at day 35. A total of 234, day-old, Arbor Acres Plus male broilers were assigned to 3 replicated (n=6) dietary treatments with 13 birds per replicate. Treatments were: a control group (T1); receiving a diet using inorganic trace mineral salts (100 ppm Zn, 6 ppm Cu, 120 ppm Mn and 0.3 ppm Se), (T2); diet containing HMTBa chelated sources of 3 trace minerals included at (50 ppm Zn, 10 ppm Cu and 60 ppm Mn) with 0.3 ppm Se in inorganic form, and (T3); diet containing chelated sources of minerals as in T2 but with added antioxidant (ethoxyquin) at 125 ppm and Se (as selenium yeast) 0.3 ppm. Birds in all groups were exposed to mild stress by increasing house temperature to 30 °C from day 21 to 35. At the end of the study, several carcase paramaters were scored and analysed statistically using Chi square test (alpha=0.05). T1, T2 and T3 birds recorded zero incidence of white striping in breast meat at (0%, 4.17% & 12.5%) respectively, with significant difference between all treatments ($P<0.05$). T1, T2 and T3 birds recorded zero incidence of woody breast tissue at (8.33%, 20.83% & 37.5%) respectively with significant difference between all treatments ($P<0.05$). The incidence of tibial head lesions was more pronounced in T1 where zero incidence of tibial head lesions was (8.33%), significantly lower ($P<0.05$) than in group T2 and T3 which had (54.17% & 58.33%) respectively. From these findings it could be deduced that including HMTBa chelated minerals and an in-feed antioxidant with organic Se could be used as an effective tool to minimise the incidence of carcase downgrades such as white striping and woody breast meat. Moreover, replacement of inorganic minerals with chelated minerals could significantly reduce the occurance of lameness in birds due to improved tibial head integrity.

Management of phosphorus in pullet and laying hen diets

A. Rogiewicz and B.A. Slominski
Department of Animal Science, University of Manitoba, Winnipeg, Manitoba, R3T 2N2, Canada;
bogdan.slominski@umanitoba.ca

Dietary phosphorus (P) over-formulation is a common practice in poultry operations. Therefore, the objectives of the current study were: (1) To determine the total and NPP contents of key feed ingredients and complete pullet and laying hen diets in the Province of Manitoba, Canada; and (2) To determine P balance and P excretion in selected pullet and laying hen operations. Three sets of samples were collected in 2015 and 2016 for the total of 350 samples of feed ingredients and 210 samples of complete feeds. The results clearly indicated a significant variation in the total and NPP contents. The means and CV for NPP contents of feed ingredients were as follows (%, as-is basis): corn 0.04, 57; DDGS 0.69, 14; wheat 0.06, 61; barley 0.09, 39; soybean meal 0.29, 0.16; canola meal 0.29, 24. The means and CV for NPP contents of diets were (%, as-is basis): pullet starter 0.0.39, 27; pullet grower 0.38, 29; pre-layer 0.37, 21; layer (phase 1) 0.43, 29; layer (phase 2) 0.37, 22; layer (phase 3) 0.38, 24; layer (phase 4) 0.40, 39; layer (phase 5) 0.37, 52. Phosphorus measurements were also conducted on the feed and excreta samples from two pullet and one laying hen barns to determine the amount of dietary P retained by the body and the amount of P excreted. Total P retention values averaged 41.8, 44.9, and 48.3% for pullets (barn A), pullets (barn B), and laying hens, respectively. When these data are expressed as actual total P content of the diets, the amount of dietary P retained by the body averaged 0.26, 0.35, and 0.30% and the amount of P excreted 0.35, 0.43, and 0.32% for pullets (barn A), pullets (barn B), and laying hens, respectively. From the amounts of P excretion, it would appear evident that the amount of supplemental inorganic P in all three diets could be significantly reduced. The results of this research would allow for the development of recommendations for the cost effective and environmentally friendly P management.

Ileal digestibility of phosphorus from soybean meal in layers

M. Lichovnikova and A. Musilova

Mendel University in Brno, Zemedelska 1, 61300 Brno, Czech Republic; lichovmartina@gmail.com

Objective of the study was to determine precaecal P digestibility (pcdP) from soybean meal (SBM). Three experimental diets were fed to ISA Brown layers housed in 6 replicate cages each treatment, 12 layers per cage. The experiment started when laying hens were at the peak of egg production (97%) in the 26th week of age and lasted 9 days. SBM was added at 34, 50 and 65% to a P free basal diet to obtain diets with 2.7, 3.7 and 4.6 g/kg P, respectively. Dietary Ca/P ratio was kept at 13.1, 9.6, 7.6 and Cr_2O_3 was used as inert marker. Feed and water were freely available. The beginning of killing layers started 8 h after switch on the lights. The digesta of the posterior half of the ileum, except the 2 cm prior to the ileo-ceco-colonic-junction, was flushed out with distilled water and lyophilized. The coefficients of pcdP of the diets were 0.380, 0.158 and 0.204. The coefficients of pcdCa were 0.642, 0.723 and 0.705. The correlation between Ca and P coefficients was -0.99. The equation for pcdP of SBM in layers was y=$0.51x^2$-3.83x+7.66, R^2=1. It is not possible to use linear slope regression for pcdP estimation in feeds for layers.

Growth performance and tibia bone measurements of chicks fed increasing dietary vitamin D3 levels

H. Sievers, M.-A. Lieboldt, I. Halle, L. Hüther, J. Frahm, J. Kluess and S. Dänicke

Friedrich-Loeffler-Institute, Institute of Animal Nutrition, Bundesallee 50, 38116 Braunschweig, Germany; henrieke.sievers@fli.bund.de

Vitamin D_3 is important for avian bone mineral metabolism through regulation of Ca- and P-homeostasis in blood. The present study investigated the effect of increasing dietary vitamin D_3 levels on growth performance and tibia bone measurements of chicks from four layer lines differing in phylogeny and performance (WLA/R11: high/low performing white layers; BLA/L68: high/low performing brown layers) during 12 weeks of rearing. At hatch, 24 one-day-old chicks (3 males/3 females) were weighed and slaughtered for tibia bone dissection. Additional 27 male and 27 female one-day-old chicks of each line were housed in six floor-range pens with one pen per sex and diet, providing 300, 1000 and 3,000 IU of vitamin D_3 (Cholecalciferol)/kg supply and a calculated Ca (9.5 g/kg)/P (6.3 g/kg)-ratio of 1.5 under *ad libitum* feeding conditions. From hatch to week 12 residual feed was recorded weekly per pen. Daily feed intake, daily weight gain and feed to gain-ratio were calculated. From every pen 3 chicks were weighed and slaughtered for analysis of dry bone weight, breaking strength (BS) and Ca-/P-content of tibia bones. There were no significant effects of increasing dietary vitamin D_3 levels on growth performance and bone measurements. Brown layer lines showed higher relative bone weights than white lines from 8 weeks of age ($P<0.05$). Male brown layer lines had higher relative bone weights and Ca-/P-content than females ($P<0.01$). BS and Ca-content increased with aging ($P<0.01$). L68 had the highest relative Ca-/P-content in the bone from 4 weeks of age ($P<0.001$) and the highest BS in week 4 and 12 of age ($P<0.05$). Increasing dietary vitamin D_3 levels had no effect on performance and tibia bone measurements of male and female chicks of four genetically diverse layer lines. Differences were primarily dependig on age, sex an genetics.

Intermittent lighting improves broiler performance

I. Rodrigues[1], M. Toghyani[1], B. Svihus[2], M. Bedford[3], R. Gous[4] and M. Choct[1]
[1]University of New England, Australia, Armidale, NSW 2351, Australia, [2]Norwegian University of Life Sciences, Universitetstunet 3, 1430 Ås, Norway, [3]AB Vista Feed Ingredients, Marlborough, SN8 4AN, United Kingdom, [4]University of KwaZulu-Natal, King George V Ave, Durban, 4041, South Africa; imendotr@myune.edu.au

Manipulation of feed retention time in the foregut via lighting management programs may further enhance performance and efficacy of exogenous enzymes. Following an adaptation starter period until d10, 624 one-day-old ROSS 308 male broilers were subjected to two different lighting programs (continuous (CL) (18L:6D) or intermittent (IL) (1L:3D:1L:3D:1L:3D:1L:3D:2L:6D)) and fed isoenergetic (on an ME basis) and isonitrogenous wheat-based diets with or without the supplementation of phytase and xylanase over 34 days. Chicks were randomly allocated to eight treatments (2×2×2) with six replicates per treatment. Data were analysed using ANOVA (SPSS Statistics, ver. 24). Means were compared using the Tukey multiple range test. At d10, after the initial adaptation period, body weight (BW) of chicks was the same for both treatments (290 vs 288 g, for CL and IL groups, respectively). At d34, CL birds were heavier than those in IL (2,224 vs 2,159 g, $P=0.08$) but presented statistically significant higher mortality-corrected FCR (FCRc) (1.390 vs 1.370, $P=0.003$). Phytase supplementation improved final BW (2,247 vs 2,137 g, $P=0.004$) and FCRc was better for supplemented animals (1.375 vs 1.286). Xylanase addition to basal diet improved FCRc (1.370 vs 1.391, $P=0.001$). There was an interaction ($P<0.05$) between phytase and xylanase supplementation on final body weight, which shows the additive/synergistic effect of these enzymes. We hypothesise intermittent lighting had an effect on feed retention in the upper gastro-intestinal tract, evidenced by the lower pH found in the crop and gizzard of birds in IL groups (data not shown), and that this enabled better digestibility of nutrients.

Effect of different post-peak feed withdrawal programmes on overall broiler breeder performance

E.H. Helander, H. Willemsen and O. Van Tuijl
Aviagen, Newbridge, Midlothian, EH28 8SZ Scotland, United Kingdom; ehelander@aviagen.com

Broiler breeder hens need energy and nutrients for maintenance, growth and egg production. After peak production, the requirements for egg production gradually decrease over time. Oversupply of energy and nutrients after peak causes increased growth and excessive breast meat deposition, leading to poorer egg production and increased egg size. Reducing feed allocation after peak is one way to prevent excess bodyweight gain, which may improve production persistency and control the egg size. Globally, feed withdrawal programs differ widely. This experiment was conducted to study the effect of 4 post-peak feed withdrawal programmes on the overall performance of modern Ross 308 broiler breeders. The feed withdrawal programmes tested were 0(A), 4(B), 8(C) and 12%(D) up to 60 weeks of age. Every treatment (Trt) consisted of 4 replicates with 150 females and 15 males per pen. Coarse mash diets for females were formulated according to Aviagen recommendations. Females received precalculated amounts of metabolizable energy/day, with peak feed being 163 g (1.91 MJ or 456 kcal/hen/day). Feed withdrawals started at 36 wks of age, and the birds were depleted at 60 wks of age. Average feed consumption of Trt A from transfer (20 wks) to depletion was 43.971 kg/hen, and was 0.726, 1.278 and 1.902 kg less for hens fed B, C and D Trt, respectively. Average body weights at 60 wks were 4,368, 4,326, 4,219 and 3,950 g for Trt A, B, C and D, respectively. Abdominal fat content measured at 60 wks of age was lower for Trt C and D ($P<0.001$). Egg sizes were very similar between the treatments throughout the trial. Total hen-week egg productions were 186.6, 186.5, 186.9, and 184.7 (NS); hatching eggs 177.1, 176.3, 177.3 and 174 (NS); and 151, 150.8, 152.9 and 150 chicks (NS) were obtained for A, B, C, and D, respectively. Post-peak feed withdrawal is an effective tool to control bodyweight gain. Due to lower feed consumption feed withdrawal improves the economic performance of the breeders.

Performance and range use of organic broilers with access to different vegetation in outdoor areas

A.L.F. Hellwing[1], J.S. Petersen[2] and S. Steenfeldt[1]
[1]Aarhus University, P.O. Box 50, 8830 Tjele, Denmark, [2]SEGES, Agro Food Park 15, 8200 Arhus N, Denmark; annelouise.hellwing@anis.au.dk

Outdoor range areas are an important part of the organic broiler production, and the question is how to make this area as attractive for the broilers as possible. The aim of the study was to investigate the influence of vegetation in outdoor range areas on performance, crop and gizzard content and range use of organic broilers. The experiment was performed on an organic farm in the summer of 2015 and 2016. The outdoor range was a grass field (F) or grass field combined with a plantation (GP). There were two replicates for each type of outdoor range. Eight hundred broilers were allocated to each replicate. Feed intake were registered daily. In 2015, 80 broilers from each replicate were weighed at 7, 14, 28, 42, 56, 70 and 84 days of age, and in 2016 at 28, 56 and 70 days of age. The range use was observed three times in 2015 and two times in 2016. Crop and gizzard content and composition were investigated in 12 weeks old broilers in 2015. From each range area, 2 male and 2 female broilers were killed in the morning and in the afternoon. The average daily gain was 40.8 and 39.3 g for GP and F, respectively in 2015 for the whole growth period, and 41.9 and 40.7 g for GP and F in 2016. The differences were statistically significant in 2015 but not in 2016. Feed intake was 125 and 133 g/day on GP and F, respectively in 2015, but feed intake per chicken or feed efficiency were not statistically different for the two types of range areas in 2015. The crop contained more material in broilers from GP than F, which was not the case for gizzard content. However, the results for crop content should be interpreted with care as outdoor range, sex, and time of day interaction were observed. The crop contained mainly feed and the gizzard mainly grass. It was observed that broilers on F remained more inside the chicken house than on GP, and broilers on GP were most often observed to stay in the area covered with trees.

Effect of grazing on the carotenoids and vitamins content in egg yolk of laying hens during storage

J. Vlčková, V. Skřivanová, M. Englmaierová and T. Vít
Institute of Animal Science, Department of Physiology of nutrition and quality of animal products, Přátelství 815, 104 00, Prague, Uhříněves, Czech Republic; vlckova.jana@vuzv.cz

The objective of this study was to compare the effect of housing system (enriched cage, free range and grazing) on the concentration of zeaxanthin, lutein, α-tocopherol, retinol and β-carotene in fresh egg yolks and in egg yolks after 21 days of storage at 18-20 °C. In each housing system was kept sixty laying hens genotype Hisex Brown. The group in free range had access only to a range without grass whereas the group from grazing had access to permanent grassland. Total two hundred and forty eggs were used to determine the vitamin and carotenoid content in the egg yolks once on the beginning of lying period (3 eggs per sample). Lyophilized pasture contained 102 mg/kg dry matter (DM) of zeaxanthin, 119.8 mg/kg DM of lutein, 44.2 mg/kg DM of α-tocopherol and 62.1 mg/kg DM of β-carotene. Laying hens in each housing system were fed identical commercial feed mixtures. Significant interactions between housing system and storage time were found out in α-tocopherol ($P \leq 0.029$) and retinol ($P \leq 0.029$). The highest content of α-tocopherol (116.2 mg/kg DM) was detected in fresh egg yolks from hens kept on grazing compared to free range and enriched cage. Likewise the highest content of retinol was in eggs from grazing without different between storage times. The significant effect ($P \leq 0.001$) of housing system was detected in all monitored parameters. The content of zeaxanthin, lutein, α-tocopherol, retinol and β-carotene had 3.42, 2.70, 1.03, 1.30 and 80 times higher values, respectively, in grazing compared to enriched cage. Zeaxanthin ($P \leq 0.035$), retinol ($P \leq 0.045$) and β-carotene ($P \leq 0.035$) were affected by storage time, when the values after 21 days decreased by 2.40, 0.60 and 0.091 mg/kg DM, respectively. Results of the study suggest that grazing produced eggs with increased concentrations of carotenoids and vitamins and may have a higher nutritional value for humans.

Drinking water temperature influences the performance of intensive broilers

D. Albiker and R. Zweifel
Aviforum Foundation, Burgerweg 22, 3052 Zollikofen, Switzerland; albiker@aviforum.ch

The influence of different drinking water temperatures (DWTs) in a broiler barn was tested in winter regarding the production performance of 4,320 Ross 308 hybrids. The as hatched one day old chicks were evenly distributed to 16 compartments and randomly allocated to four different DWTs, allowing 4 compartment replications. They were slaughtered at the age of 37 days. The barn temperature (BT) started at 34 °C and continuously decreased to an average of 23.7 °C. DWTs were achieved by a short water pipe (SP: Ø 10 m), a long water pipe (LP: Ø 48 m), heating the water up to around 25 °C (T25) and up to around 30 °C (T30). SP led to a minimal (32 to 20 °C), LP to a maximal (33 to 22 °C) warming up of the water by BT. T25 decreased from 33 to 24 °C, T30 from 33 to 27 °C. The actual DWT depended on the BT and the water consumption. It differed between 1 to 2 °C at the beginning to up to 7 °C at the end of the trial. All the groups received the same feed (starter: 12.8 MJ ME, 21% CP; grower: 12.5 MJ ME, 19% CP). During the starter phase, SP ate and weighed significantly less than T25 and T30 (SP: 250 g/a&d, 254 g BW, LP: 250 g/a&d, 264 g BW, T25: 272 g/a&d, 270 g BW, T30: 271 g/a&d, 269 g BW). During the whole trial, T30 ate significantly less than the other groups (SP: 3,581 g/a&d, LP: 3,547 g/a&d, T25: 3,600 g/a&d, T30: 3,473 g/a&d). The BW of T30 was significantly lower at the end of the trial as well and showed the highest FCR (SP: 2,297 g, 1.586, LP: 2,297 g, 1.571, T25: 2,309 g, 1.586, T30: 2,202 g, 1.605). Mortality was not influenced by the water temperature, nor was the number of germs (aerobic mesophilic germs, *Escherichia coli* and enterococcus) in the water. T30 showed a significantly higher water to feed ratio in the last week than SP and T25 (SP: 1.78, LP: 1.87, T25: 1.79, T30: 1.94). The best performance results were achieved at a DWT of 25 °C. In conclusion, the closer the DWT to the BT during the whole production cycle, the better the broiler performance. The DWT should not be higher than the BT.

Water temperature difference can influence on the broiler performance

Y. Paek[1], S.M. Kim[1], C.I. Lim[2] and K.S. Ryu[2]
[1]National Research Institute of Agricultural Science, 300 Nongsangmyung-Ro, Wansan-Gu, 54875 Jeonju, Korea, South, [2]Chonbuk National University, Animal Science, 567 Baeje-Daero, Duckjin-Gu, Jeonju 54896, Korea, South; dlacjsdlr@naver.com

A study was conducted to know the effect of water supply with three temperature regime on the growth performance of broiler chicken during summer season. Three hundred sixty (360) Cobb broiler chickens of 7 days of age were distributed into three treatments having eight replications with 15 birds in each. The duration of the trial was 8th day to 35 days old. The water temperature groups were: (1) High house temperature (28.1 °C) with tap water temperature 24.3 °C; (2) House and tap water temperature (24.3 °C); and (3) House temperature 25.5 °C and Cold tap water temperature (15.4 °C). Commercial broiler chicken diet was supplied to all three treatments birds. Significantly ($P<0.5$) highest final body weight (2,165.94 g/bird) and weight gain(1,964.77 g/bird) were observed in cold water supply group compared to other two groups. Although significantly highest feed intake was recoded in cold water supply group compared to high temperature group but no significant difference was observed in feed conversion ratio. Based on the findings of the growth performance it can be concluded that cold water supply to the broiler chicken has a positive impact on growth performance.

Effect of feed and water treatments during a transport period (6 h) on broiler chick development

M.H. Priester[1], C.W. Van Der Pol[1], D.M. Lamot[2], H.J. Wijnen[1], G. Aalbers[1] and I.A.M. Van Roovert-Reijrink[1]
[1]HatchTech B.V., P.O. Box 256, 3900 AG Veenendaal, the Netherlands, [2]Cargill Animal Nutrition Innovation Center Velddriel, Veilingweg 23, 5334 LD, the Netherlands; mpriester@hatchtech.nl

Currently, chicks that had access to feed and water in a hatcher do not always have access to feed and/or water during handling and transport. Objective of the current trial was to investigate if a transport period of 6 h without feed and water leads to BW loss and reduced development, and if providing gel had the same effect on chick development as water. After pull time from HatchCare (a hatcher with feed and water), the following treatments were applied: a pre-starter and water, only a pre-starter, a pre-starter and gel (89% water and starch), or no feed and water. At pull, 9 cradles per treatment with 90 Ross 308 chicks (n=3,240) from a prime parent flock were placed back in HatchCare, in which optimal body temperatures were ensured. BW gain and feed and gel intake were measured per cradle after 6 h. Per treatment, for 5 randomly chosen chicks per cradle, the individual BW, yolk free body mass (YFBM), and organ weights (intestine, stomach, heart and liver) as a percentage of YFBM were determined (n=40 per treatment, per time point). After 6 h, chicks provided with a pre-starter and water gained BW (+2.51 g, $P<0.001$) whereas chicks provided with only a pre-starter, a pre-starter and gel or without a pre-starter and water lost weight (-1.65, -0.60, and -2.09 g, respectively). When providing pre-starter with gel, feed intake was lower than when providing pre-starter and water (0.94 vs 1.49 g). YFBM was highest ($P<0.001$) in chicks provided with a pre-starter and water or gel. Relative organ weights were not affected by treatment. In the current trial, providing gel besides a pre-starter resulted in lowered BW loss in comparison to feeding only feed or no feed and water during a 6 h period after pull, but it did not result in BW gain as pre-starter and water did.

Gastrointestinal taste perception in broilers

Z. Uni and S. Cheled-Shoval
Hebrew University, Animal Science, Faculty of Agriculture, Rehovot 76100, Israel; zehava.uni@mail.huji.ac.il

Taste perception play an essential role in nutritional evaluation and has been shown to affect feed choice, acceptance and intake. In pigs and other mammals, nutrient sensing in the gastrointestinal tract (GIT) occur through taste components (i.e. taste receptors (TRs)) which are expressed in the GIT and affect processes such as gastric movement, digestion functionality and GIT hormonal secretion. However, knowledge on taste perception and taste pathways in the chicken GIT is limited and therefore our research aim is to study taste perception in poultry. For that, a 2-alternative-forced-choice test was develop for broilers and the detection of different representative tastants *in vivo* was done. In addition, taste-related genes expression in the broiler's GIT was studied. Results shows *in vivo* detection for umami, sweet and bitter tastants in hatchlings (n=105). Moreover, expression of bitter taste receptor genes (T2R1, T2R2, T2R7) umami (T1R1 and T1R3) and their downstream protein genes (TRPM5, α-gustducin and PLCβ2) in the upper (palate, tongue and stomach) and lower GIT (small intestine, cecum and colon) is exhibited in poultry embryos (n=10) and in mature broilers (n=10). Bitter molecule (quinine) administration decreased the expression of two bitter TRs and brush border enzymes (PepT1 and SGLT-1) in the intestine. Taste gene expression and their expression alterations imply on the involvement of taste pathways for sensing amino acids and bitter compounds in the chicken GIT, although the exact mechanism is yet to be revealed. These suggest the future possibility of affecting broiler feeding behavior and gastrointestinal physiology via specific tastant sensing mechanisms in the GIT.

Relationship between satiety hormones and fermentation products in fasted broiler chickens

G. González-Ortiz[1], D. Solà-Oriol[2], M. Martínez-Mora[2], J.F. Pérez[2] and M.R. Bedford[1]
[1]AB Vista, Woodstock Ct, 3, SN8 4AN, Marlborough, United Kingdom, [2]Universitat Autònoma de Barcelona, Department de Ciencia Animal i dels Aliments, 08193 Bellaterra, Spain; gemma.gonzalez@abvista.com

The objective of the present study was to evaluate the relationship between peptide tyrosine-tyrosine (PYY), cholecystokinin (CCK) and short chain fatty acids (SCFA) in the blood and caecal digesta, and whether a period of fasting influences this relationship upon reintroduction of feed. A total of 72 37-days-old broiler male chickens (Ross 308) were randomly distributed into two experimental groups, with continuous illumination. Group 1 (n=36) were fasted for 12 h, and Group 2 (n=36) were feed *ad libitum*. Immediately prior to reintroduction of feed to the fasted animals (T_{0h}), 12 animals per group were sacrificed. Blood samples were collected for PYY and CCK analyses. Short-chain fatty acids were analysed in blood and caecal digesta. At 3 h (T_{3h}) and 6 h (T_{6h}) after feed reintroduction, the sample collection procedures were repeated. Statistical comparisons were performed using a two-way (sampling time and feed restriction) ANOVA (JMP 12). For PYY and CCK, the interactions of main factors were significant ($P<0.05$). PYY and CCK levels remained constant with time for the *ad libitum* fed animals. For fasting animals the highest PYY concentration was observed at T_{0h} and it gradually reduced at T_{3h} and T_{6h}. CCK levels were lower at T_{0h} compared with the later time points, and refeeding birds resulted in values similar to those observed in the *ad libitum* birds by T_{6h}. Lower concentrations of butyrate in caeca were observed in fasted animals at T_{0h}, and this increased once the animals had access to feed. PYY and CCK satiety hormone levels responded in an inverse manner to one another after a fasting challenge of 12 h, and both returned to normal levels after 6 h re-feeding. These changes were not accompanied by changes in SCFA in either the caeca digesta or blood.

Performance, behaviour and plumage of brown layers with intact beaks

D. Albiker and R. Zweifel
Aviforum Foundation, Burgerweg 22, 3052 Zollikofen, Switzerland; albiker@aviforum.ch

Haylage as an occupation (OC), a high crude fibre (HCF, 6.5% CF) and a control diet (LCF, 3.5% CF) were tested regarding the performance, plumage quality and feather pecking (FP) of 2,860 LB and BN hybrids with intact beaks. The hens were evenly distributed to 8 pens. 4 pens faced east, 4 pens west. 2 pens on each side were either allocated to HCF or LCF. Half of each diet group received OC. This allowed 4 replications of each treatment without considering their combination. Light intensity and barn climate were also observed. With HCF or LCF and with or without OC, average laying performance (90%), feed consumption (129.8 g), FCR (2.293), BW (1,978 g) and egg weight (63 g) showed no important differences between treatments after 12 laying periods (LP). The plumage deteriorated and cannibalism increased from LP 8 onwards due to the change of protein content (CP) in the diet during the second phase (from 16.0% CP to 15.3% CP). The diet and OC had the same influence on the feather scoring of the back (LCF 1.98; HCF 1.82, OCM 1.86; no OCM 1.90). Both diets resulted in a similar cannibalism rate (HCF 11.3%; LCF 11.6%). With HCF, severe (s) and mild (m) FP was less frequent than with LCF (HCF: 0.89% s, 4.22% m; LCF 1.74% s, 5.78% m), as HCF hens were supported by a more humid barn climate. From LP 1 to 9, cannibalism with OC was 1.9% lower than without (OC 2.5%). This difference minimized down to 0.6% after 12 LPs, the OC losing its positive effect. With OC, sFP was lower than without until LP 8 (OC 0.13%, no OC 0.45% per LP). From LP 8 to LP 12, FP occurred as often with or without OC (OC 3.1%, no OC 2.9% per LP). On the dryer side of the barn, sFP occurred 0.32% more often per LP. In conclusion, the protein content must cover the hens' needs. A dry barn climate led to FP and cannibalism. HCF did not improve the welfare of hens with intact beaks while OC was able prolong it. OC should be changed regularly to keep up the hens' interest. The reduction of the light intensity stabilized the FP level. Further research on how to optimize the light management is recommended.

Measuring feather appetite and digestibility and the effect on artificial feather pecking motivation

K.M. Prescilla, G.M. Cronin, S.Y. Liu, K.M. Hartcher and M. Singh
University of Sydney, Poultry Research Foundation, 425 Werombi Road, Camden 2570, Australia;
kevin.prescilla@sydney.edu.au

There is evidence to suggest a link between nutritional factors, severe feather-pecking, and feather eating behaviour. The aims of this experiment were to investigate feather appetite and digestiblity in ISA Brown birds, and to determine the effect of feather consumption on feather pecking. Sixty individually housed, 56-week-old ISA Brown hens were randomly allocated to one of three dietary treatments: commercial diet (Control), commercial diet and pelleted feed containing 15% ground feathers (Ground), and commercial diet and 20 whole feathers (Whole), presented in two separate feeders to allow for choice feeding. Each bird was presented with 10 semi-plume feathers mounted on an artificial substrate and bird behaviour was observed daily over 14 days, where the latency to peck was recorded. Feed intake was higher ($P<0.01$) in the Ground treatment (125.9 g/d) when compared to Control (111.8 g/d) and Whole (114.7 g/d) treatments and average ground feather appetite was ~5.5% (range 0.2-10.6% of total intake). However, whole feather consumption was low when presented in feeders. Although overall protein digestibility was unaffected by treatment, increased feed intake and protein consumption allowed Ground treatment birds to digest more protein (26.7 g, $P<0.001$) than Control and Whole treatments over 48 hours (18.3 g and 17.3 g respectively). Ileal amino acid digestibility was unaffected by ground feather inclusion, but Ground treatment birds digested more cysteine ($P<0.01$) and methionine ($P<0.001$) content than Control and Whole treatments. Latency to peck at artificially mounted feathers was higher for Ground treatment birds ($P<0.001$) suggesting decreased feather pecking motivation. ISA Brown birds may be capable of digesting ground feathers, and increased digestion of cysteine and methionine may be related to feather pecking motivation (latency to peck). This has implications for feather pecking behaviour which will be investigated in further studies.

Assessment of dual energy X-ray absorptiometry for measuring *in vivo* body composition in broilers

C.A. Gonçalves, N.K. Sakomura, E.P. Silva, N.T. Ferreira, R.M. Suzuki, W.J. Alves and L.G. Pacheco
UNESP, Univ Estadual Paulista, Animal Science, Via de Acesso Prof. Paulo Donato Castellane s/n, 14884-900, Brazil;
lavinesp.lab@gmail.com

Dual energy X-Ray absorptiometry (DXA) is a non-invasive technique used for measuring body composition. This study aimed to evaluate DXA in substitution to chemical analysis (CHE) and calibrate a model that estimates *in vivo* body composition of broilers. A total of 720 Cobb500 broilers (360 males and 360 females) was distributed in a factorial arrangement 3 (protein levels) ×2 (sexes) with six replications of 20 birds. To change broilers body composition, diets were formulated according to Brazilian Tables, with 3,100 kcal ME/kg and varying the crude protein (CP) on 70, 100, and 130% of the requirement. At 7, 14, 28, 42, 56 and 77 days of age, one bird per replicate was alive scanned on Discovery® Wi DXA to determine bone mineral content (BMC), fat, lean and total mass. Birds were slaughtered and the body feather-free was processed for chemical analysis of fat, ash and protein. Based on the values obtained, it was developed the following prediction equations: $\text{Protein(g)}=e^{(-2.60837+1.10137\times\ln(\text{Lean}))}$, $R^2=0.99$; $\text{Ash(g)}=e^{(0.67199+0.96387\times\ln(\text{BMC}))}$, $R^2=0.98$; $\text{Fat(g)}=e^{(-3.24453+0.35802\times\ln(\text{Fat mass})+0.89557\times\ln(\text{Lean}))}$, $R^2=0.96$, for females, and $\text{Protein(g)}=e^{(-2.54693+1.09328\times\ln\text{Lean})}$, $R^2=0.99$; $\text{Ash(g)}=e^{(0.63735+0.96876\times\ln(\text{BMC}))}$, $R^2=0.98$; $\text{Fat(g)}=e^{(-2.89916+0.45512\times\ln(\text{Fat mass})+0.75153\times\ln(\text{Lean}))}$, $R^2=0.94$, for males. All equations showed significant parameters (intercept and independent variables) ($P<0.01$). We concluded that DXA is a brand new methodology to assess *in vivo* body composition and the generated equations supports and improves the evaluation of body composition of broilers.

Feed density and body weight influenced body composition and behaviour of broiler beeder pullets

J. De Los Mozos[1], A.I. García-Ruiz[1] and M.J. Villamide[2]
[1]*Nutreco, Trouw Nutrition R&D, Ctra. CM 4004 km 10.5, 45950 Casarrubios del Monte, Spain,* [2]*E.T.S.I. Agrónomos, Departamento de Producción Agraria, Universidad Politécnica de Madrid, 28040 Madrid, Spain; j.delosmozos@trouwnutrition.com*

Dietary strategies to reduce restriction have been addressed to reduce hunger feeling and stereotypic behaviour in broiler breeders. Ross308 breeder female chicks allocated in 20 pens (150 birds/pen) were used in a 2×2 factorial design in which 2 diets (control vs 15% diluted) and 2 target body weights (TBW; recommended vs 15% heavier) were tested. Diluted diet was formulated using highly fibrous raw materials as wheatbran, sunflower and rapeseed meal, cereals straw and soya hulls. From 0 to 21 days of age all birds were fed the same diet. A 4 phase feeding program was followed, starter (0-3 wks), grower-1 (4-7 wks), grower-2 (8-17 wks) and pre-lay (18-19 wks). Results showed that, independently of the diet, time spent eating was greater ($P<0.05$) in pullets reared to heavier TBW. Animals fed diluted diets had lower breast meat yield but no differences were found in carcass crude protein content. Carcass fat content at 19 wks was reduced in animals on recommended TBW when fed a diluted diet. However, feeding diluted diets to high TBW group provoked an increase ($P<0.05$) in fat carcass composition similar to that observed for the control diet. Animals fed diluted diets were less motivated to eat after one day without feed, under daily restricted and ad-libitum regimen ($P<0.05$). Analysis of behaviour parameters as fixed elements pecking ($P<0.1$) or feeding ($P<0.05$) corroborated the results of the feeding rate test. Results of this trial confirm that stress derived from the feed restriction might be reduced through high TBW and diluted diets.

Validation of a method of tibia breaking strength as an indicator of skeletal integrity in broilers

K. Yupanqui[1], R. Riveros[1], J. Inca[1], F. Prado[2] and C. Vilchez[1]
[1]*Universidad Nacional Agraria La Molina, Nutrición Animal, Av. La Molina s/n La Molina, Lima 12, Peru,* [2]*Universidade Estadual Paulista, Zootecnia, Via de Acesso Prof. Paulo Donato Castellane s/n Jaboticabal, 14870 Jaboticabal, Brazil; faprado91@gmail.com*

The study aimed to validate a methodology of breaking strength of the diaphysis of tibia as an indicator of skeletal integrity in 21 day-old broilers chicken. Forty Cobb 500 male chickens were raised in battery cages (13.5 birds/m2) for 21 days. It was considered two phases of isonutritive feeding: pre-starter for the first to tenth day (0.90% Ca, 0.45% Pd) and starter from eleven to twenty one days (0.84% Ca, 0.42% Pd). At 21 day, the chickens were slaughtered to remove the right tibia of each bird to measure the following morphometric variables: tibia weight, tibia length, average diaphysis diameter, volume and mineralization indexes (bone density, shape index, Seedor modified index, Quetelet index and robustness index) and breaking strength (BS). The resistance was measured by direct transverse pressure at the mid-diaphysis of the tibia. Data were subjected to analysis of phenotypic Pearson correlation using the SPSS statistical program. The result showed not significant correlation ($P>0.05$) between BS and tibia volume, density, shape index, robustness index. There were a significant ($P<0.05$) positive correlations between tibia BS with length, average diaphysis diameter and Quetelet index and highly significant ($P<0.01$) positive correlation with tibia weight and the modified Seedor index. In conclusion, the methodology of breaking strength of tibia diaphysis can be used as a reliable indicator of skeletal integrity.

Comparative study of methodologies to evaluate egg freshness

M. Umar Faruk[1], M. Yamashita[2], S. Gerard[1] and F. Cisneros-Gonzalez[3]
[1]DSM Nutritional Products, 1 Blvd d'Alsace, 68128, Village-Neuf, France, [2]Kyoto Womens University, 35 Kitahiyoshi-cho, 605-8501, Japan, [3]DSM Nutritional Products, Wurmisweg, 4303 Kaiseraugst, Switzerland; murtala.umar-faruk@dsm.com

This work evaluated three different methodologies to assess egg Yolk Index (YI) as an indicator of egg freshness. YI was measured using: (1) automated digital device based on laser beam technology; and (2) two different manual methods on a total of 796 eggs stored under different storage temperature (4, 15 and 22 °C) and time (0, 7, 14, 21 and 28 days). Eggs were collected from 162, 31-weeks-old Lohmann Brown laying hens. The manual methods were adapted from Sharp and Powell. In method 1, yolk height (mm) was determined by measuring the total height of the egg and the support and substracting support height from the total height. Method 2, measures the height by direct yolk puncture using stem of a vernier caliper. Manual method 1 calculates $YI=((H-h)\times2)/(D1+D2)$ where H=height of yolk h=height of support, D1=Diameter of yolk, D2= Diameter of yolk (at 90° of D1). Manual method 2 calculates $YI=(H\times2)/(D1+D2)$. Haugh unit was obtained from the digital testing machine. Results indicated a significant effect of temperature and storage time on the YI and HU. Irrespective of the method used, a significantly lower values of YI and HU were observed on day 28 after storage. Eggs stored at 4 °C maintained higher YI across all storage days compared to 15 and 22 °C. Using linear regression, it was observed that YI and HU fitted linearly at 4 °C irrespective of the storage time. However, at higher temperatures (15 and 22 °C), HU showed a lower linearity compared to YI. A significant correlation ($r^2=0.865$) was observed between the two manual methods. Thus, making them to be alternatives. The digital method of measuring yolk index did correlated ($r^2=0.97$) with both manual methods and is thus considered as a viable alternative to manual measurement. Finally, YI is more sensitive when eggs are stored for a longer period at higher temperatures compared to the HU.

The ileal digestibility of amino acids and fatty acids in chickens fed freeze-dried pasture herbage

M. Englmaierová, E. Skřivanová, M. Skřivan and M. Marounek
Institute of Animal Science, v.v.i., Department of Nutrition Physiology and Animal Product Quality, Pratelstvi 815, 104 00, Prague 10, Uhrineves, Czech Republic; englmaierova.michaela@vuzv.cz

This experiment was focused on the determination of effect of a basal diet or diets supplemented with freeze-dried pasture herbage (FDPH) at 20 or 40 g/kg on the ileal digestibility of amino acids and fatty acids in chickens. A total of 48 one-day-old chickens (Ross 308) were used in this experiment. At 28 days of age, the chickens were individually relocated to balance cages (16 chickens per group) and were fed the basal or experimental diets until 36 days of age. The dominant species in the pasture herbage which was harvested in May were *Lolium perenne*, *Fescua pratensis* and *Trifilium pratense*. Freeze-dried pasture herbage contained especially linolenic acid and its addition increased content of this fatty acid in experimental diets. The supplementation of diet with 40 g/kg of FDPH increased the ratio of unsaturated to saturated fatty acid from 2.67 to 2.86. The content of amino acids in diet was also influenced by FDHP. Concentrations of most amino acids were lower in the diets with FDPH than in the basal feed. The effect of FDPH on digestibility was variable. In chickens with 40 g/kg of FDPH in diet, apparent ileal digestibility of amino acids varied from 0.39 (cysteine) to 0.91 (methionine). The variability in the ileal digestibility of fatty acids was less pronounced (from 0.65 to 0.89). The ileal digestibility of amino acids and fatty acids decreased in a dose-dependent manner in chickens fed FDHP. This suggests that the pasture herbage contains anti-nutritional factors that inhibit proteolysis and lipolysis.

Determining broiler system nutritional strategies to reduce environmental impact in two regions

C.W. Tallentire, S.G. Mackenzie and I. Kyriazakis
Newcastle University, AFRD, Newcastle upon Tyne, NE1 7RU, United Kingdom; c.w.tallentire@newcastle.ac.uk

The environmental impacts associated with broiler production arise mainly from the production and consumption of feed. The aim was to develop a tool for formulating broiler diets designed to target and reduce individually specific environmental impact categories in two contrasting regions (the UK and US). Using linear programming, least cost broiler diets were formulated for each region, using the most common genotype specific to each region. The environmental impact of the systems was defined using 6 categories calculated through a LCA method: global warming potential (GWP), fresh water eutrophication potential (FWEP), marine eutrophication potential (MEP), terrestrial acidification potential (TAP), non-renewable energy use (NREU) and agricultural land use (ALU). Diets were then formulated for each region to minimise each impact category, without compromising bird performance. The diets formulated for environmental impact objectives increased the cost of the diets in most cases by between 20 and 30% (the cost increase limit), with the exception of the least GWP (+16%) and the least NREU (+4%) diets in the UK, and the least TAP diet in the US (+14%). The degree of flexibility to reduce simultaneously several environmental impact categories in the UK and the US differed due to the different feed ingredients available to each region. The results suggested there was potential to minimise several impact categories simultaneously by reducing the impact of one impact category compared to least cost, through diet formulation in the UK; this was shown to a greater and lesser extent in the least FWEP and the least NREU diet formulations respectively. In the US, there was no way to minimise one impact category through diet formulation without increasing other impact categories caused by the system. Employing a multi-criteria approach to diet formulation methodologies, which account for environmental impact, will be important for improving the sustainability of animal production.

The effect of pre-peak feeding strategies on early laying performance

J. De Los Mozos, A. Navarro-Villa, C. Torres and A.I. García-Ruiz
Nutreco, Trouw Nutrition R&D, Ctra. CM 4004 km 10.5, 45950, Spain; j.delosmozos@trouwnutrition.com

This study aimed to assess contrasting nutritional alternatives to support early production in layers through feeding strategies that promote total nutrient intake during the pre-peak phase. Transition from rearing to production is a period in which pullets undergo many changes (transport, environment, diet, etc.) and the rapid increase in egg production is not always synchronized with a sufficient feed intake. Trial design consisted of 3 Diets (Standard; Dense; Low Energy) × 3 Premixes (Control; Metabolic [Organic minerals and higher levels of Vitamin E, C, D and B]; Gut [which included a synergetic blend of a phenolic compound with slow release C12, medium chain fatty acids and target release organic acids; as intestinal health modulator]) during the pre-peak phase (15 to 24 weeks). Each of the 9 dietary treatments were randomly distributed across the experimental units (n=7), each of them consisting of 22 collectively caged Isa Brown hens. From 25 to 32 weeks of age, the Standard diet with the Control premix was used in all the previous treatments. Pullets were selected at 15 weeks to be 100 g BW lighter than recommended by the breeding company. Pullets were light stimulated at 15 weeks of age to provoke early production in hens with low BW. Results from 15 to 24 weeks of age, showed that a pre-peak diet based on Low Energy, compared to the Standard and Dense diets, increased CP, Ca and P intake 5% on average ($P<0.05$). Moreover, egg production (89.19 vs 86.45 and 86.58%; $P<0.05$) and egg mass (51.41 vs 49.66 and 49.53 g/d; $P<0.05$) of the whole period of production studied was higher with the Low Energy Pre-peak diet. Gut Premix tended to improve egg production ($P=0.071$) compared to the other premixes. A positive interaction was observed when a Low Energy diet with the Gut Premix was fed, obtaining the highest egg production ($P<0.05$) and egg mass ($P=0.058$). As conclusion, the usage of Low Energy diets improved production in early phases. Moreover, its effect might be higher with the addition of the Gut Premix.

Providing a post-hatch feed has beneficial effects on the performances of male broilers

S. Leleu[1], E. Delezie[1] and S. De Vos[2]
[1]ILVO (Institute for Agricultural and Fisheries Research), Animal Sciences Unit, Scheldeweg 68, 9090 Melle, Belgium, [2]INVE Belgium R&D, Oeverstraat 7, 9200 Baasrode, Belgium; saskia.leleu@ilvo.vlaanderen.be

A post-hatch feed is provided to one-day-old chicks in order to obtain an improved fulfilment of their needs. The objective of this trial was to investigate the effect of a post-hatch feed on the production performances in a trial with male broilers. One-day-old chicks were fed a post-hatch feed (P, mash/pellet mixture) or a common starter (S, coarse mash) on a carton plate. 12 pens (=replicates) with 30 male broilers per pen received either P or S (30 g/chick). Then, after 24 h, the feed mangers with the starter feed were placed into the pens. In this trial, a three-phase feeding scheme was used with a starter (0-12 d, mash), grower (12-26 d, pellet) and finisher (26-35 d, pellet) diet. At the end of each period, body weight (BW) and feed intake (FI) were measured per pen, and body weight gain (BWG) and feed conversion ratio (FCR) were calculated. Significant differences were found in the starter and grower period for most of the parameters. In the whole trial period (0-35 d), the mean BW of P fed broilers was 4.9% higher (2,504 vs 2,388 g; $P<0.001$). They had a 5.0% higher BWG (69.5 vs 66.3 g) and a 3.8% higher FI (102.2 g vs 98.5 g) compared to the S fed broilers (all significant, $P<0.001$). Also their FCR was significantly better (1.470 vs 1.486; significant, $P<0.05$). Thus, it can be concluded that when 30 g of a post-hatch feed is provided per chick at arrival in the stable, significantly improved production performances are obtained.

Estimation of choline requirement for broiler chickens feeding corn-soybean meal basal diet

S. Honarbakhsh[1], M. Zaghari[2], M. Shivazad[2], M. Pahlevan-Hassan[2] and H.R. Ghalamkari[3]
[1]University of Tehran, Department of Animal and Poultry Science, College of Aburaihan, Pakdasht, Tehran, 33916-53755, Iran, [2]University of Tehran, Department of Animal Science, College of Agriculture and Natural Resources, Karaj, Alborz, 31587-11167, Iran, [3]Sepahan Daneh Parsian Co., Isfahan, 81655-668, Iran; honarbhk@ut.ac.ir

The aim of present study was to estimate choline requirement for broiler chickens feeding corn-soybean meal basal diet. One hundred and twelve male broiler chickens (Ross 308) reared in a 42 days' trial. Four dietary treatments included added levels of choline (0, 400, 800 and 1,200 mg/kg) to the basal diet. Choline content of basal diets of starter, grower and finisher, respectively, were 1,118, 1,218 and 1,318 mg/kg, but the other nutrients were the same as Ross 308 broiler nutrition specifications. Birds received water and experimental feed *ad libitum*. Treatments replicated 7 times each included four chickens in a completely randomized design. Performance and carcass characteristics were compared statistically by using the GLM procedure of SAS. Duncan's multiple range test was used to determine differences among treatment means at $P<0.05$. Performance (body weight gain and feed conversion ratio) and carcass characteristics (breast, thighs, liver and abdominal fat fractional weight) had no significant difference. In conclusion it is not necessity to add excess sources of choline to corn-soybean meal diets with high levels of metabolizable energy (3,000 to 3,200 kcal/kg) and crude protein (23 to 19.5%) because of same performance ($P>0.05$) of birds which received different levels of choline.

Effects of using dietary metformin on broiler breeder hens performance and plasma metabolites

R. Taherkhani

Payame Noor University, Department of Animal Science, Faculty of Agriculture, Tehran, 19395-3697, Iran; r_taherkhani@pnu.ac.ir

High blood glucose in broiler breeders could results in development of lipotoxicity and impair their reproduction performance. This experiment was carried out to evaluate if metformin (a human glucose lowering agent) could lower broiler breeder blood glucose and prevent lipotoxicity associated depression of egg production in broiler breeder hens. The experiment was conducted using a completely randomized design with factorial arrangement. The factors were consisted of feeding regimen (restrict and *ad libitum*) and the levels of metformin (0, 5 and 10 g/kg of diet). A total of 48 Cobb 500 broiler breeder were used in the experiment. The experiment was started at 27 wk and lasted to 39 wk of age. Each of dietary treatments was replicated 8 times and each hen was assumed as an experimental unit. Daily egg production, feed intake, and body weight were recorded and blood samples were taken for determining blood glucose and triacylglycerol. Results have shown that 5 mg metformin/kg of diets suppressed feed intake and also blood glucose both in feed satiated and feed restricted birds. Reduced blood glucose was associated with improved egg production along with lower plasma triacylglycerol and also lower abdominal fat pad in birds received 5 mg metformin in their feed. Our results indicated that reducing blood glucose in broiler breeder hen may improve their reproduction performance.

N-acetyl-L-cysteine improves performance of chronic cyclic heat stressed finisher broilers

J. Michiels[1], M. Majdeddin[1,2,3], E. Rosseel[1], J. Degroote[1,2], P. Vermeir[1] and A. Golian[3]

[1]Ghent University, Department of Applied Biosciences, Valentin Vaerwyckweg 1, 9000 Gent, Belgium, [2]Ghent University, Laboratory for Animal Nutrition and Animal Product Quality, Department of Animal Production, Coupure Links 653, 9000 Gent, Belgium, [3]Ferdowsi University of Mashhad, Centre of Excellence in the Animal Science Department, P.O. Box: 91775-1163, Mashad, Iran; joris.michiels@ugent.be

Heat stress is known to induce oxidative stress. The most abundant endogenous intracellular antioxidant is the tripeptide glutathione (GSH), for which cysteine is the limiting amino acid. It was hypothesized that dietary supplementation of N-acetyl-L-cysteine (NAC), as a source of cysteine, could improve performance of heat stressed finisher broilers by enhancing GSH availability. Four levels of NAC; 0, 500, 1000 and 2,000 mg/kg, were added to a finisher diet with a ratio dig M+C to dig LYS of 0.73 (d25-41). Dietary treatments were replicated in 8-9 pens with 20 male Ross308 birds each. A chronic cyclic heat stress model (34 °C, 50-60% rh for 7 h daily) was initiated at d28. Two birds per pen were sampled on d29 (acute heat stress) and d41 (chronic heat stress) between 4 and 7 h of the heat stress period that day, to determine levels of GSH in liver and heart. Final BW, growth and feed conversion in the finisher phase were all improved by supplemental N-acetyl-L-cysteine ($P<0.05$). ADG was 88.2, 92.2; 93.7 and 97.7 g/d, and F:G ratio equalled 2.21, 1.91, 1.84 and 1.80 for 0, 500, 1000 and 2,000 mg/kg NAC treatments, respectively. GSH levels in liver and heart were numerically increased in birds supplemented with 1000 and 2,000 mg/kg NAC at both sampling days (+3.9-7.8%). Irrespective of treatment, GSH in liver (organ of synthesis) was positively and negatively correlated with sampling time on d26 and d39, respectively ($P<0.05$), in opposite to heart (sink organ for GSH) ($P<0.05$). In conclusion, GSH metabolism was affected by heat stress. However, NAC supplementation had little influence on this, in contrast to clear performance improvements.

Caffeine induces changes in hematological, biochemical, and cardiovascular parameters in quail

M. Kamely
Tarbiat Modares University, Poultry Science, 14115-336 Tehran, Iran; m.kamely@gmail.com

Caffeine is one of the most widely consumed pharmacological substances in the world, found in food and drinks. The effects of caffeine on human health and cardiovascular disease have been the focus of much debate, and much research has been performed in various animal models, including some in chicken. Previously, we reported caffeine to cause blood biochemistry changes and heart failure in young chickens (Gallus gallus). Recently, quails are increasingly used as a more economical and convenient bird model. Here, a trial was conducted to assess the suitability of adult quail as an avian model for caffeine research. Fifty six, 20-week-old laying Japanese quails (Coturnix coturnix japonica) were randomly divided into two groups (28 birds each with 7 replicates): one with no added caffeine, and one which was placed on a 15 mg/kg body weight/day caffeine in drinking water. On d 21, 14 birds per group were sacrificed and analysed for physiological and hematological parameters. Results showed that body weight, liver, heart and spleen relative weight were not affected by caffeine ($P>0.05$), whereas a significant right ventricular hypertrophy was found ($P<0.01$). Hematocrit and hemoglobin also significantly increased in caffeine group ($P<0.001$). Plasma alanine aminotransferase and aspartate aminotransferase activity were highly decreased in caffeine group ($P<0.001$). The levels of plasma total protein, albumin, globulin, glucose, triglyceride, cholesterol and low density lipoprotein were significantly higher in caffeine group ($P<0.001$). Overall, the results obtained were consistent with those reported in chicken and those in mammals, demonstrating the suitability of quail as an animal model for caffeine research.

Embryonic feeding of silver nanoparticles for improving post hatch status of broilers

A. Goel, S. Bhanja, M. Mehra, N. Nayak, J. Rokade, A. Mandal and S. Majumdar
Central Avian Research Institute, PHM Section, Central Avian Research Institute, Izatnagar, 243122 Bareilly, India; genesakshat@gmail.com

Silver nanoparticles (SN) are known for its unique physical, chemical as well as biological properties, enabling them to penetrate inside the cells and exhibits anti-inflammatory response. In this study, three different doses T1, T2 and T3 (12.5, 25 and 50µg per egg, respectively) of SN were administered in 18 day embryonated eggs at the broad end, utilising *in ovo* techniques to explore their effects on the post hatch status of broiler chickens. Sham control (T4) and un-injected control (T5) were kept for comparison. A total of 50 eggs per treatment were used for injection. Each treatment group had four replicates of 8 birds each with equal sexes and the chicks were reared up to 42 day of age following standard nutritional practices. Hatchability parameters, weekly body weight gain, immunity parameters including organ weight, *in vivo* cell mediated immune response to Phytohaemagglutinin-P (PHA-P) and humoral immune response to Sheep red blood cells (SRBC) were studied at 3rd week post hatch in 8 birds per treatment. Hatchability parameters such as egg weight (55.82-57.97 g), chick weight (37.59-38.70 g), ratio (66.07-67.90%) and hatchability (82.35-86.84%) were similar in all the treatment groups. Similarly, body weight gain was similar in all the treatment groups during the whole experiment in both male and female broilers. No variation was seen in the weight of thymus however the bursa and spleen weight was significantly increased ($P<0.05$) in T2 (0.291 mg/kg; 0.217 mg/kg) and T3 (0.306 mg/kg; 0.199 mg/kg) in comparison to T5 (0.245 mg/kg; 0.168 mg/kg) treatment. *In vivo* immune response to PHA-P and SRBC was also increased ($P<0.05$) in T1 (0.718 mm; 10.0) in comparison to T4 (0.54 mm; 8.38) and T5 (0.54 mm; 8.13) treatment. From the above study it can be concluded that *in ovo* supplementation of SN modulates the post hatch immune response without effecting the hatchability, growth and development in broilers.

Author index

A

Aalbers, G.	289
Abdallh, M.E.	233
Abdelrahman, W.	198
Abdel-Wareth, A.A.A.	176, 198
Abdollahi, M.R.	47, 242
Abdulla, J.	153
Adam, M.	200
Adedokum, S.A.	168
Adeleye, O.O.	226
Adeola, O.	160
Ader, P.	231, 243
Adrizal, A.	172
Agboola, J.O.	168
Aggrey, S.E.	191
Agostini, P.	254
Agostini, P.S.	173, 188
Airlang, A.	184
Albiker, D.	288, 290
Albrecht, A.	174
Alfonso, C.	190
Alfonso-Carrillo, C.	217
Ali Arabi, H.	282
Alijosius, S.	160
Alleno, C.	217
Alves, M.C.	199
Alves, W.J.	291
Amalia Nurhuda, G.	167
Amerah, A.M.	242, 267
Ammari, F.	170
Anderson, K.E.	277
Andrianova, E.N.	159, 190, 193, 221, 228
Angga, W.A.	153
Annisa, A.	153
Antonissen, G.	130
Antoszkiewicz, Z.	215
Aoun, C.	196
Applegate, T.J.	252
Arbe Ugalde, X.	284
Ardi, A.	167
Arsenault, R.J.	200, 264
Arturo-Schaan, M.	151
Asadian, N.	229
Asad, S.	284
Ashoori, A.	253
Assunção, P.S.	239
Attamnangkune, S.	237
Auclair, E.	201, 260
Auer, B.	188
Avila, E.	169
Awaad, M.H.H.	205
Ayllón Ramos, S.	209
Azhar, M.R.	162

B

Bachanek, I.	265
Baeyen, S.	251
Bailey, C.	200
Baker, K.T.	264
Barbe, F.B.	270, 278, 279
Barbosa, A.C.	178
Barea, R.	203
Barekatain, R.	177
Barnard, L.P.	157, 236
Barnes, J.	192
Barroeta, A.C.	118, 152, 164, 226
Bartelt, J.	219, 220, 224
Bashir, Y.	241
Batonon-Alavo, D.I.	179
Bébin, K.	151, 222
Beckers, S.	231, 232
Beckers, Y.	251
Bedford, M.R.	161, 162, 234, 263, 281, 286, 290
Bekker, M.S.	284
Belalcazar, A.	236
Belloir, P.	184, 185
Berg Kehlet, A.	196
Bernadet, M.D.	260
Bernard, N.	168, 283
Bhanja, S.	297
Bhuiyan, M.M.	233
Biasato, I.	163, 271
Biasibetti, E.	163, 271
Bieber, M.	240
Bierman, K.	231, 232, 245, 249
Bindelle, J.	251
Blake, D.P.	275
Blanch, A.	196, 197, 259, 261
Blanch, M.	192, 213
Bliznikas, S.	160, 222
Bonato, M.A.	201
Bonnin, E.	262
Boonyoung, P.	263
Borges, L.L.	201
Borin, H.	217
Boros, D.	159
Bortoluzzi, C.	252
Bourdillon, A.	168
Bourdonnais, A.	283
Bourgueil, E.	170
Bouwhuis, G.J.	253
Braun, U.	178
Brévault, N.	225
Briens, M.	279, 280
Brinch, K.S.	257
Broekner, C.	154
Brøgger Pedersen, M.	238

Rodriguez-Sanchez, R.	152	Schollenberger, M.	230
Roekngam, N.	263	Sedeghi, M.	229
Rogiewicz, A.	284	Selle, P.H.	163, 172
Rokade, J.	297	Senkoylu, N.	283
Romeo, A.	281, 282	Sereno, A.	271
Romero, L.	245, 248	Shachnev, Y.D.	228
Rose, S.P.	153, 157, 162	Sharma, N.K.	183
Rosseel, E.	296	Shevyakov, A.N.	228
Roth, N.	212	Shimao, R.	218
Rouault, M.	197	Shivazad, M.	295
Roura, E.	29, 191, 213	Shokri, P.	280
Rousseau, X.	246	Shtylla Kika, T.	193
Roux, J.F. Le	217	Siegert, W.	189
Royer, A.F.B.	178	Sievers, H.	285
Rozhkov, O.A.	193	Sifri, M.	113
Ruangpanit, Y.	237	Silva, E.P.	291
Rudeaux, F.	196	Sims, M.	261
Rutkowski, A.	156, 157, 158, 235, 249	Singh, M.	151, 291
Ryan, W.J.	239	Sirri, F.	165
Ryu, K.S.	213, 215, 228, 288	Sirukhi, M.	198
		Sizer, D.	239
S		Skjoet-Rasmussen, L.	259
Sacopta, D.	206	Skřivan, M.	293
Sacranie, A.	180, 248	Skřivanová, E.	293
Sacy, A.S.	270, 278, 279	Skřivanová, V.	287
Saiz, A.	224	Slominski, B.A.	284
Saki, A.A.	240, 253, 282	Smeets, N.	204, 233
Sakkas, P.	223, 275	Smit-Heinsbroek, A.	169
Sakomura, N.K.	91, 291	Smith, S.	223
Salari, S.	194	Smulikowska, S.	265
Salgado, H.	171	Soares, N.	231, 232
Salih, A.	200	Socoliuc, R.P.	166
Salobir, J.	256	Soica, C.	256
Sánchez, F.	217	Soisuwan, K.	184
Sandvang, D.	259	Solà-Oriol, D.	226, 290
Sang, B.	274	Sol, C.	205
Santiago, R.	169	Sommerfeld, V.	230
Santomà, G.	136	Soumeh, E.A.	267
Santos, J.S.	178	Sozcu, A.	199
Saracila, M.	166, 220, 256	Sparagano, O.	193
Saravanan, S.	241	Starkl, V.	273
Saremi, B.	174, 179, 191	Steenfeldt, S.	287
Sarker, M.S.K.	213	Stefanova, I.L.	221
Sarsour, A.H.	192, 206, 242	Stef, L.	207
Sasyte, V.	160	Steyl, P.D.	208
Sauerwein, H.	219, 220, 224	Stringhini, J.H.	178, 199, 239
Saulnier, L.	262	Styrishave, T.	259
Savary, R.K.	36	Sulyok, M.	273
Scarsella, E.	175	Sun, L.H.	280
Schallier, S.	171, 272	Suzuki, R.M.	291
Schaumberger, S.	273, 274	Svihus, B.	59, 152, 286
Schiavone, A.	163, 271	Svitkin, V.S.	221
Schierle, J.	225	Świątkiewicz, S.	164, 194
Schlagheck, A.	196	Swick, R.A.	166, 177, 182, 183
Schokker, D.	269	Syafwan, S.	172
Scholey, D.	281	Syed, B.A.S.	214